Seismicity Patterns, their Statistical Significance and Physical Meaning

Edited by
Max Wyss
Kunihiko Shimazaki
Akihiko Ito

Springer Basel AG

Reprint from Pure and Applied Geophysics
(PAGEOPH), Volume 155 (1999), No. 2-4

Editors:

Max Wyss
Geophysical Institute
University of Alaska
Fairbanks
USA
email: max@giseis.alaska.edu

Kunihiko Shimazaki
Earthquake Research Institute
University of Tokyo
Tokyo
Japan
email: niko@eri.u-tokyo.ac.jp

Akihiko Ito
University of Utsunomiya
Utsunomiya
Tochigi
Japan
email: ito@acc.utsunomiya-u.ac.jp

A CIP catalogue record for this book is available from the Library of Congress, Washington
D.C., USA

Deutsche Bibliothek Cataloging-in-Publication Data

Seismicity patterns, their statistical significance and physical meaning / ed.
by Max Wyss ... - Springer Basel AG, 1999
    (Pageoph topical volumes)
    Aus: Pure and applied geophysics : Vol. 155. 1999
    ISBN 978-3-7643-6209-6   ISBN 978-3-0348-8677-2 (eBook)
    DOI 10.1007/978-3-0348-8677-2

© 1999 Springer Basel AG
Originally published by Birkhäuser Verlag in 1999
Printed on acid-free paper produced from chlorine-free pulp

ISBN 978-3-7643-6209-6

9 8 7 6 5 4 3 2 1

Contents

Pure appl. geophys. 155 (1999) 203–205
0033–4553/99/040203–03 $ 1.50 + 0.20/0

Pure and Applied Geophysics

Introduction

Seismicity Patterns, their Statistical Significance and Physical Meaning

The topic of seismicity patterns is one in which significant advances in the understanding of seismotectonics and earthquake hazard are being made, and in which controversies have recently developed. This special issue grew out of a workshop attended by forty-one participants on May 11/12, 1998 in Nikko, Japan. The collection of papers we ultimately generated does not cover all aspects of problems related to seismicity patterns, but a fair part of them. In a few cases, authors wrote papers especially for this workshop, however, in general, they submitted papers they fortuitously worked on at the time. Nevertheless, this collection of papers shows what issues currently top the priority list in this field of research.

The workshop was sponsored by the Geophysical Institute of the University of Alaska, Fairbanks (UAF), in order to develop contacts within the international community of experts in this field, who may participate in collaborative research at the International Arctic Research Center (IARC) in the future. The IARC was established as a part of the common agenda between Japan and the Untied States and is located at UAF. It is a research facility that will house thirty scientists and which opened its doors in December 1998.

This two-day meeting of minds with opposing views took place in the oldest hotel in Japan, the Kanaya Hotel, within a few stones throw of the grave of Iyeyasu Tokugawa, the first Shogun, amidst the serene beauty of the waterfall rich Nikko mountain area. In addition to thirty-four formal lectures, two extended evening discussions, each led by an advocate and an opponent, were conducted on the topics of "Foreshocks, their Recognition and Properties" and "Can Earthquakes be Predicted," respectively.

The topic of *foreshocks* and *moment release increase* was discussed in six papers from an observational and modeling point of view. K. Yamaoka, T. Ooida and Y. Ueda reported that in May 1993 they interpreted a seismicity rate increase as a potential foreshock sequence and estimated the probable occurrence time of the main shock by fitting a *time to failure curve* to it. In real time they estimated the expected magnitude as 5, and installed a strong-motion instrument that recorded the M 5.1 earthquake which occurred ten days after the expected date, at about 5 km from it. Studies estimating the time of main-shock failure from the regionally observed increase of moment release as a function of time, including the correct prediction of the 1996 M 7.9 Delaroff Islands earthquake, were summarized by S. Jaumé and L. Sykes.

P. Reasenberg demonstrated that in Cascadia earthquakes are four times more likely to be foreshocks than in California. Many speakers emphasized the regional differences in all earthquake parameters, and it was generally understood that basic models of the earthquake occurrence must be modified for regional application. The idea that the *focal mechanisms of foreshocks* may differ from that of background activity was advocated by Y. Chen and identified by M. Ohtake as possibly the thus far most neglected property of foreshocks, in efforts to identify them. S. Matsumura proposed that focal mechanism patterns of small earthquakes may differ characteristically near locked fault segments into which fault creep is advancing.

Considerable discussion was devoted to the status of the *seismic gap hypothesis* because M. Wyss argued that the occurrence of the M 7.9, 1986, Andreanof Islands earthquake was a confirmation of Reid's rebound theory of earthquakes and thus of the time predictable version of the gap hypothesis, whereas Y. Kagan believed he could negate this view by presenting a list of nine earthquake pairs with $M > 7.4$, moment centroid separation of less than 100 km, and time difference less than about 60% of the time he estimated it would take plate motions to restore the slip of the first event. Most participants did not accept this table as conclusive evidence supporting the idea that the energy for very large main shocks can be drawn from the same volume within short-time intervals, because, when high-quality data are available, it can be shown that rupture areas of very large neighboring shocks usually abut, or, if they overlap, the overlapping segments are locations of slip deficiency. K. Shimazaki presented evidence that showed that the slip distribution in the previous event is important to estimate the occurrence time of the next, a result that again was pointed out by B. Shaw in his rupture models.

Several authors discussed evidence and models for *seismic quiescence*. J. Zschau illustrated that in an elastic layer overlaying a viscous one, slow stress changes (e.g., due to creep) propagate to considerably longer distances than rapid ones (e.g., coseismic stress changes). N. Kato and T. Hirasawa presented a model in which, applying constitutive laws of failure, creep events on a megathrust could start and stop without generating a main shock, thus potentially explaining swarms and seismic quiescence episodes not preceded or followed by a main shock.

Several papers dealt with statistical models for seismicity patterns, such as the point process model by Y. Ogata, while other papers dealt with the statistical evaluations of observations. For example, V. Kossobokov established that the *M 8 hypothesis* passed the ongoing real-time test at the 99% confidence level, if $M > 7.9$ events were considered, but failed it (85% confidence level) when $M > 7.5$ earthquakes were included.

The test of the *swarm hypothesis* by D. Rhoades and F. Evison is an example of rigorous testing of a hypothesis. Nonetheless, several attendees criticized it because a single excellent success generated such a large bonus that it alone could validate their hypothesis. The consensus was that a hypothesis cannot be validated on the basis of one event.

Computer models of earthquake generation and interaction were based on constitutive laws of friction (J. Dieterich, B. Shaw) and on pore pressure variations (J. Miller, T. Yamashita). Both types of models produce many aspects of earthquake sequences realistically, although different models concentrated on different parameters. It seems that these models have recently become more realistic and useful for understanding the earthquake generation process. J. Dieterich proposed that his model could be tested quantitatively.

The *b*-value of the *frequency-magnitude relation* was discussed by several authors, with a controversy developing because Y. Kagan argued that the factor of two differences in *b*-values, shown by S. Wiemer over 10-km distances along the San Andreas fault, must be due to catalog errors. He proposed that *b*-values are always approximately equal to 1, because he observed constant *b*-values in the Flinn-Engdahl regions of the world.

The new problem of self-organized criticality (SOC) in seismicity received insufficient discussion at the workshop. However, papers authored by C. Sammis and S. W. Smith and S. Jaumé and L. Sykes in this volume summarize the current status of SOC models.

Although everyone realizes that most analyses of seismicity patterns are sensitive to errors in the earthquake catalogs, relative few search quantitatively for changes in reporting. R. Zúñiga and S. Wiemer demonstrated new problems that can exist in catalogs.

From the vigorous debate at the workshop, and from the papers collected in this special issue, it is evident that seismicity patterns are currently a controversial topic. It seems that all participants at the workshop believe that considerable information regarding the generation of large earthquakes may be locked in seismicity patterns, and that we are making progress in learning how to extract it and put this knowledge to use.

Max Wyss
Geophysical Institute
University of Alaska
Fairbanks
Alaska
U.S.A.
E-mail: max@giseis.alaska.edu

Akihiko Ito
University of Utsunomiya
Utsunomiya
Tochigi
Japan
E-mail: ito@acc.utsunomiya-u.ac.jp

Kunihiko Shimazaki
Earthquake Research Institute
University of Tokyo
Tokyo
Japan
E-mail: niko@eri.u-tokyo.ac.jp

Pure appl. geophys. 155 (1999) 207–232
0033–4553/99/040207–26 $ 1.50 + 0.20/0

Rethinking Earthquake Prediction

LYNN R. SYKES,[1] BRUCE E. SHAW[1] and CHRISTOPHER H. SCHOLZ[1]

Abstract—We re-examine and summarize what is now possible in predicting earthquakes, what might be accomplished (and hence might be possible in the next few decades) and what types of predictions appear to be inherently impossible based on our understanding of earthquakes as complex phenomena. We take predictions to involve a variety of time scales from seconds to a few decades. Earthquake warnings and their possible societal uses differ for those time scales. Earthquake prediction should not be equated solely with short-term prediction—those with time scales of hours to weeks—nor should it be assumed that only short-term warnings either are or might be useful to society. A variety of "consumers" or stakeholders are likely to take different mitigation measures in response to each type of prediction. A series of recent articles in scientific literature and the media claim that earthquakes cannot be predicted and that exceedingly high accuracy is needed for predictions to be of societal value. We dispute a number of their key assumptions and conclusions, including their claim that earthquakes represent a self-organized critical (SOC) phenomenon, implying a system maintained on the edge of chaotic behavior at all times. We think this is correct but only in an uninteresting way, that is on global or continental scales. The stresses in the regions surrounding the rupture zones of individual large earthquakes are reduced below a SOC state at the times of those events and remain so for long periods. As stresses are slowly re-established by tectonic loading, a region approaches a SOC state during the last part of the cycle of large earthquakes. The presence of that state can be regarded as a long-term precursor rather than as an impediment to prediction. We examine other natural processes such as volcanic eruptions, severe storms and climate change that, like earthquakes, are also examples of complex processes, each with its own predictable, possibly predictable and inherently unpredictable elements. That a natural system is complex does not mean that predictions are not possible for some spatial, temporal and size regimes. Long-term, and perhaps intermediate-term, predictions for large earthquakes appear to be possible for very active fault segments. Predicting large events more than one cycle into the future appears to be inherently difficult, if not impossible since much of the nonlinearity in the earthquake process occurs at or near the time of large events. Progress in earthquake science and prediction over the next few decades will require increased monitoring in several active areas.

Key words: Earthquakes, earthquake prediction, earthquake precursors, physics of earthquakes.

Introduction

National programs to predict earthquakes have been underway for more than 30 years in Japan, the former Soviet Union and China. Most work on prediction in the United States, however, did not commence until after 1978 with the establish-

---
[1] Lamont-Doherty Earth Observatory and Department of Earth and Environmental Sciences, Columbia University, Palisades NY 10964, USA. Fax: 914-365-8150; E-mail: sykes@ldeo.columbia.edu

ment and funding of the National Earthquake Hazards Reduction Program
(NEHRP). In fact, earthquake prediction has been a relatively small component of
NEHRP, which also includes funds for fundamental studies of earthquakes,
earthquake engineering, estimates of risk to people and the built environment,
insurance and various measures to reduce the loss of life and property in earthquake (all of which arguably belong in a well-balanced national program). Nevertheless, successes and failures of predictions worldwide and the recognition that the
earthquake process is an example of a complex system dictate the need for a
thorough review of what predictions are achievable now on various time scales,
what might be accomplished in the next few decades, and what is likely to be
inherently unknowable, i.e., unpredictable.

A number of recent articles, e.g., MAIN (1997), GELLER *et al.* (1997), GELLER
(1997a–d) and EVANS (1997), carry strongly worded titles like *Long Odds on
Prediction* and *Earthquakes Cannot be Predicted.* They make claims regarding the
earthquake process that we think are either incorrect or misleading. For example,
articles like those of GELLER (1997a,b) and MAIN (1997), which were published as
either opinion pieces in newspapers or short comments intended for more general
scientific audiences, equate earthquake prediction with short-term prediction without stating so explicitly. Even for longer scientific articles in more specialized
journals (e.g., GELLER, 1997d; GELLER et al., 1997) one must read well beyond the
bold pronouncements of either the titles or abstracts to ascertain that by earthquake prediction they mean short-term prediction. All of the above articles argue
that reliable short-term prediction, is inherently difficult if not impossible and that
very high accuracy is needed for mitigation measures either to be taken or to be
cost effective. The present extreme divergence of views relative to the feasibility of
earthquake prediction and whether the study of earthquakes is, in fact, a science,
are similar to those that existed about continental drift from 1920 to 1965. Many
earth scientists, especially in the U.S. and U.K., do not regard prediction as "worth
working on." Our paper is an attempt to challenge these views and to enlarge this
debate.

The notion that several different kinds of prediction might be possible, each
with its own time scale, is central to current debates about earthquake prediction.
A short-term prediction of a few days to weeks, based on some earthquake process
with a short time scale (e.g., nucleation), is distinct from a long-term prediction
based on a longer-term process (e.g., stress buildup due to plate motions). These
different kinds of predictions may have very different chances for success.

The public perception in many countries and, in fact, that of many earth
scientists is that earthquake prediction means short-term prediction, a warning of
hours to days. They typically equate a successful prediction with one that is 100%
reliable. This is in the classical tradition of the oracle. Expectations and prepara-
tions to make a short-term prediction of a great earthquake in the Tokai region of
Japan have this flavor. We ask instead are there any time, spatial and physical

characteristics inherent in the earthquake process that might lead to other modes of prediction and what steps might be taken in response to such predictions to reduce losses? In this context we examine briefly what has been learned about the earthquake generation process, complexity and chaotic behavior, long-term prediction, and changes in rates of small and moderate-size earthquakes and in chemistry prior to large events. We argue that an important middle ground exists between very accurate predictions of hours to weeks and estimates of long-term seismic potential that are typically made for centuries or millennia.

Terms for Expressing Earthquake Predictions and Potential

We use four categories of earthquake prediction (long-term, intermediate-term, short-term and immediate alert) keyed to various warning times (Table 1). Each category has a known or inferred scientific basis (or bases) and some mitigation measures that might be taken to reduce loss of life and property. These categories and their associated warning times are similar to those advocated by WALLACE *et al.* (1984), except that we add the category immediate alert. All of these categories of prediction involve time windows that are shorter than the average repeat times of large earthquakes along a given fault segment or segments. They are examples of time-varying probabilities in contrast to estimates of long-term potential, which assume that large earthquakes occur randomly in time. We take 30 years to mark the transition from predictions to estimates of earthquake potential since time-varying probabilities probably have modest additional societal value for longer periods. From a tectonic viewpoint, however, the transition from one to another is better thought of as occurring at a certain fraction of the average repeat times of large shocks, which, of course, vary with long-term slip rates of faults.

We use the term "large earthquakes" very specifically to mean events that rupture the entire downdip width of the seismogenic zone of faults, the part of faults that is capable of building up stresses and releasing them suddenly (Fig. 1). Large shocks cause most of the cumulative damage and strong shaking worldwide. Also, our discussions are aimed mainly at very active faults—those of long-term slip rates of 10 mm/year or greater—where the repeat times of large earthquakes are typically decades to several hundred years.

Complexity, Chaos and Predictability of Natural Systems

Earthquakes, and the faults on which they occur, are thought to be an example of a complex physical system that exhibits chaotic behavior. Yet such a characterization does not preclude useful predictions. Many other natural hazards including floods, severe storms, wildfires, and climatic changes associated with El Niño also

Pure appl. geophys.,

Table 1

Warning times, scientific bases, feasibility and mitigation measures for various types of earthquake predictions and estimates of long-term potential

Term	Warning time	Scientific bases	Feasibility	Some mitigation measures
Immediate alert	0 to 20 seconds	speed of electro-magnetic waves $\gg$ speed of seismic waves	good	warning to take cover immediately, close valves in refineries, scram reactors, shut off gas and electricity, produce maps of shaking for quick emergency response
Short-term prediction	hours to weeks	accelerating aseismic slip, foreshocks for some events	unknown	deploy more emergency services and put on higher alert status, warn people to keep in safest places for period of prediction, education
Intermediate-term prediction	1 month to 10 years	changes in seismicity, strain, chemistry, fluid pressure	fair for areas of intensive monitoring and past study	strengthen structures and lifelines, improve emergency services, enact laws to reduce losses, education, prepare to react to short-term prediction, more instrumentation and technical studies
Long-term prediction	10 to 30 years	time remaining in cycle of large shocks, increase in regional shocks	good for well-studied faults of high long-term activity	make time-dependent hazard and risk estimates, site new critical and expensive facilities to minimze losses, education, increase instrumentation to detect possible intermediate-term precursors
Long-term potential	>30 years	long-term rate of activity, plate tectonic setting	very good	make time-independent hazard and risk estimates, building codes, land use planning

exhibit chaotic behavior. Weather, the classical example of chaotic behavior does have predictable elements. Atmospheric forecasts are made routinely for the next several days. The advent of satellite imagery greatly improved forecasts and warnings of hurricanes on a time scale of hours to days. Knowledge and monitoring of parameters affecting El Niño now furnish bases for making forecasts of departures from average climatic conditions for the following six months to two years. SHUKLA (1998) claims that wind and rain patterns in certain regions of the tropics are so strongly determined by the temperature of the underlying sea surface that they do not show sensitive dependence on the initial conditions of the atmosphere, as is the case for much of the earth's atmosphere. Nevertheless, the sensitive dependence on initial conditions for most of the atmospheric system indicates that accurate short-term predictions for specific dates, months and years ahead of time probably are impossible, e.g., whether it will rain on a given day in New York City a year from now or whether the temperature will be higher than normal on that date.

Volcanoes may also be a complex, chaotic system. Nevertheless, many volcanoes that have been active in the past 10,000 years cycle through decades to centuries of dormancy followed by months to years of activity of various types including increased strain, tilt, emission of gases, temperature and flow of water, shallow earthquakes and volcanic tremor. These increases can and have been used to make predictions on time scales of months to years that the chance of a damaging eruption is greater than in the long period of dormancy between major eruptions. Successful predictions on a time scale of days have been made, such as

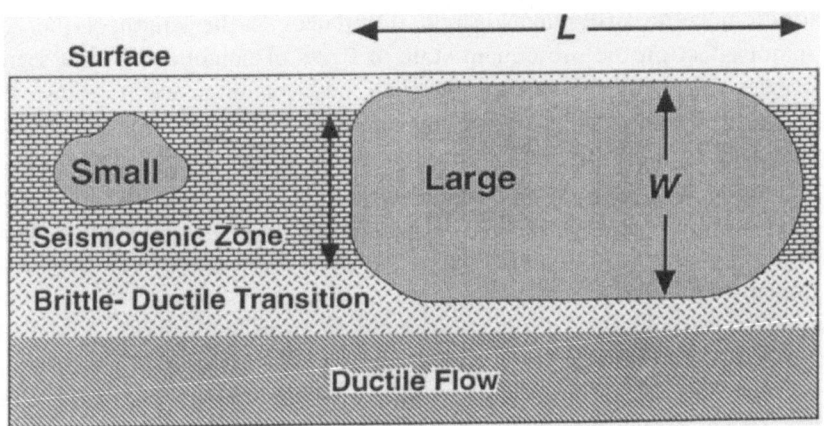

Figure 1
Two types of earthquakes—small and large, after PACHECO *et al.* (1992). L is rupture length along strike of fault; W is its downdip width. Small shocks do not rupture entire downdip width, W, that is capable of breaking in earthquakes, i.e., the seismogenic zone. Large events rupture entire downdip width, W. Small shocks are capable of growing in two dimensions, L and W whereas large earthquakes can become yet larger only by growing in one dimension, L.

that for the major Pinatubo eruption of 1991 in the Philippines, on a short-enough time scale to permit evacuation and saving of many lives. Nevertheless, the duration and size of an eruption have proven difficult to ascertain in most cases. Monitoring is a key to greater understanding of the complex physics of volcanoes and to the prediction of eruptions on various time scales.

As with large earthquakes along a given fault segment, it appears to be inherently difficult, if not impossible, to predict beyond the next major eruption cycle into the future. Fortunately, prediction beyond one cycle into the future is of minimal societal value for either major volcanic eruptions or large earthquakes.

Chaos and Determinism

Chaos does not mean complete unpredictability or randomness, the common nonscientific usage of the word. It does mean that predictability eventually will be lost over long enough time scales. Key questions are the reasons for the complexity, the time scales over which predictability is lost and when do most nonlinear and complex changes occur during the cycle of large shocks along a given fault segment.

Earthquakes occur when stress exceeds the strength of a portion of a fault. During intervals between earthquakes, stress gradually increases as a result of plate tectonic loading. It is then suddenly released as the fault slips during an earthquake. The stress condition of a fault system is not uniform, but rather a complex distribution governed by the history of earthquakes, especially large earthquakes, on the fault system. We argue below that most of the complexity of the earthquake process arises from the sensitivity of the stress distribution to the details of the slip distributions in large earthquakes. Slight differences in the length of the rupture have a major effect on the subsequent state of stress of neighboring fault segments. The processes operating in the time intervals between large events, however, are usually far less complex. Thus, in the case of earthquakes, chaos and nonlinearity arise mainly during unstable sliding in large events. Even though small shocks are more numerous, large to great earthquakes account for most of the total slip that occurs in earthquakes as well as most of the seismic moment and energy release. It is mainly during large earthquakes that the system gets "stirred up" in a sensitive and complex way.

This point is illustrated using measurements from the simple "slider-block" physical model shown in Figure 2a which consists of a very long, one-dimensional elastic system of blocks dragged slowly at a constant rate. Motion of the blocks is resisted by friction along their bases. When friction weakens (decreases) sufficiently according to either slip or slip-rate relationships, a chaotic sequence of events ensues and one or more blocks slide. This is shown in Figure 2b where we follow the manner in which two identical faults (such as the one shown in Fig. 2a) with nearly identical initial conditions at time $T = 0$ differ as time progresses. Both faults evolve slowly at the same loading rate. Eventually an event occurs on one of them,

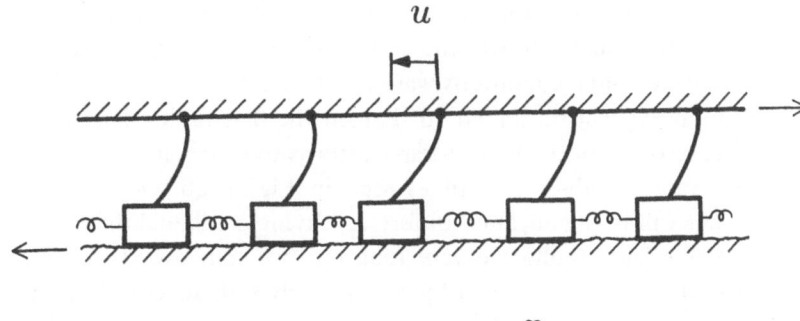

(a)

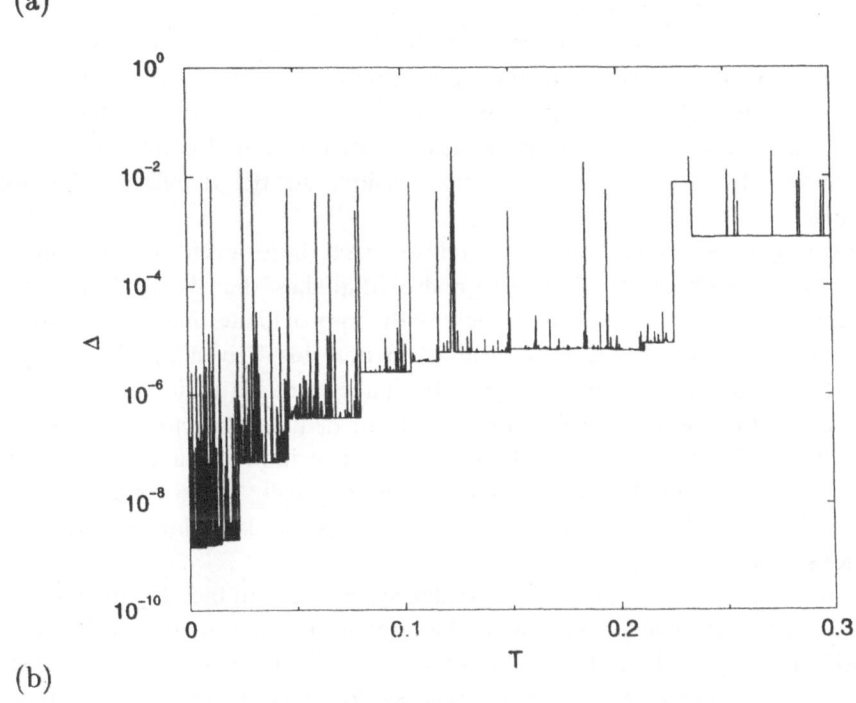

(b)

Figure 2

Chaotic behavior in a simple model of an earthquake fault. (a) Model consists of a very long, one-dimensional chain of blocks of equal mass. Each block is joined to its nearest neighbors by horizontal coupling springs of equal strength. Inclined pulling springs of equal strength attach blocks to upper plate, which is moving to the right at constant velocity. A frictional force that depends only on the slip history of blocks is present along contact between blocks and lower plate. (b) Chaotic behavior in two simple identical models of faults with nearly identical initial conditions is exhibited by exponential divergence as a function of time, T. Only a very small random initial difference between the two exists at $T = 0$. Both faults are then loaded at the same rate and evolve separately. Vertical axis shows average of absolute value of the differences between the two faults as a function of time, $\Delta = \int |u(x) - u'(x)| \, dx / \int dx$, where x is distance along each fault and u and u' are the displacements along the first and second faults, respectively. Note that on this log-linear plot the log of the separation increases roughly in a linear fashion, showing on average, the exponential growth in Δ as a function of time for two faults with nearly identical initial configurations.

and the differences between the two systems, Δ as defined in Figure 2, increase by the net slip of that event. Shortly thereafter, a corresponding event occurs on the other fault since its initial conditions were nearly identical to the first. Typically, the corresponding event will be similar to that on the first fault. Hence, Δ, a measure of the difference between the two systems, returns to nearly its value before the two events. The events produce the spikes seen in Figure 2b. Occasionally, a corresponding event will be significantly different, leaving a residual difference or step in Δ. Since the simulated faults are very long and we sum over their entire lengths, many large events (in which tens of blocks typically slide together) occur during the average repeat time, $T = 1.0$, of a particular point on one of the faults.

Several important phenomena are illustrated in Figure 2b. First, most of the net increase Δ (i.e., when the spikes do not descend) occurs in large events (i.e., the largest spikes). Large events cause most of the divergence between the two systems. Second, the divergence grows roughly exponentially with time (since log Δ increases approximately linearly with time). This exponential divergence is the hallmark of chaotic behavior. Third, the system is very sensitive to initial conditions; its state immediately after a large event is very sensitive to the details of the initial conditions.

Predicting slider-block events beyond the next large event is very difficult. Fortunately, in applying this finding to real earthquakes, that which we really care most about from a societal point of view is the approximate time and location of the next large shock. Here determinism is on our side. Given the dominance of large events in the model and of large earthquakes along a major fault, it is very important for long-term prediction to measure or deduce as accurately as possible the distribution of slip in large earthquakes. This distribution mainly sets the stage for the next large event(s). It is this that we choose to call "initial conditions" from a practical point of view since we have no hope of knowing detailed initial conditions thousands of years ago.

Far from being random, chaotic, complex systems are, in fact, highly correlated. Thus, despite that which will always be limitations in our knowledge of the distribution of slip in the last large earthquake and of the dynamics of the system, we can use the knowledge we now possess to make probabilistic statements concerning the future extending to the time of the next large event. While the uncertainty of these statements remains an open question, prediction, in this sense is possible.

Uniformity and Complexity in the Earthquake Process

Several aspects of the earthquake process indicate that it is less complex than most aspects of the circulation of the atmosphere. Unlike the atmosphere, which moves significantly even on short time scales, faults remain stationary over periods of tens of thousands of years. Faults do not change their configuration significantly even over several cycles of large events since the displacement in the largest shocks is at most meters to tens of meters. Relative plate motion, such as that between the North American and Pacific plates in California, is remarkably uniform on a time scale of years to a few million years. Complexity in the earthquake process results mainly from the following two effects: 1) from initial conditions resulting from the distribution of slip (and hence in stress drop) in the last large shock along a fault segment and 2) changes in stress along that segment produced by nearby large earthquakes. The second can delay or advance the timing of the next large event along the fault segment under consideration (DENG and SYKES, 1997). We have a reasonable chance, however, of being able to calculate those changes in stress generated by nearby shocks and hence to improve intermediate- and long-term predictions.

Are Earthquakes a Self-organized Critical Process?

In their classic paper on self-organized critical processes BAK et al. (1988) used the analogy of grains of sand being added slowly to a sandpile (Fig. 3). For large earthquakes, as for large avalanches on a sandpile, it is important to distinguish regional and global effects. Large sand avalanches can occur at various azimuths on the pile at any time (a "global" effect). Nevertheless, once a large slide occurs at a given azimuth on the sandpile (Fig. 3b) considerable time is needed to restore that segment to the angle of repose through the slow addition of grains from above such that a large avalanche can reoccur at the same place. In contrast, small avalanches (Fig. 3c) can occur at any time along a given azimuth since they affect only a small part of the slope at that azimuth.

Small earthquakes in a given region and large shocks globally manifest a similar behavior and follow the Gutenberg-Richter frequency (N)-magnitude (M) relationship, $\log N = a - bM$, where the slope, or "b value," is a constant of about 1.0. Large to great shocks along a major fault system are akin to large avalanches along a given azimuth of the sandpile (Fig. 3b). This effect can be seen in the pattern of earthquakes in the San Francisco Bay area. A broad neighboring area in which shear stress was lowered in the 1906 earthquake was very quiet for events of $M \geq 5$ for about 70 years after 1906 (e.g., SYKES and JAUMÉ, 1990). Shocks of that size were about 10 times more numerous from 1883 to 1906. Thus, the greater San Francisco area can be regarded as at or close to a self-organized critical (SOC) state from 1883 to 1906 as manifested by the frequent occurrence of moderate to large earthquakes. This effect is illustrated for the sandpile in Figure 3d. Most of the greater San Francisco area was dropped below a SOC state for decades after 1906 akin to the azimuth of the sandpile in Figure 3b that recently experienced a large avalanche. A portion of the area affected by the 1906 shock became more active for moderate-size events from 1978 until just before the Loma Prieta earthquake of

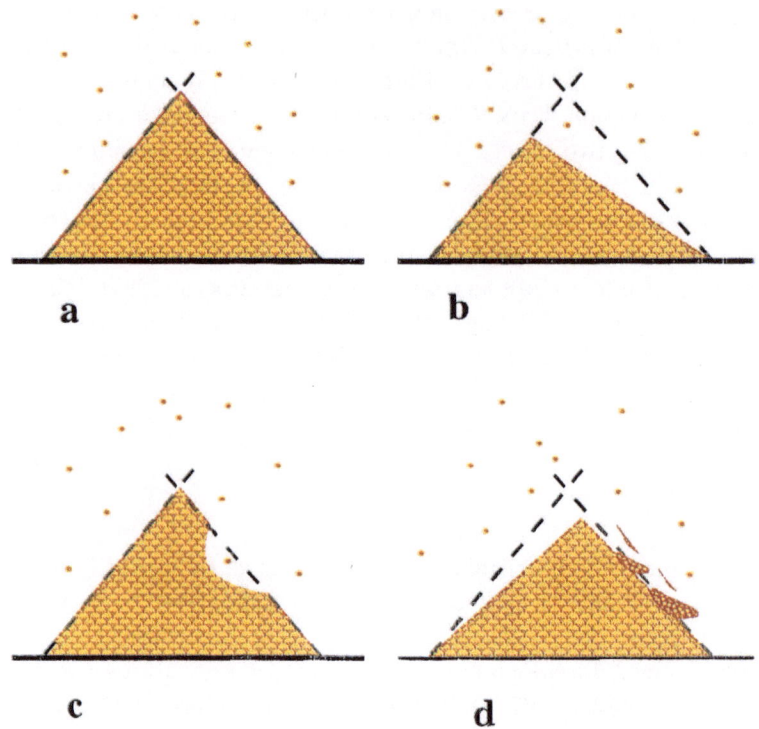

Figure 3

Grains of sand (small dots) being added slowly to a sandpile. (a) All sides of the sandpile have reached the angle of repose whereby additions of sand result in instabilities, i.e., avalanches, of various sizes. (b) A large avalanche has taken place along one small range of azimuths of pile taking that zone out of a self-organized critical state and making it incapable of being the site of a large avalanche for a long time (until grains of sand are added to it to bring its slope back to the angle of repose; large avalanches can still occur at any time along other azimuths. (c) A small avalanche occurs along one azimuth but does not affect its entire downdip slope; small avalanches can still occur along other portions of slope either up or downdip of that small avalanche. (d) Moderate-size avalanches occur as a given azimuth approaches or reaches a state of instability prior to a large avalanche. Large and small avalanches correspond to large and small earthquakes in Figure 1.

1989 (SYKES and JAUMÉ, 1990; JAUMÉ and SYKES, 1996, 1999; SYKES, 1996). In fact, five years before the Loma Prieta earthquake SYKES and NISHENKO (1984) proposed that the increase in regional moderate-size shocks from 1978 to 1983 may represent a long-term precursor to a large event on the nearby section of the San Andreas fault. The rate of moderate-size events in the greater San Francisco area returned to a low level from 1990 until the writing of this paper in early 1999 and hence may be regarded as out of a SOC state for large events, i.e., those of $M \geq 7$.

TRIEP and SYKES (1997) found that a large region of Asia to the north of the Himalayas was considerably more active for shocks of $M \geq 7$ in the decades before the giant (M_w 8.7) Assam earthquake of 1950 than in the decades since. The increased activity before 1950 and the fact that the b-value during that period remained about 1.0 even up to events of $M > 8$ led them to conclude that the region was in a SOC state before 1950 and dropped out of that state in the following decades. The eventual return to a SOC state for a region that has been in such a stress "shadow" for many decades may be regarded as a long-term precursor.

Many critics of earthquake prediction argue that the constancy of the b value, i.e., the slope of the log frequency-magnitude relationship, over a far-reaching range of sizes of earthquakes implies that no spatial or temporal scale exists for the earthquake process. The b value does change, however, in going (Fig. 1) from small to large earthquakes (PACHECO et al., 1992; MAIN, 1996; TRIEP and SYKES, 1997). Most importantly, however, the Gutenberg-Richter relationship applies "in the large," i.e., either to the entire world or to regions larger than the size of the rupture zones of large earthquakes. It breaks down for fault segments comparable in size to the rupture zones of individual large earthquakes. The strongest evidence for this is that the observed rate of occurrence of large shocks along individual segments of fast-moving faults is considerably higher than the rate extrapolated from the occurrence of small shocks using the Gutenberg-Richter relationship with a b value of about 1.0. Hence, the often-repeated argument invoking the Gutenberg-Richter relationship as a justification for self-similarity among earthquakes, the existence of a SOC state at all times and places, and the lack of earthquake predictability is not correct at the scale of individual fault segments. Those arguments only apply to the less interesting case (at least for prediction) of earthquakes in a very large region such as the entire world, a continent or all of California.

Is the Strain Buildup and Release Theory of Reid Still Valid?

In his famous paper interpreting the geodetic data collected before and after the 1906 earthquake, REID (1910) proposed that the next 1906-type event in the San Francisco Bay area along the San Andreas fault would recur about the time that stress was restored to the level just before the 1906 shock. He proposed making geodetic measurements of deformation to ascertain that approximate time, i.e., to make a long-term prediction. His strain buildup/relief theory, as subsequently interpreted in a plate tectonic framework, is the basis of much of seismology including seismic gap theories and many long-term earthquake predictions. He assumed that stresses build up slowly, by what we now recognize as plate motion.

KAGAN and JACKSON (1991) challenge this view. They claim that once aftershocks are removed from consideration that clustering, not quasi-periodicity, characterizes the occurrence of all other earthquakes. Neither they, we, nor others, however, claim that large shocks occur strictly periodically along a given segment of an active fault. Rather, what is at issue is whether large shocks can recur soon after a previous large event along the same fault segment or whether a considerable waiting time is required for stresses to slowly re-accumulate.

In their study, KAGAN and JACKSON (1991), use data sets that consist largely of small events as defined by us in Figure 1. In their analysis of the global Harvard catalog of seismic moments of shallow earthquakes they take a lower cutoff magnitude, M_w, of 6.5, claiming "these shocks are large enough to be plate-rupturing." A significant number of those events occurred at subduction zones. Their choice of M 6.5 as being a large event, i.e., one that breaks the entire seismogenic (downdip) width of a fault, is not correct since the transition from small to large events occurs at about M 7.5 for subduction zones and M 7.0 for many intracontinental regions (TRIEP and SYKES, 1997). Kagan and Jackson use an even lower cutoff, $M = 1.5$, in their analysis of the Calnet (California) catalog.

Kagan and Jackson state correctly that quasi-periodic behavior would result in a deficit of pairs with short interval times. Statistical statements at high levels of confidence can rarely be made for the repeats of large shocks that rupture the same or a similar fault segment. Since the time intervals between such events is typically greater than 50 to 100 years, the number of known and well-described repeats of large shocks of that type is small. Probably the best sequence of this type, however, is the long record of historically great earthquakes along the Nankai trough of Japan. Those events are reasonably described as quasi-periodic; none have reruptured the same part of the plate boundary in a brief time (years to a few decades) compared to the average repeat time of about 100 to 200 years.

A relatively new set of data has become available on repeats of small shocks along parts of the San Andreas and Calaveras faults in California. ELLSWORTH (1995), NADEAU and MCEVILLY (1997) and NADEAU and JOHNSON (1998) use a correlation of waveforms from densely-spaced stations to demonstrate that families of repeating small events of similar magnitude occur along the same fault patches to within the accuracy of their relative locations, 3 to 30 m. Events they examined were of magnitude 0.2 to 5. Those of $M > 3$ have calculated source dimensions larger than 100 m, i.e., larger than the uncertainties in the relative locations of events in individual sequences. Thus, the events in each of those sequences ruptured the same or nearly the same fault patch. The events of $M < 3$ have calculated source dimensions of about 1 to 10 m. Nevertheless, the identified shocks in each of those sequences also probably break the same fault patch since creep, not earthquakes, account for a large amount of the plate motion along those fault segments, and the patches with earthquakes appear to be relatively isolated spatially.

The number of small earthquakes breaking each fault patch, up to 13, and the number of patches are large enough that various models of recurrence can be tested with high statistical confidence. Events associated with a given patch are usually of similar size and occur quasi-periodically in time. Thus, their temporal probability distributions are decidedly non-Poissonian, i.e., non-random. Events along the same fault patch do not recur without sufficient time for stress to re-accumulate.

Temporal clustering does occur for patches in close proximity to one another where one event triggers the occurrence of a nearby earthquake soon thereafter. A similar clustering of adjacent events in time is well known for large shocks along major plate boundaries.

We think that the examples of clustering cited by KAGAN and JACKSON (1991) also involve nearby events, not those that rerupture the same or nearly the same fault segment. Thus, their study does not invalidate the strain buildup hypothesis of REID (1910). In contrast, the work on small repeating earthquakes strongly supports Reid's hypothesis as well as it indicates that events along the same fault segment often occur quasi-periodically. ROSENDAHL et al. (1994) find a similar behavior for large avalanches on an actual sandpile of the type illustrated in Figure 3. These results suggest that long-term prediction has promise.

The rupture patterns of large shocks along many active plate boundaries often differ from one sequence to another. A single segment may rupture in one event while two or more segments may break together in a subsequent large earthquake. For example, the 1906 California earthquake ruptured about 430 km of the San Andreas fault. Only about 50 km of that zone near its southeastern end reruptured in 1989; two segments some 90-km long ruptured in 1838 (TUTTLE and SYKES, 1992). The next rupture of a fault segment or segments is controlled strongly by the distribution of slip (and stress drop) in the last large event(s). For example, the 1989 shock ruptured that portion of the 1906 fault break that experienced the smallest displacement in 1906. Thus, it was along that segment that stresses were restored the soonest to their pre-1906 level. That segment was widely recognized as more likely to rupture in the few decades after 1983 than segments to the north of San Francisco that experienced larger slip in 1906 (LINDH, 1983; SYKES and NISHENKO, 1984; SCHOLZ, 1985; SYKES, 1996; HARRIS, 1998a,b). Hence, long-term predictions could be made for the next cycle of large shocks along the 1906 rupture zone once the distribution of slip in 1906 was ascertained approximately. Presumably, better knowledge of that distribution both in depth and along strike would have led to more accurate long-term or even intermediate-term predictions. Predicting more than one cycle ahead, however, appears to be inherently difficult, if not impossible, since considerable nonlinear behavior associated with a fault segment occurs at the time of large to great earthquakes.

Time Scales and Physical Bases for Predictions

Table 1 describes various types of earthquake predictions, their known or inferred scientific basis (or bases) and examples of some mitigation measures that might be taken for each. We also present our assessment of the feasibility in principle of being able to make various predictions, assuming that they apply to an active region which has been studied extensively and is well instrumented. To the extent possible, we endeavor to use physical understanding of various precursory effects in separating and defining the types and time scales of predictions. We strive to employ either knowledge (or inference) relative to what is now possible, what may be possible and what is likely to be inherently unknowable.

Long-term Prediction

We use 10 to 30 years as the warning time for long-term predictions and have in mind very active faults for which individual fault segments are characterized by average repeat times of about 50 to 300 years. Predictions of this type are based on knowledge of the average repeat time and variations in individual repeat times for each segment, the size of its last large shock and the time that has elapsed since it occurred. The physical bases for this type of prediction are the slow buildup of stress, the loading rate for each fault segment, and the timing of the warning interval with respect to the approximate time remaining in the cycle of large earthquakes. Predictions of this type are usually probabilistic in nature to allow for observed differences in individual repeat times and uncertainties in the parameters used in the calculations.

In the United States a consensus exists for 30-year predictions of this kind for very active faults of northern and southern California. Since the first US Government report of this type a decade ago (WORKING GROUP ON CALIFORNIA EARTHQUAKE PROBABILITIES, 1988), follow-up studies have been published for northern and southern California in 1990 and 1995, respectively. The fact that these are consensus documents and that they have been reviewed by governmental panels such as the National Earthquake Prediction Council and the California Earthquake Prediction Council appear to be major factors in their use and acceptance by the public, large corporations and several governmental organizations. It is clear that a variety of stakeholders (i.e., "customers") exists even for long-term predictions. A major electrical utility in California used the reports in selecting the site for a new multi-billion dollar generating facility. The results of the 1990 report for the greater San Francisco Bay area were featured in an insert in local newspapers focussing on earthquake risks and how citizens could prepare for the next large earthquake in the Bay area. Over a million copies of that insert were distributed. Responses to long-term predictions are likely to differ from those of the owner of a single-frame home to those of a corporation planning a major investment for a facility with a projected lifetime of decades.

The most successful long-term forecasts for large shocks in California were those for the Loma Prieta earthquake of 1989, which are summarized in HARRIS (1998a,b). Long-term predictions for either 20- or 30-year duration made in the 1980s for the next Parkfield earthquake do not expire until 2003 or later. Nevertheless, the intermediate-term prediction of BAKUN and LINDH (1985) that the next Parkfield earthquake would occur before 1993 was not correct. Two explanations have been put forward to explain that failure: 1) the 1934 and 1966 Parkfield events did not break the same segments of the San Andreas fault and 2) stresses, including perhaps fluid pressures, at Parkfield were perturbed significantly by the nearby Coalinga earthquake of 1983. Perhaps equally as successful as the Loma Prieta forecasts were predictions (WORKING GROUP ON CALIFORNIA EARTHQUAKE PROBABILITIES, 1988) that certain segments of the San Andreas fault, such as that to the north of San Francisco where slip in 1906 was high and the Carrizo segment where slip was as great as 9 m in the great shock of 1857, had a low probability of rupturing in large events in the subsequent 30 years. Large shocks have not occurred along any of the segments assigned a low probability in the 1988 document.

The accuracy of long-term predictions probably can be improved through better knowledge of initial conditions, i.e., slip in the last large shock and better modeling of fault interactions. We think that it is unlikely, however, that forecasts based on these types of data can be improved to better than about 10% of the average repeat time of large shocks along a given fault segment. Nevertheless, that would be about 8 years for the Loma Prieta segment of the San Andreas fault. It would be longer, of course, for faults with lower long-term slip rates.

Seismic Gap Theory

A crude form of long-term prediction used in the absence of quantitative data called seismic gap theory simply states that large events along a specific plate boundary segment will be widely separated in time. We do not want to suggest, however, that seismic gap theory is more advanced than a first, primitive step towards prediction. Most work on seismic gap theory was published before 1982. Considerable subsequent work (e.g., SYKES and NISHENKO, 1984) has focused on making time-varying probabilistic predictions for fault segments along the main plate boundary in California, offshore western Canada, southern Alaska and the Aleutians. Those calculations took into account the rate of loading for each fault segment, the size and date of its last large event, its average repeat time and their standard deviation, which simple seismic gap theory does not. Many of the newer calculations not only were more quantitative but also they took into account pronounced gradients in slip in great earthquakes, such as from about 2 to 6 m along the rupture zone of the 1906 California shock, which gap theory could not. The application of simple gap theory in about 1981 merely yielded the result that the entire 1906 zone had not ruptured in many decades. LINDH (1983) and SYKES and NISHENKO (1984) proposed that reloading had brought stresses along the southeastern end of the 1906 rupture zone close to their pre-1906 levels whereas stresses along segments to the north of San Francisco where slip was highest in 1906 were still far below their pre-1906 levels. Similarly, that portion of the rupture zone of the 1968 Tokachi-oki earthquake where slip was smallest in 1968 broke in the great thrust earthquake of December 1994 (TANIOKA et al., 1996). Thus, we conclude that time-varying, long-term probabilistic predictions that include information treating pronounced gradients in slip in the last great earthquake represent an advance with respect to simple gap theory. The accuracy of the newer probabilistic methods, however, does require more information about initial conditions, such as the distribution of slip in the last large shock, particularly for great events that rupture several fault segments.

Intermediate-term Prediction

We take the warning time for this type of prediction to be one month to ten years. Changes in chemistry, fluid pressure, seismicity and strain have been observed which have time scales of this period.

The rate of occurrence of moderate-size events along nearby faults is known to have increased in the last 5 to 25 years of the seismic cycle of numerous large to great earthquakes (SYKES and JAUMÉ, 1990; JAUMÉ and SYKES, 1996, 1999; TRIEP and SYKES, 1997). By our definitions these changes include both long- and intermediate-term precursors. The physical mechanism of these increases in seismicity appears to be the return of stresses to levels that existed prior to the last large to great shock in the neighborhood of the fault segment under consideration. If so, the warning time is expected to scale with the average repeat time for that fault segment. Simple computer simulations of earthquakes along a one-dimensional fault show increased organization and an increase in the frequency of moderate-size events prior to the occurrence of large to great events (PEPKE *et al.*, 1994).

Two earthquakes near Lake Elsman, California, of about M 5.3 occurred in the 1.5 years before the 1989 shock. The Loma Prieta segment and the adjacent Peninsular segment of the San Andreas fault to the northwest were characterized by very low levels of activity of that size in the 70 years after the 1906 earthquake. The Lake Elsman shocks were interpreted by some as being on the San Andreas fault and hence as possible foreshocks to a larger event. In retrospect we now understand that they occurred on a nearby subparallel fault to the one that ruptured in 1989. Hence, they neither reflected the nucleation of slip on the 1989 rupture zone itself nor were they short-term precursors. Instead, they more likely represent an intermediate-term precursor that reflects the return of stresses to pre-1906 levels close to the fault segment that ruptured in 1989. While their number was small, their occurrence led to increased anxiety among a number of earth scientists about an approaching larger shock. Better knowledge of fault geometries on a scale of kilometers and of which of those faults are either likely or unlikely to produce large earthquakes could help distinguish moderate-size events that occur on nearby faults, and hence are not likely to represent nucleation of rupture in a large event, from those that are located on the main fault itself (SYKES, 1996). The occurrence of moderate-size events of this type may provide a means to improve intermediate-term predictions at the year to decade level for places where precise locations of events are possible. A capability to locate earthquakes at the kilometer or better level so as to distinguish one nearby fault from another does not exist for the shallow plate boundaries at subduction zones.

The Joshua Tree earthquake of April 1992 occurred close to the southernmost portion of the San Andreas fault, a segment that has not ruptured in a large to great event in about 300 years. Its occurrence led to the issuance of a short-term warning for a larger nearby shock in the next few days. Such an event did not occur during that time frame. Aftershocks, however, did migrate northward in the next two months to near the epicenter of the large Landers earthquake of M 7.2. Thus, in retrospect, the Joshua Tree sequence appears to have been causally related to the coming Landers event and may be regarded as an intermediate-term precursor.

Wyss and his colleagues (e.g., WYSS et al., 1996) report precursory quiescence for small earthquakes along parts of the coming rupture zones of many large earthquakes. These reports of quiescence are not in conflict with increased levels of moderate-size shocks before large earthquakes since reports of quiescence are associated with small shocks along the rupture zones of coming large events while the latter involve moderate-size events in a considerably larger neighboring area. The time scales of quiescence are about 0.5 to 3 years and hence are intermediate-term precursors by our definition. The physical mechanism of quiescence probably differs from that controlling the increased frequency of moderate-size regional events. Quiescence of this type may reflect either dilatancy hardening of materials along parts of a fault zone or slip weakening late in the seismic cycle (SCHOLZ, 1988, MAIN and MEREDITH, 1991).

Chemical changes in ground water have been monitored at numerous sites in Japan for up to 20 years. An anomalous change in radon in the few months before the 1978 Izu-Oshima earthquake and in the concentration of chloride and in sulfate ions in the five months before the 1995 Kobe earthquake (TSUNOGAI and WAKITA, 1995; WAKITA, 1996) are two of the clearest examples of precursory changes in chemistry. Changes in radon and in water level have been observed in some wells in Japan and California although not in others (WAKITA, 1996; ROELOFFS and QUILTY, 1997). It is clear, however, that some wells are more responsive to earth tidal stresses and changes in barometric pressure than others, which may explain the presence of precursory signals at some but not other nearby wells (ROELOFFS and QUILTY, 1997). Anomalous changes in earth strain in the months before some large earthquakes generally have been reported at single observation points, which is to be expected given their sparse deployment. For geochemical, hydrological and strain changes to be accepted widely as precursors by the scientific community, multiple observations for individual earthquakes are needed. To "catch" multiple examples of changes in one or more parameters will require a substantially greater concentration of observations than presently exist except in a few places. In the United States such a concentration of observations exists only at Parkfield. To our knowledge, however, changes in chemistry are not being monitored there.

We think the prospects for intermediate-term prediction are better than for short-term prediction. A prediction of five years would permit a number of measures to be taken that would not be possible for short-term predictions alone. These include strengthening of critical structures and lifelines, stockpiling more emergency supplies, improving emergency response and conducting disaster drills (SYKES, 1996). Predictions of that type are not possible today but might be possible in about twenty years if programs to investigate intermediate-term precursors were pursued actively. Undertaking serious mitigation measures would require a scientific and technical consensus about intermediate-term prediction of the type that exists today for long-term predictions in California. Such warnings need not be at a confidence level greater than say 95% for at least certain mitigation measures to be undertaken; however, they probably would need to be at a level higher than 50% for many potential "users" of those warnings to do so.

Short-term Prediction

Long-term and intermediate-term predictions will not satisfy everyone as being "real" earthquake predictions. Can we do any better? Is short-term prediction with a lead time of weeks or hours impossible as a number of authors contend? This issue does not depend on the SOC nature of seismicity. It depends on whether or not there is a precursory phase of the earthquake instability that can be detected confidently with instruments. Frictional instability theory indicates that the stick-slip instability that results in earthquakes must be preceded by an aseismic nucleation stage (e.g., DIETERICH and KILGORE, 1996). Its likely existence means that short-term prediction is possible in principle but we don't know whether or not it can be accomplished in practice.

The most commonly observed phenomena that are a likely manifestation of this nucleation process are foreshocks. This is simply a result of the fact that the only continuously recording instruments that are widely deployed are seismometers and their effective pass band is limited to periods less than one hour. Nucleation, however, is inherently an aseismic process to which foreshocks typically are incidental. For many large earthquakes, foreshocks are either absent or of very small size, making them unreliable for prediction purposes. Hence, the question of whether short-term prediction is possible depends on whether or not a means can be devised to detect nucleation directly, and if the form of the nucleation is predictive of the size as well as the time and the place of the subsequent event. These are questions to which we do not know the answers. The size of the nucleation phase of large earthquakes is hotly debated (e.g., ABERCROMBIE *et al.*, 1995; DODGE *et al.*, 1996) and, we think, unresolved. Its depth of occurrence and dimensions probably will determine whether short-term precursors can be detected or not.

Strain measurements give the best signal-to-noise ratio for earth deformations with periods from about an hour to months. It is in this period range that slow nucleation of rupture is likely to occur. GPS is usually superior for longer periods and seismometers for shorter periods. We think that greater emphasis needs to be given to the detection and study of aseismic nucleation, not foreshocks, for short-term prediction. State-of-the-art strain measurements are not being made with sufficient density except in a few places.

It is often stated that it is not possible to ascertain if a magnitude say 2 or 4 earthquake will be the particular small event that develops into a large event. During the early and middle part of the cycle of stress buildup to large events, however, a small event along a fault segment is unlikely to grow into a large earthquake along the same segment. Even in the later part of the cycle of stress buildup, it may be that it is only small shocks in certain locations, such as near a large asperity, that have the possibility of either being foreshocks or continuing their rupture process to become large events.

Unlike intermediate-term predictions, insufficient time is available in response to short-term predictions to take many mitigation measures, such as retrofitting buildings and lifelines (SYKES, 1996). The complexities of the nucleation phase of rupture in large events are likely to make short-term prediction approximately 100 days in advance inherently impossible.

Immediate Alert

Seismic waves, especially more damaging shear and surface waves, travel at wave velocities considerably less than the speed of electromagnetic waves. Several urban areas in California and elsewhere are located 50 to 150 km from faults that are capable of generating large to great earthquakes. Warning times of seconds to tens of seconds are possible if information pertaining to strong shaking close to the rupture zones of large shocks is detected, transmitted to an analysis center and used to generate an immediate warning of impending strong shaking in adjacent areas. Such a system is under development in southern California (KANAMORI *et al.*, 1997). It will be used first to rapidly ascertain the locations of strong shaking in future large events and to infer likely or possible levels of damage and loss of life. These will greatly aid rapid emergency response, something that is often delayed many hours (and sometimes days) as a result of inferior information on the loci of strong shaking, damage, injury and loss of life.

Such systems also could be used for immediate alerts that could permit the shutting down of critical valves in refineries, the insertion of control rods in nuclear reactors, the stoppage or slowdown of trains, and other measures that could be implemented in seconds to tens of seconds. These types of mitigation measures would need to be preplanned and initiated nearly automatically. Issuing immediate alerts appears to be doable in principle but must be developed and tested. While an immediate alert would not be useful for a city very close to the rupture zone of a large earthquake, nevertheless, sensors still could provide for rapid assessment of the disaster and rapid emergency response.

Responses to Earthquake Warnings

Many of the recent critics of earthquake prediction argue that high reliability is a prerequisite for engaging in [short-term] prediction. MAIN (1997) states "otherwise a programmed evacuation could not take place." GELLER *et al.* (1997) state that predictions need to be reliable, accurate and short-term "to justify the cost of response." Many scientists equate prediction with short-term prediction in the erroneous belief that only it can result in significant mitigation of either damage or loss of life. Others argue that short-term prediction is the public's desire or expectation.

We agree that evacuation of population centers is an extreme measure that should only be undertaken when the threat is dire and the benefits outweigh the costs. The Los Angeles area, however, could not be evacuated in hours and probably not in a few days. Even given a perfectly reliable short-term prediction, deaths and injuries resulting from an attempt to evacuate the region would likely surpass those resulting from the earthquake itself. Many things can be done in response to short-term earthquake warnings that are far less drastic than evacuating cities. Considerable literature exists in the social sciences addressing various responses to hazard warnings.

In the United States forecasts of natural hazards such as those for severe storms are made in more than one category. For example, a "hurricane watch" is a forecast of lower probability than a "hurricane warning." Those engaged in emergency response, governmental officials and the public generally respond differently to those two types of forecasts. If applied to weather forecasting, the high accuracy demanded by GELLER *et al.* (1997), MAIN (1997) and others would preclude many one- to four-day forecasts and would result in the omission of lower-level warnings for severe storms.

Many different users or potential users exist for warnings of natural hazards. There is no single user of hazard warnings, i.e., a generalized "public," but rather many different stakeholders or "customers," each with different assets at risk and different "costs" associated with taking measures to reduce loss of property and/or life. For example, NASA moved a multi-billion dollar space shuttle from its launching pad back into its assembly building in response to a watch issued for hurricane Georges in 1998. Others merely waited to see if the watch would be either cancelled or upgraded to a hurricane warning.

Scientifically-based predictions for many natural hazards are likely to be probabilistic statements. The U.S. public has, in fact, become accustomed to forecasts that give the probability of rain tomorrow or a few days from now. Past claims that the public cannot understand or deal with probability no longer seem valid.

We urge earth scientists to take the lead in educating one another and the public about that which can be accomplished now, what might be accomplished in the next few decades and what may well remain unknowable for various time scales. We may not be able to deliver on highly-accurate short-term predictions. Nevertheless, our inability to produce the spectacular should not prevent us from engaging in mitigation measures that are more modest, and to be proud of those accomplishments. Clearly, debate is needed relative to the benefits and costs of predictions of various reliability and for those of different warning times.

Summary

We assert that earthquake prediction involves several different time scales: immediate (a few to tens of seconds), short-term (hours, days, weeks), intermediate-term (one month to one decade) and long-term (10 to 30 years). Precursors or possible precursors in each of these categories appear to have different scientific bases. We emphasize that which is possible now (long-term prediction in a few well-studied areas), what might be achievable in a few decades given a vigorous program of monitoring and study (intermediate-term predictions on a time scale of several years to a decade) and what now appears to be inherently impossible (predictions of large shocks more than a cycle in advance or of short-term predictions of events far into the future). While earthquakes are the culmination of complex physical processes, that does not mean that predictions of all types are impossible. We argue that a variety of responses is possible for predictions of various warning times which occupy a middle ground between doing nothing and evacuating cities. In addition, a variety of potential users (i.e., consumers or stakeholders) exist for prediction information. Each has different assets (including possibly their lives) at risk for different warning times. The mitigation measures they may take are likely to vary.

GELLER et al. (1997) claim that "the obvious ideas [in prediction] have been tried and rejected for over 100 years." In fact, only a few individual earth scientists, like Imamura in Japan, worked seriously on prediction prior to 1965. Nearly all work regarding prediction worldwide has been carried out during the last 30 years. Reid's proposal in 1910 for long-term prediction based on geodetic measurements was not instituted in the United States for more than 60 years. Parkfield is the only area in the U.S. monitored by a dense array of sensors which measure a variety of physical and hydrological parameters. This is despite the recommendation in 1965 by a Presidential panel for similar arrays and a multiplicity of types of measurements along major faults in California, Alaska and the Aleutians to better understand the earthquake process and to record possible precursors to earthquakes. The

U.S. National Earthquake Prediction Evaluation Council (NEPEC) warned in 1986 of the danger of "putting all of the U.S. eggs in the Parkfield basket" and recommended intense monitoring of about 10 segments of very active, major faults in California and Alaska. Recently, this was done in part for the Hayward fault but for no others recommended in 1986. Unless dense monitoring of a variety of physical and chemical parameters is undertaken in many other areas in the U.S. the chance of "catching" a large event and its possible precursors in the next twenty years is unlikely.

Considerable progress has been made in the last 20 to 30 years in understanding the earthquake process, including the plate tectonic bases of earthquakes, the nature of frictional processes, precursory slip in the lab and the possible role of fluids at depth in fault zones as well as in determining rates of strain buildup using GPS and other instruments, developing geological techniques to detect large paleoseismic (pre-historic) events, determining long-term rates of fault movement, modeling the evolution of stresses in California, developing instruments that have detected so-called slow earthquakes, discovering guided waves along fault zones, detecting quiescence of small shocks in the immediate vicinity of coming large earthquakes and delineating increased rates of moderate-size shocks in larger neighboring regions before large earthquakes. Long-term (30 year) forecasts have become accepted by many in the scientific and policy communities for many faults in California. Much remains to be done, however, in understanding the earthquake rupture process itself, detecting the nucleation of slow slip at depth along faults and other slow changes in deformation, detecting temporal changes in fluid pressure at depth in fault zones, and ascertaining why some subduction zones experience shocks as large as M_w 9 while others lack historic earthquakes larger than M_w 7.

We do not claim that earthquake prediction is easy; the subject is still in its infancy for all time scales. GELLER *et al.* (1997) state that the scientific question of whether prediction is either inherently impossible or just fiendishly difficult can be addressed using a Bayesian approach. They claim that each failed attempt at prediction lowers the *a priori* probability for the next attempts. The recent proof of Fermat's Last Theorem after hundreds of years of failed attempts is an interesting counter-proof. Geller *et al.*'s argument might hold more weight if nothing of significance had been learned in the meantime (or the learning curve is negative). But we think significant progress in understanding earthquakes has been made in recent decades. We would agree that many reported precursors that have been identified retrospectively are a product of either environmental noise or low signal-to-noise ratio. We do not, however, put all precursory observations in those two categories.

We favor a stepped approach that starts with estimating long-term potential and then moves to probability-based long-term prediction and thence to intermediate-term predictions. In most cases attempts to make a more specific prediction should be reserved for those fault segments that are identified as having moderate to high probability in the less specific category of warning. Resources will always be scarce enough that choices must be made carefully about fault segments to receive increased levels of funding for monitoring and studies. That is best done by progressively increasing investments as a fault segment is categorized as being likely to rupture in a large shock with increasingly shorter warning times. Thus, we think it is a mistake to jump from long-term to short-term prediction without attempting intermediate-term prediction. We sense that little attention is being given to intermediate-term prediction in the United States.

Of the various categories of predictions that we consider in Table 1, we know the least about the feasibility of short-term prediction. Whether it will be feasible depends on the nature and scale of the aseismic nucleation phase of large earthquakes. In addition to some of the possible mitigation measures that we list in Table 1, increased education concerning pre-event response options will be needed at each stage of prediction for it to be even potentially useful to various "consumers" of that knowledge.

We stress the need to determine the distribution of slip in large to great earthquakes along given fault segments as accurately as possible. It is these "initial conditions" as well as the effect of nearby large shocks on the distribution of stress that set the stage for the occurrence of the next large event along that segment. Monitoring networks and studies to determine that distribution of slip may well need to be tailored differently for areas of complex transpressional fault bends, such as those in California near Loma Prieta and San Gorgonio Pass, than for relatively straight fault segments. It is this legacy of accurate determinations of slip in the last large events that we can pass on to future generations so that they can better estimate the timing, location and size of future earthquakes in those areas.

While it is desirable to understand the physics of the earthquake process, it should be remembered that viable mechanisms of continental drift were not proposed until decades after data supporting drift became available. Hence, certain types of predictions may become not only possible but accepted before their full physical and/or chemical bases are established.

Some earthquakes are likely to be more predictable than others. The Haicheng earthquake in northeastern China in 1975 is an example of an event with regional long-term anomalies, which led to increased monitoring, followed by a well-defined foreshock sequence and the issuance of a short-term warning. Large earthquakes which involve thrust faulting at subduction zones may be more difficult to predict on time scales shorter than about 10 years since slip typically starts at a depth of some 50 km, considerably deeper than along transform faults and faults in other continental settings. Monitoring of thrust faults at subduction zones is usually more difficult since those zones are largely located at sea. Predictions for specific time intervals are likely to be more difficult for faults with low long-term slip rates, especially those in intraplate areas.

We are concerned that earthquake prediction is not regarded as a subject worthy of serious scientific study by many in the United States. Prediction is typically regarded as too "hot to handle" and too controversial. As a student one of us (LRS) was told by a professor in the late 1950s that continental drift was impossible and that serious young scientists should engage in other topics. Similar views are conveyed to many students today regarding earthquake prediction. Scientists working on predictions also must contend with the occupational hazard that many seers and psychics make predictions that are not scientifically based — often very specific, dire predictions. Being more specific and dramatic, they are often given greater emphasis by the mass media than scientifically-based ones. Nevertheless, we think there is room for creative and superior work on the science of earthquakes and on prediction.

Prediction does involve a commitment to monitoring and research on a time scale of decades, a time scale that far exceeds that of typical research grants and contracts in the U.S. Many Japanese scientists seem to take a longer-term view of monitoring and the prospects for prediction. Some who claim that prediction is impossible argue for increased funding for fundamental research, which may represent a nostalgia for the "good old days" when funding for basic research was easier to obtain than in the post cold-war era. Earthquake studies involve a continuum from very basic research to specific societal applications. We see room for excellent work throughout that continuum. Justification of funding for earthquake studies should emphasize the importance of both curiosity-driven research and benefits to society through the reduction of earthquake losses.

A recurrent proposal has been to cease work on prediction and earthquake research and to allocate all funding for earthquakes into better construction of buildings. Infrastructure of a major city, however, is typically built over more than 50 years. While better building codes and their enforcement can be applied to new structures at small additional costs, the greatest threats to life and property are usually associated with a legacy of older structures that were not built to modern standards. Roughly half a trillion dollars would be needed to bring older structures up to present codes in each of the several major U.S. urban areas, an expense that is unlikely to be forthcoming for many of them in the foreseeable future. The capability to make a reliable intermediate-term prediction with say a 5-years time window could provide a rationale for focusing more resources on a major urban area for which such a forecast had been issued. Many older structures could be strengthened on a time scale of a few years. We do note that the City of Los Angeles is making major efforts to strengthen or retire structures that do not meet current codes. Nevertheless, it will take decades to bring the built environment up to certain standards of earthquake resistance even if a given country focuses solely on that topic. We favor balanced national programs that include the science of earthquakes, better delineation of hazards and risks, monitoring, prediction, engineering and mitigation measures of a variety of types.

Acknowledgements

We thank K. Jacob, W. Menke and M. Wyss for their critical reading of the manuscript and J. Deng and S. Jaumé for discussions. This work was supported by the Southern California Earthquake Center. SCEC is funded by NSF cooperative agreement EAR-8920136 and USGS cooperative agreement 14-08-001-A0899 and 1434-HQ-97AG01718. Lamont-Doherty Earth Observatory contribution no. 5928; SCEC contribution no. 467.

REFERENCES

ABERCROMBIE, R. E., AGNEW, D. C., and WYATT, F. K. (1995), *Testing a Model of Earthquake Nucleation*, Bull. Seismol. Soc. Am. *85*, 1873–1878.

BAK, P., TANG, C., and WIESENFELD, K. (1988), *Self-organized Criticality*, Phys. Rev. A *38*, 364–374.

BAKUN, W. H., and LINDH, A. G. (1985), *The Parkfield, California, Earthquake Prediction Experiment*, Science *229*, 619–624.

DENG, J., and SYKES, L. R. (1997), *Stress Evolution in Southern California, and Triggering of Moderate, Small and Microearthquakes*, J. Geophys. Res. *102*, 24,411–24,435.

DIETERICH, J. H., and KILGORE, B. (1996), *Implications of Fault Constitutive Properties for Earthquake Prediction*, Proc. Nat. Acad. Sci. USA *93*, 3787–3794.

DODGE, D. A., BEROZA, G. C., and ELLSWORTH, W. L. (1996), *Detailed Observations of California Foreshock Sequences: Implications for the Earthquake Initiation Process*, J. Geophys. Res. *101*, 22,371–22,392.

ELLSWORTH, W. L., *Characteristic earthquakes and long-term earthquake forecasts: implications of central California seismicity*. In *Urban Disaster Mitigation: The Role of Science and Technology* (eds. Cheng, F. Y., and Sheu, M. S.) (Elsevier 1995) pp. 1–14.

EVANS, R. (1997), *Assessment of Schemes for Earthquake Prediction*, Geophys. J. Int. *131*, 413–420.

GELLER, R. S. (1997a), *Predicting Earthquakes is Impossible*, Los Angeles Times, Feb. 2.

GELLER, R. S. (1997b), *Earthquakes: Thinking about the Unpredictable*, EOS Trans. Amer. Geophys. Union *78*, 63–66.

GELLER, R. S. (1997c), *Predictable Publicity*, Seismol. Res. Lett. *68*, 477–480.

GELLER, R. S. (1997d), *Earthquake Prediction: A Critical Review*, Geophys. J. Int. *131*, 425–450.

GELLER, R. S., JACKSON, D. D., KAGAN, Y. Y., and MULARGIA, F. (1997), *Earthquakes Cannot be Predicted*, Science *275*, 1616–1617.

HARRIS, R. A. (1998a), *The Loma Prieta, California, Earthquake of October 17, 1989—Forcasts*, U.S. Geol. Survey Prof. Paper *1550-B*, 1–28.

HARRIS, R. A. (1998b), *Forecasts of the Loma-Prieta, California, Earthquake*, Bull. Seismol. Soc. Am. *88*, 898–916.

JAUMÉ, S. C., and SYKES, L. R. (1996), *Evolution of Moderate Seismicity in the San Francisco Bay Region, 1850 to 1993: Seismicity Changes Related of the Occurrence of Large and Great Earthquakes*, J. Geophys. Res. *101*, 765–789.

JAUMÉ, S. C., and SYKES, L. R. (1999), *Evolving Toward a Critical Point: A Review of Accelerating Seismic Moment/Energy Release prior to Large and Great Earthquakes*, Pure appl. geophys, this issue.

KAGAN, Y. Y., and JACKSON, D. D. (1991), *Long-term Earthquake Clustering*, Geophys. J. Int. *104*, 117–133.

KANAMORI, H., HAUKSSON, E., and HEATON, T. (1997), *Real-time Seismology and Earthquake Hazard Mitigation*, Nature *390*, 461–464.

LINDH, A. G. (1983), *Preliminary Assessment of Long-term Probabilities for Large Earthquakes along Selected Fault Segments of the San Andreas Fault System in California*, U.S. Geol. Survey Open-File Rep. *83–63*, 1–5.

MAIN, I. (1996), *Statistical Physics, Seismogenesis and Seismic Hazard*, Rev. Geophys. *34*, 433–462.

MAIN, I. (1997), *Long Odds on Prediction*, Nature *385*, 19–20.

MAIN, I. G., and MEREDITH, P. G. (1991), *Stress Corrosion Constitutive Laws as a Possible Mechanism of Intermediate-term and Short-term Seismic Quiescence*, Geophys. J. Int. *107*, 363–372.

NADEAU, R. M., and JOHNSON, L. R. (1998), *Seismological Studies at Parkfield VI: Moment Release Rates and Estimates of Source Parameters for Small Repeating Earthquakes*, Bull. Seismol. Soc. Am. *88*, 790–814.

NADEAU, R. M., and MCEVILLY, T. V. (1997), *Seismological Studies at Parkfield V: Characteristic Microearthquake Sequences as Fault-zone Drilling Targets*, Bull. Seismol. Soc. Am. *87*, 1463–1472.

PACHECO, J. F., SCHOLZ, C. H., and SYKES, L. R. (1992), *Changes in Frequency-size Relationship from Small to Large Earthquakes*, Nature *355*, 71–73.

PEPKE, S. L., CARLSON, J. M., and SHAW, B. E. (1994), *Prediction of Large Events on a Dynamical Model of a Fault*, J. Geophys. Res. *99*, 6769–6788.

REID, H. F. (1910), *The Mechanics of the Earthquake*. Vol. 2 of *The California Earthquake of April 18, 1906. Report of the State Earthquake Investigation Commission* (Carnegie Institution of Washington Publication 87).

ROELOFFS, E., and QUILTY, E. (1997), *Water Level and Strain Changes Preceding and Following the August 4, 1985 Kettleman Hills, California, Earthquake*, Pure appl. geophys. *149*, 21–60.

ROSENDAHL, J. M., VEKIC, M., and RUTLEDGE, J.E. (1994), *Predictability of Large Avalanches on a Sandpile*, Phys. Rev. Lett. *73*, 537–540.

SCHOLZ, C. H. (1985), *The Black Mountain Asperity; Seismic Hazard of the Southern San Francisco Peninsula, California*, Geophys. Res. Lett. *12*, 717–719.

SCHOLZ, C.H. (1988), *Mechanisms of Seismic Quiescences*, Pure appl. geophys. *126*, 701–718.

SHUKLA, J. (1998), *Predictability in the Midst of Chaos: A Scientific Basis for Climate Forecasting*, Science *282*, 728–731.

SYKES, L. R. (1996), *Intermediate- and Long-term Earthquake Prediction*, Proc. Nat. Acad. Sci USA *93*, 3732–3739.

SYKES, L. R., and JAUMÉ, S. C. (1990), *Seismic Activity on Neighboring Faults as a Long-term Precursor to Large Earthquakes in the San Francisco Bay Area*, Nature *348*, 595–599.

SYKES, L. R., and NISHENKO, S. P. (1984), *Probabilities of Occurrence of Large Plate Rupturing Earthquakes for the San Andreas, San Jacinto and Imperial Faults, California*, J. Geophys. Res. *89*, 5905–5927.

TANIOKA, Y., RUFF, L., and SATAKE, K. (1996), *The Sanriku-oki Earthquake of December 28, 1994 (M_w 7.7): Rupture of a Different Asperity from a Previous Earthquake*, Geophys. Res. Lett. *23*, 1465–1468.

TRIEP, E., and SYKES, L. R. (1997), *Frequency of Occurrence of Moderate to Great Earthquakes in Intracontinental Regions*, J. Geophys. Res. *102*, 9923–9948.

TSUNOGAI, U., and WAKITA, H. (1995), *Precursory Chemical Changes in Ground Water: Kobe Earthquake, Japan*, Science *269*, 61–63.

TUTTLE, M. P., and SYKES, L. R. (1992), *Re-evaluation of Several Large Historic Earthquakes in the Vicinity of Loma Prieta and Peninsular Segments of the San Andreas Fault, California*, Bull Seismol. Soc. Am. *82*, 1802–1820.

WAKITA, H. (1996), *Geochemical Challenge to Earthquake Prediction*, Proc. Nat. Acad. Sci USA *93*, 3781–3786.

WALLACE, R. E., DAVIS, J. E., and MCNALLY, K. C. (1984), *Terms for Expressing Earthquake Potential, Prediction and Probability*, Bull. Seismol. Soc. Am. *74*, 1819–1825.

WORKING GROUP ON CALIFORNIA EARTHQUAKE PROBABILITIES (1988), *Probabilities of Large Earthquakes Occurring in California on the San Andreas Fault*, U.S. Geol. Survey Open-File Rep. *88–398*, 1–62.

WYSS, M., SHIMAZAKI, K., and URABE, T. (1996), *Quantitative Mapping of a Precursory Seismic Quiescence to the Izu-Oshima (M 6.5) Earthquake, Japan*, Geophys. J. Int. *127*, 735–743.

(Received November 12, 1998, revised January 12, 1999, accepted January 22, 1999)

Pure appl. geophys. 155 (1999) 233–258
0033–4553/99/040233–26 $ 1.50 + 0.20/0

❙ Pure and Applied Geophysics

Is Earthquake Seismology a Hard, Quantitative Science?

Y. Y. KAGAN[1]

Abstract—Our purpose is to analyze the causes of recent failures in earthquake forecasting, as well as the difficulties in earthquake investigation. We then propose that more rigorous methods are necessary in earthquake seismology research. First, we discuss the failures of direct earthquake forecasts and the poor quantitative predictive power of theoretical and computer simulation methods in explaining earthquakes. These failures are due to the immense complexity of earthquake rupture phenomena and lack of rigor in the empirical analysis of seismicity. Given such conditions, neither "holistic," interdisciplinary analysis of geophysical data nor greater reliance on the currently available results of earthquake physics is likely to work without revising scientific methodology. We need to develop more rigorous procedures for testing proposed patterns of earthquake occurrence and comparing them to predictions of theoretical and computer modeling. These procedures should use methods designed in physics and other sciences to formulate hypotheses and carry out objective validation. Since earth sciences study a unique object, new methods should be designed to obtain reliable and reproducible results. It is likely that the application of sophisticated statistical methods will be needed.

Key words: Crisis in earthquake seismology, earthquake occurrence, statistical methods, hypotheses testing, fractals, earthquake source models.

1. Introduction

We define earthquake seismology as the study of earthquake source and occurrence. Thus, the problems of elastic wave propagation and the study of earth structure are outside the scope of this paper. It is widely accepted that failure has dogged the extensive efforts of the last 30 years to find 'reliable' earthquake prediction methods, the efforts which culminated in the Parkfield prediction experiment (ROELOFFS and LANGBEIN, 1994, and references therein) in the U.S.A. and the Tokai experiment in Japan (MOGI, 1995). EVANS (1997), GELLER *et al.* (1997), JORDAN (1997), SCHOLZ (1997), SNIEDER and VAN ECK (1997), and WYSS *et al.* (1997) discuss various aspects of earthquake prediction and its lack of success. Jordan (1997) commented that "The collapse of earthquake prediction as a unifying theme and driving force behind earthquake science has caused a deep crisis."

[1] Institute of Geophysics and Planetary Physics, University of California, Los Angeles, CA 90095-1567, U.S.A. E-mail: ykagan@ucla.edu; WEB: http://scec.ess.ucla.edu/ykagan.html

Similar difficulties and failures exist in other earth sciences, however earthquake seismology is a special case because its predictions are so important to seismic hazard reduction. Moreover, in other disciplines, empirical models or a common sense approach seem at least to be weakly predictive in a qualitative way for observations. Hurricanes, flooding, volcanic eruptions and other geophysical catastrophes can be forecast with increasing accuracy as prediction lead time diminishes. By contrast, in earthquake seismology the question remains whether current models have even qualitative predictive power which is assumed to be a property of even 'soft' sciences. For example, for long-term predictions based on the seismic gap hypothesis, it is not clear whether the predictions are better or worse than those made by the null hypothesis, i.e., by random choice (NISHENKO and SYKES, 1993; JACKSON and KAGAN, 1993).

The difference in predictive capability is due to the fact that other geophysical catastrophes involve large mass transport of fluids, which is a relatively slow process. Manifestations of such a transport can be monitored and a warning can be issued, usually a few days or hours, in advance. Earthquakes represent a propagation of fracture with a velocity of km/s, thus if their preparation stage cannot be monitored or, as some evidence suggests (KAGAN, 1997b), is absent, the prediction can only be a statistical one.

As we argue below, current research in earthquake seismology is in a fact-gathering, phenomenological stage, accompanied by attempts to find empirical correlations between the observed parameters of earthquakes. Theoretical models have not been successful in explaining, let alone quantitatively predicting, these phenomena. MARGENAU (1950, p. 28) calls this stage of scientific development "*correlation science.*" In other words, earthquake seismology is not a *normal* science in KUHN's (1970) sense of the word: there is no widely accepted *paradigm* explaining the basic properties of earthquake occurrence.

In a recent paper (KAGAN, 1997b) temporal earthquake prediction, interpreted more traditionally, has been reviewed. Therein, I argue that, earthquake prediction as understood by the majority of geophysicists and the general public—the prediction of specific individual events—is not possible. However, earthquake rates can be predicted either in a time-independent mode (seismic hazard) or as a temporally dependent rate. The major problem, presently unknown, is whether we can achieve magnitude-dependent prediction, i.e., a prediction of magnitude distribution of future earthquakes statistically different from the GUTENBERG-RICHTER (1944) law.

Here we do not focus on the question of earthquake predictability in a narrow sense (KAGAN, 1997b), but rather on a broader question, why recent research in earthquake seismology has not achieved the success of other sciences. Despite an increase in the quality and quantity of data collected, there has been no major breakthrough; no significant progress has been made in understanding seismicity and earthquake occurrence.

Two approaches are often proposed as a solution to this crisis: (1) "holistic" analysis of geophysical data (JORDAN, 1997), (2) further reliance on earthquake physics (BEN-ZION et al., 1999). As stated below, these approaches alone are unlikely to produce fruitful results since the major problem is methodological: earthquake research needs a significantly higher level of rigor. If we replace one paradigm by another and continue to conduct our research as before, relatively little will change. Thus, we argue that the major challenge facing earthquake seismology is not a paradigm shift but the application of more rigorous methods of analysis. Moreover, new methods for hypotheses verification need to be developed. The development of many sciences such as medical, biological, agricultural research, etc., especially in the second half of the 20th century, has been characterized by their transformation from qualitative "soft" disciplines to quantitative ones. This transformation has been accompanied by the application of increasingly sophisticated statistical methods.

The above distinction between mathematically based hard sciences and descriptive soft sciences can be traced back to the beginning of the scientific method in Western civilization (SANTILLANA, 1961, pp. 214–222). Whereas astronomy, and physics in general, used the Pythagorean-Platonic paradigm of an abstract mathematical model applied to observed phenomena, the Aristotelian scheme of descriptive categorization and teleological understanding dominated biological sciences. It is interesting to note that whereas biology is now moving towards mathematization, geology is still largely a descriptive science, plate tectonics being an analog of the 19th century Darwinian theory.

Below, we discuss a broader definition of prediction, summarize the failure of present research methods, and attempt to explain their causes. In the final section we consider more rigorous methods that can be applied to verify hypotheses of earthquake patterns.

2. Prediction

Quantitative prediction is the aim of every science. BEN-MENAHEM (1995, p. 1217) says: "[T]he ultimate test of every scientific theory worthy of its name, is its ability to predict the behavior of a system governed by the laws of said discipline." By quantitative prediction, in a broad sense, we mean any new numerical information on earthquake occurrence, be that the distribution of size, location, time, or other properties. ENGELHARDT and ZIMMERMANN (1988, p. 203) define *retrodiction* as a "prediction" of a past state, and *codiction* as a "prediction" of a simultaneous state. Thus, on the basis of a given hypothesis, we can "predict" past earthquake occurrence or the present state of earthquake-related variables in a certain region. Because of the stochastic nature of the earthquake process (KAGAN, 1992, 1994), the prediction must be statistical.

"The practical demand for the increased use of factually and hypothetically founded prognoses requires a complete turn around in the geologist's accustomed point of view. This exposes him to a danger he does not encounter in making retrodictions: each prognosis will at some time be proven right or wrong, and he who made it must reckon with the possibility that the course of events or the very measures he proposed may "refute" it. The backward-looking geologist is safe from such dangers, for the ambiguity of abductive inference and the impossibility of "direct" proof make it impossible to disprove 'once and for all' certain retrodictions or abductive codictions, possibly setting long years of research at nought" (ENGEL-HARDT and ZIMMERMANN, 1988, p. 230). By *abduction* the authors mean a logical determination of an initial state of a natural system, if the final state and the controlling laws are known.

The citation above points out the challenges that exist in earthquake seismology. In other scientific disciplines the failure of a model to explain and predict phenomena can be established by carrying out an appropriate experiment. The long-term recurrence time for large earthquakes makes it difficult to evaluate earthquake prediction methods. At the same time, occurrence of an unpredicted earthquake, especially in regions covered by dense observational networks, demonstrates our deficient understanding of earthquake process. Some of these unpredicted events are destructive, thus dramatizing for the general public the fundamental failure of the science.

3. Failures

3.1. Earthquake Prediction: Failures and Difficulties

3.1.1. Seismic hazard—limited success. Earthquake hazard estimates are presently based on earthquake geology, i.e., a description of active faults (COBURN and SPENCE, 1992; YEATS et al., 1997). Failures are obvious—most large earthquakes in California in the 20th century (1952 Kern County, 1971 San Fernando, 1989 Loma Prieta, 1992 Landers, 1994 Northridge) occurred on faults that had not been properly identified earlier. Similarly, several earthquakes of unexpectedly large magnitude occurred recently in Spitak, Armenia (1988), Netherlands (1992), Latur, India (1993), and North Sakhalin (1995).

Although geology, paleoseismology in particular, has achieved significant progress in assessing past seismicity, the problems of the non-uniqueness of its estimates are still very serious (WELDON, 1996). While estimates for seismic hazard due to intermediate ($M\,5-M\,7$) earthquakes in seismically active regions may differ by a factor of about 2–3, the difference may be as large as a few orders of magnitude for large ($M\,7-M\,8$) and very large ($M > 8$) earthquakes. The uncertainty in the maximum magnitude size is especially pronounced at intracontinental areas with a low level of contemporary seismicity (KAGAN, 1997b).

Standard estimates of earthquake size distribution are based on the characteristic earthquake model (KAGAN, 1997a,b). The model assumes that a maximum earthquake depends on the length of a causal tectonic fault or fault segment (SCHWARTZ and COPPERSMITH, 1984). Although no quantitative model has been proposed to identify faults and no critical testing of the hypothesis has been carried out by its proponents, the characteristic model is generally accepted in hazard estimates. KAGAN (1997a,b) demonstrates that this model greatly underestimates the size of the largest earthquakes, and the prospective (forward) testing of the model reveals its failure to describe regional magnitude-frequency relation (KAGAN, 1997b).

Several recent investigations (TANIOKA and GONZALEZ, 1998; SCHWARTZ, 1999) obtained maps of slip distribution for large earthquake pairs. These maps suggest that the slip pattern is highly non-uniform, and the slip distribution for subsequent earthquakes significantly differs from a previous event. This pattern contradicts the central assumption of the characteristic earthquake model (SCHWARTZ and COPPERSMITH, 1984): that slip distribution during characteristic events is nearly identical.

Recently, repeating microearthquakes have been studied in the Parkfield area (NADEAU and McEVILLY, 1997; NADEAU and JOHNSON, 1998, and references therein). These event sequences exhibit certain properties of characteristic earthquakes: regularity of occurrence, nearly identical wave forms and illustrate many interesting features of earthquake occurrence. However, these micro-earthquakes are not characteristic earthquakes in a strict sense: these events do not release most of the tectonic deformation on a fault segment as the real characteristic earthquakes are assumed to do (SCHWARTZ and COPPERSMITH, 1984). As with large characteristic earthquakes, attempts to predict these microearthquakes using their quasi-periodicity have not yet succeeded (R. Nadeau, private communication, 1997).

3.1.2. Long-term prospective predictions—unsuccessful or inconclusive results. Recently several experiments were carried out to forecast long-term earthquake potential. Most of these attempts were based on the seismic gap model which was first proposed by FEDOTOV (1968). SYKES (1971) reformulated it on the basis of plate tectonics and McCANN *et al.* (1979) developed a map of circum-Pacific seismic zones, including a forecast for each zone. NISHENKO (1991) based his prediction of seismic activity in the Pacific zones on a new seismic gap model which considers the recurrence time and characteristic earthquake magnitude specific to each plate boundary segment. These authors apparently did not anticipate the necessity for rigorously testing the model performance. However, it was a great service to science that they produced a prediction that can (admittedly with some difficulty) be tested, i.e., the seismic gap model was formulated as a falsifiable hypothesis.

The testing of the MᴄCᴀɴɴ *et al.* (1979) and Nɪsʜᴇɴᴋᴏ (1991) forecasts by Kᴀɢᴀɴ and Jᴀᴄᴋsᴏɴ (1991, 1995) showed that no version of the seismic gap hypothesis has yet evidenced a significant statistical advantage over a reasonable null hypothesis (Kᴀɢᴀɴ, 1997b). There is a possibility that due to the long-term clustering property of earthquakes, the gap model performs significantly worse than the Poisson (null) hypothesis, although Nɪsʜᴇɴᴋᴏ and Sʏᴋᴇs (1993) dispute this conclusion.

Two prediction experiments which generated immense public interest—Tokai, Japan (Isʜɪʙᴀsʜɪ, 1981; Moɢɪ, 1995) and Parkfield, California (Bᴀᴋᴜɴ and Lɪɴᴅʜ, 1985)—are also based on the seismic gap model. However, in contrast to the investigations mentioned above, these experiments were not planned as falsifiable statements. Despite the significant cost of these prediction experiments, the seismic gap hypothesis which was their basis was not tested rigorously either before or during the experiments. As Kᴀɢᴀɴ (1997a) indicates, even if an earthquake similar to the expected Parkfield event were to occur, no statistically significant conclusions could be drawn. Meanwhile, damaging earthquakes for which no specific long-term forecasts were issued, occurred elsewhere in Japan (1993 Okushiri Island, 1995 Kobe) and California (1992 Landers, 1994 Northridge). Although one may expect that unanticipated earthquakes may shake some remote areas, these events occurred in places covered by one of the densest observational networks, and caused significant damage. Thus the absence of any reasonable estimate of seismic hazard for these regions seems especially embarrassing.

The *M* 8 algorithm was proposed about 15 years ago by using a pattern-recognition technique (Kossobokov *et al.*, 1997) to predict long-term earthquake probabilities. The prospective testing of the model was carried out in 1992–1997 for large ($M \geq 7.5$) earthquakes in the Pacific area. Preliminary test results indicated that the algorithm performs no better than the null Poisson hypothesis (*ibid.*). Although final test results show improvement in the hypothesis' performance (Kossobokov *et al.*, 1998, 1999), the model's predictive skill for $M \geq 7.5$ earthquakes is apparently low.

The only prediction experiment planned from the beginning as a technically rigorous test of the hypothesis is the precursory swarm model formulated by Evɪsᴏɴ and Rʜᴏᴀᴅᴇs (1993, 1997). In these experiments all necessary parameters of the earthquake forecast and statistical criteria for acceptance/rejection of the hypothesis were specified before the tests were started. Both tests revealed that the research hypothesis (the precursory swarm model) does not exhibit a statistically significant advantage over the null hypothesis, i.e., the test results are negative. Tests of the modified model are ongoing.

An often expressed opinion (see, for example, Gusᴇv, 1998) is that since the tectonic deformation responsible for earthquakes is slow, we need decades or even centuries to accumulate sufficient data to test earthquake occurrence hypotheses. This may be true for certain aspects of models, however as discussion in this subsection testifies, it is possible to design experiments so that testing of long-term forecasts is accomplished on a considerably shorter timescale, i.e., in a few years.

3.1.3. Short-term prediction. Several algorithms have been proposed to use foreshock-mainshock-aftershock patterns for short-term earthquake prediction (KAGAN and KNOPOFF, 1987; REASENBERG and JONES, 1989; MICHAEL and JONES, 1998). Appropriate models for short-term clustering are very important even for studying long-term effects. Seismicity rate increases by a factor of many thousands in the wake of earlier earthquakes, whereas long-term rate changes are less than one order of magnitude (KAGAN, 1994). Thus we cannot see these long-term signals in the presence of stronger short-term effects.

However, geophysicists are primarily concerned with deterministic short-term earthquake prediction—the search for immediate earthquake precursors that would allow reliable forecasting on a time scale of days and hours. A recent debate on the VAN predictions (GELLER, 1996; KAGAN and JACKSON, 1996) shows clearly that the geophysical community lacks unambiguous methods to test even short-term predictions. Although there is no consensus on this issue, an extensive search for deterministic short-term precursors has been successful (GELLER *et al.*, 1997; WYSS *et al.*, 1997). Failure to find any reliable earthquake "precursors" (KAGAN, 1997b) may indicate that predicting individual earthquakes is inherently impossible.

3.2. Earthquake Phenomenology

A significant number of publications concerning earthquake phenomenology exists: for example, according to the INSPEC database, a paper on earthquake size distribution (magnitude-frequency relation and its variants) is published about once a week. A vast diversity exists in the description of earthquake size distributions, spatial patterns of earthquake focal zones, temporal regularities of earthquake occurrence, and temporal interrelation between earthquakes (KAGAN, 1994, 1997b).

In my opinion, most (90% or more) of the seismicity patterns published, even in the best geophysical journals, are partly or wholly artifacts. It is impossible, of course, to substantiate this definitely, but the discussion in the previous subsections (see also more in KAGAN, 1994, 1997b) demonstrates that efforts and predictions of many experienced researchers proved fruitless when tested against a future earthquake record. These investigations were carried out in the "mainstream" of earthquake seismology, thus subject to special attention and criticism. Because no unifying theoretical principle exists, a significant part of the phenomenological research explores specific problems, and thus is outside the normal cross-validation and result duplication which is normative in other quantitative sciences. Such diversity of research directions and lack of requirements for objective model testing results in lower standards.

Are large earthquakes quasi-periodic or clustered in time? What is the size distribution of the largest earthquakes in a particular area? Is the distribution characteristic or Gutenberg–Richter? Is the geometry of earthquake faults or short-term earthquake temporal behavior scale-invariant or are there several different length and timescales (see, for example, OUILLON et al., 1996, and references therein)? The differences in opinion are so great to date that no common ground has been found and, most importantly, the debate has continued for many years without resolution, i.e., the research seems largely ineffective.

3.3. Theory

The term *earthquake physics* usually denotes the physical processes and conditions that govern the occurrence of sudden slip on earthquake faults, e.g. a slip initiation and arrest, dynamic rupture, as well as all supporting processes and conditions which lead to an earthquake, including fault zone structure, tectonic loading, stress evolution and redistribution within the crust, and the operative boundary conditions. No comprehensive theory of earthquake generation process or fundamental equations exist. Thus, earthquake physics represents a set of ideas, some of them simple, based on observations of earthquake deformation and common sense taken from engineering disciplines such as rock mechanics, friction physics, fracture mechanics, and materials science, and from recent developments in statistical physics and nonlinear sciences.

Since there are no current comprehensive reviews of earthquake physics, for a more specific discussion see SCHOLZ (1990); NEWMAN et al. (1994, in particular, GABRIELOV and NEWMAN, 1994); KAGAN (1994); KNOPOFF et al. (1996); MAIN (1996); SCHOLZ (1998); BEN-ZION et al. (1999). Existing physical models of earthquakes are often mutually contradictory. Lack of agreement on the fundamental principles of earthquake rupture and seismicity patterns modeling is demonstrated by the papers cited above.

3.3.1. Elastic rebound theory. The elastic rebound theory was first formulated by REID (1910) following his analysis of the 1906 San Francisco earthquake. SCHOLZ (1997, p. 18) states with reference to the rebound theory: "... the next earthquake is likely to happen when the strain released in the previous one has been restored." In recent years the rebound model has become the basis for several hypotheses, such as earthquake recurrence, earthquake cycle, and seismic gap models (EVANS, 1997). These models, in turn, have been used in long-term prediction efforts. Tests of the seismic gap hypothesis (section 3.1.2) show that its performance is not better than the Poisson model. Researchers who feel that the recurrence hypothesis is correct and that the failure of the seismic gap models can be explained by inappropriate applications of the seismic cycle model, will do well to follow the example of McCANN et al. (1979) and NISHENKO (1991) and propose a testable, falsifiable prediction of future seismicity. Such a forecast should be made for a sufficiently large area, to be verifiable in a reasonable time span.

The problem with the Reid model is whether stress concentration and release is local, i.e., involving 10–15 km along a fault trace, or, as California data suggest (JACKSON *et al.*, 1997) there is no "strain concentration" around faults expected to slip in the near future, and the old concentrations of strain rate are after-slip of events (including those long past). This finding would suggest that strain builds up on a broad regional scale and is randomly (stochastically) released somewhere. The presence of large foreshocks or aftershocks (some of them comparable in size with a mainshock) indicates that stress is not fully released even by a very large earthquake. Geodetic measurements of deformation along known tectonic faults should help to resolve this problem.

3.3.2. Dilatancy-diffusion model. The dilatancy-diffusion hypothesis (an increase in rock volume prior to failure) was proposed in the late 1960s–early 1970s (SCHOLZ *et al.*, 1973; GELLER, 1997; SCHOLZ, 1997, and references therein) to explain the variation in elastic wave velocities, as well as other geophysical anomalies observed before earthquakes. Subsequent measurements indicated that the original observations may have been artifacts. Thus the dilatancy model was largely abandoned by the late 1970s. SCHOLZ (1997) suggests that this rejection may have been premature: "[T]hese results were accepted by consensus as deeming the failure of both the theory and of that particular form of earthquake precursor, neither of which has been seriously investigated since." We argue below, that such lack of critical testing and rigorous verification characterizes not only the dilatancy hypothesis but all others as well.

3.3.3. Rock friction in laboratory and earthquake models. Various forms of rock friction law are offered to explain and predict earthquake rupture and occurrence (SCHOLZ, 1998, and references therein). The original friction law was formulated from laboratory experiments and invoked to explain earthquakes *in situ*. However, the conditions in the laboratory experiments are radically different from the conditions *in situ*. The laboratory experiments are conducted in a block with (usually) a single polished fault, contained in an apparatus. Thus the rock friction law is based on the mechanics of man-made, engineered objects. Three clearly defined, highly different characteristic geometrical scales can be distinguished in such objects: (1) exterior or macroscopic, which applies to an apparatus and the sample size; (2) interior or microscopic, which describes the defects and intergranular boundaries of the object material; and (3) the irregularities of the object surface.

The factors connected with the latter scale are usually taken into account by the introduction of frictional forces (SCHOLZ, 1998). The analysis of fault geometry (KAGAN, 1994) indicates that earthquakes do not occur on a single (possibly fractally wrinkled) surface, but on a fractal structure of many closely correlated faults (GABRIELOV and NEWMAN, 1994). The total number of these small faults can be very large, practically infinite.

Similarly, scales (1) and (2) are usually separated in materials science and in engineering applications by introducing effective properties of the material (LEARY,

1997). In earthquake studies we consider the propagation of a rupture through rock material which, during millions of years of its tectonic history, has been subjected to repeated deformation. Material properties of rocks are scale-invariant (CHELIDZE, 1993; LEARY, 1997; TURCOTTE, 1997). Even if natural rock specimens are used in laboratory experiments, these specimens are carefully selected for internal defects to be small compared to the size of the sample. Thus the internal structure of a specimen resembles the uniform material properties characteristic of engineered materials.

Most of the mechanical models of earthquake occurrence are based on Euclidean geometrical forms that are plane surfaces of rupture, whereas the geometry of natural tectonic forms is fractal (CHELIDZE, 1993; MAIN, 1996; KAGAN, 1994). This is only one reason to doubt the relevance of laboratory experiments for earthquake prediction.

3.3.4. Computer seismicity simulation. Ideas from the engineering practice of fracture mechanics and rock friction are used in many computer models of earthquake occurrence, like various modifications of block-friction models (KNOPOFF *et al.*, 1996). Hundreds of papers treating computer simulation of earthquakes have been published in physics and geophysics journals in the last decade. As was mentioned above, these models are not based on the fundamental laws of physics, and the mechanical and geometrical properties of these constructs differ radically from natural fault systems. These models must be justified and confirmed by carefully comparing their results with earthquake data.

Several concepts taken from solid state physics (critical phenomena) or self-organized criticality have been proposed to model earthquake phenomena, although as in mechanical models, their relevance to earthquake problems is unclear, partly because a comparison of the results drawn from these theories with the experimental data is inconclusive. Although the ideas borrowed from nonlinear mechanics, critical point phenomena, and self-organized criticality explain in general the scaling properties of earthquakes and provide new tools with which to describe seismicity, these methods are still of limited value (KAGAN, 1994; MAIN, 1995, 1996).

Two reasons explain the ineffectiveness of such research: (1) considerable theoretical work is conducted in a corroborative vacuum: there is very little confrontation between experimental facts and results of simulations; (2) as discussed in section 3.2, the seismological observational evidence is inadequate and insufficient: the fundamental regularities of earthquake occurrence are still in doubt. Such a diversity of opinion makes it relatively easy to "confirm" simulation results, using one or another of the contradictory empirical "observations." Thus the major difficulty we face in modeling earthquake occurrence arises not only because of deficient models, but also from the lack of reliable observational results to verify the models' predictions.

The results of synthetic models are usually compared to the power-law distribution for event size and the value of the distribution's exponent. However, earthquake size distribution is the least informative of all power laws that govern earthquake occurrence: the power-law size distribution can be obtained from a variety of models (KAGAN, 1994); the fact that events in a model follow a fractal distribution, even with a correct exponent value, does not prove that the model can describe seismicity.

Short-term earthquake clustering, the major feature of a shallow earthquake occurrence, is not reproduced by most models, casting doubt on the applicability of their results to real earthquakes. Some models generate either aftershocks or foreshocks, although evidence exists (KAGAN, 1994; MICHAEL and JONES, 1998) that both aftershocks and foreshocks manifest the same clustering process. Therefore they should be explained by the same mechanism. In addition, during a rupture, each earthquake can be regarded as a dynamic fusion of many infinitesimal events (KAGAN, 1982); a model which generates a clustered earthquake sequence should explain a complex internal structure of each earthquake as well.

There is increasing information regarding the complexity of earthquake source time release; currently an earthquake catalog with source time functions is available on the WEB (RUFF and MILLER, 1994). Such data may soon allow one to obtain a statistical description of the source function complexity. When such results become available, quantitative testing of theoretical models can be carried out pertaining to whether these models correctly predict statistics of seismic moment release complexity, multiplicity of earthquake sources, etc.

Since most simulations do not generate foreshocks and aftershocks, only temporal interaction for main events can be investigated. Mainshock sequences can be fairly well approximated by a Poisson process which indicates that these earthquakes are largely statistically independent. There is little information in sequences of independent events; as in the power-law size distribution, such time series can be produced by a variety of mechanisms.

3.3.5. Laboratory fracture experiments. In a laboratory, a crack develops instabilities which makes its propagation highly chaotic and unpredictable (XU et al., 1997; KALIA et al., 1997; MARDER, 1998; SHARON and FINEBERG, 1999). The instabilities and sensitive dependence on initial conditions are due to crack propagation, especially at a speed close to elastic wave velocity. Stress and fracture conditions in laboratory specimens differ significantly from those in earthquake fault zones: the boundary effects are controlled by the laboratory researcher. Therefore, fracture can self-organize only at spatial scales substantially smaller than those of the specimen. In the fault zones stress, rock mechanical properties and fault geometry are self-organized as large-scale self-similar patterns develop.

Consequently, even if a good predictive capability can be achieved in modeling laboratory ruptures, transferring this capability to real earthquakes is completely different. As explained, in the laboratory: (1) the spatial domain of fracture is well defined; (2) the boundary conditions are well known and controlled. Conversely, in the earth's crust: (1) fracture can occur virtually anywhere; (2) the boundary conditions are unknown primarily because the boundaries themselves cannot be defined. Therefore, the relevance of laboratory results to actual earthquakes and the predictive power of these results must be demonstrated.

3.3.6. Stress triggering. Stress in the earth's interior where earthquakes occur is not practically measurable. Moreover, stress is a tensor, not a scalar, with significant complications ensuing. In stress diagrams (KAGAN, 1994; KING et al., 1994; DENG and SYKES, 1997) stress tensor invariants (or components) form a complicated mosaic, even when only moderate and large earthquakes are accounted for. This mosaic should become increasingly more complex if we calculate stress for smaller earthquakes.

Stress, earthquake size, and other parameters of seismicity have a power law or a fractal distribution with an exponent value of less than one (KAGAN, 1994). This results in extreme stochastic variability of seismicity parameters. The statistical variability can easily be confused with regional or temporal variations of seismicity. The statistical properties of power-law variables are not well known and are the subject of current intensive research (SAMORODNITSKY and TAQQU, 1994; ADLER et al., 1998). These properties significantly differ from more familiar stochastic quantities with a finite second moment, such as Gaussian variables. Neither the average nor the correlations are defined for such variables: one should use quantiles/percentiles and codifferences, respectively (*ibid.*). One may argue that in the real earth the stress singularities are smoothed out one way or another, and as soon as the stress is finite, all statistical moments would also be finite. However we do not yet know how to handle these singularities. Thus the results would strongly depend on how the stress infinities near the earthquake focal zone are treated. These stresses in focal zones of earlier earthquakes trigger the next events and define the way earthquake rupture propagates and stops.

To verify the ability of the stress triggering patterns inferred from observations to predict future earthquake activity, it is important to make forward predictions. These predictions must be compared with the actual seismicity record. The predictions should also be tested against a null hypothesis which uses known statistical properties of earthquake occurrence to specify time, space, and focal mechanism for future events (KAGAN and JACKSON, 1994). Only when the stress triggering method outperforms the empirical based algorithm can the predictive ability of the method be considered verified (cf. HARDEBECK et al., 1998).

3.3.7. Theory—critical review. The major problem with earthquake physics is that—given the lack of appropriate testing—it is not known which models (if any) can reproduce the fundamental properties of earthquake occurrence quantitatively. Presently we are uncertain whether theoretical or computer models of rupture are even approximately correct, the predictions of these models are tested only qualitatively and never in an actual prospective mode. Even if some models are successful in engineering applications, i.e., within the narrow range of laboratory experiments, we still must show that these constructs are useful in fundamentally understanding the phenomena.

At the present time, numerical earthquake models have shown no predictive capability exceeding or comparable to the empirical prediction based on earthquake statistics. Even if a theoretical or physical model exhibits some predictive skill (DIETERICH, 1994), we should always question whether the predictive power derives from a deeper theoretical understanding, or from the earthquake statistics which are imbedded in the model. A model may have a large number of adjustable parameters both obvious and implied to simulate a complicated pattern of seismic activity successfully. However, the model may not have a theoretical predictive capability. Thus we should request that model's prediction must outperform an empirically based one.

Earthquake physics as practiced now employs mathematical tools that were designed before 1870. Substantial progress has been made in mathematics subsequently with the appearance of set theory, topology, group theory, and the theory of stochastic processes, to mention just some new mathematical disciplines. The necessity of new mathematical approaches to earthquake seismology problems is highlighted by scale-invariant, fractal properties of seismicity, discussed above. These properties do not only mean that empirical distributions can be approximated by a straight line in a log–log plot. The scale-invariant geometry of earthquake faults signifies, for example, that the geometrical and mechanical characteristics of these objects would radically differ from those of solid bodies with more familiar Euclidean forms (MANDELBROT, 1983).

PENROSE (1989, p. 125) asks whether the Mandelbrot set (MANDELBROT, 1983, pp. 188–189) is computable: is there any computer procedure that in a finite number of steps would decide that an arbitrary point in a complex plane belongs to the set? The answer proves to be "*no*" (BLUM *et al.*, 1998, p. 55); even if we use the *real-number* arithmetic operations, no algorithm can decide whether a point is in the set. BLUM *et al.* (1998) determine that the reason for the "undecidability" of the Mandelbrot set and many similar complex mathematical objects, is that their boundary has a fractal Hausdorff dimension.

As discussed earlier in this section, earthquake occurrence in all variables is characterized by the scale-invariance and fractal dimensions. Earthquake fault system, for example, is a fractal object (KAGAN, 1982, 1994), thus its boundary cannot be computed in a continuum limit (BLUM *et al.*, 1998). Moreover, the stress at the fractal boundary should be nowhere differentiable function (see subsection 3.3.6), thus it is possible that the calculation of earthquake rupture criteria for points close to a "fault-tip" cannot effectively be carried out. It seems probable, therefore, that the application of new mathematical tools would be necessary to create a comprehensive physical theory of earthquake occurrence.

The diversity of earthquake models testifies to the lack of a common disciplinary matrix (KUHN, 1970). These theoretical models have not yielded conclusions that can be verified or rejected by comparing them to actual earthquake data, i.e., the hypotheses are not falsifiable (POPPER, 1980). Currently, these theoretical studies have contributed no identifiable new knowledge regarding actual earthquakes. Hence earthquake physics fails to function as a *theoretical science* (MARGENAU, 1950). One can argue that the physics of earthquakes does not yet exist. As a result, we have no clear idea which research directions are worth pursuing.

4. Causes of Failures: Lack of Rigor, Hypotheses are not Falsifiable

POPPER (1980) introduced the notion of falsifiability for two essential reasons: (1) to reject the hypotheses which contradict observational evidence and (2) for demarcation—to identify models and hypotheses which are unscientific. As we discussed above, the models, hypotheses, and experiments employed in earthquake seismology are often not falsifiable even in principle. For example, the earthquake recurrence hypothesis (section 3.3.1) has never been formulated in terms amenable to rigorous testing. How do we test the elastic rebound theory and its derivative models? Does it predict that earthquake rupture is not repeated on the same fault plane at short time intervals? What are the formal tolerance limits? How is the rupture zone defined? Since slip varies along the fault surface (if the surface can be unambiguously defined), how do we draw boundaries? To verify the earthquake recurrence model we should formulate a formal hypothesis and a null hypothesis and test both of them against independent data.

Below we briefly list several common deficiencies of earthquake occurrence models and data processing methods (some of these are proposed for case studies by WYSS, 1991 and WYSS and DMOWSKA, 1997):

(1) Non-uniqueness of earthquake occurrence models. In the previous section we mentioned that theoretical models of seismicity are not specific enough to be tested against real data.

(2) Drawing conclusions from extremely small samples (e.g., a few earthquakes in one or a few small regions). Moreover, most studies and predictions are retrospective. Such sampling obviously introduces biases, as the selection process searches for patterns, and almost any pattern can be found if enough data are sifted. The mathematical discipline called Ramsey theory (GRAHAM and SPENCER, 1990) states that a surprisingly broad range of patterns can be found in even a small amount of data. Retrospective prediction "successes" may also be due to selection bias.

(3) The number of adjustable parameters in a statistical model is often comparable to the effective number of data points, i.e., data are overinterpreted or overfitted. Hidden degrees of freedom and systematic effects must be carefully

investigated before the results of statistical studies can be presented. A probably apocryphal statement attributed to Enrico Fermi holds, that with four adjustable parameters one can approximate an elephant.

(4) *Ad hoc* and *ex post facto* adjustment of prediction hypotheses, after they failed in original tests.

(5) Failure to employ statistical methods properly, if at all. Retroactive adjustment of parameters to produce the best correlation must be taken into account when evaluating the statistical significance of retrospective tests (MULARGIA, 1997).

(6) Use of inappropriate null hypotheses in statistical tests (KAGAN, 1997b; STARK, 1997). Earthquakes are known to cluster in space and time (foreshock-mainshock-aftershock sequences). Inappropriate "strawman" null hypotheses (e.g., uniform randomness in space and time) can be rejected with an apparently high level of confidence even when the working hypothesis has no merit (KAGAN and JACKSON, 1996).

(7) Non-reproducibility of results. A case history is the usual method for investigating seismicity; the tendency is to collect the maximum possible information from different scientific disciplines and integrate this information into a complete, holistic picture of earthquake occurrence. The problem with these descriptive results is poor reproducibility. Even if a few studies seem to produce similar results, it is not clear whether this replication is caused by similar prior assumptions and selection bias (i.e., the reproducibility is illusory), or an underlying physical cause. Many papers have been published, for instance, to confirm the seismic gap model using particular case histories (NISHENKO and SYKES, 1993; JACKSON and KAGAN, 1993, and references therein). However, forward tests of this model (section 3.1.2) demonstrate that it lacks predictive power.

The difficulties described above can be ignored if applying current methodology such as an extensive descriptive, phenomenological analysis of earthquake case histories would result in success. In the Parkfield and Tokai prediction experiments major efforts were made to accumulate and integrate geologic, geodetic, and seismic information for an in-depth analysis of the data. However, both experiments have apparently failed. One may argue that both experiments were designed to detect short-term precursors, and since no large earthquake occurred in the respective areas, the experiments are still continuing. Nonetheless, for what duration should monitoring of putative short-term precursors be continued to be considered productive? KAGAN (1997a) argues that the return time for a $M \approx 6$ earthquake in the Parkfield area may exceed 100 years. For the seismic gap hypothesis or the $M8$ method, the results of the test are obtained in a few years (subsection 3.1.2). Thus experiments can be designed to provide an answer much sooner than the Tokai and Parkfield efforts. Nevertheless, despite failure in the Parkfield and Tokai tests, the seismological community still accepts the premises on which these experiments are based, with the exception of a few investigators. A likely failure of the seismic gap models (section 3.1.2) has not generally been acknowledged, and has not led to critical large-scale tests of the earthquake rebound models (section 3.3.1).

5. Recommendations and Conclusions

The failures described in the previous sections imply that a review of the foundations of earthquake science is necessary. Whereas valid scientific reasons for disagreement on earthquake predictability may exist (WYSS et al., 1997), most researchers agree that the rules for seismicity pattern investigations as well as reporting of the results must be significantly tightened. However, implementing more rigorous rules for data analysis is likely to be difficult and contentious. Let us consider a few methodological criteria relevant to the problems of earthquake seismology.

5.1. Simple versus Complex Models

Two explanatory approaches for seismic patterns can be proposed: (1) Introduce more *ad hoc* factors influencing seismicity, in effect introducing as many degrees of freedom in a model as the number of features the hypothesis seeks to explain. As shown in the previous section, the success of this approach is small to non-existent. (2) Treat most of the observed features as a random phenomenon. Only major features should be analyzed by using extensive statistical data. These simple patterns of earthquake occurrence may characterize the universal properties of seismicity, and such investigations are reproducible.

KIRCHHERR et al. (1997, see also LI and VITANYI, 1997) consider how one should treat simple versus complex models, accounting for the Occam's razor principle and Bayesian reasoning, from the point of view of Kolmogorov's complexity. They suggest that each model should be assigned a prior probability equal to its complexity, with these probabilities used in the Bayesian posterior evaluation of observational evidence. According to this rule, simple models have a strong initial advantage in explaining phenomena. Such a methodology corresponds to usual scientific practice: simple hypotheses are not rejected unless there is compelling evidence against them. Thus, although the earth is very complex (ORESKES et al., 1994), unless a complicated hypothesis is based on a properly verified understanding of basic relations, a complex model would lack predictive power and would remain a descriptive, phenomenological tool.

5.2. Reductionist versus Emergent Phenomena Approach

JORDAN (1997) argues that earthquake seismology may not succeed using the reductionist program, which explains the complexity of natural phenomena by a few fundamental equations. Following ANDERSON's (1972) arguments, Jordan suggests that due to the complexity of geosystems, new "emergent" laws must be found as explanatory constructs for earthquake science. There is no doubt that the earth is an extremely complicated system; however, some observational regularities of earthquake rupture and occurrence, such as the power-law size distribution, are seen in relatively simple systems, such as a rock specimen or man-made objects (MAIN, 1996).

Calculations of molecular dynamics (KALIA et al., 1997; MARDER, 1998; SHARON and FINEBERG, 1999) demonstrate that the basic properties of tensile fracture can effectively be derived from simple laws. Similarly, precise laboratory measurements of fault propagation demonstrate multiple branching of fault surfaces. These simulations reproduce the fractal character of a fracture. Moreover, calculating the total energy balance in laboratory fracture experiments (SHARON and FINEBERG, 1996, 1999) establishes that most elastic energy goes into creating new surface. Although the conditions during tensile fracture differ from those of shear failure in earthquakes, the above result may bear on the problem of the heat paradox for earthquake faults (SCHOLZ, 1996). One should anticipate that these computer simulations, even if extended to shear fracture, would need to be combined with continuum-style models of material rupture. The new, yet unknown, emergent effects may appear during a transition from one to another model.

5.3. Seismicity and Turbulence

Tectonic earthquakes and rupture in a laboratory exhibit scale-invariance over a very broad range of distances $10^{-3}-10^5$ m. Similar scaling, even over an extended distance range, is observed for another large-scale deformation of matter—turbulence of fluids (KAGAN, 1992, 1994). Both phenomena share stochastic scale-invariance, have hierarchical space-time structures, and multiple power-law dependencies. Turbulence is characterized by an energy transport cascade from large-scale to smaller structures (MANDELBROT, 1983; MOLCHAN, 1997). KAGAN (1973) proposes a stochastic model where time-space-magnitude features of seismicity are represented as a cascade in energy dimension from the largest to the smallest earthquakes. This model, using very few assumptions, reproduces major statistical properties of seismicity, suggesting that the dissipation of energy in earthquakes follows the pattern similar to turbulent cascades: large-scale tectonic deformation energy is dissipated on creating new fractal surfaces of earthquake faults (see the previous subsection).

Although the exact mechanism of both seismicity and turbulent motion is unknown, the ubiquity of their manifestations makes one question whether this complexity is due to the deep topological properties of space-time itself (see also discussion on computability of the Mandelbrot set and other fractals in subsection 3.3.7). Thus, although its fundamental equations are still unknown, earthquake seismology may be a branch of mathematical physics, like turbulence of fluids.

Comparing seismicity to turbulence indicates that earthquake science may have a profound problem: the Navier–Stokes formula in fluid dynamics is well known, but has not facilitated our understanding of turbulence. Thus, we cannot construct theoretical solutions from the first principle. The best hypotheses which fit the turbulence data are essentially statistical hierarchical (fractal) models. Although dynamical systems and fractal notions have provided new tools with which to describe the turbulence (as well as for seismicity characterization), these tools are still of limited use (LORENZ, 1993; NEWMAN et al., 1994; TURCOTTE, 1997).

GLEICK (1987) provides an eloquent and entertaining account of the failure of theoretical physics to explain turbulence. His story regarding Heisenberg (GLEICK, 1987, pp. 121, 329) reflects a general feeling among mathematical physicists that turbulence may be insolvable. The systems which theoretical physics analyzes are isolated and closed, whereas in earth sciences we must deal with systems which are widely open (ORESKES et al., 1994). If the analogy of seismicity to turbulence is even partially true, this suggests that prospects of predicting earthquake dynamics and seismicity patterns purely by theoretical insight are low.

5.4. Methodology in Medical Research

We argue above that for earthquake research to succeed, we need more rigorous methods of data analysis. Research in the health sciences could be used as an example of such a methodological transition, since a human organism is complex and many types of experiments are not possible with human subjects. This is similar to the circumstances in the geosciences. Because significant effort has been made to develop scientific methods in medicine, we may benefit from their experience and insight. During the second half of the 20th century, medical researchers realized that without carefully controlled trials, experimental results would be of limited use. SHAPIRO and SHAPIRO (1997, p. 73) even claim that the effect of the medications of a prescientific era was largely psychological: "Despite the problem of bias and the hazards of interpretation, we propose that the available data support the overall hypothesis that the history of medical treatment up to the era of scientific medicine is largely the history of the placebo effect" (however, see also STERNBERG, 1998, who partially disagrees with this view).

Medical researchers found that single-blind experiments (when patients do not know who is getting a medication and who is on a placebo) are insufficient to exclude a bias. The experimenters should not know this either. Such double-blind experiments have been found to be reproducible and since the 1970s became a standard research method. In addition to the double-blind method, SHAPIRO and SHAPIRO (1997, p. 210) propose other general rules for the statistical analysis of medical data. Most of these rules are applicable to analyzing earthquake data as well (cf. our list in section 4): "Well-controlled studies usually include . . . a priori specification of hypotheses; independent and dependent variables; specification of statistical procedures to be used; adequate sample size; the use of measures with demonstrated reliability and validity, to control for errors in measurement; assessment of statistical power; clear differentiation between a priori hypotheses and post *hoc* hypotheses; the use of adjusted probability values based on the number of variables tested, to avoid errors; and subsequent replication of results."

Reviews of statistical trials in medical research (such as NOWAK, 1994; TAUBES, 1998) make for fascinating reading: a seemingly slight change in methodology or data selection may lead to completely different conclusions. As the above publications testify, the development of a new scientific methodology which began around the 1930s (SHAPIRO and SHAPIRO, 1997, p. 153), continues today in the health sciences. Such history testifies that the major source of errors and biases is usually not technically insufficient methods, but neglect of the basic statistical rules for gathering and processing data. We cannot adopt all the critical methods employed in medical and biological research, since many such methods are specific to these disciplines. No experiments are possible with real earthquakes, and the earth is a unique object of study; thus our ability to make controlled tests is even more limited than in medicine. However, we need to develop methods which yield reproducible, objective results.

5.5. Discussion

The need for sophisticated statistical treatment of data emerged as medical research moved from strong, easily observable effects to more subtle ones (SHAPIRO and SHAPIRO, 1997). Similarly, in earthquake seismology short-term clustering is an obvious feature of shallow seismicity, whereas long-term effects (long-term clustering or quasi-periodicity) must be studied in the presence of a much stronger signal. Thus without an efficient and faithful model of short-term clustering, in most cases no practical result can be obtained for long-term effects. Such a model is needed to explain the short-term, time-space-focal mechanism regularities of earthquake sequences.

With the exception of a few problems such as wave propagation and its influence on rupture, no conclusions can be obtained from the fundamental equations, i.e., no deductive conclusion is yet possible to explain real earthquake patterns. Thus the problems in earthquake seismology must be solved inductively, and each proposed seismicity pattern should be rigorously tested for its predictive skill. A starting point might be to require that all claims of successful earthquake predictions outperform the following two "trivial" (null) hypotheses: (1) predictions assuming regular earthquake occurrence, based on the long-term average of past seismicity, and (2) predictions based on extrapolating the recent seismicity record, such as aftershock sequences of strong earthquakes or the earthquake history of recent years. These null hypotheses correspond to so-called "climatological" and "persistence" forecasts in meteorology (MURPHY, 1996). Seismic gap model validation (KAGAN and JACKSON, 1991, 1995) and testing of VAN predictions (KAGAN and JACKSON, 1996) offer examples of such tests applied to seismicity.

Therefore, the following methodology can be proposed for testing of earthquake prediction schemes (KAGAN, 1997b):

1. Case history investigations: these studies should satisfy the criteria formulated, for example, by the IASPEI group (WYSS, 1991; WYSS and DMOWSKA, 1997, pp. 13–15; WYSS and BOOTH, 1997). However, case histories of "successful prediction" of one or several earthquakes do not demonstrate a method has predictive skill, since such success may be due to chance or a selection bias. Since seismicity is characterized by extreme randomness, it is possible in principle to select almost any pattern from abundant data. For example, the lack of agreement over whether the 1989 Loma-Prieta earthquake was predicted meaningfully (U.S. GEOLOGICAL SURVEY STAFF, 1990; HARRIS, 1998) demonstrates (1) a forecast must be formally and rigorously specified in advance, and (2) a track record of a prediction method is needed—no reliable conclusion can be reached on the basis of one or even several events.

2. Rigorous large-scale retrospective statistical testing, using a control sample, i.e., the data that were not considered in formulating the working hypothesis and evaluating adjustable parameters for the model. The null hypothesis should be formulated and tested against the same data. However, tectonic and geological conditions always vary and the long-term, long-range clustering property of seismicity makes it difficult, if not impossible, to use an earthquake catalog in different time intervals or in different regions as a control sample. As experience in medical research (section 5.4) suggests, even meticulous efforts to control the bias often fail. Test of the $M8$ algorithm (KOSSOBOKOV et al., 1997, p. 228; 1999) again demonstrates that the results of the a posteriori forecast could be significantly better than those of the real-time prediction.

3. Forward, prospective prediction testing, during which no adjustment of parameters is allowed and all relevant possible ambiguities in the data or the interpretation technique are specified in advance (EVISON and RHOADES, 1993, 1997; KOSSOBOKOV et al., 1997). Such testing should be carried out for empirical and theoretical predictions of earthquake occurrence. A track record of the predictions should be established to verify the method's predictive power. Only the results of properly executed forward testing are to be accepted as a final verdict of a method performance.

Because of practical and theoretical considerations, many such tests involve the occurrence of large earthquakes. To carry out these tests in a reasonably short time, the test space should be extended to large seismic regions such as circum-Pacific (see section 3.1.2), or global seismicity. For smaller earthquakes, seismicity in one region is usually dominated by a few major earthquakes. Thus we cannot separate the effects of random fluctuations in effectively small samples from possible regularities of earthquake occurrence. This would necessitate that even for small events, prediction tests must be carried out in several different regions.

5.6. Conclusions

The above discussion suggests the insufficiency of two commonly proposed solutions for the crisis in earthquake seismology: (1) the holistic, interdisciplinary integration of geophysical data and the search for new "emergent" regularities, or (2) the application of earthquake physics methods. These methods may be necessary to make progress in the understanding of earthquakes, however they will remain unsuccessful if their performance is not properly validated. Without critical testing, the holistic approach erodes into descriptive phenomenology, its constructs deprived of predictive power. Similarly, without confronting real data and controlled verification of its performance, theoretical or computer modeling breaks down into speculation and computer manipulation.

The "fifth force" controversy (FRANKLIN, 1993) demonstrates the difference in attitudes between geophysicists and theoretical physicists in establishing scientific truth. Whereas the latter have a critical and rigorous approach to formulating and testing new hypotheses, and enforce a strong conformance to these norms, the earth science community lacks effective rules to verify a hypothesis. A method to enforce discipline in formulating and justifying hypotheses in geophysics needs to be found and discussed. Higher standards for research in earthquake seismology must be enforced. Authors should adopt a more rigorous style of scientific investigation, and reviewers and editors of geophysical journals should reject manuscripts which do not satisfy the above requirements.

Summarizing the discussion, we conclude that the only quantitative, reproducible knowledge pertinent to earthquake occurrence available currently is statistical. The regularities, discovered by OMORI (1984), GUTENBERG and RICHTER (1944) and others (see more in KAGAN, 1994), have withstood the test of time and are confirmed, albeit qualitatively, by recent developments in nonlinear dynamics and critical phenomena. The advances in statistical analysis of seismological data, and new understanding of the scaling properties of seismicity, including possible universality of major properties of earthquake occurrence (KAGAN, 1994, 1997b; MAIN, 1995, 1996), provide a unique opportunity to evaluate seismic hazard and to estimate the short- and long-term rate of future earthquake occurrence, i.e., to predict earthquakes statistically.

Stress accumulation and release models may yield new understanding of the earthquake processes and eventually allow us to predict earthquakes more reliably. The advances in space geodetic measurements—Global Positioning System (GPS), Synthetic Aperture Radar Interferometry (InSAR)—should greatly increase our understanding of the earthquake process (JACKSON *et al.*, 1997; MEADE and SANDWELL, 1996). New powerful observational tools and interpretive techniques allow us to obtain time-space distribution of moment release for large earthquakes

(RUFF and MILLER, 1994; TANIOKA and GONZALEZ, 1998; SCHWARTZ, 1999—see additional discussion in subsections 3.1.1 and 3.3.4). Seismic moment tensor solutions for small ($m \geq 3.5$) and moderate size earthquakes (ZHU and HELMBERGER, 1996, and references therein) are becoming routinely available. These and other new methods and techniques may soon change a perspective in earthquake research.

Finally, I would like to answer the question in my title: "Is earthquake seismology a hard, quantitative science?" It is clear from the above discussion, that the answer is "not yet," although with application of more rigorous methods for proposing and validating scientific hypotheses, we may reach this goal.

Acknowledgments

I appreciate partial support from the Southern California Earthquake Center (SCEC). SCEC is funded by NSF Cooperative Agreement EAR-8920136 and USGS Cooperative Agreements 14-08-0001-A0899 and 1434-HQ-97AG01718. This work has greatly benefited from stimulating discussions with D. D. Jackson and D. Sornette of UCLA, Y. Ben-Zion of USC, R. Geller of Tokyo University, F. Mulargia of Bologna University, P. B. Stark of UC Berkeley, M. Wyss of Univ. of Alaska, I. Main of Univ. of Edinburgh, M. Ohtake of Tohoku Univ., V. I. Keilis-Borok and V. G. Kossobokov of Russian Academy of Science. I am grateful to three anonymous reviewers and to the editor M. Wyss for their very useful remarks. The SCEC contribution number is 454.

Note Added in Proof

The recent *Nature* WEB debate on earthquake prediction (see http://helix. nature.com/debates/earthquake/equake_frameset.html) confirmed several of the ideas and conclusions discussed in this paper. Scarcely anyone proposed that deterministic type reliable prediction of individual earthquakes is feasible in the near future. It is especially interesting to note the opinions of the debators on seismic gap and earthquake recurrence models. Scholz (week 6 contribution) in effect claims that none of the results based on the seismic gap hypothesis are ready to be tested. This statement is surprising: according to the GEOREF database more than 375 publications, many of them published in the last few years, cite as their subject the *seismic gap model*. If this model, despite its claimed great potential value for evaluation of the intermediate- and long-term seismic hazard, cannot be tested after nearly 30 years of development, general scientific methods in earthquake seismology need to be significantly revised.

REFERENCES

ADLER, A., FELDMAN, R., and Taqqu, M. S. (eds), *A Practical Guide to Heavytails: Statistical Techniques for Analyzing Heavy Tailed Distributions* (Birkhäuser, Boston 1998).

ANDERSON, P. W. (1972), *More is Different*, Science *177*, 393–396.

BAKUN, W. H., and LINDH, A. G. (1985), *The Parkfield, California, Earthquake Prediction Experiment*, Science *229*, 619–24.

BEN-MENAHEM, A. (1995), *A Concise History of Mainstream Seismology—Origins, Legacy, and Perspectives*, Bull. Seismol. Soc. Am. *85*, 1202–1225.

BEN-ZION, Y., SAMMIS, C., and HENYEY, T. (1999), *Perspectives on the Field of Physics of Earthquakes*, Seismol. Res. Lett., in press.

BLUM, L., CUCKER, F., Shub, M., and Smale, S., *Complexity and Real Computation* (Springer, New York 1998).

CHELIDZE, T. (1993), *Fractal Damage Mechanics of Geomaterials*, Terra Nova *5*, 421–437.

COBURN, A. W., and SPENCE, R., *Earthquake Protection* (Wiley, New York 1992).

DENG, J. S., and SYKES, L. R. (1997), *Stress Evolution in Southern California and Triggering of Moderate-, Small-, and Micro-size Earthquakes*, J. Geophys. Res. *102*, 24,411–24,435.

DIETERICH, J. (1994), *A Constitutive Law for Rate of Earthquake Production and its Application to Earthquake Clustering*, J. Geophys. Res. *99*, 2601–2618.

ENGELHARDT, VON W., and ZIMMERMAN, J., *Theory of Earth Science* (Cambridge University Press, Cambridge 1988).

EVANS, R. (1997), *Assessment of Schemes for Earthquake Prediction: Editor's Introduction*, Geophys. J. Int. *131*, 413–420.

EVISON, F. F., and RHOADES, D. A. (1993), *The Precursory Earthquake Swarm in New Zealand: Hypothesis Tests*, New Zealand J. Geophys. *36*, 51–60 (Correction p. 267).

EVISON, F. F., and RHOADES, D. A. (1997), *The Precursory Earthquake Swarm in New Zealand: Hypothesis Tests II*, New Zealand J. Geol. Geophys. *40*, 537–547.

FEDOTOV, S. A. (1968), *On the Seismic Cycle, Feasibility of Quantitative Seismic Zoning and Long-term Seismic Prediction*. In *Seismic Zoning of the USSR*, pp. 121–150, Nauka, Moscow (in Russian); English translation: Israel program for scientific translations, Jerusalem, 1976.

FRANKLIN, A., *The Rise and Fall of the "Fifth Force": Discovery, Pursuit, and Justification in Modern Physics* (American Inst. Physics, New York 1993).

GABRIELOV, A. M., and NEWMAN, W. I., *Seismicity modeling and earthquake prediction: A review*. In *Nonlinear Dynamics and Predictability of Geophysical Phenomena* (Newman, W. I., Gabrielov, A., and Turcotte, D. L., eds.), Geoph. Monogr. *83* (Washington, American Geophysical Union 1994) pp. 7–13.

GELLER, R. J. (ed.) (1996), *Debate on "VAN,"* Geophys. Res. Lett. *23*(11), 1291–1452.

GELLER, R. J. (1997), *Earthquake Prediction: A Critical Review*, Geophys. J. Int. *131*, 425–450.

GELLER, R. J., JACKSON, D. D., KAGAN, Y. Y., and MULARGIA, F. (1997), *Earthquakes Cannot be Predicted*, Science *275*, 1616–1617.

GLEICK, J., *Chaos, Making a New Science* (Viking, New York 1987).

GRAHAM, R. L., and SPENCER, J. H. (1990), *Ramsey Theory*, Scientific American *263*(7), 112–117.

GUSEV, A. A. (1998), *Earthquake Precursors: Banished Forever?* EOS Trans. AGU *79*, 71–72.

GUTENBERG, B., and RICHTER, C. F. (1944), *Frequency of Earthquakes in California*, Bull. Seismol. Soc. Am. *34*, 185–188.

HARDEBECK, J. L., NAZARETH, J. J., and HAUKSSON, E. (1998), *The Static Stress Change Triggering Model: Constraints from two Southern California Aftershock Sequences*, J. Geophys. Res. *103*, 24,427–24,437.

HARRIS, R. A. (1998), *Forecasts of the Loma-Prieta, California, Earthquake*, Bull. Seismol. Soc. Am. *88*, 898–916.

ISHIBASHI, K., *Specification of a soon-to-occur seismic faulting in the Tokai District, Central Japan, based upon seismotectonics*. In *Earthquake Prediction: An International Review* (Simpson, D. W. and Richards, P. G., eds.), Ewing Monograph Series, *4*, pp. 297–332 (Am. Geophys. Union, Washington 1981).

JACKSON, D. D., and KAGAN, Y. Y. (1993), *Reply [to Nishenko and Sykes]*, J. Geophys. Res. *98*, 9917–9920.

JACKSON, D. D., SHEN, Z.-K., POTTER, D., GE, B.-X., and SUNG, L.-Y. (1997), *Southern California Deformation*, Science *277*, 1621–1622.

JORDAN, T. H. (1997), *Is the Study of Earthquakes a Basic Science?* Seismol. Res. Lett. *68*(2), 259–261.

KAGAN, Y. Y. (1973), *A Probabilistic Description of the Seismic Regime*, Izv. Acad. Sci. USSR, Phys. Solid Earth, 213–219 (English translation).

KAGAN, Y. Y. (1982), *Stochastic Model of Earthquake Fault Geometry*, Geophys. J. Roy. Astr. Soc. *71*, 659–691.

KAGAN, Y. Y. (1992), *Seismicity: Turbulence of Solids*, Nonlinear Sci. Today *2*, 1–13.

KAGAN, Y. Y. (1994), *Observational Evidence for Earthquakes as a Nonlinear Dynamic Process*, Physica D *77*, 160–192.

KAGAN, Y. Y. (1997a), *Statistical Aspects of Parkfield Earthquake Sequence and Parkfield Prediction Experiment*, Tectonophysics *270*, 207–219.

KAGAN, Y. Y. (1997b), *Are Earthquakes Predictable?* Geophys. J. Int. *131*, 505–525.

KAGAN, Y. Y., and JACKSON, D. D. (1991), *Seismic Gap Hypothesis: Ten Years After*, J. Geophys. Res. *96*, 21,419–21,431.

KAGAN, Y. Y., and JACKSON, D. D. (1994), *Long-term Probabilistic Forecasting of Earthquakes*, J. Geophys. Res. *99*, 13,685–13,700.

KAGAN, Y. Y., and JACKSON, D. D. (1995), *New Seismic Gap Hypothesis: Five Years After*, J. Geophys. Res. *100*, 3943–3959.

KAGAN, Y. Y., and JACKSON, D. D. (1996), *Statistical Tests of VAN Earthquake Predictions: Comments and Reflections*, Geophys. Res. Lett. *23*, 1433–1436.

KAGAN, Y. Y., and KNOPOFF, L. (1987), *Statistical Short-term Earthquake Prediction*, Science *236*, 1563–1567.

KALIA, R. K., NAKANO, A., OMELTCHENKO, A., TSURUTA, K., and VASHISHTA, P. (1997), *Role of Ultrafine Microstructures in Dynamic Fracture in Nanophase Silicon Nitride*, Phys. Rev. Lett. *78*, 2144–2147.

KING, G. C. P., STEIN, R. S., and LIN, J. (1994), *Static Stress Changes and the Triggering of Earthquakes*, Bull. Seismol. Soc. Am. *84*, 935–953.

KIRCHHERR, W., LI, M., and VITANYI, P. (1997), *The Miraculous Universal Distribution*, Math. Intelligencer *19*, 7–15.

KNOPOFF, L., AKI, K., ALLEN, C. R., RICE, J. R., and SYKES, L., eds. (1996), *Earthquake Prediction: The Scientific Challenge, Colloquium Proceedings*, Proc. Nat. Acad. Sci. USA *93*, 3719–3837.

KOSSOBOKOV, V. G., HEALY, J. H., and DEWEY, J. W. (1997), *Testing an Earthquake Prediction Algorithm*, Pure appl. geophys. *149*, 219–232.

KOSSOBOKOV, V. G., HEALY, J. H., and DEWEY, J. W. (1998), *Testing an Earthquake Prediction Algorithm: M 7.5 + Earthquakes in Circum-Pacific, 1992–1997*, XXVI General Assembly European Seismol. Commis., Abstracts, Tel Aviv, Israel, p. 42.

KOSSOBOKOV, V. G., ROMASHKOVA, L. L., KEILIS-BOROK, V. I., and HEALY, J. H. (1999), *Testing Earthquake Prediction Algorithms: Statistically Significant Advance Prediction of the Largest Earthquakes in the Circum-Pacific, 1992–1997*, Phys. Earth Planet. Inter. *111*, 187–196.

KUHN, T. S., *The Structure of Scientific Revolutions*, 2nd ed. (Univ. Chicago Press, Chicago 1970).

LEARY, P. C. (1997), *Rock as a Critical-point System and the Inherent Implausibility of Reliable Earthquake Prediction*, Geophys. J. Int. *131*, 451–466.

LI, M., and VITANYI, P., *An Introduction to Kolmogorov Complexity and its Applications*, 2nd ed. (Springer, New York 1997).

LORENZ, E. N., *The Essence of Chaos* (Univ. Washington Press, Seattle 1993).

MAIN, I. G. (1995), *Earthquakes as Critical Phenomena—Implications for Probabilistic Seismic Hazard Analysis*, Bull. Seismol. Soc. Am. *85*, 1299–1308.

MAIN, I. G. (1996), *Statistical Physics, Seismogenesis, and Seismic Hazard*, Rev. Geophys. *34*, 433–462.

MANDELBROT, B. B., *The Fractal Geometry of Nature*, 2nd ed. (W. H. Freeman, San Francisco 1983).

MARDER, M. (1998), *Computational Science—Unlocking Dislocation Secrets*, Nature *391*, 637–638.

MARGENAU, H., *The Nature of Physical Reality: A Philosophy of Modern Physics* (McGraw-Hill, New York 1950).

McCann, W. R., Nishenko, S. P., Sykes, L. R., and Krause, J. (1979), *Seismic Gaps and Plate Tectonics: Seismic Potential for Major Boundaries*, Pure appl. geophys. *117*, 1082–1147.

Meade, C., and Sandwell, D. T. (1996), *Synthetic Aperture Radar for Geodesy*, Science *273*, 1181–1182.

Michael, A. J., and Jones, L. M. (1998), *Seismicity Alert Probabilities at Parkfield, California, Revisited*, Bull. Seismol. Soc. Am. *88*, 117–130.

Mogi, K. (1995), *Earthquake Prediction Research in Japan*, J. Phys. Earth *43*, 533–561.

Molchan, G. M. (1997), *Turbulent Cascades: Limitations and a Statistical Test of the Lognormal Hypothesis*, Physics of Fluids *9*, 2387–2396.

Mulargia, F. (1997), *Retrospective Validation of Time Association*, Geophys. J. Int. *131*, 500–504.

Murphy, A. H. (1996), *General Decompositions of MSE-based Skill Scores—Measures of Some Basic Aspects of Forecast Quality*, Mon. Weather Rev. *124*, 2353–2369.

Nadeau, R. M., and McEvilly, T. V. (1997), *Seismological Studies at Parkfield. V. Characteristic Microearthquake Sequences as Fault-zone Drilling Targets*, Bull. Seismol. Soc. Am. *87*, 1463–1472.

Nadeau, R. M., and Johnson, L. R. (1998), *Seismological Studies at Parkfield VI: Moment Release Rates and Estimates of Source Parameters for Small Repeating Earthquakes*, Bull. Seismol. Soc. Am. *88*, 790–814.

Newman, W. I., Gabrielov, A., and Turcotte, D. L., eds., *Nonlinear Dynamics and Predictability of, Geophysical Phenomena*, Geoph. Monogr. *83* (American Geophys. Union, Washington 1994).

Nishenko, S. P. (1991), *Circum-Pacific Seismic Potential: 1989–1999*, Pure appl. geophys. *135*, 169–259.

Nishenko, S. P., and Sykes, L. R. (1993), *Comment on "Seismic Gap Hypothesis: Ten Years After" by Y. Y. Kagan and D. D. Jackson*, J. Geophys. Res. *98*, 9909–9916.

Nowak, R. (1994), *Problems in Clinical Trials go far Beyond Misconduct*, Science *264*, 1538–1541.

Omori, F. (1894), *On the After-shocks of Earthquakes*, J. College Sci., Imp. Univ. Tokyo 7, 111–200 (with Plates IV–XIX).

Oreskes, N., Shrader-Frechette, K., and Belitz, K. (1994), *Verification, Validation, and Confirmation of Numerical Models in the Earth Sciences*, Science *263*, 641–646, also see Science *264*, 331.

Ouillon, G., Castaing, C., and Sornette, D. (1996), *Hierarchical Scaling of Faulting*, J. Geophys. Res. *101*, 5477–5487.

Penrose, R., *The Emperor's New Mind: Concerning Computers, Minds, and the Laws of Physics* (Oxford Univ. Press, New York 1989).

Popper, K. R., *Logic of Scientific Discovery*, 2nd ed. (Hutchinson, London 1980).

Reasenberg, P. A., and Jones, L. M. (1989), *Earthquake Hazard After a Mainshock in California*, Science *243*, 1173–1176.

Reid, H. F. (1910), *The California Earthquake of April 18, 1906, Vol. 2: The Mechanics of the Earthquake*, Carnegie Institution of Washington, Washington, D.C.

Roeloffs, E., and Langbein, J. (1994), *The Earthquake Prediction Experiment at Parkfield, California*, Rev. Geophys. *32*, 315–336.

Ruff, L. J., and Miller, A. D. (1994), *Rupture Process of Large Earthquakes in the Northern Mexico Subduction Zone*, Pure appl. geophys. *142*, 101–172 (see also WEB site http://www.geo.lsa.umich.edu/SeismoObs/STF.html).

Samorodnitsky, G., and Taqqu, M. S., *Stable non-Gaussian Random Processes: Stochastic Models with Infinite Variance* (Chapman and Hall, New York 1994).

Santillana de, G., *The Origins of Scientific Thought; from Anaximander to Proclus, 600 B.C. to 500 A.D.* (Univ. Chicago Press, Chicago 1961).

Scholz, C. H., *The Mechanics of Earthquakes and Faulting* (Cambridge Univ. Press, Cambridge 1990).

Scholz, C. H. (1996), *Faults Without Friction?* Nature *381*, 556–557.

Scholz, C. (1997), *Whatever Happened to Earthquake Prediction?* Geotimes *42*(3), 16–19.

Scholz, C. H. (1998), *Earthquakes and Friction Laws*, Nature *391*, 37–42.

Scholz, C. H., Sykes, L. R., and Aggarwal, Y. P. (1973), *Earthquake Prediction: A Physical Basis*, Science *181*, 803–810.

Schwartz, D. P., and Coppersmith, K. J. (1984), *Fault Behavior and Characteristic Earthquakes: Examples from Wasatch and San Andreas Fault Zones*, J. Geophys. Res. *89*, 5681–5698.

SCHWARTZ, S. Y. (1999), *Non-characteristic Behavior and Complex Recurrence of Large Subduction Zone Earthquakes*, J. Geophys. Res., in press.

SHAPIRO, A. K., and SHAPIRO, E., *The Powerful Placebo: From Ancient Priest to Modern Physician* (J. Hopkins Univ. Press, Baltimore 1997).

SHARON, E., and FINEBERG, J. (1996), *Microbranching Instability and the Dynamic Fracture of Brittle Materials*, Physical Review B *54*, 7128–7139.

SHARON, E., and FINEBERG, J. (1999), *Confirming the Continuum Theory of Dynamic Brittle Fracture for Fast Cracks*, Nature *397*, 333–335.

SNIEDER, R., and VAN ECK, T. (1997), *Earthquake Prediction: a Political Problem?* Geologische Rundschau *86*, 446–463.

STARK, P. B. (1997), *Earthquake Prediction: The Null Hypothesis*, Geophys. J. Int. *131*, 495–499.

STERNBERG, E. M. (1998), *Unexpected Effects*, Science *280*, 1901–1902.

SYKES, L. R. (1971), *Aftershock Zones of Great Earthquakes, Seismicity Gaps, and Earthquake Prediction for Alaska and the Aleutians*, J. Geophys. Res. *76*, 8021–8041.

TAUBES, G. (1998), *The (Political) Science of Salt*, Science *281*, 898–907, see also Science *281*, 1961–1963.

TANIOKA, Y., and GONZALEZ, F. I. (1998), *The Aleutian Earthquake of June 10, 1996 (M_w 7.9) Ruptured Parts of both the Andreanof and Delarof Segments*, Geophys. Res. Lett. *25*, 2245–2248.

TURCOTTE, D. L., *Fractals and Chaos in Geology and Geophysics*, 2nd ed. (Cambridge Univ. Press, Cambridge 1997).

U.S. GEOLOGICAL SURVEY STAFF (1990), *The Loma-Prieta, California, Earthquake—An Anticipated Event*, Science *247*, 286–293.

WELDON, R. J. (1996), *Why We Cannot Predict Earthquakes with Paleoseismology, and What We (Paleoseismologists) Really Should Be Doing*, EOS Trans. AGU 77(46), Fall AGU Meet. Suppl., p. F53.

WYSS, M., ed., *Evaluation of Proposed Earthquake Precursors* (American Geophys. Union, Washington 1991).

WYSS, M., and BOOTH, D. C. (1997), *The IASPEI Procedure for the Evaluation of Earthquake Precursors*, Geophys. J. Int. *131*, 423–424.

WYSS, M., ACEVES, R. L., and PARK, S. K., GELLER, R. J., JACKSON, D. D., KAGAN, Y. Y., and MULARGIA, F. (1997), *Cannot Earthquakes be Predicted? (Technical comments)*, Science *278*, 487–490.

WYSS, M., and DMOWSKA, R., eds. (1997), *Earthquake Prediction—State of the Art*, Pure appl. geophys. *149*, 1–264.

XU, X.-P., NEEDLEMAN, A., and ABRAHAM, F. F. (1997), *Effect of Inhomogeneities on Dynamic Crack Growth in an Elastic Solid*, Modelling and Simulation in Materials Science and Engineering *5*, 489–516.

YEATS, R. S., SIEH, K., and ALLEN, C. R., *The Geology of Earthquakes* (Oxford Univ. Press, New York 1997).

ZHU, L. P., and HELMBERGER, D. V. (1996), *Advancement in Source Estimation Techniques Using Broadband Regional Seismograms*, Bull. Seismol. Soc. Am. *86*, 1634–1641.

(Received August 19, 1998, revised January 18, 1999, accepted January 22, 1999)

To access this journal online:
http://www.birkhauser.ch

Pure appl. geophys. 155 (1999) 259–278
0033–4553/99/040259–20 $ 1.50 + 0.20/0

How Can One Test the Seismic Gap Hypothesis? The Case of Repeated Ruptures in the Aleutians

MAX WYSS[1] and STEFAN WIEMER[2]

Abstract—The test that KAGAN and JACKSON (1991, 1995) applied to the seismic gap hypothesis did not bring us closer to understanding the generation of large earthquakes. On the contrary, it led some to the conclusion that the rebound theory of earthquake generation should be rejected. We disagree with this point of view and argue that a global test of the simplified gap hypothesis cannot be done because it cannot account for differences in the slip history of fault segments and tectonic differences between separate plate boundaries. Kagan and Jackson did show, however, that the original gap hypothesis was oversimplified and should be refined. We propose that consideration of all the facts, including slip history and seismicity patterns in the Andreanof Islands, show that the concept of seismic gaps and the elastic rebound theory are correct for that segment of the plate boundary. The coseismic slip in the M_w 8.7 earthquake that broke this plate boundary segment in 1957 was only 2 m, as published before the repeat earthquake of 1986 (M_w 8), and thus, using a plate convergence rate of 7.3 cm/year, the return time in this cycle was expected to be less than 30 years, unless substantial aseismic creep occurs. This supports the time predictable model of mainshock recurrence. In addition, KISSLINGER *et al.* (1985) and KISSLINGER (1986) noticed a seismic quiescence in the subsequent source volume before the 1986 earthquake and attempted to predict it. The specific parameters he estimated were not entirely correct although his interpretation of the observed quiescence as a precursor was. We conclude that the 1986, M_w 8, Andreanof earthquake was not an example that disproves the seismic gap hypothesis. On the contrary, it shows that the hypothesis that plate motions reload plate boundaries after most of the elastic energy is released in great ruptures was correct in this case. This suggests that great earthquakes occur preferably in mature gaps. We believe the testing of the seismic gap hypothesis by algorithm on a global scale is an example that illustrates that overly simplified tests can lead to erroneous conclusions. To make progress in the actual understanding of the physics of the process of great earthquake ruptures, one must consider all the facts known for case histories.

Key words: Seismic gap, seismotectonics, earthquake prediction.

Introduction

The main issue addressed in this paper is the misuse of a simple algorithm applied worldwide to test the seismic gap hypothesis and, indirectly, the elastic

---

[1] Geophysical Institute, University of Alaska, Fairbanks Alaska, U.S.A. E-mail: max@giseis.alaska.edu.
[2] Seismological and Volcanological Research Department, Meteorological Research Institute, Tsukuba, now at Eidgenössiche Technische Hochschule, Zürich, Switzerland.

rebound theory. Although hypothesis testing is important for accepting or rejecting new ideas, it can also lead to erroneous results if applied indiscriminately to cases in which the preconditions differ. We demonstrate that a case history approach better advances our understanding of the rupture process of great earthquakes than global hypothesis testing by simple algorithm.

We also touch on the side issue that the analysis of seismicity patterns, such as seismic quiescence and increased moment release, can aid assessment of time-dependent seismic hazard in a gap. Although not yet well understood, these phenomena have been recognized before the subsequent mainshock as precursors in two great Aleutian earthquakes, and they have been used in attempting to predict these earthquakes. These facts further support our idea that a case history approach leads to improved understanding of the earthquake process, as well as to the possibility of predicting them.

Competing Hypotheses

The seismic gap hypothesis was developed after seismologists understood that great earthquakes occur along the boundaries of plates and that often their aftershock areas abut (FEDOTOV, 1965; SYKES, 1971). The concept of the existence of an earthquake cycle, in which it takes hundreds of years to build up the elastic strain that is eventually released within seconds in great earthquakes, originated with the observations following the 1906 San Francisco earthquake (REID, 1910). The third element important to the current concept of how great earthquakes occur is the recognition that the shear stress along plate boundaries is low (BRUNE et al., 1969; GAMAR and BERNARD, 1997; HENYEY and WASSERBURG, 1971; LACHEN-BRUCH and SASS, 1980; SIBSON, 1980; WANG et al., 1995; ZOBACK and BEROZA, 1993; ZOBACK et al., 1987). Based on the cited facts, most seismologists believe that it is reasonable to formulate the basic seismic gap hypothesis as follows:

> The energy for large and great earthquakes along plate boundaries is accumulated by plate motions, from low to maximum levels. This process takes decades to centuries to load a plate boundary segment. Great earthquakes are more likely in "loaded" segments, called "seismic gaps," than in segments recently unloaded by great ruptures.   (1)

Because we also know that some faults, including plate boundaries, exhibit physical discontinuities, the idea of a characteristic earthquake that breaks repeatedly the same fault segment (e.g., AKI, 1984; SCHWARTZ and COPPERSMITH, 1984) was added to the basic gap hypothesis. The segmentation is evident along many faults that are accessible for geologic inspection, and the multiple, complex nature of the energy release in great earthquakes has been documented numerous times (e.g., CHRISTENSEN and BECK, 1994; HARVEY and WYSS, 1986; WYSS and BRUNE,

1967). Thus, it is clear that segments of faults can repeatedly break by themselves (as the Parkfield earthquakes (BAKUN and McEVILLY, 1984)), as well as join with neighboring segments in complex larger ruptures (e.g., THATCHER, 1990). This generally accepted phenomenon of segmentation leads to the next step in defining seismic gaps:

> Seismic gaps are likely to rupture in one or a small number of large, gap-filling earthquakes.   (2)

In addition, several facts are known about the slip along plate boundaries. (A) It is known that aseismic creep along plate boundaries releases some of the accumulated strain energy (e.g., DAVIES and BRUNE, 1971). However, it is uncertain how much creep occurs and how heterogeneous its distribution is. Recent evidence suggests that more creep occurs than we had hitherto assumed (e.g., HEKI et al., 1997; FREYMUELLER et al., 1999; NISHIMURA et al., 1998)). (B) Further, we know from numerous case studies of coseismic surface displacements (e.g., SIEH et al., 1992; WALD et al., 1996), paleoseismic displacement mapping (ATWATER, 1992; SIEH, 1984) and teleseismic moment release analyses (e.g., CHRISTENSEN and BECK, 1994; HAESSLER et al., 1992), that the coseismic slip in great earthquakes varies strongly along the rupture surface. Therefore, the picture that virtually all seismologists have of plate boundaries is one of complex resistance to slip (asperities versus creeping segments) and complex slip history. This translates into the recognition that seismic gaps are not monolithic and may join in ruptures with neighboring segments beyond the gap boundary. Acknowledging this complexity leads to weakening of the adherence to statement (1), which could be formulated as follows:

> Because of the unknown creep and coseismic slip history, it is not certain that a seismic gap will ever be filled, or that the rupture stops at its edge.   (3)

It is standard for geological problems that varied scenarios exist. However, this variability is not convenient for testing one simple hypothesis with a single algorithm for the entire globe.

The hypothesis that all earthquakes, even the greatest ones, occur at random with a Poissonian distribution, predates all knowledge of tectonics and earthquake rupture mechanisms. It has long served earthquake engineers to estimate locally the probability of exceedence of a certain ground motion. It also has been used as the standard null hypothesis to evaluate the success of earthquake prediction schemes. Most seismologists regard the assumption that great earthquakes occur at random not as a serious physical model, but just as the null hypothesis in which one professes to be ignorant of tectonic facts. However, recently the proposal has been made that the hypothesis of temporal random or clustered occurrence of great earthquakes should be considered seriously (GELLER, 1997; KAGAN, 1997). This view could be formulated as follows:

The energy released in large and great earthquakes along plate boundaries is only a fraction of the total elastic energy stored at any time. Therefore, great earthquakes occur at random, and the probability of another great earthquake is the same, or increased, directly following a great earthquake, as it was before the event.

The fact that large ruptures often are multiple events, starting as a small earthquake, makes it difficult to predict earthquakes within short time windows (BRUNE, 1979). Exaggerating this difficulty, some proponents of the randomness hypothesis expanded their model as follows:

> Because small quakes can develop into a large energy release, and because energy is always available along plate boundaries, great earthquakes occur at random and earthquakes cannot be predicted (GELLER, 1997).

Testing the seismic gap hypothesis against the null hypothesis on a global basis, KAGAN and JACKSON (K&J) (1991, 1995) claimed to have shown the gap hypothesis to be invalid. Although these authors argue that they have formally only addressed the gap hypothesis as applied by McCANN et al. (1979), their difference in opinion goes deeper. In recent publications they, and the others, extend this result to claim that great earthquakes are clustered, or occur at random and are unpredictable (GELLER et al. 1997; KAGAN, 1997; MAIN, 1997). However, MC-CANN et al. (1979) applied a crude first approximation to the problem by classifying the state of plate boundary segments in three classes only (the most recent rupture having occurred less than 30, between 30 and 100 and more than 100 years ago). That this represents a simplification was clear to most seismologists at the time, and was stated by the authors in various articles (SYKES and NISHENKO, 1984). NISHENKO (1991) elaborated the hazard assessment for the Circum-Pacific area by considering local interevent intervals. K&J repeated the crucial shortcoming of the early gap work, which was to endeavour to evaluate the hypothesis along all plate boundaries simultaneously, instead of evaluating the seismic hazard as well as the hypothesis on a case by case or regional basis, as THATCHER (1990) did for the Circum-Pacific.

We argue in this paper that the case history and regional approach should be used to advance our knowledge of earthquake generation because of the strong differences in local conditions. In general, one must expect significant differences between earthquakes in extensional stress regimes, along strike-slip faults, in subduction zones, in continental collision areas, and in intra-continental settings. In the different settings, the orientation of the principal stresses, as well as the ambient stress level, differ widely. In addition the regularity of earthquakes along specific fault segments may be interrupted due to the complex interaction with neighboring segments, as is the case in the Circum-Pacific area (THATCHER, 1990).

Making Progress in Understanding the Earthquake Generating Mechanism

Imagination is the force with which humans launch new insight into the physics of processes surrounding us. However, critical testing of hypotheses thus generated must keep us from accepting erroneous concepts. In some fields of seismology, testing of hypotheses was not practiced extensively until several authors pointed out this shortcoming. In the late 1980s, IASPEI's subcommission on earthquake prediction initiated an effort to evaluate earthquake prediction hypotheses (WYSS, 1991a, 1997b; WYSS and BOOTH, 1997) with the aim of improving the rigor in developing hypotheses in this field. In the 1990s, tests were proposed for the seismic gap hypothesis, which brought to light that this hypothesis was not well formulated, and which triggered a debate about its validity (KAGAN and JACKSON, 1991, 1995; NISHENKO and SYKES, 1993). Also, rigorous testing of the precursory swarm hypothesis (EVISON and RHOADES, 1993) led to the rejection of that hypothesis and to a subsequent new formulation.

Often, proponents and critics of a hypothesis hold different ideas of how it should be tested. This is, for example, the case for testing the M 8 algorithm (HABERMANN and CREAMER, 1994; KEILIS-BOROK et al., 1990; KOSSOBOKOV et al., 1990, 1997; MINSTER and WILLIAMS, 1996). Nevertheless, hypothesis testing and tight definitions of the hypotheses most often bring advances and a clearer understanding of the problem at hand. However, it is possible that an overly simplified test may reject a hypothesis that correctly describes the physics of the process. We suggest that this is the case with the test of the seismic gap hypothesis by K&J. It only failed the seismic gap hypothesis as applied by MCCANN et al. (1979), but it did not fail the elastic rebound theory (REID, 1910), and thus, it did not fail the idea that fault segments recently broken in maximum magnitude earthquakes are unlikely to rupture soon, because the strain energy must be accumulated again. Below we will demonstrate how K&J reach incorrect results in the case of the M_w 8, 1986, Andreanof Island earthquake.

Low Stress Levels along Plate Boundaries

The ideas that the seismic gap hypothesis is incorrect (KAGAN and JACKSON, 1991, 1995) and that earthquakes cannot be predicted (GELLER, 1997) are based on the proposals that large earthquakes occur at random or that they cluster (KAGAN, 1997). The latter idea requires the assumptions (1) that the stress level is very high in all seismogenic areas, such that enough elastic strain energy for a second large earthquake is still available immediately after a first one, and (2) that the critical stress level for failure in that volume is changed, such that the fault can fail at a lower stress in the second event. If the ambient stress along plate boundaries were about 3 kbar, as necessary for fracturing rocks in the laboratory, then the average stress drop of 30 bars observed in large earthquakes (KANAMORI and ANDERSON, 1975) would lower the stress level by only 1%. Hence, one might imagine that great earthquakes may be repeated at any time, if one assumes that a mechanism exists by which the failure stress along the freshly ruptured fault has been appropriately lowered. If, however, stress levels along plate boundaries are approximately 100 bars or less, then stress drops are a substantial fraction of the total, possibly nearly complete, and hence strain energy must be accumulated again by tectonic processes, before another great rupture can follow a first one. Because this is an important question, we briefly review the evidence for low stress levels along plate boundaries.

The first hint that stress levels are low derives from the fact that earthquakes are concentrated along faults. Faults are clearly zones of low resistance to faulting, along which coseismic slip occurs in thousands of large earthquakes over millions of years.

Second, many authors (BRUNE et al., 1969; HENYEY and WASSERBURG, 1971; LACHENBRUCH and SASS, 1980) pointed out that, if friction along the San Andreas Fault was high and the work during coseismic slip was done by shear stresses larger than about 100 bar, then a heat-flow anomaly should be observed across the fault. This is not the case, and the hot springs, which would be necessary if the heat was removed by ground-water flow, do not exist.

Third, the orientation of the greatest principal stress within a few kilometers of the San Andreas Fault is nearly perpendicular to the fault, which means that the shear component parallel to the fault is very low at all times (JONES, 1988; WYSS and LU, 1995; ZOBACK and BEROZA, 1993; ZOBACK et al., 1987).

Fourth, the ubiquitous local perturbations of the directions of principal stresses that have been demonstrated by many techniques can easily be explained by local slip on faults, if the ambient stress only slightly exceeds the perturbing stress. GAMAR and BERNARD (1997) have shown this explicitly in an analysis of shear-wave polarization in the vicinity of the Erzincan earthquake, M 6.8, 1992, on the North Anatolian Fault. They inferred a local perturbation of the principal stress directions, which can be explained as due to the redistribution of stress by the slip in the 1939, M 8, earthquake along that fault, if the ambient stress level is about equal to the stress drop due to the M 8, 1939, earthquake (GAMAR and BERNARD, 1997), which means that the ambient stress is less than 100 bars in this area.

Fifth, the fact that the peak accelerations measured by strong motion instruments rarely exceed 0.5 g suggests that the ambient stress levels are generally low.

Sixth, the fact that stress drops for large and great earthquakes exceeding 100 bars are virtually never observed, also suggests that stress levels are below, or at that value. If stress levels were at 3 kbars, an occasional stress drop exceeding 1 to 2 kbar would likely be observed.

Seventh, stress indicators in the Cascadia and Nankai trough subduction zones indicate that these megathrust zones are weak (WANG et al., 1995), analogous to results obtained for the strike slip regimes.

The above remarks apply to plate boundaries. In the interior of continents, where faults are less well developed, several lines of evidence exist that suggest stresses may be higher. For plate boundaries, we conclude that stress levels are low and comparable to the stress drops observed for earthquakes. It follows that great earthquakes cannot be repeated after short intervals and that the amount of energy available depends on the slip history, as well as the time elapsed. Therefore, the rebound theory (REID, 1910), the time predictable model (SHIMAZAKI and NAKATA, 1980) and the seismic gap hypothesis (FEDOTOV, 1965) are likely to be essentially correct, along plate boundaries.

Case History of the Andreanof Segment of the Aleutian Plate Boundary

The volcanic island chain of the Aleutians, with the trench along its southern front, is the location of the subduction of the Pacific under the North American plate. One of the characteristics of this plate boundary is the enormous size of the megathrust earthquakes occurring along it (Fig. 1). More than half of the 3500-km long subduction zone ruptured in just three great earthquakes that took place within eight years of each other. The source parameters of these three earthquakes, which are among the six largest of this century, are given in Table 1.

Although one might assume that these three earthquakes were similar (because of their size and their common tectonic setting), they are in fact quite different because the style of subduction changes from east to west. In the east, the Prince

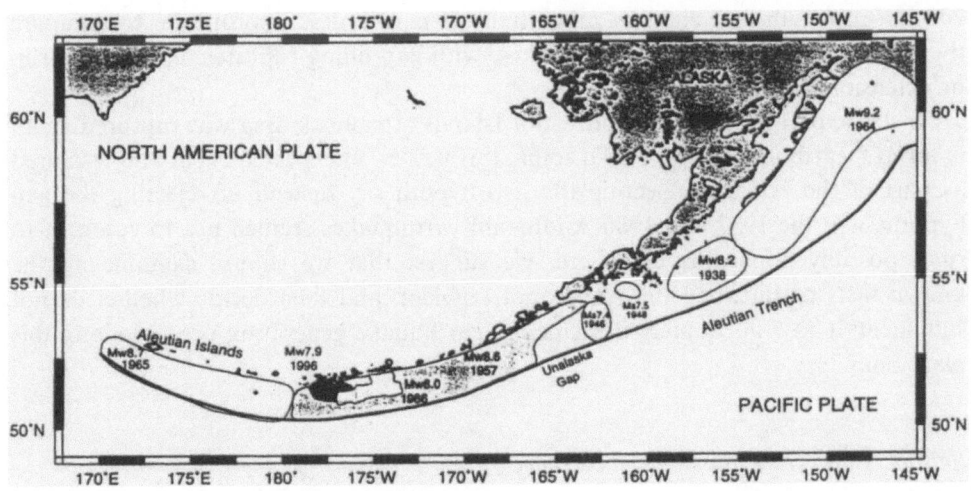

Figure 1

Map of the Aleutian Islands with aftershock areas of great earthquakes. Two *M* 8 earthquakes have re-ruptured the western part of the great 1957 rupture in 1986 and 1996.

Table 1

Source parameters of the Great Aleutian earthquakes

Location	Year	M_w	D (m)	L (km)	W (km)
Andreanof Islands	57	8.6	2	1200	75
Prince William Sound	64	9.2	10–20	700	300
Rat Islands	65	8.7	5–7	600	100
Andreanof Islands	86	8.0	2.4	270	50
Delarof Islands	96	7.9	2	130	30

William Sound earthquake of 1964 M_w 9.2) ruptured over an unusually wide plate interface of 300 km (PAGE, 1968) and an exceptionally large coseismic slip of 20 m (CHRISTENSEN and BECK, 1994) because in its source volume bits of continental crust, brought by the Pacific plate, are docking against the North American plate. In contrast, in the central Aleutians, the subducting plate is oceanic, and the 1957 Andreanof Island earthquake (M_w 8.7) ruptured a 75-km wide interface (BOYD and NABELEK, 1988) and developed an unusually small displacement of only about 2 m (JOHNSON and SATAKE, 1993; WAHR and WYSS, 1980), although it was 1200 km long (RUFF *et al.*, 1985; SYKES, 1971). In the west, the 1965 Rat Island earthquake (M_w 8.7) produced partial strike-slip motion, in addition to the thrusting component, and was thus again different than the other two ruptures.

The style of rupture along the Aleutian arc is thus varied. One cannot simply expect the same, narrow rules governing repeat ruptures to apply throughout the Aleutians. It makes even less sense to evaluate by one algorithm M_w 9 Aleutian earthquakes together with Central American megathrust ruptures that rarely exceed 200 km in length and eight in magnitude. It is even less appropriate to compare these, and other megathrust earthquakes, with gap-filling ruptures along strike-slip or extensional plate boundaries.

In 1986, part of the 1957 Andreanof Islands aftershock area was ruptured again in an M 8 earthquake (Fig. 1). To some, this was an unexpected event and was used as part of the evidence rejecting the gap hypothesis. Instead of rejecting the gap hypothesis if the 1957 and 1986 Andreanof earthquakes seemed not to conform to rules possibly applicable elsewhere, we suggest that we should examine all the known facts pertinent to the Andreanof segment, and then decide whether or not statements 1 to 3 adequately describe the earthquake generating process along this plate boundary.

Seismic Gap versus Random Occurrence

The segment of the 1957 Aleutian break was classified as recently ruptured, thus considered safe from earthquakes in the near future by MCCANN *et al.* (1979). Therefore, K&J regarded it a failure of the seismic gap hypothesis, when in 1986 the M_w 8 Andreanof earthquake re-ruptured part of the same segment (Fig. 1). We agree that it was a failure of the forecast by MCCANN et al., nonetheless we argue that it was clearly supporting the seismic gap hypothesis as formulated in statements 1 through 3 above, based on the additional facts detailed below.

The Slip History

Information regarding the slip history of the fault segment in question is important, since the seismic gap concept, and more basically the idea that earthquakes 'return,' is based on the hypothesis that a large percentage of the locally available elastic strain is released in great earthquakes, and that this energy must be built up again by plate motions before another major rupture can occur. SHIMAZAKI and NAKATA (1980) presented data that support this time-predictable model, however MULARGIA and GASPERINI (1995) argued that the case was not strong. The notion that slip deficiencies along parts of great ruptures must be made up in additional, possibly large, ruptures is generally accepted (SYKES and NISHENKO, 1984; TANIOKA et al., 1996; SYKES et al., 1999).

In the case of the Andreanof segment, limited information was available to estimate the coseismic displacement of 1957 from tide-gauge observations. Two tide gauges, which were operated over decades, were located within the 1957 rupture segment. They recorded subsidence of the islands by about 0.15 m during the 1957 rupture (Fig. 2). WAHR and WYSS (1980) used these observations to estimate that the coseismic slip in the 1957 earthquake was only about 2 m, although the major asperity may have slipped about twice as much (JOHNSON et al., 1994). This is a surprisingly small amount, when compared to slip in other great events of the M_w 9 class, especially the 1964 Prince William Sound earthquake (Table 1). However, varying the dip of the fault plane and its width, within acceptable bounds, could not produce a model with a substantially larger displacement (WAHR and WYSS, 1980). Thus, these authors pointed out that there must be large differences in recurrence times along the Aleutian plate boundary. If one assumes that no aseismic creep reduces the accumulated displacement, then one must conclude that a 2-m slip is re-accumulated in 27 years, since the plate convergence is about 7.3 cm/year at that location (DEMETS et al., 1990). That means that, after the 1957 earthquake, another rupture repeating the slip of 1957 could be expected near 1984 plus or minus several years, based on the time predictable model.

The Quiescence History

In addition to the seismic slip, information about seismicity patterns was available which suggested the approach of a great earthquake in the Andreanof segment in the early 1980s. C. KISSLINGER (personal communication, 1983) remarked to the senior author of this paper "my people have nothing to do, there are no earthquakes occurring." By "my people" he was referring to his employees, whose job it was to locate earthquakes recorded by the Adak seismograph network, operated by him. When asked where the seismic quiescence was occurring, he replied "everywhere." When told that, based on the quiescence hypothesis, a quiescence with a spatial extent of 150 km would suggest that an M_w 8 earthquake was to be expected, he argued we should not advocate this because the quiescence hypothesis was not firmly established.

In the years following this discussion, Kisslinger documented the seismic quiescence, added other tectonic information and formulated a prediction of a large earthquake in the Adreanof segment of the Aleutians (KISSLINGER, 1986; KISSLINGER et al., 1985). The earthquake that followed did not strictly conform to the prediction as formulated. Kisslinger had opened too short a time window of only half a year. If he had assessed our ignorance of durations of precursory

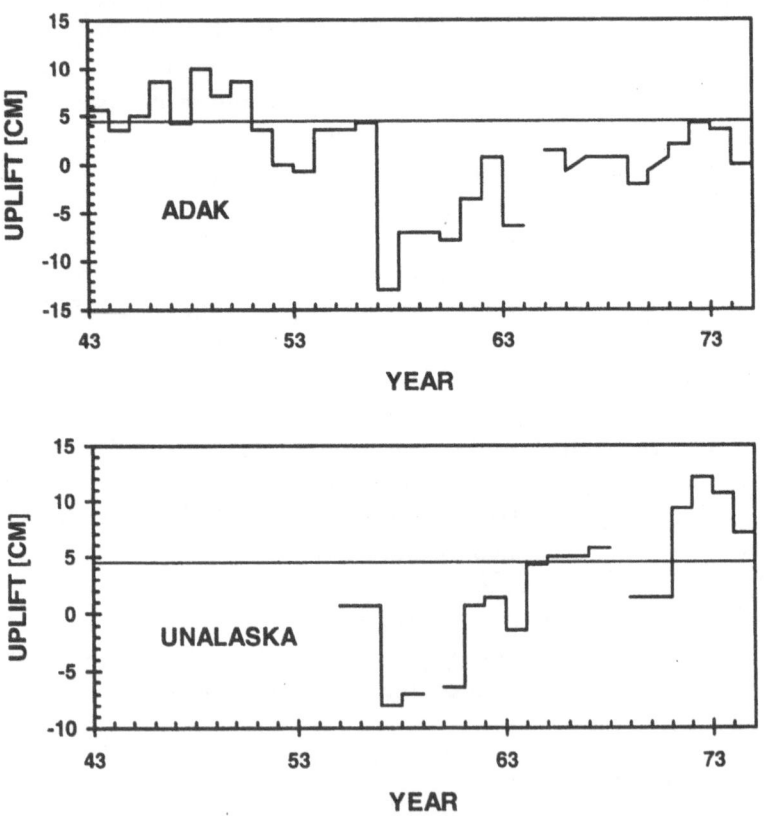

Figure 2

Mean annual sea level at Adak and Attu Islands as a function of time, showing the small coseismic subsidence that constrains the model of coseismic slip at the time of the $M_w = 8.7$, 1957 earthquake to about 2 m.

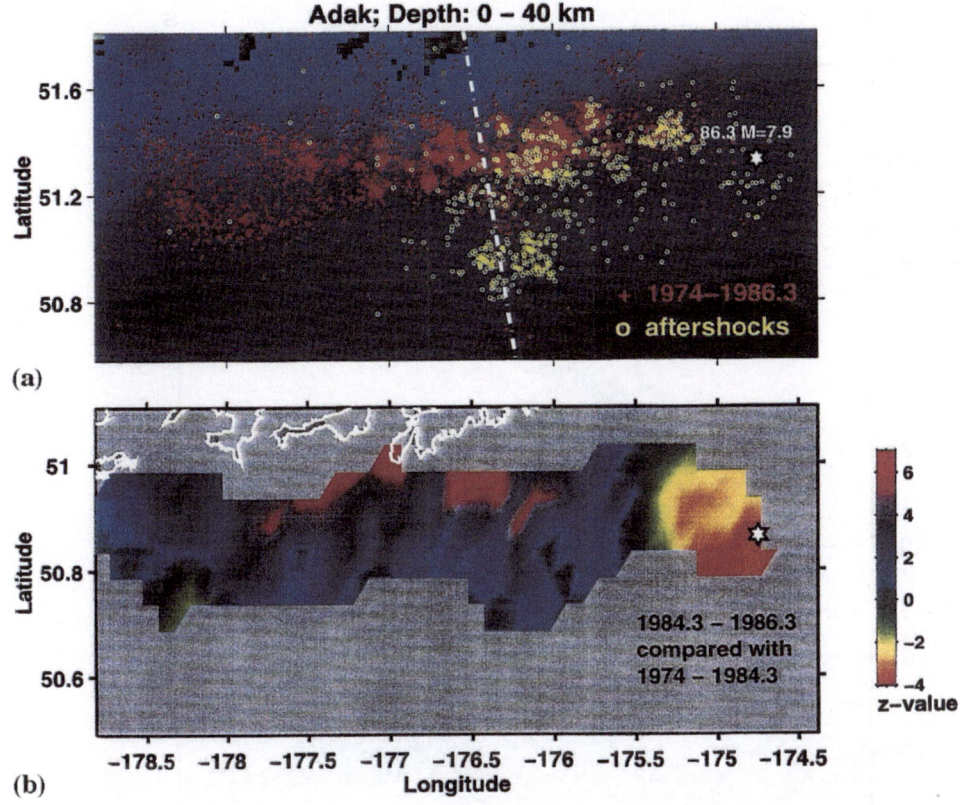

Figure 3

(a) Epicenter map of the Adak region from the Adak network catalog for depths < 40 km. Aftershocks to the 1986 (M_w 8) Andreanof earthquake are plotted in yellow. The dashed line defines the location of the cross section shown in Figure 4. (b) Map of the distribution of Z values in the Adak region. The Z value compares the background seismicity rate (1974–1984.3) with the final two years before the main shock. The 300 nearest earthquakes to each node of a densely spaced grid were sampled. High Z values (blue-pink) indicate a decrease or quiescence in the period 1984.3–1986.3 as compared to the background period. Note the increased seismicity rate at the hypocenter of the 1986 mainshock.

quiescence more realistically, he would have opened a one- or two-year window and been correct with his prediction. He also underestimated the magnitude of the 1986 earthquake. He estimated the dimensions of the quiescence as 150 km, hence an M_w 8 event should have been anticipated if one assumes that the size of the quiet volume indicates the size of a future rupture.

The computer tools which we recently developed (WIEMER, 1996) to search for seismicity rate changes map the quiescence before the 1986 Andreanof earthquake as located mainly above the megathrust (Fig. 4b), and in the western 150 km of the rupture (Fig. 3b), while the seismicity rate in the hypocentral volume increased (Fig. 5b). No information regarding rate changes is available in the easternmost portion

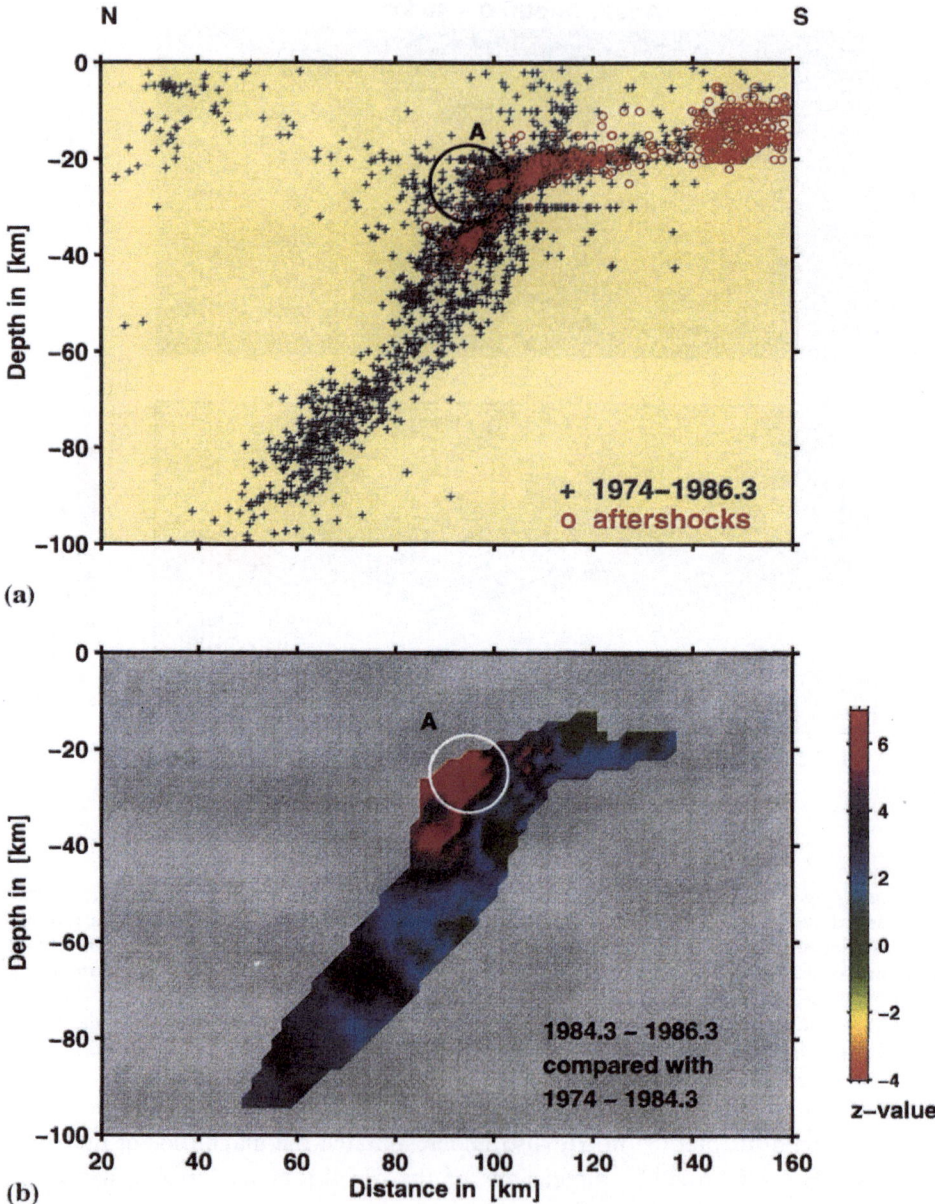

Figure 4
Cross section perpendicular to the plate boundary, following the profile defined in Figure 3. (a)
Hypocenter distribution. Events in the period 1974–1986.3 are plotted as blue +, aftershocks to the
1986 Andreanof mainshock as red 'O'. (b) Distribution of the Z value comparing the seismicity in the
final two years before the mainshock with the 10 years of background. The circle delineates the volume
for which the seismicity rate is shown in the first frame of Figure 5.

of the 1986 rupture, because this area is outside the coverage of the local Adak seismic network. This quiescence is statistically highly significant and it is the most significant anomaly in the data set of the Adak network. In addition, the quiescence was confirmed by an independent data set of analog records from station Adak (WYSS, 1991b), and the rates returned to the pre-quiescence background shortly after the 1986 mainshock (KISSLINGER and KINDEL, 1994). Therefore, we conclude that the seismic quiescence, noticed during the early to mid-1980s by Kisslinger, was a real phenomenon; that it was highly significant, even unique, in the data, and that it was correctly interpreted before the fact as a precursor to a large earthquake.

There are two facts pertaining to the mapped quiescence (Figs. 3 and 4) that suggest it was caused by precursory fault creep as envisioned in the model by KATO *et al.* (1998). The quiescence is seen to be concentrated in the wedge above the megathrust (Fig. 4b), and the hypocentral volume itself manifested an increase of the seismicity rate synchronously with the quiescence elsewhere (Fig. 5b). This would result if the hypocentral volume was an asperity in which no precursory creep occurred, but in which stress was accumulated at a faster than normal rate, due to precursory creep on the neighboring fault segment.

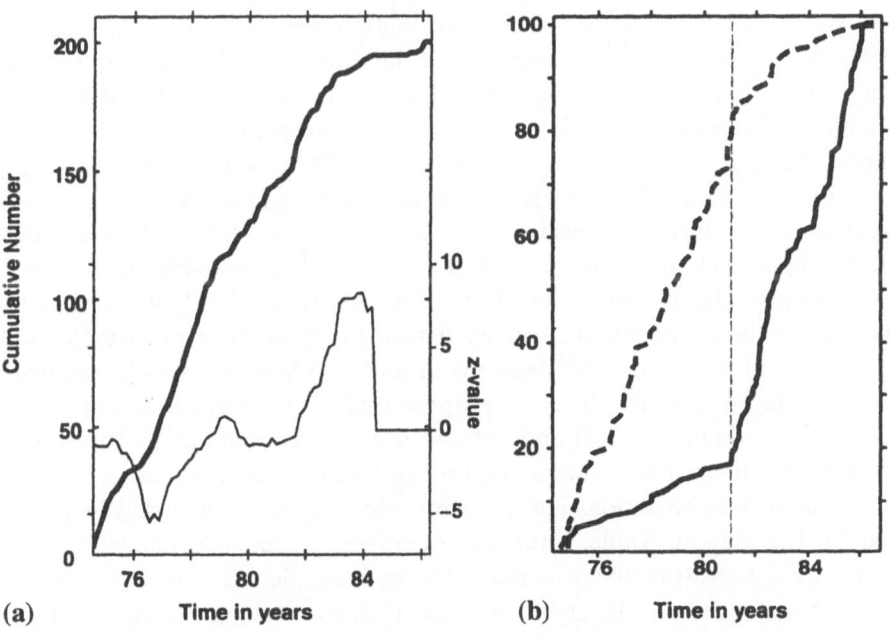

Figure 5

Cumulative number of earthquakes as a function of time showing the precursory quiescence and seismicity rate increase in two separate segments of the 1986 great rupture. (a) Volume A marked in Figure 3. The thick line represents the cumulative number as a function of time, the thin line the LTA(t) (e.g., WIEMER and WYSS, 1994) value computed for a window length of two years. (b) Cumulative number plots for the area of highest Z value (dashed) and the lowest Z value at the epicenter location (solid).

Now that we know that the 1986, M_w 8, earthquake was definitely preceded by a quiescence precursor, it seems likely that the precursory quiescences reported after the fact for other Aleutian earthquakes (HABERMANN, 1981) were also real, although they were based on inferior teleseismic data. It is therefore reasonable to assume that precursory seismic quiescence will occur again in the Aleutians, and thus, it would be irresponsible not to monitor the Aleutians for quiescence in real time.

Discussion and Conclusions

The Main Issue: Artificial Intelligence versus Common Sense

The artificial intelligence as exercised by the simple algorithm used by K&J suggests that the occurrence of the 1986 Andreanof, M_w 8, earthquake is part of the evidence opposing the validity of the seismic gap hypothesis, and thus implies that great earthquakes do not occur in gaps along plate boundaries. This conclusion is wrong. We illustrate that, based on the information concerning the 1957 coseismic slip (WAHR and WYSS, 1980) known before this earthquake, the energy to repeat the 1957 slip was most likely restored by the mid to late 1980s. In addition, the observation of seismic quiescence was correctly interpreted before the fact by Kisslinger (KISSLINGER, 1986; KISSLINGER et al., 1985) as a precursor to a large earthquake. We conclude that the common sense analysis of the Andreanof segment as a case history, considering all available facts, led to the conclusion, before the mainshock, that a large to great earthquake was possible, or even likely, in this segment in the mid-1980s. The facts of the Andreanof 1957/1986 case history firmly support the elastic rebound theory (REID, 1910), the time predictable model (SHIMAZAKI and NAKATA, 1980), and the seismic gap hypothesis as formulated in statements 1 through 3. We therefore propose that time-dependent seismic hazard should be assessed mostly based on local case histories, considering all known facts, and that global tests of hypotheses can lead to incorrect conclusions.

Since subduction conditions and coseismic slip vary as dramatically as demonstrated by the data in Table 1 for the Aleutians, as also documented for the Aleutians by JOHNSON (1998), one may have to conclude that one cannot test the seismic gap hypothesis on the global data set. Unfortunately, great earthquakes do not occur frequently enough that hypotheses of their prediction could be tested statistically in subregions. Thus, for any hypothesis for which parameters are strongly dependent on local conditions, meaningful statistical tests have yet to be designed. This does not mean that we cannot progress toward ultimate acceptance of a hypothesis. However, it does mean that we must formulate our hypothesis in a complex way, such that we can account for regional differences.

A case history that shows that Reid's model operates not only along plate boundaries is the 1975 (M 7.2) Kalapana, Hawaii, earthquake. In this case, we know that the previous significant rupture of the same fault surface occurred in 1868 (WYSS, 1988), and that geodetic measurements between 1900 and 1976 revealed that the compressive strain of about 4×10^{-4} that was accumulated before the 1975 earthquake (SWANSON et al., 1976) was released as extension during the rupture (TILLING et al., 1976; WYSS and KOVACH, 1988). As in the case of the 1986 Andreanof earthquake, the Kalapana earthquake was not formally correctly predicted, although it was anticipated. SWANSON et al.'s paper was in press at the time of the earthquake and explicitly contained the expectation that a large earthquake should occur in the Kalapana area because of the observed large strain accumulation.

Negative case histories that would show two or more great (large) earthquakes generated in short sequence (an order of magnitude shorter than expected from the time-predictable model) from the same source volume do not exist (e.g., THATCHER, 1990). In all cases in which source volumes of two large earthquakes separated by short intervals overlapped, and detailed information on the slip distribution existed, it was found that the slip in the first event was deficient in the segment in which the overlap occurred (e.g., TANIOKA et al., 1996).

The study of case histories should include a detailed examination of the extent of past ruptures (SYKES and NISHENKO, 1984), the interevent times (NISHENKO, 1991) and the amount of slip (as well as its variation along the rupture) of each local rupture (SHIMAZAKI and NAKATA, 1980; SYKES and NISHENKO, 1984). In this manner, one will be able to construct a common sense local scenario for great earthquake ruptures.

Important contributions to estimating the strain buildup and the release in creep will likely come from geodetic measurements by GPS and SAR interferometry. This will take time, however modern geodetic networks will ultimately capture strain cycles. Once many case histories of strain accumulation over decades, followed by its release in tens of seconds during earthquake ruptures, have been measured, we will know more about this problem.

The Side Issue: Precursory Seismicity Patterns

Beyond the pattern of seismic gaps, two precursory patterns, those of quiescence and increasing moment release, are emerging as helpful in assessing the time-dependent seismic hazard locally. Both of these patterns have been recognized before and after the fact in several cases in the Aleutians and California, and should therefore be employed in the future to attempt to anticipate large and great earthquakes.

Seismic quiescence was mapped in quantitive detail in several outstanding case histories with modern tools (WIEMER and WYSS, 1994; WYSS et al., 1998, 1996) in addition to many earlier case histories based on high quality data (WYSS, 1997a).

Along the San Andreas Fault, one earthquake was predicted correctly (WYSS and BURFORD, 1987), and in the Aleutians one earthquake was anticipated but not correctly predicted (KISSLINGER, 1988) based on seismic quiescence. Although there are issues that remain to be solved, such as the percentage of false alarms and missed events, seismic quiescence is now at a state where it should be monitored in seismically active areas, especially since viable quantitative models based on constitutive laws of failure in the laboratory (DIETERICH, 1994; KATO et al., 1998) may explain the observations in certain cases.

Increased moment release is also a phenomenon that can be understood with failure criteria developed in other applications in earth sciences (VARNES, 1989). After the fact, this phenomenon was documented for parts of California (JAUMÉ and SYKES, 1999, 1996; SYKES and JAUMÉ, 1990), and it led to the correct prediction of the 1996 Delarof (M_w 7.9) earthquake in the Aleutian Islands (BUFE et al., 1994; BUFE and VARNES, 1996; NISHENKO et al., 1996). Therefore, increased moment release should also be monitored in real time, and targeted on sesimic gaps or other locations which could form a case history.

We conclude that recent attempts to evaluate hypotheses by algorithms have led to many advances in earthquake prediction research, such as sharpening definitions of predictions and hypotheses, but that it can also lead to erroneous rejection of a hypothesis that can be demonstrated to be supported by case histories. Since great earthquakes occur relatively infrequently, it may be very difficult to statistically test the effectiveness of models concerning their generation. Therefore, we advocate that both testing by algorithm as well as evaluating by common sense based on case histories should be exercised to advance our knowledge of the physics of earthquake failure processes.

Acknowledgements

We thank C. Estabrook, Y. Y. Kagan, S. McNutt, D. Christensen and K. Satake for comments on the manuscript. This work was supported by the USGS under grant number 1434-HQ-97-GR-03073, by the GeoForschungs Zentrum, Potsdam, the Japanese Meteorological Agency and the Wadati Foundation at the Geophysical Institute of the University of Alaska, Fairbanks.

REFERENCES

AKI, K. (1984), *Asperities, Barriers, Characteristic Earthquakes and Strong Motion Prediction*, J. Geophys. Res. *89*, 5867–5872.
ATWATER, B. F. (1992), *Geologic Evidence for Earthquakes during the Past 2000 Years along the Copalis River, Southern Coastal Washington*, J. Geophys. Res. *97*, 1901–1919.

BAKUN, W. H., and MCEVILLY, T. V. (1984), *Recurrence Models and Parkfield, California, Earthquakes*, J. Geophys. Res. *89*, 3051–3058.

BOYD, T. M., and NABELEK, J. L. (1988), *Rupture Process of the Andreanof Islands Earthquake of May 7, 1986*, Bull. Seismol. Soc. Am. *78*, 1653–1673.

BRUNE, J. N. (1979), *Implications of Earthquake Triggering and Rupture Propagation for Earthquake Prediction Based on Premonitory Phenomena*, J. Geophys. Res. *84*, 2195–2198.

BRUNE, J. N., HENYEY, T. L., and ROY, R. F. (1969), *Heat Flow, Stress, and Rate of Slip along the San Andreas Fault, California*, J. Geophys. Res. *74*, 3821–3827.

BUFE, C. G., NISHENKO, S. P., and VARNES, D. J. (1994), *Seismicity Trends and Potential for Large Earthquakes in the Alaska-Aleutian Region*, Pure appl. geophys. *142*, 83–99.

BUFE, C. G., and VARNES, D. J. (1996), *Time-to-failure in the Alaska-Aleutian Region: An Update*, EOS *77*, F456.

CHRISTENSEN, D. H., and BECK, S. L. (1994), *The Rupture Process and Tectonic Implications of the Great 1964 Prince William Sound Earthquake*, Pure appl. geophys. *142*, 29–53.

DAVIES, G. F., and BRUNE, J. N. (1971), *Regional and Global Fault Slip Rates from Seismicity*, Nature *229*, 101–107.

DEMETS, C., GORDON, R. G., ARGUS, D. F., and STEIN, S. (1990), *Current Plate Motions*, Geophys. J. Int. *101*, 425–478.

DIETERICH, J. (1994), *A Constitutive Law for Rate of Earthquake Production and its Application to Earthquake Clustering*, J. Geophys. Res. *99*, 2601–2618.

EVISON, F. F., and RHOADES, D. A. (1993), *The Precursory Swarm in New Zealand: Hypothesis Test*, New Zealand J. Geol. Geophys. *36*, 51–60.

FEDOTOV, S. A. (1965), *Regularities of the Distribution of Strong Earthquakes in Kamchatka, the Kurile Islands and Northeastern Japan*, Tr. Inst. Fiz. Zemli Akad. Nauk SSSR *36*, 66.

FREYMUELLER, J., COHEN, S., and FLETCHER, H. (1999), *Deformation in the Kenai Peninsula*, J. Geophys. Res., submitted.

GAMAR, F., and BERNARD, P. (1997), *Shear-wave Anisotropy in the Erzincan Basin and its Relationship with Crustal Strain*, J. Geophys. Res. *102*, 20,373–20,393.

GELLER, R. J. (1997), *Earthquake Prediction: Are Further Efforts Warranted*, Geophys. J. Int. *147*, 111–999.

GELLER, R. J., JACKSON, D. D., KAGAN, Y. Y., and MULARGIA, F. (1997), *Earthquakes Cannot be Predicted*, Science *275*, 1616–1617.

HABERMANN, R. E. (1981), *The Quantitative Recognition and Evaluation of Seismic Quiescence: Applications to Earthquake Prediction and Subduction Zone Tectonics*, University of Colorado, Boulder, 253 pp.

HABERMANN, R. E., and CREAMER, F. (1994), *Catalog Errors and the M 8 Earthquake Prediction Algorithm*, Bull. Seismol. Soc. Am. *84*, 1551–1559.

HAESSLER, H., DESCHAMPS, A., DUFUMIER, H., FUENZALIDA, H., and CISTERNAS, A. (1992), *The Rupture Process of the Armenian Earthquake from Broad-band Teleseismic Body Wave Records*, Geophys. J. Int. *109*, 151–161.

HARVEY, D., and WYSS, M. (1986), *Comparison of a Complex Rupture Model with the Precursor Asperities of the 1975 Hawaii $M_s = 7.2$ Earthquake*, Pure appl. geophys. *124*, 957–973.

HEKI, K., MIYAZAKI, S., and TSUJI, H. (1997), *Silent Fault Slip Following an Interplate Thrust Earthquake at the Japan Trench*, Nature *386*, 595–598.

HENYEY, T. L., and WASSERBURG, G. J. (1971), *Heat Flow near Major Strike-slip Faults in California*, J. Geophys. Res. *76*, 7924–7946.

JAUMÉ, S., and SYKES, L. R. (1999), *Evolving towards a Critical Point: A Review of Accelerating Seismic Moment/Energy Release prior to Large and Great Earthquakes*, Pure appl. geophys., this issue.

JAUMÉ, S. C., and SYKES, L. R. (1996), *Evolution of Moderate Seismicity in the San Francisco Bay Region, 1850 to 1993: Seismicity Changes Related to Occurrence of Large and Great Earthquakes*, J. Geophys. Res. *101*, 765–790.

JOHNSON, J. M. (1998), *Heterogeneous Coupling along the Alaska-Aleutians as Inferred from Tsunami, Seismic and Geodetic Inversions*, Advances in Geophysics *39*, 1–116.

JOHNSON, J. H., and SATAKE, K. (1993), *Source Parameters of the 1957 Aleutian Earthquake from Tsunami Waveforms*, Geophys. Res. Lett. *20*, 1487–1490.

JOHNSON, J. M. et al. (1994), *The 1957 Great Aleutian Earthquake*, Pure appl. geophys. *142*, 3–28.

JONES, L. M. (1988), *Focal Mechanisms and the State of Stress on the San Andreas Fault in Southern California*, J. Geophys. Res. *93*, 8869–8891.

KAGAN, Y. Y. (1997), *Are Earthquakes Predictable?*, Geophys. J. Int. *131*, 505–525.

KAGAN, Y. Y., and JACKSON, D. D. (1991), *Seismic Gap Hypothesis: Ten Years after*, J. Geophys. Res. *96*, 21,419–21,431.

KAGAN, Y. Y., and JACKSON, D. D. (1995), *New Seismic Gap Hypothesis: Five Years after*, J. Geophys. Res. *100*, 3943–3960.

KANAMORI, H., and ANDERSON, D. L. (1975), *Theoretical Basis of Some Empirical Relations in Seismology*, Bull. Seismol. Soc. Am. *65*, 1073–1095.

KATO, N., OHTAKE, M., and HIRASAWA, T. (1998), *Possible Mechanism of Precursory Seismic Quiescence: Regional Stress Relaxation due to Preseismic Sliding*, Pure appl. geophys. *150*, 249–267.

KEILIS-BOROK, V. I., KNOPOFF, L., KOSSOBOKOV, V. G., and ROTVAIN, I. (1990), *Intermediate-term Prediction in Advance of the Loma Prieta Earthquake*, Geophys. Res. Letts. *17*, 1461–1464.

KISSLINGER, C. (1986), *Seismicity Patterns in the Adak Seismic Zone and the Short-term Outlook for a Major Earthquake*, Meeting of the National Earthquake Prediction Evaluation Council. U.S. Geol. Survey, Open-file Rept. 86–92, Anchorage, Alaska, pp. 119–134.

KISSLINGER, C. (1988), *An Experiment in Earthquake Prediction and the 7 May 1986 Andreanof Islands Earthquake*, Bull. Seismol. Soc. Am. *78*, 218–229.

KISSLINGER, C., McDONALD, C., and BOWMAN, J. R. (1985), *Precursory Time-space Patterns of Seismicity and their Relation to Fault Processes in the Central Aleutian Islands Seismic Zone*, IASPEI, 23rd General Assembly, Tokyo, Japan, 32 pp.

KISSLINGER, K., and KINDEL, B. (1994), *A Comparison of Seismicity Rates near Adak Island, Alaska, September 1988 through May 1990 with Rates before the 1982 to 1986 Apparent Quiescence*, Bull. Seismol. Soc. Am. *84*, 1560–1570.

KOSSOBOKOV, V. G., HEALY, J. H., and DEWEY, J. W. (1997), *Testing an Earthquake Prediction Algorithm*, Pure appl. geophys. *149*, 219–248.

KOSSOBOKOV, V. G., KEILIS-BOROK, V. I., and SMITH, S. W. (1990), *Localization of Intermediate-term Earthquake Prediction*, J. Geophys. Res. *95*, 19,763–19,772.

LACHENBRUCH, A. H., and SASS, J. H. (1980), *Heat Flow and Energetics of the San Andreas Fault Zone*, J. Geophys. Res. *85*, 6185–6222.

MAIN, I. G. (1997), *Long Odds on Prediction*, Nature *385*, 19–20.

McCANN, W. E., NISHENKO, S. P., SYKES, L. R., and KRAUSE, J. (1979), *Seismic Gaps and Plate Tectonics: Seismic Potential for Major Plate Boundaries*, Pure appl. geophys. *117*, 1082–1147.

MINSTER, J. B., and WILLIAMS, N. P. (1996), *M 8 Intermediate-term Earthquake Prediction Algorithm: Performance Update for M > 7.5, 1985–1996*, EOS 77, F456.

MULARGIA, F., and GASPERINI, P. (1995), *Evaluation of the Applicability of the Time- and Slip-predictable Earthquake Recurrence Models to Italian Seismicity*, Geophys. J. Int. *120*, 453–473.

NISHENKO, S. P. (1991), *Circum-Pacific Seismic Potential: 1989–1999*, Pure appl. geophys. *135*, 169–259.

NISHENKO, S. P. et al. (1996), *1996 Delarof Islands Earthquake—A Successful Earthquake Forecast/Prediction?*, EOS 77, F456.

NISHENKO, S. P., and SYKES, L. R. (1993), *Comment on "Seismic Gap Hypothesis: Ten Years after" by Y. Y. Kagan and D. D. Jackson*, J. Geophys. Res. *98*, 9909–9916.

NISHIMURA, T. et al. (1998), *Source Model of the Co- and Post-seismic Deformation Associated with the 1994 far off Sanriku Earthquake (M 7.5) Inferred from Strainmeter and GPS Measurements*, Tohoku Geophys. J. *35*, 15–32.

PAGE, R. (1968), *Aftershocks and Microaftershocks of the Great Alaska Earthquake of 1964*, Bull. Seismol. Soc. Am. *58*, 1131–1168.

REID, H. F. (1910). *The Mechanics of the Earthquake*, The California Earthquake of April 18, 1906, Report of the State Earthquake Investigation Commission. Carnegie Institution of Washington, Washington, D.C.

RUFF, L. R., KANAMORI, H., and SYKES, L. R. (1985), *The 1957 Great Aleutian Earthquake (abstract)*, EOS *66*, 298.

SCHWARTZ, D. P., and COPPERSMITH, K. J. (1984), *Fault Behaviour and Characteristic Earthquakes: Examples from the Wasatch and San Andreas Fault Zones*, J. Geophys. Res. *89*, 5681–5698.

SHIMAZAKI, K., and NAKATA, T. (1980), *Time-predictable Recurrence Model for Large Earthquakes*, Geophys. Res. Letts. *7*, 279–282.

SIBSON, R. H. (1980), *Power Dissipation and Stress Levels on Faults in the Upper Crust*, J. Geophys. Res. *85*, 6239–6247.

SIEH, K. E. *et al.* (1992), *Near-field Investigations of the Landers Earthquake Sequence, April to July 1992*, Science *260*, 171–176.

SIEH, K. E. (1984), *Lateral Offsets and Revised Dates of Large Prehistoric Earthquakes at Pallett Creek, Southern California*, J. Geophys. Res. *89*, 7641–7670.

SWANSON, D. A., DUFFIELD, W. A., and FISKE, R. S. (1976), *Displacement of the South Flank of Kilauea Volcano: The Result of Forceful Intrusion of Magma into the Rift Zones*, U.S. Geol. Surv. Prof. Pap. *963*, 1–39.

SYKES, L. R. (1971), *Aftershock Zones of Great Earthquakes, Seismicity Gaps, Earthquake Prediction for Alaska and the Aleutians*, J. Geophys. Res. *76*, 8021–8041.

SYKES, L. R., and JAUMÉ, S. C. (1990), *Seismic Activity on Neighbouring Faults as a long-term Precursor to Large Earthquakes in the San Francisco Bay Area*, Nature *348*, 595–599.

SYKES, L. R., and NISHENKO, S. P. (1984), *Probabilities of Occurrence of Large Plate Rupturing Earthquakes for the San Andreas, San Jacinto, and Imperial Faults, California*, J. Geophys. Res. *89*, 5905–5927.

SYKES, L. R., SHAW, B. E., and SCHOLZ, C. H. (1999), *Rethinking Earthquake Prediction*, Pure appl. geophys., this volume.

TANIOKA, Y., RUFF, L., and SATAKE, K. (1996), *The Sanriku-oki, Japan Earthquake of December 28, 1994 (M_w 7.7): Rupture of a Different Asperity from a Previous Earthquake*, Geophys. Res. Letts. *23*, 1465–1468.

THATCHER, W. (1990), *Order and Diversity in the Modes of Circum-Pacific Earthquake Recurrence*, J. Geophys. Res. *95*, 2609–2624.

Tilling, R. I. *et al.* (1976), *Earthquake and Related Catastrophic Events, Island of Hawaii, November 29, 1975: A Preliminary Report*, Geological Survey Circular 740, 33 pp.

VARNES, D. J. (1989), *Predicting Earthquakes by Analyzing Accelerating Precursory Seismic Activity*, Pure appl. geophys. *130*, 661–686.

WAHR, J., and WYSS, M. (1980), *Interpretation of Postseismic Deformation with a Viscoelastic Relaxation Model*, J. Geophys. Res. *85*, 6471–6477.

WALD, D. J., HEATON, T. H., and HUDNUT, K. W. (1996), *The Slip History of the 1994 Northridge, California, Earthquake Determined from Strong-motion, Teleseismic, GPS, and Levelling Data*, Bull. Seismol. Soc. Am. *86* (B1), S49–S70.

WANG, K., MULDER, T., ROGERS, G. C., and HYNDMAN, R. D. (1995), *Case for Very Low Coupling Stress of the Cascadia Subduction Fault*, J. Geophys. Res. *100*, 12,907–12,918.

WIEMER, S. (1996), *Analysis of Seismicity: New Techniques and Case Studies*, Dissertation Thesis, University of Alaska, Fairbanks, Alaska, 151 pp.

WIEMER, S., and WYSS, M. (1994), *Seismic Quiescence before the Landers (M = 7.5) and Big Bear (M = 6.5) 1992 Earthquakes*, Bull. Seismol. Soc. Am. *84*, 900–916.

WYSS, M. (1988), *A Proposed Source Model for the Great Kau, Hawaii, Earthquake of 1868*, Bull. Seismol. Soc. Am. *78*, 1450–1462.

WYSS, M. (1991a), *Evaluation of Proposed Earthquake Precursors*, Am. Geophys. Union, Washington, 94 pp.

WYSS, M. (1991b), *Reporting History of the Central Aleutians Seismograph Network and the Quiescence Preceding the 1986 Andreanof Island Earthquake*, Bull. Seismol. Soc. Am. *81*, 1231–1254.

WYSS, M. (1997a), *Nomination of Precursory Seismic Quiescence as a Significant Precursor*, Pure appl. geophys. *149*, 79–114.

WYSS, M. (1997b), *Second Round of Evaluations of Proposed Earthquake Precursors*, Pure appl. geophys. *149*, 3–16.

WYSS, M., and BOOTH, D. C. (1997), *The IASPEI Procedure for the Evaluation of Earthquake Precursors*, Geophys. J. Int. *131*, 423–424.

WYSS, M., and BRUNE, J. N. (1967), *The Alaska Earthquake of 28 March 1964: A Complex Multiple Rupture*, Bull. Seismol. Soc. Am. *57*, 1017–1023.

WYSS, M., and BURFORD, R. O. (1987), *A Predicted Earthquake on the San Andreas Fault, California*, Nature *329*, 323–325.

WYSS, M., HASEGAWA, A., WIEMER, S., and UMINO, N. (1999), *Quantitative Mapping of Precursory Seismic Quiescence before the 1989, M 7.1, off-Sanriku Earthquake, Japan*, Annali di Geophysica, in press.

WYSS, M., and KOVACH, R. L. (1988), *Comments on "A Single Force Model for the 1975 Kalapana, Hawaii, Earthquake," by H.K. Eissler and H. Kanamori*, J. Geophys. Res. *93*, 8078–8082.

WYSS, M., and LU, Z. (1995), *Plate Boundary Segmentation by Stress Directions: Southern San Andreas Fault, California*, Geophys. Res. Letts. *22*, 547–550.

WYSS, M., SHIMAZAKI, K., and URABE, T. (1996), *Quantitative Mapping of a Precursory Quiescence to the Izu-Oshima 1990 (M 6.5) Earthquake, Japan*, Geophys. J. Int. *127*, 735–743.

ZOBACK, M. D., and BEROZA, G. C (1993), *Evidence for Near-frictionless-faulting in the 1989 (M 6.9) Loma Prieta, California, Earthquake and its Aftershocks*, Geology *21*, 181–185.

ZOBACK, M. D., et al. (1987), *New Evidence on the State of Stress of the San Andreas Fault System*, Science *238*, 1105–1111.

(Received August 31, 1998, revised/accepted January 27, 1999)

 To access this journal online:
http://www.birkhauser.ch

Pure appl. geophys. 155 (1999) 279–306
0033–4553/99/040279–27 $ 1.50 + 0.20/0

|Pure and Applied Geophysics

Evolving Towards a Critical Point: A Review of Accelerating Seismic Moment/Energy Release Prior to Large and Great Earthquakes

STEVEN C. JAUMÉ[1] and LYNN R. SYKES[2]

Abstract—There is growing evidence that some proportion of large and great earthquakes are preceded by a period of accelerating seismic activity of moderate-sized earthquakes. These moderate earthquakes occur during the years to decades prior to the occurrence of the large or great event and over a region larger than its rupture zone. The size of the region in which these moderate earthquakes occur scales with the size of the ensuing mainshock, at least in continental regions. A number of numerical simulation studies of faults and fault systems also exhibit similar behavior. The combined observational and simulation evidence suggests that the period of increased moment release in moderate earthquakes signals the establishment of long wavelength correlations in the regional stress field. The central hypothesis in the critical point model for regional seismicity is that it is only during these time periods that a region of the earth's crust is truly in or near a "self-organized critical" (SOC) state, such that small earthquakes are capable of cascading into much larger events. The occurrence of a large or great earthquake appears to dissipate a sufficient proportion of the accumulated regional strain to destroy these long wavelength stress correlations and bring the region out of a SOC state. Continued tectonic strain accumulation and stress transfer during smaller earthquakes eventually re-establishes the long wavelength stress correlations that allow for the occurrence of larger events. These increases in activity occur over longer periods and larger regions than quiescence, which is usually observed within the rupture zone of a coming large event. The two phenomena appear to have different physical bases and are not incompatible with one another.

Key words: Accelerating seismic moment/energy, earthquake forecasting, critical point hypothesis, self-organized criticality, stress correlations.

Introduction

Earthquakes do not occur randomly in space and time. Seismology has long recognized this fact and has included this aspect of seismicity directly into its specialized language; we speak of foreshocks and aftershocks, clusters and swarms, precursory activity and quiescence. Some of these patterns, such as the temporal

[1] Queensland University Advanced Centre for Earthquake Studies, The University of Queensland, Brisbane 4072 Queensland, Australia, E-mail: jaume@shake2.earthsciences.uq.edu.au
[2] Lamont-Doherty Earth Observatory and Department of Earth and Environmental Sciences, Columbia University, Palisades, New York 10964, U.S.A., E-mail: sykes@ldeo.columbia.edu

decay of the occurrence of aftershocks, follow well-defined empirical laws and a number of plausible physical mechanisms have been proposed to explain their nature. Other patterns, such as the apparent acceleration of seismic moment/energy release prior to large and great earthquakes that we review here, have been proposed more recently and their physical mechanism still remains to be fully explained.

In this paper we review the evidence that at least some $M \geq 6.5-7.0$ earthquakes are preceded by a period of increased occurrence of moderate (generally $M \geq 5.0$) earthquakes in the region surrounding the oncoming large earthquake. In particular, we focus on those cases where the rate of seismic moment or energy release appears to accelerate in the years to decades prior to the large earthquake, and then decrease to a lower level afterwards. We describe what we believe are the major defining features of these accelerating moment release sequences and discuss what implications they hold for the nature of the earthquake process. Finally, we also review models of the earthquake process that we believe shed some light on the physics underlying this seismicity pattern.

Historical Background

The hypothesis that the rate of earthquake activity, particularly of moderate earthquakes, changes in the period between large and great earthquakes has existed for some time. Workers such as WILLIS (1924), IMAMURA (1937), GUTENBERG and RICHTER (1954), and TOCHER (1959) noted changes in the rate of moderate-sized earthquakes before and after great earthquakes in Japan and California. FEDOTOV (1968) was perhaps the first to describe a "seismic cycle" in the occurrence of smaller earthquakes between great plate rupturing events in the Kurile-Kamchatka region. MOGI (1969) further described the evolution of seismicity in the period between great earthquakes. In general, these workers found that, following the aftershock sequence, there is a relative quiescence of earthquake activity (approximately 50% of the interval between great earthquakes) which built up to a higher background level before the next great earthquake.

A number of scientists have also studied the spatial association between earthquakes occurring during the years and decades before great earthquakes. KELLEHER and SAVINO (1975) reviewed seismicity prior to a number of $M \geq 7.8$ earthquakes in the circum-Pacific region. They found that pre-mainshock seismicity clusters around the edges of the oncoming great earthquake rupture zone. MOGI (1969) noted that, in southwest Japan, the earthquakes during the period of increased seismicity concentrate around the periphery of the oncoming rupture zone and that rupture zone itself was relatively quiescent; he referred to such a sequence as a "doughnut pattern." MOGI (1981) also noted that the occurrence of moderate earthquakes in the San Francisco Bay region prior to the great 1906 earthquake appears to fit this pattern.

In general, these early workers suggested that changes in the level of driving stress during the earthquake cycle was the cause of the observed seismicity changes. The great earthquakes that define the cycle release large amounts of accumulated tectonic stress, which is subsequently reacquired through plate motion. The variations in the rate of regional seismicity is seen as a response to this cycle of stress accumulation and release.

Of particular interest to us are the observations of changes in the rate of seismic activity in the San Francisco Bay region between the great 1906 earthquake and the 1989 Loma Prieta earthquake. It was noted by GUTENBERG and RICHTER (1954) that the rate of moderate earthquakes in the several decades following 1906 was considerably less than in the 50 years before the great 1906 earthquake. TOCHER (1959) suggested that a series of $M \geq 5.0$ in the mid-1950s had ended the quiescence following the 1906 earthquake. ELLSWORTH et al. (1981) showed that, at an $M \geq 5$ level, the seismicity rate in the San Francisco Bay region at the latitudes of the 1906 surface rupture was significantly higher since 1955 compared with the preceding 50 years. SYKES and NISHENKO (1984) pointed out that the post-1955 $M \geq 5$ earthquakes were concentrated in the southern part of the San Francisco Bay region. They postulated that this may represent a long-term precursor to an $M \sim 7$ earthquake in this region.

An important but sometimes overlooked observation of ELLSWORTH et al. (1981) was that, in contrast with the changes in the rate of moderate-sized earthquakes, the rate of smaller earthquakes had remained approximately constant since the 1930s. In a prescient paragraph in their discussion, they suggested that the equilibrium of the San Andreas fault system is controlled by long wavelength forces which are modulated by great earthquakes on the San Andreas fault and tectonic strain accumulation, but that these stress modulations do not disturb the details of the balance between individual parts of the system.

Accelerating Seismic Moment/Energy Release Before Large and Great Earthquakes

Characteristics of Accelerating Seismic Moment/Energy Release Sequences

In the following sections we review cases where there is evidence that accelerating seismic moment/energy release occurred prior to large and great earthquakes, in addition to a small number of cases where this seismicity pattern is known not to have occurred. In this section we first elucidate what we believe are the major characteristics of these sequences, before reviewing specific cases that exemplify these characteristics. The physical interpretation of these features will be covered in a later section.

Based on our review, we find four major characteristics that describe this seismicity pattern:

1. The rate of seismic moment or Benioff strain (the square root of seismic energy) release in moderate earthquakes accelerates prior to the occurrence of the large or great earthquake. In many of these cases this acceleration can be modeled using a power-law time-to-failure relationship to estimate the time and in some cases the magnitude of the oncoming event.
2. The moderate earthquakes involved in the accelerating sequence occur primarily outside the rupture zone of the oncoming large or great earthquake.
3. The change in rate of earthquake occurrence is limited to earthquakes with magnitudes within about 2.0 units of the magnitude of the mainshock.
4. The size of the region in which the moderate earthquakes participating in the accelerating sequence occur scales with the size of the oncoming mainshock.

In our view any physical model vetted to explain this seismicity pattern would have to make predictions consistent with these four observations.

In the sections immediately following we first review cases from the San Francisco Bay region, where accelerating moment/energy release prior to large and great earthquakes was first clearly recognized. We then review other cases in California followed by Alaska, where most of the work on this topic has been conducted. We follow this by reviewing cases outside of California/Alaska. In addition we briefly note other areas were the spatial and temporal patterns of pre-mainshock seismicity appear to match some of the features elucidated above, but which have not been examined for the occurrence of accelerating seismic moment or Benioff strain. At the end we also note two cases where this seismicity pattern clearly did not occur before a large or great earthquake.

Accelerating Moment/Energy Release before Large Earthquakes in the San Francisco Bay Region

On October 18, 1989 an $M = 7.0$ earthquake struck the southern Santa Cruz mountains of California. This was the first $M \geq 7.0$ earthquake to occur in the region of the great 1906 San Francisco earthquake, and its occurrence prompted a number of workers to re-examine the seismic history of the San Francisco Bay region.

The apparent acceleration in the rate of seismic moment release in the San Francisco Bay region prior to the 1989 earthquake was recognized by two groups. As reported by BUFE and VARNES (1993), D. Varnes conducted a time-to-failure analysis of seismic moment in the San Francisco Bay region 1 year prior to the event. Independently, SYKES and JAUMÉ (1990) recognized accelerating moment release prior to both the 1989 Loma Prieta earthquake and the $M \geq 6.8$ San Francisco Bay region earthquakes of 1868 and 1906, plus a 1948 $M = 6.0$ earthquake in southern California.

SYKES and JAUMÉ (1990) found that the rate of seismic moment release in $M \geq 5.0$ earthquakes accelerated prior to the San Francisco Bay region earthquakes of 1868 ($M \sim 7$), 1906 ($M = 7.9$), and 1989 ($M = 7.0$). Like ELLSWORTH et al. (1981), they found that there were no significant changes in the rate of occurrence of smaller earthquakes. They found that an exponential increase in the rate of seismic moment release (expressed as cumulative seismic moment) better fit the data in the San Francisco Bay area than a simple linear increase. They attempted to fit the rate-dependent failure equation of VOIGHT (1989) to the cumulative seismic moment data, but were unable to make accurate estimates of the failure time. They also note that the moderate earthquakes participating in the acceleration of moment release appear to occur mostly or totally outside the rupture zone of the oncoming mainshock, even in the case where some of these events occurred very close to the mainshock hypocenter (i.e., the 1988 and 1989 Lake Elsman earthquakes prior to the 1989 Loma Prieta earthquake). Finally, they noted that the seismicity preceding the great 1906 earthquake covered a wider region than that before the events of 1868 and 1989, suggesting that the region of accelerating moment release scales with the magnitude of the oncoming mainshock (Fig. 1).

BUFE and VARNES (1993) focused their work on modeling the accelerating seismicity sequences in the San Francisco Bay area. They applied a power-law time-to-failure relationship to the changes in the rate of seismicity preceding the large and great earthquakes of 1868, 1906, and 1989. They quantified seismicity as cumulative event count, cumulative Benioff strain, and cumulative seismic moment. They used a time-to-failure function of the form:

$$d\Omega/dt = k/(t_f - t)^n \tag{1}$$

where Ω is some measurable quantity describing the rate of seismicity (event count, seismic moment, or Benioff strain), t_f is the time of a large earthquake, t is the time of the last measurement of Ω, and k and n are constants. This formulation was first used by VARNES (1989) to model accelerating seismic moment/energy during 11 foreshock sequences to estimate the time (and in some cases the magnitude) of the mainshock. BUFE and VARNES (1993) show that equation (1) is equivalent to equation (20) of VOIGHT (1989).

BUFE and VARNES (1993) found that, in general, using cumulative Benioff strain release for Ω in equation (1) leads to more accurate predictions of the time of the oncoming mainshock (Fig. 2) than using cumulative seismic moment release. It also allows the magnitude of the mainshock to be estimated. They note that seismic moment is preferable on theoretical grounds (i.e., cumulative Benioff strain becomes unbounded as one goes to smaller and smaller magnitudes), but using cumulative moment as Ω in equation (1) leads to predictions of t_f that are consistently late and that the magnitude of the mainshock cannot be estimated because Ω becomes infinite at t_f.

BUFE and VARNES (1993) had the most success in modeling the accelerating sequences prior to the $M \sim 7$ earthquakes in 1868 and 1989. They found that they could retrospectively predict the time and magnitude of these events to within 2 years and 0.5 magnitude units. They were unable to accurately model the pre-1906 sequence without fixing either t_f or the exponent n. They found that the time of the

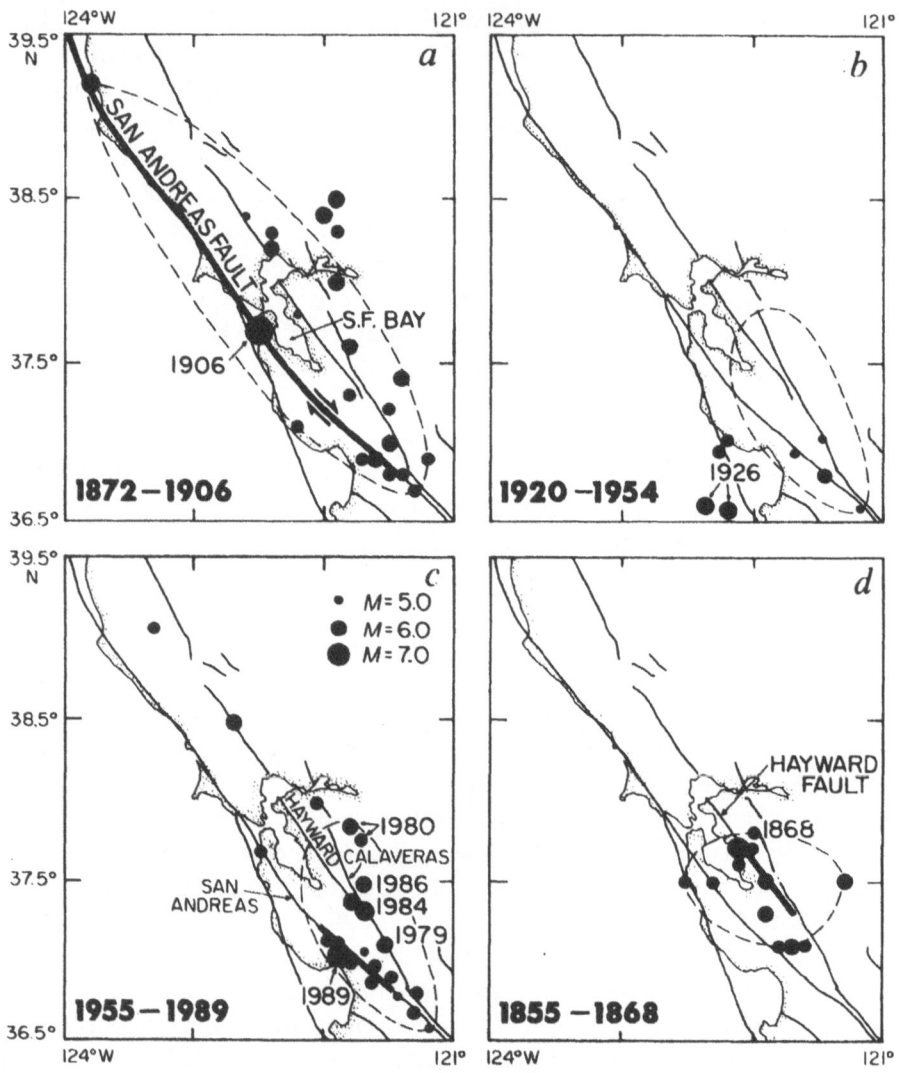

Figure 1

Moderate earthquakes ($M \geq 5.0$) in the San Francisco Bay region during four equal time periods. The area in which earthquakes participating in the accelerating moment release prior to the mainshocks of 1868 ($M \sim 7$), 1906 ($M = 7.8$), and 1989 ($M = 7.0$) are outlined with a dashed line. Note that these areas scale with the magnitude of the oncoming mainshock. Figure 1 from SYKES and JAUMÉ (1990).

Reprinted by permission from Nature 348, 595–599. Copyright (1990) Macmillan Magazines Ltd.

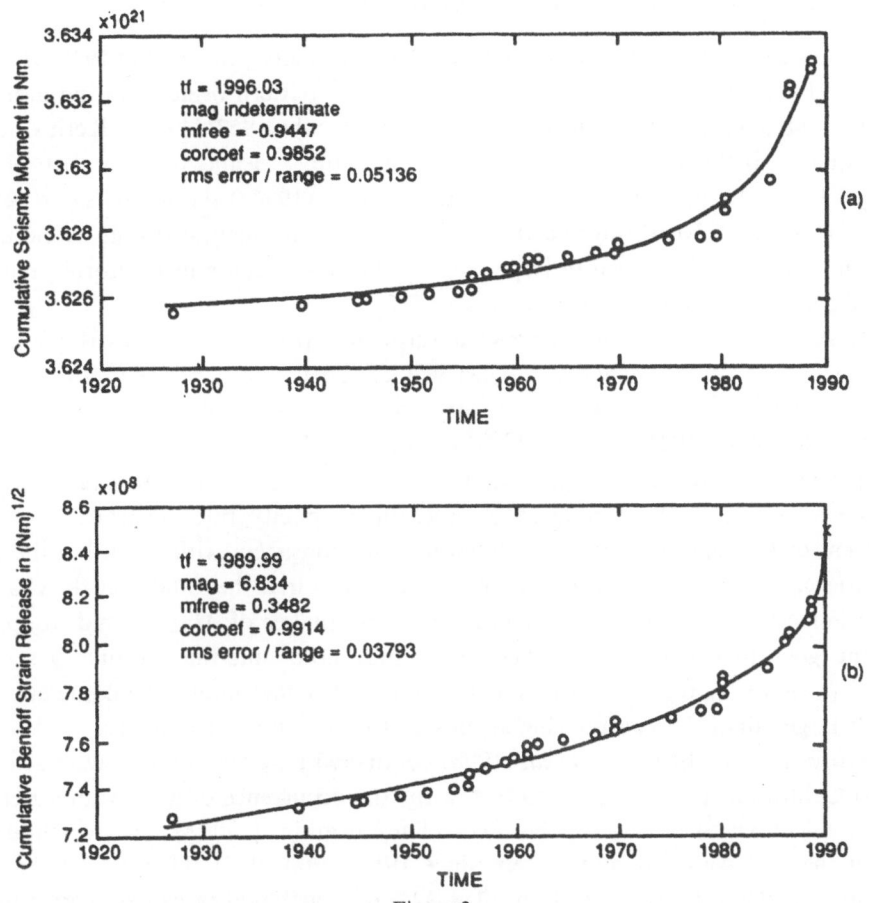

Figure 2

Fits to equation (1) using seismicity in the San Francisco Bay region prior to the 1989 Loma Prieta earthquake. Figure 3 from BUFE and VARNES, J. Geophys Res. *98*, 9871–9883, 1993, published by the American Geophysical Union.

1989 Loma Prieta earthquake could be accurately predicted using moderate earthquakes within 70 km of its epicenter, but that events from a wider region were needed to accurately estimate its magnitude.

BUFE and VARNES (1993) also attempted to model the entire seismic cycle between 1906-type events. They assumed that the pre- and post-1906 seismicity was part of the same sequence, with a period of unknown length missing. Their model, although speculative, predicted a recurrence time of 1906-type events of 230 to 270 years, consistent with paleoseismic results (e.g., HEINGARTNER and SCHWARTZ, 1996). Finally, they suggested that the seismic cycle in the San Francisco Bay region consisted of a long-term acceleration of seismicity leading to the next 1906-type event, punctuated by shorter cycles of acceleration prior to $M \sim 7$ earthquakes. They also suggested that this sequence may be scale-invariant.

Accelerating Moment/Energy Release before Large Earthquakes in California

SYKES and JAUMÉ (1990) pointed out that seismicity prior to the 1948 $M = 6.0$ Desert Hot Springs earthquake appeared to fit the pattern of accelerating moment release, and suggested that this may also be true for the 1952 $M = 7.5$ Kern County earthquake. BUFE *et al.* (1993) report accelerating seismic moment before the $M = 7.3$ 1992 Landers earthquake and BUFE *et al.* (1994a) report the same before the $M = 6.7$ 1994 Northridge earthquake. These results suggested that accelerating seismic moment release before large earthquakes also occurs in California outside the San Francisco Bay region.

A systematic review of changes in seismicity prior to large earthquakes in California and off its Mendocino coast was conducted by KNOPOFF *et al.* (1996). They examined changes in the rate of moderate earthquakes prior to $M \geq 6.8$ earthquakes from 1941 to 1993. They found that all 11 large earthquakes during this period were preceded by increases in the rate of $M \geq 5.1$ earthquakes. Although KNOPOFF *et al.* (1996) examined changes in seismicity rates instead of seismic moment or Benioff strain rates, their findings are consistent with several of the four features elucidated at the beginning of this section. First, they find that the changes in seismicity rates are only well-defined for earthquakes of $M \geq 5.1$ and disappear as one goes to smaller magnitudes. Also, they note that the precursory seismic activity occurs in regions with dimensions up to a few hundred kilometers; i.e., much larger than the rupture dimensions of the $M \geq 6.8$ earthquakes.

More recently, BOWMAN *et al.* (1998) conducted a systematic review of seismicity in California specifically aimed to test for the occurrence of accelerating Benioff strain release before large earthquakes. They examined all $M \geq 6.5$ earthquakes along the San Andreas fault system since 1950 (total of 8). They found that the cumulative Benioff strain prior to all 8 $M \geq 6.5$ earthquakes can be better fit by equation (1) than by a linear increase in Benioff strain. As part of this study they also searched for an optimal "critical region," i.e., a circular region surrounding the epicenter in which the cumulative Benioff strain of the pre-mainshock seismicity best fits equation (1). The results appear to weakly scale with the size of the oncoming mainshock, but the narrow range of mainshock magnitudes considered (6.5 to 7.5) made this observation tenuous. Therefore BOWMAN *et al.* (1998) also analyzed seismicity prior to a few other earthquakes where accelerating moment or Benioff strain is known or suspected to have occurred. These events span a magnitude range from 4.8 to 8.6, and the results appear to support the concept that the region of accelerating seismic release scales with the magnitude of the oncoming mainshock, with $\log R \propto 0.44\, M$, with R being the radius of the critical region.

An important aspect of the work of BOWMAN *et al.* (1998) is that they noted their optimization procedure would also pick out "patterns" given a random set of earthquakes. Therefore, in order to test the significance of their results, they utilized their procedure on a set of 1000 randomly generated synthetic catalogs. They found that an individual random earthquake catalog had close to a 50% chance of generating a spurious acceleration, but that the null hypothesis (i.e., that the Benioff strain accelerations are due to chance) for all eight California cases could be rejected at better than 99% confidence.

Accelerating Moment/Energy Release before Large Earthquakes in the Alaska-Aleutian Subduction Zone

BUFE et al. (1990) were the first to point out that accelerating moment release, similar to that seen along the San Andreas fault system, appeared to be occurring since 1975 in the Alaska Peninsula-Shumagin Islands segments of the Alaska-Aleutian subduction zone. JAUMÉ and ESTABROOK (1992) further established the existence of accelerating moment release in this region, plus documented accelerated moment release in the Kodiak Island segment prior to the $M = 9.2$ 1964 Prince William Sound earthquake.

JAUMÉ (1992) examined seismic moment release rates before three $M > 8.5$ earthquakes in the Alaska-Aleutian subduction zone (1957 Central Aleutians, 1964 Prince William Sound, and 1965 Rat Islands earthquakes). HOUSE et al. (1981) had pointed out that $6.0 \leq M \leq 7.0$ earthquakes clustered at both ends of the 1957 rupture in the decade before that event. JAUMÉ (1992) found that the rate of moment release appeared to accelerate prior to these events at one or both edges of the oncoming rupture zone. He found, however, that only in the case of the $M = 9.2$ 1964 Prince William Sound earthquake did the regional moment release rate change from a higher rate before the event to a considerably lower rate thereafter.

BUFE et al. (1992) showed that the cumulative Benioff strain prior to the 1957 Central Aleutians earthquake could be modeled using equation (1) to estimate the time and magnitude of that earthquake. They also extended the results of BUFE et al. (1990) to include the Unimak Island segment to the west of the Shumagin Islands. They noted that the accelerating Benioff strain sequence appeared to be in an early stage, but that their modeling results suggested a failure time for an $M > 7.5$ earthquake around 1997.

BUFE et al. (1994b) systematically examined cumulative Benioff strain for $M > 5.2$ earthquakes along the Alaska-Aleutians subduction zone, using the segmentation of NISHENKO and JACOB (1990). They identified three regions of apparent accelerating Benioff strain release: the earlier identified Shumagin Islands-Alaska Peninsula segment, the Delarof Islands segment (Fig. 3) and the Kommandorski Islands segment. They modeled the Benioff strain curves using equation (1) to estimate the time and magnitude of potential oncoming large earthquakes. For the first two segments noted above, they estimated failure times in the interval 1994 through 1996, with magnitudes in the range 7.3 to 8.2. For the Kommandorski Islands segment the expected failure time was relatively poorly constrained, being in the range of 1995 through 2003 with a magnitude range of 7.5 to 8.5.

On 10 June 1996, an $M = 7.9$ earthquake occurred in the Delarof Islands segment, fulfilling one of the predictions of BUFE et al. (1994b). We have re-examined the seismicity in the Delarof Islands segment prior to this event and find that most of the earthquakes contributing to the acceleration modeled by BUFE et al. (1994b) occur outside or on the edge of the western part of the rupture zone (Fig. 4). Following the 1996 Delarof Islands event, BUFE et al. (1996) updated the results of BUFE et al. (1994b) to include more recent moderate earthquakes. Results for the Kommandorski Islands segment changed little, but they again combine the Unimak Island segment with the Shumagin Islands-Alaska Peninsula (as in BUFE et al., 1992), yielding an expected failure time of 1996 through 1998 and magnitude 7.7–8.7. We note, however, that a large to great earthquake did not occur in that area during the period (1994 through 1996) of the first prediction by BUFE et al. (1994b).

Accelerating Moment/Energy Release before Earthquakes Outside California and Alaska

Although most studies of accelerating moment/energy release have been conducted in plate boundary regions, BREHM and BRAILE (1998) have applied the technique of BUFE and VARNES (1993) to earthquakes in the New Madrid Seismic

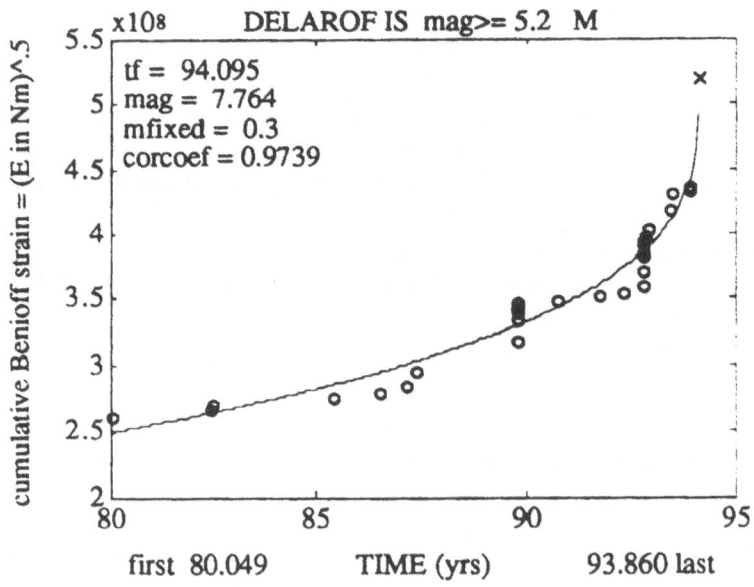

Figure 3

Fit of equation (1) to cumulative Benioff strain release prior to the 1996 Delarof Islands earthquake. Part of Figure 4 from BUFE et al., Pure appl. geophys. *142*, 83–99, 1994, published by Birkhäuser Verlag.

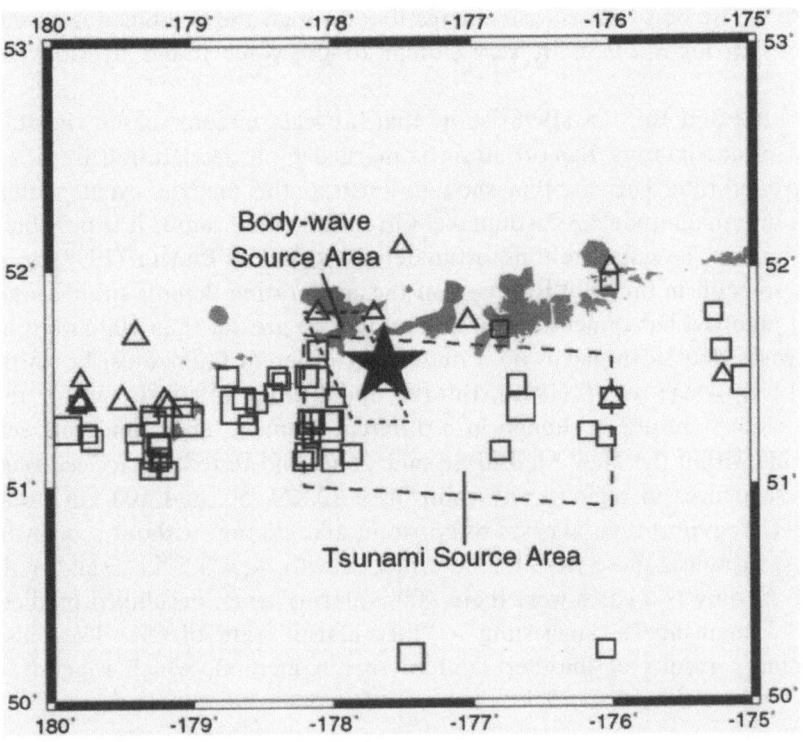

Figure 4

Seismicity ($M \geq 5.5$) in the region of the 10 June 1996 $M = 7.9$ Delarof Islands earthquake. Large star is the epicenter of the 10 June 1996 earthquake. Squares represent seismicity during preceding ten years; triangles represent seismicity during the decade preceding the 9 March 1957 $M = 8.7$ Central Aleutians earthquake. Source areas estimated from tsunami and body-wave analysis are also shown (S. Schwartz, pers. comm.).

zone in the central United States. Surprisingly, they found they can model accelerating Benioff strain sequences before mainshocks as small as $m_b = 3.5$ that yield reasonable predictions of time and magnitude. Of 26 post-1979 mainshocks ranging from $m_b = 3.5$ to 5.2, 16 could be modeled using equation (1). Accelerating Benioff strain sequences were also found before three earlier events, including the 1895 $M = 6.2$ (NUTTLI, 1979) earthquake.

Similar to BOWMAN et al. (1998), BREHM and BRAILE (1998) conducted a search for the circular region in which pre-mainshock Benioff strain best fits equation (1). They also found that the size of the region scales with the magnitude of the oncoming event. They suggest that $\log R \propto 0.5 \log M_0$ (seismic moment) provides a reasonable fit to the data, which translates to $\log R \propto 0.75\, M$; i.e., different from the $\log R \propto 0.44\, M$ scaling suggested by BOWMAN et al. (1998). We have taken the radius versus magnitude information from Table 1 in BOWMAN et al. (1998) and Table 1 in BREHM and BRAILE (1998) and plotted them together in

Figure 5. The best-fitting least-squares line through the combined data set has the relationship $\log R \propto 0.36\,M$, very similar to the value found by BOWMAN *et al.* (1998).

BREHM and BRAILE (1998) show that, at least in some cases, the start of the period of accelerating Benioff strain is marked by a deceleration from a constant background rate. The case they show to illustrate this includes events within a 5-km radius of a magnitude 4.3 earthquake. Given this small radius it is possible all these events are on the same fault; unfortunately BREHM and BRAILE (1998) do not show a map to confirm this. But for many of the accelerating Benioff strain sequences the region involved has dimensions of several tens to greater than 100 km; it is clear in these cases that earthquakes on a number of different faults must be participating.

Like BOWMAN *et al.* (1998), BREHM and BRAILE (1998) tested for the significance of their results, although in a different manner. They randomly selected 20 locations within the New Madrid Seismic zone, and tested for accelerating Benioff strain sequences in regions with radii of 5, 10, 20, 50, and 100 km (total of 100 tests). They report several cases of apparent acceleration without a mainshock, but in most instances these were for mainshocks with $m_b < 3.5$, i.e., smaller than their cutoff. In only two cases were there "false-alarms" (i.e., unfulfilled predictions) for $m_b \geq 3.5$ mainshocks, suggesting a "false-alarm" rate of 2%. They also report preliminary results of another random search method, which suggests a "false-alarm" rate of less than 33%.

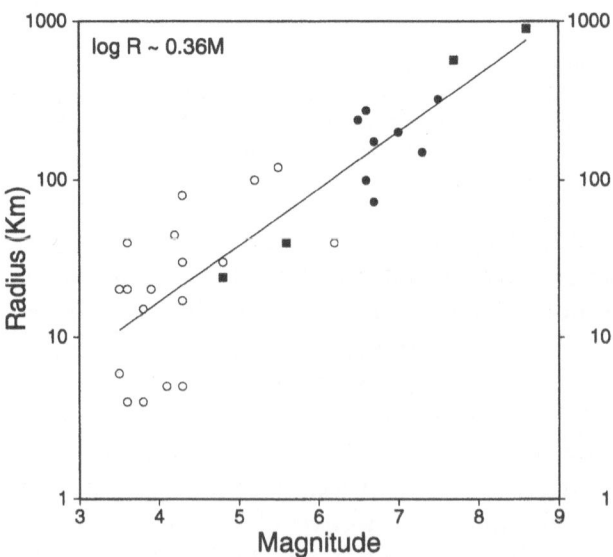

Figure 5

Size of "critical region" relative to magnitude of oncoming mainshock. Filled circles are the eight post-1950 California events and filled squares are the additional four events analyzed by BOWMAN *et al.* (1998). Open circles are the nineteen New Madrid Seismic zone events analyzed by BREHM and BRAILE (1998). The solid line is the best least-squares fit to the combined data set.

Other work outside of California and Alaska includes VARNES and BUFE (1996), who, using data from FRANKEL (1982), found that cumulative square root of moment accelerated before an $m_b = 4.8$ earthquake in the Virgin Islands. Like the cases above, this sequence was modeled using equation (1) to yield an accurate time and magnitude estimate of the mainshock. TRIEP and SYKES (1997) noted that earthquakes of $M > 7.0$ in Asia during the period 1926–1950 cluster near the rupture zone of the 1950 $M = 8.6$ Assam earthquake. They also remark that the rate of $M > 7.0$ earthquakes in Asia was 3.3 times greater during 1900–1957 than during 1958–1994; i.e., that the moment release rate over a large part of Asia decreased following the great 1950 earthquake. BOWMAN et al. (1998) used both the Virgin Islands and Assam data sets to extend the magnitude range of their study.

Another region that may potentially exhibit this phenomenon is Japan. The "doughnut pattern" described by MOGI (1969, 1979, 1981) for the seismicity in southwestern Japan before the great Nanki trough earthquakes of 1944 and 1946, is consistent with the spatial and temporal patterns of activity described here. MOGI (1980) also recognized a similar pattern before the great 1923 Kanto earthquake and MOGI (1985) also points out that damaging earthquakes also occurred frequently in Tokyo in the decades leading up to the great earthquake of 1703. To our knowledge these sequences have not yet been assessed with respect to the accelerating moment/energy release model, but we suspect that examining them in this respect will prove a fruitful line of inquiry.

Large Earthquakes without a Preceding Acceleration in Moment/Energy Release

Although the purpose of this study is to review accelerating moment/energy release before large and great earthquakes, any physical explanation of this phenomenon must also explain cases where it does not occur. While attention has not been focused on these cases, two such occurrences are known to us.

First, BUFE et al. (1994), in their systematic review of moment release rates in the Alaska-Aleutian subduction zone, found no moment release rate acceleration in the vicinity of the 1986 $M = 8.1$ Andreanof Islands earthquake. This case is of particular interest because the 1986 earthquakes lies within the rupture zone of the 1957 Central Aleutians earthquake, which was preceded by accelerating moment at both ends, and just east of the 1996 Delarof Islands earthquake, which was preceded by accelerating moment at its western end.

Another case is the 1988 Tennant Creek earthquakes in the Northern Territory of Australia. Three $M > 6.0$ earthquakes ($M = 6.2$, 6.4, and 6.7; equivalent to a single $M = 7.0$ earthquake) occurred on adjacent fault segments within 12 hours of each other. What is interesting to us is that this region of central Australia had no record of $M \geq 5.0$ earthquakes within 500 km from Tennant Creek before 1987 (BOWMAN, 1992). This area has been well monitored since 1965 when the Warramunga seismic array was installed 30 km east of the surface ruptures. There was an extended (~ 1 year) foreshock sequence to the 1988 earthquakes, but these events were clustered in what was the nucleation zone of the mainshocks, and not distributed across the surrounding region.

Statistical Tests of Accelerating Moment/Energy Release before Large and Great Earthquakes

There have been three attempts to statistically test the significance of the accelerating moment/energy release patterns we describe in this paper. One test was made by BOWMAN *et al.* (1998) and another by BREHM and BRAILE (1998), which we reviewed in the section above. The other test was executed by GROSS and RUNDLE (1998). They used the DNAG catalog (ENGDAHL and RINEHART, 1991) from 1960 through 1985 and earthquakes with $m_b \geq 5.0$. They fit the cumulative Benioff strain data up to 1985 using either equation (1) or the log-periodic version of SORNETTE and SAMMIS (1995), and tested for "predictions" of $m_b \geq 6.5$ earthquakes up to 1995 in arbitrary zones of the northern Western Hemisphere. They find that a Poisson model better explains the observed seismicity than either of the power-law time-to-failure models.

One criticism we have of GROSS and RUNDLE (1998) is their use of m_b as a measure of an earthquake's size. m_b is known to saturate at $M \sim 6.0$, underestimating the size of larger earthquakes. This could potentially lead to "false negative" results; i.e., not recognizing an accelerating moment/energy release sequence because the size of the largest events were underestimated. In addition, use of m_b in constructing cumulative Benioff strain curves would be very susceptible to artificial changes in earthquake catalogs (HABERMANN, 1987). In our view, use of longer period measures of magnitude, such as M_s or M_w, are preferred.

Physical Mechanisms of Accelerating Seismic Moment/Energy Release Sequences

Early models of the physical process underlying accelerating seismic energy/moment release generally referred to fault and soil creep or damage mechanics processes (e.g., SYKES and JAUMÉ, 1990; BUFE and VARNES, 1993). These authors noted the similarity between the acceleration in the rate of seismic moment release and acceleration in other measures of material deformation seen before failure in many different types of natural and man-made materials. BUFE and VARNES (1993) specifically referred to nucleation/crack propagation and damage mechanics models when deriving equation (1).

JAUMÉ and SYKES (1996) and DENG and SYKES (1997) have examined the occurrence of moderate-sized earthquakes in California in the context of evolutionary stress models that include the effects of large and great earthquakes ($M \geq 7.0$)

plus strain accumulation on major faults. They find that moderate earthquakes occur preferentially in regions of positive stress (quantified as a Coulomb failure function); i.e., in regions where stress has been increased by prior earthquakes and strain accumulation and/or where they have overriden a decrease in stress produced by a previous earthquake. JAUMÉ and SYKES (1996) found that the timing and location of the moderate earthquakes that contributed to the acceleration in moment release prior to the 1989 Loma Prieta earthquake is consistent with their stress evolution model.

Workers such as HUANG et al. (1998), SALEUR et al. (1996), SAMMIS et al. (1996), and SORNETTE and SAMMIS (1995) have begun developing models of the earthquake process based upon the concept that an earthquake represents a type of critical point. In this hypothesis, a large or great earthquake is a consequence of the progressive ordering of a fault system under the influence of many small-scale changes. At the "critical point" correlations exist at all scales and the system moves globally and abruptly. An important feature of this hypothesis is that the accelerating moment/energy release patterns we review in this paper are an expected consequence of systems that either have embedded discrete scale invariance (e.g., a fractal distribution of fault sizes, HUANG et al., 1998) or where such a hierarchy appears spontaneously from the physics of a heterogeneous non-hierarchical system.

The critical point models have a number of attractive features, particularly when viewed against the four characteristics of accelerating moment/energy release sequences described at the beginning of this paper. These models predict power-law time-to-failure behavior of measures of deformation should occur close to the critical point. In addition, log-periodic corrections to equation (1) are also predicted, leading to an equation of the form (SORNETTE and SAMMIS, 1995):

$$\epsilon(t) = A + B(t_f - t)^m \left[1 + C \cos\left(2\pi \frac{\log(t_f - t)}{\log \lambda} + \Psi \right) \right],$$ (2)

where $A + B(t_f - t)^m$ is the integrated form of equation (1). SORNETTE and SAMMIS (1995) show that equation (2) yields a prediction of the time of the Loma Prieta earthquake that is closer to the actual time of occurrence and has a smaller uncertainty than using equation (1). However, a similar fit to the Kommandorski Islands acceleration noted by BUFE et al. (1994) yields a predicted time (1996.3 ± 1.1) that has since passed.

In these critical point models the power-law behavior of seismicity results from the emergence of long-range correlations in the stress field preceding the large or great earthquake. This is consistent with the observation that the earthquakes contributing to the accelerating seismicity occur outside the rupture zone of the oncoming event, and that the area over which they occur scales with the size of oncoming large event. The critical point model is also consistent with the observation that the seismicity rate changes before large and great earthquakes are confined to moderate magnitude events. SALEUR et al. (1996) show that even when the system is very close to a critical point in a small local region, it can still be far away from criticality at larger scales. The results of BREHM and BRAILE (1998) suggest that, in some cases, one can observe this approach to criticality even for relatively small earthquakes.

Discussion

Accelerating Moment/Energy Release and the Wavelength of the Stress Field

One aspect that repeatedly arises during our review of accelerating moment/energy release sequence is the notion of *scale*; spatial scales in particular. The region participating in the accelerating seismicity sequences appear to scale with the size of the mainshock; existing data support a relationship of $\log R \propto 0.5\ M$. Changes in seismicity rates appear to only occur for earthquakes above a certain magnitude; i.e., only for earthquake ruptures above a certain length scale. Time scales also appear to be relevant. BREHM and BRAILE (1998) note that the length of the accelerating sequence scales with magnitude in the New Madrid Seismic zone. HUANG et al. (1998) find in their model that precursory activity for events smaller than the system size occur on shorter time scales than for system spanning events. When one examines Figures 6 and 8 in BOWMAN et al. (1998) one finds that accelerating Benioff strain before the events they examined also appears to scale in time; i.e., over the course of a decade or more before the larger events (e.g., 1906 $M = 7.9$ San Francisco, 1989 $M = 7.0$ Loma Prieta, etc.) but only years to months before the smaller events (e.g., 1968 $M = 6.5$ Borrego Mountain, 1994 $M = 6.7$ Northridge, etc.).

This leads us to the view, consistent with the critical point hypothesis, that the accelerating seismic moment/energy release sequences reviewed here are related to changes in the length scale of the earthquake process with time. That such changes can take place is supported by the work of TRIEP and SYKES (1997), who found that the frequency-magnitude (i.e., b value) statistics of large earthquakes in the broad plate boundary deformation zone of Asia changed dramatically during this century. Earthquakes from 1958 through 1994 have a change in scaling near $M_w = 7.0$, from a b value about 1.0 below to a b value of 2.4 to 3.0 above $M_w = 7.0$, similar to other continental regions. However, during an earlier period (1900–1957) they find that the frequency-magnitude distribution can be described with a b value near 1.0 up to $M_w = 8.2$. TRIEP and SYKES (1997) suggest that, prior to 1957, most of Asia was in a self-organized critical state (SOC). They attribute this to stress changes caused by the great 1950 Assam earthquake, together with slow aseismic slip below the brittle zone.

Building upon this observation, SYKES *et al.* (1997) hypothesized that, although the crust may be in a SOC state over large time and space scales, the occurrence of a great earthquake moves its surrounding region away from a SOC state. Tectonic loading progressively re-establishes a SOC state before the occurrence of the next great earthquake. Observed increases in the rate of moderate-sized earthquakes herald the re-establishment of SOC.

Both the critical point model of HUANG *et al.* (1998), etc. and the model of SYKES *et al.* (1997) imply that a change in the length scale of the underlying stress field drive the observed acceleration in seismic moment/energy release. If so, this should be reflected in the frequency-magnitude scaling of earthquakes in the affected region. To test this hypothesis, we examine the frequency-magnitude distribution of earthquakes during earlier and later periods of several observed accelerating moment/energy release sequences. The cases we examine are three in which accelerating moment/energy release appears to have led to a large or great earthquake (i.e., 1989 $M = 7.0$ Loma Prieta, 1992 $M = 7.3$ Landers, 1996 $M = 7.9$ Delarof Islands) and one where an ongoing accelerating sequence has not yet ended in a large or great event (i.e., the Unimak Island-Shumagin Islands-Alaska Peninsula segments of the Alaska-Aleutian subduction zone). We choose to examine modern examples because earthquake catalogs for these cases are more likely to be complete than for earlier examples. In this initial survey we do not attempt to decluster the earthquake catalogs to remove aftershocks. We note that such declustering preferentially removes smaller events from the catalog (e.g., FROHLICH and DAVIS, 1993); here we are interested in the larger events in the magnitude-frequency distribution and thus would not expect declustering to make a significant difference.

We start with the Loma Prieta case. We use the University of California at Berkeley earthquake catalog because it appears to be complete down to magnitude 3 since 1942 (see JAUMÉ and SYKES, 1996, for a more detailed discussion of this earthquake catalog). Initially we look at earthquakes within 100 km from the 1989 epicenter, and split the catalog into equal earlier (1942–1965) and later (1966–1989) periods (Fig. 6). During the earlier period the b value appears to be linear up to a magnitude (M_L in this case) of 5.2 to 5.3, but thereafter the b value appears to increase and the largest event is 5.6. During the later time period the b value is linear up to a larger magnitude (5.7 to 6.2), and, if the Loma Prieta mainshock is included in the distribution, there is no apparent change in b value. Basically, the largest events in the distribution only occur near in time to the Loma Prieta mainshock. We also examined a larger area (200 km from the Loma Prieta epicenter) as suggested by BOWMAN *et al.* (1998), and get similar results.

In the Landers case, we use the catalog of the Southern California Seismic Network available online from the Southern California Earthquake Center. We examine a region within 150 km from the Landers epicenter and split the catalog into 1972–1982 and 1982–1992 time periods, based upon the results of BOWMAN

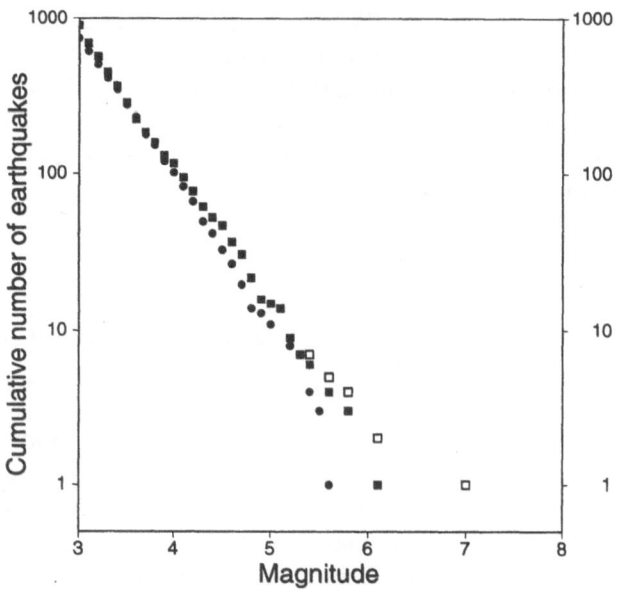

Figure 6

Changes in the frequency-magnitude distribution of earthquakes within 100 km of the Loma Prieta epicenter. Circles—earthquakes during 1942/01/01 through 1965/11/24; closed squares—earthquakes from 1965/11/25 through 1989/10/17. Open squares—same as dark squares but including Loma Prieta mainshock.

et al. (1998). The results are dissimilar to the Loma Prieta case in that the *b* value appears linear up to the largest magnitude (5.7) during 1972–1982, but similar in that the largest events (up to 6.6) all occur in the latter period (Fig. 7).

A difficulty in examining the Alaska-Aleutian cases is finding a suitable earth-quake catalog. Ideally we need a catalog that is both complete to relatively small magnitudes and far enough back in time, plus does not have any saturation problems in magnitude (i.e., like with m_b as discussed above). The Harvard Centroid Moment Tensor catalog does not have saturation problems, but is only complete to $M \sim 5.8$ and only goes back to 1977. The M_s catalog of JAUMÉ (1992) goes back further in time but also is not complete to small magnitudes. The Preliminary Determination of Epicenter (PDE) catalog available online from the U.S. Geological Survey is complete to smaller magnitudes and goes back to 1973. We compared both m_b and M_s in the PDE catalog to M_w in the Harvard catalog, and find that m_b appears to correlate well with M_w at smaller magnitudes and M_s correlates with M_w at larger magnitudes. We find that using the larger of either m_b or M_s is a better representation of M_w than either alone. Therefore, in examining the Delarof Islands and Unimak-Shumagin-Alaska Peninsula cases we use the PDE catalog, looking at earthquakes with depths of 50 km and less and using the largest assigned magnitude.

Based upon Figure 4 in BUFE and VARNES (1994), we split the Delarof Islands seismicity in 1980–1988 and 1988–1996 time segments (Fig. 8). As expected, the largest magnitude events ($M > 6.4$) occur in the latter period, but in this case the seismicity rate appears to increase at all magnitudes. This also appears to be the case when the Harvard CMT catalog is used.

In the Unimak-Shumagin-Alaska Peninsula segments, BUFE and VARNES (1994) and BUFE et al. (1996) find that the apparent acceleration begins about 1970, thus we are able to use the entire PDE catalog (i.e., from 1973). Initially, we examined each segment separately, but found the Unimak Island and Shumagin Islands segments were similar to one another and dissimilar to the Alaska Peninsula segment. The combined Unimak-Shumagin segment appears similar to the Loma Prieta and Landers examples, except that the deviation in seismicity rate appears to be approximately at magnitude 6.0 instead of 5.0 (Fig. 9). $M > 6.3$ earthquakes all occur during 1985–1996. The Alaska Peninsula case is more complex. The earthquakes in this segment do not appear to form a continuous distribution in magnitude; instead there appears to be a continuous distribution up to near magnitude 6.0 and then a separate distribution for events with $M = 6.7–6.9$. We do note that, as in the other cases explored, the largest event(s) occur in the later time period, whether viewed as one or two magnitude distributions. Nevertheless, only one event of $M > 6$ occurred in each time period.

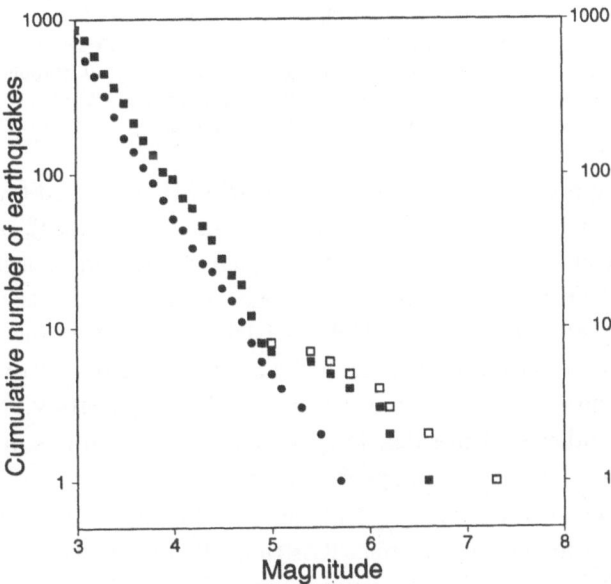

Figure 7

Changes in the frequency-magnitude distribution of earthquakes within 150 km of the Landers epicenter. Circles—earthquakes from 6/28/1972 through 6/27/1982; closed squares—earthquakes from 6/28/1982 through 6/27/1992. Open squares—same as dark squares except including Landers mainshock.

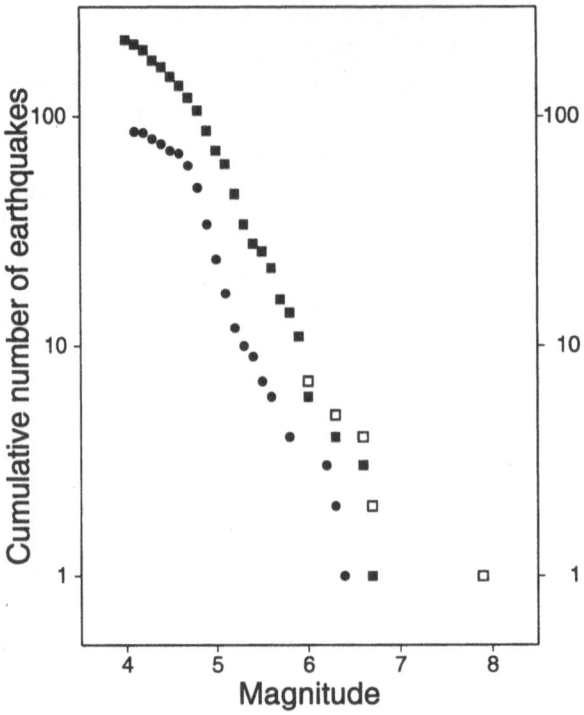

Figure 8

Changes in the frequency-magnitude distribution of earthquakes in the Delarof Islands segment of the Alaska-Aleutian subduction zone. Circles—earthquakes from 6/10/1980 through 6/9/1988; closed squares—earthquakes from 6/10/1988 through 6/9/1996. Open squares—same as dark squares except including Delarof Islands mainshock.

Although the procedure used above can obviously be refined (particularly with respect to potential problems with the earthquake catalog used), at first pass the results appear consistent with the hypothesis that accelerating seismic moment/energy release is the product of increasingly longer correlations in the stress field. The one possible exception to this is the Delarof Islands case, where there appears to be an overall change in seismicity rate; in this sense it is more similar to pre-large event seismicity patterns seen in some block slider models (e.g., SHAW *et al.*, 1992).

Is there any other evidence that long wavelength correlations in the stress field herald a large or great earthquake? The best evidence for this comes from simulation models of the earthquake process. In an early attempt to simulate the fault system in southern California, RUNDLE (1988) speculated that such long wavelength stress correlations would appear to be necessary to drive the large earthquakes in his model, since the model only contained nearest-neighbor stress interactions. Since that time a number of other authors have found that long wavelength correlations in stress appear to be a necessary condition for the occurrence of a large model earthquake, even though the types of simulation models vary greatly; i.e., Burridge-Knopoff block slider models (SCHMITTBUHL *et al.*, 1996), a detailed model of the Parkfield segment (BEN-ZION, 1996), a map-like configuration of southern California faults (WARD, 1996), and a heterogeneous cellular automaton model (STEACY and McCLOSKEY, 1998). This consistency across a wide range of model types suggests this is a common feature of many-element models with large events.

Other Issues

Several issues with respect to accelerating moment/energy sequences are still unsettled. The first we address is: Why does cumulative Benioff strain appear to produce more accurate predictions than cumulative moment? As noted above, cumulative Benioff strain is an unphysical quantity. Restricting the magnitudes used in the Benioff strain sums to within 2.0 magnitude units of the magnitude of the event one is attempting to predict appears to be a useful practical measure, both because the seismicity changes are confined to moderate magnitudes and it avoids problems with catalog incompleteness. But as far as we know, there is no theoretical justification for using this quantity. A resolution to this problem would be very useful. We note that the number of cases where Benioff strain gives more accurate estimates of time to large events is small.

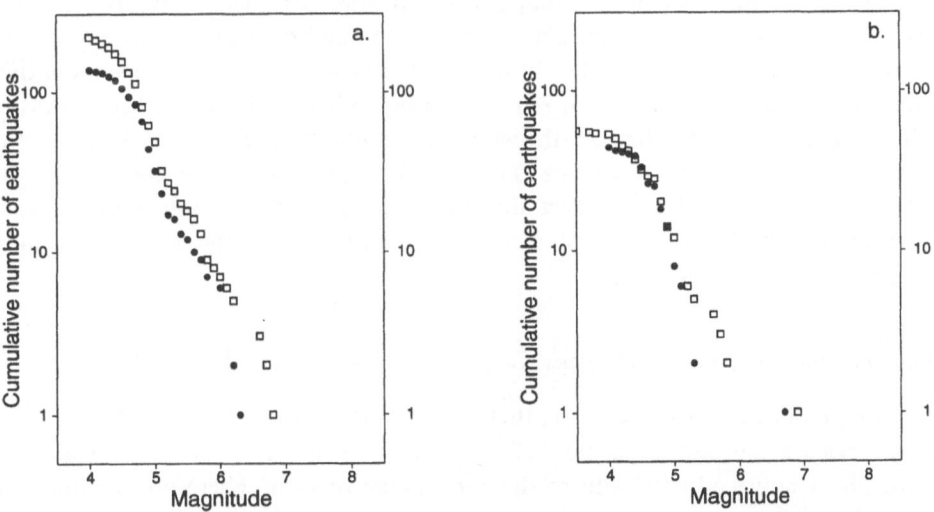

Figure 9

Changes in the frequency-magnitude distribution of earthquakes in the (a) Unimak-Shumagin Islands and (b) Alaska Peninsula segments of the Alaska-Aleutian subduction zone. Circles—earthquakes from 1973 through 1984; squares—earthquakes from 1985 through 1996.

Another issue regards the "background" frequency-magnitude distribution of seismicity from which the accelerating sequence departs. SYKES *et al.* (1997) suggested that accelerating moment/energy release represents a return to a SOC state following a great earthquake. The Loma Prieta (Fig. 6) and Unimak-Shumagin (Fig. 9a) cases fit this model very well. But in the Landers case (Fig. 7) the "background" appears to already be in a SOC state, as found by SORNETTE *et al.* (1996) for the entire southern California region; during the moment/energy acceleration the region appears "super-critical." The Delarof Islands case (Fig. 8) appears to be most consistent with a change in seismicity rate at all magnitudes, although the largest events only occur during the later period.

The difference between northern and southern California may be a function of the time since the last great earthquake. In southern California the time since the last great earthquake on the San Andreas ranges from 141 to around 300 years, depending upon location. In northern California the previous great earthquake was 83 years before the 1989 Loma Prieta event. Thus the "background" stress state in southern California may simply be more evolved from the last great earthquake than in northern California. Another difference between northern and southern California is the much greater contribution of thrust/reverse faulting events to the frequency-magnitude statistics in the latter region. Thus one is seeing the state of a coupled strike-slip/compression system in southern California, as opposed to a primarily strike-slip system in northern California.

The Unimak-Shumagin case (Fig. 9a) seems simply to be a "scaled-up" version of those in California. The Alaska Peninsula (Fig. 9b) and Delarof Islands cases (Fig. 8) are more difficult to comprehend. If one considers only the $M < 6.0$ seismicity in the Alaska Peninsula segment it resembles California closely; but where do the $M > 6.5$ earthquakes fit in? For the Delarof Islands case we note that it lies just to the west of the 1986 $M = 8.1$ Andreanof Islands event. Stress transfer along strike after a subduction thrust event would be expected to increase the driving stress on adjacent segments. Thus the frequency-magnitude statistics of the 1988–1996 time period may reflect this stress change, which would both increase the total number of events and also likely help increase the long wavelength component of the stress field.

What are the Necessary Conditions for Accelerating Moment/Energy Release?

As noted in our review of cases, there are examples where accelerating moment/ energy release has not occurred before a large or great earthquake. Thus this cannot be a universal condition of the earthquake process. Therefore the question arises: What features of the earthquake process and/or the fault system in which earthquakes occur are necessary to give rise to this phenomenon? Based upon our review, we tentatively put forth some potential answers to this question, in hopes of stimulating further work on this topic.

The first condition, which was remarked upon by SYKES and JAUMÉ (1990), is that a certain degree of heterogeneity in the fault system seems to be required. It is possible that a hierarchical distribution of fault lengths (for continental regions) or asperity sizes (for subduction zones) is necessary, as suggested by HUANG et al. (1998). This may explain why there was no acceleration observed before the great 1986 Andreanof Islands earthquake; i.e., the rupture zone contained a number of small (much less than the width of the fault zone) asperities but no large ones, as suggested by moment release distributions derived from body waves (DAS and KOSTROV, 1990). Conversely, moment release in the 1996 Delarof Islands earthquake appears to be concentrated in one large asperity west of the epicenter (KISSLINGER and KIKUCHI, 1997; SCHWARTZ, 1996).

The second condition regards the density of faults and/or asperities. As remarked by SALEUR et al. (1996), accelerating moment/energy release appears to be a cooperative phenomenon in a many-body system. If the "cooperation" arises as a result of elastic interactions between the elements of the system, increasing the distance between elements will decrease the interactions and eventually each element would act more or less independently. This may explain the lack of regional seismicity preceding the 1988 Tennant Creek, Australia earthquakes. Early aftershocks are all concentrated in the immediate vicinity of the rupture zones of the large events, although triggered events at greater distances start occurring nearly two years later (BOWMAN, 1992) and continue at least through May 1996 (AGSO, 1996). We speculate that the faults that ruptured in the Tennant Creek earthquakes are a relatively isolated set and not part of a regional system.

A final necessary condition appears to be the presence of an earthquake large enough to influence the stress state of a significant part of the fault system. The minimum size of such an event would likely depend upon the fault system in question. Based upon our review, we expect that the *minimum* size of such an earthquake is one where the rupture length exceeds its downdip width; probably by a factor of two or more. Earthquakes whose rupture length is considerably greater than the width are more efficient in decreasing regional shear stress. This is clearly illustrated in Figure 8 of KING et al. (1994). A rupture with a length/width (L/W) ratio of 1 increases and decreases regional stress in nearly the same proportions; for an event with a L/W ratio of 6 shear stress is decreased in a broad region parallel to the fault but only increased near the rupture tips. In effect, the efficiency at which a large earthquake changes the regional stress state is as much a function of its rupture length as its total moment.

If all or some of the points stated above are necessary conditions for the existence of an accelerating moment/energy release seismicity sequence before a large or great earthquake, it is clear that a study such as that conducted by GROSS and RUNDLE (1998) would likely yield inconclusive results, even using a different set of earthquake magnitudes and time windows. A challenge would be to determine which fault systems would be expected to generate such patterns *a priori*; i.e., using information other than the moment/energy release rates. We believe the best course of action may be to first construct numerous simulation models of fault systems, based upon different real world examples. Hopefully this will make clear what the necessary preconditions are for the seismicity pattern we review here, and guide us in making predictions of the future behavior of real fault systems.

Conclusions

There are a growing number of cases reported where the occurrence of a large or great earthquake is preceded by an accelerated rate of moderate-size earthquakes occurring in its surrounding region. The rate of moment and/or energy release in these sequences can be modeled using a power-law time-to-failure relationship; in some cases an accurate prediction of the time and magnitude of the oncoming event can be made. At least one great earthquake (the 10 June 1996 $M = 7.9$ Delarof Islands event) has been predicted before its occurrence using this methodology, and evidence suggests the time and magnitude of several other $M \geq 7.0$ earthquakes could also have been predicted. The locations of these events, however, could only have been specified as within a broad region, significantly greater than the rupture length of the large or great event.

The moderate earthquakes that make up these accelerating sequences generally occur outside the rupture zone of the oncoming event and often on separate faults altogether. These sequences are therefore not part of the rupture nucleation process, but are cooperative phenomena of the elements of the regional fault system. The size of the participating region appears to scale with the magnitude of the oncoming rupture, at least in continental regions. Only earthquakes with magnitudes within ~ 2.0 magnitude units of the mainshock magnitude appear to participate; the rate of occurrence of smaller events does not change. In addition, the larger events during the accelerating sequence occur preferentially during the latter half of the sequence.

The model that is most consistent with these observations is one where a large or great earthquake is considered to be analogous to a critical point. At this critical point correlations exist at all scales within the fault system, and an event spanning a significant part of the system is possible. The accelerating moment/energy release sequence is therefore a consequence of the ordering of the fault system, where the regional seismicity is responding to the establishment of progressively longer wavelengths in the stress field. The occurrence of a large event decorrelates the regional stress field at longer wavelengths, but leaves the stress field rough at short wavelengths.

The existence of such a seismicity pattern appears to require a certain regional fault system structure and density. Simulation models using a hierarchical distribution of fault and/or asperity sizes match this pattern well, but other types of fault distributions may also support this pattern. The prescence of earthquake events spanning a large part of the fault system also appear to be a prerequisite, since only these events are capable of effecting the stress field at long wavelengths.

Acknowledgments

We have benefited from discussions with the participants of the Seismicity Patterns Conference in Nikko, Japan, particularly J. Rundle and Y. Kagan. We thank C. Sammis, M. Wyss and an anonymous reviewer for helpful comments. Earthquake data used in this study are available online from the U.S. Geological Survey, the Southern California Earthquake Center, and Harvard University. S. Schwartz kindly provided the tsunami and body-wave source areas for the 10 June 1996 Delarof Is. earthquake. The GMT system (WESSEL and SMITH, 1991) was used to produce many figures. This research was supported by the Australian Research Council and the Southern California Earthquake Center. SCEC is funded by NSF Cooperative Agreement EAR-8920136 and USGS Cooperative Agreements 14-08-0001-A0899 and 1434-HQ-97AG01718. The SCEC contribution number for this paper is 468. Lamont-Doherty contribution number 5929.

REFERENCES

AUSTRALIAN GEOLOGICAL SURVEY ORGANIZATION (1996), Monthly Report on Australian Earthquakes *96/5A*.
BEN-ZION, Y. (1996), *Stress, Slip, and Earthquakes in Models of Complex Single-fault Systems Incorporating Brittle and Creep Deformations*, J. Geophys. Res. *101*, 5677–5706.
BOWMAN, J. R. (1992), *The 1988 Tennant Creek, Northern Territory, Earthquakes: A Synthesis*, Aust. J. Earth Sci. *39*, 651–699.
BOWMAN, D. D., OUILLON, G., SAMMIS, C. G., SORNETTE, D., and SORNETTE, A. (1998), *An Observational Test of the Critical Earthquake Concept*, J. Geophys. Res. *103*, 24,359–24,372.
BREHM, D. J., and BRAILE, L. W. (1998), *Intermediate-term Prediction Using Precursory Events in the New Madrid Seismic Zone*, Bull. Seismol. Soc. Am. *88*, 564–580.
BUFE, C. G., and VARNES, D. J. (1993), *Predictive Modeling of the Seismic Cycle in the Greater San Francisco Bay Region*, J. Geophys. Res. *98*, 9871–9983.
BUFE, C. G., JAUMÉ, S. C., NISHENKO, S. P., SYKES, L. R., and VARNES, D. J. (1990), *Accelerating Moment Release in the Alaska Subduction Zone: Precursor to a Great Thrust Earthquake?* (Abstract), EOS, Trans. AGU *71*, 1451–1452.
BUFE, C. G., NISHENKO, S. P., and VARNES, D. J. (1992), *Clustering and Potential for Large Earthquakes in the Alaska-Aleutian Region* (Extended Abstract), Proc. Wadati Conf. on Great Subduction Earthquakes, University of Alaska, 129–132.
BUFE, C. G., VARNES, D. J., and NISHENKO, S. P. (1993), *A Nonlinear Time- and Slip-predictable Model for Foreshocks* (Abstract) EOS, Trans. AGU 1993 Fall Meeting Suppl. *74*, 437.
BUFE, C. G., NISHENKO, S. P., and VARNES, D. J. (1994a), *Seismicity Trends and Potential for Large Earthquakes in the Alaska-Aleutian Region*, Pure appl. geophys. *142*, 83–99.
BUFE, C. G., VARNES, D. J., and NISHENKO, S. P. (1994b), *Long-term Seismicity Patterns and Pre-earthquake Failure Processes* (Abstract) EOS, Trans. AGU 1994 Fall Meeting Suppl. *75*, 434.
BUFE, C. G., VARNES, D. J., and NISHENKO, S. P. (1996), *Time-to-failure in the Alaska-Aleutian Region: An Update* (Abstract), EOS, Trans. AGU, 1996, Fall Meeting Suppl. *77*, F456.

DAS, S., and KOSTROV, B. V. (1990), *Inversion for Seismic Slip Rate History and Distribution with Stabilizing Constraints: Application to the 1986 Andreanof Islands Earthquake*, J. Geophys. Res. *95*, 6899–6913.

DENG, J., and SYKES, L. R. (1997), *Evolution of the Stress Field in Southern California and Triggering of Moderate-size Earthquakes: A 200-year Perspective*, J. Geophys. Res. *102*, 9859–9886.

ELLSWORTH, W. L., LINDH, A. G., PRESCOTT, W. H., and HERD, D. G., *The 1906 San Francisco earthquake and the seismic cycle*. In *Earthquake Prediction: An International Review* (eds. Simpson, D. W., and Richards, P. G.) (AGU, Washington, D. C. 1981) pp. 126–140.

ENGDAHL, E. R., and RINEHART, W. A., *Seismicity map of North America project*. In *Neotectonics of North America* (Geol. Soc. of Am., Boulder, Colo. 1991) pp. 21–27.

FEDOTOV, S. A., *The seismic cycle, quantitative seismic zoning, and long-term seismic forecasting*. In *Seismic Zoning of the USSR* (ed. Medvedev, S.) (Idatel'stvo "Nauka", Moscow 1968) pp. 133–166.

FRANKEL, A. (1982), *Precursors to a Magnitude 4.8 Earthquake in the Virgin Islands: Spatial Clustering of Small Earthquakes, Anomalous Focal Mechanisms, and Earthquake Doublets*, Bull. Seismol. Soc. Am. *72*, 1277–1294.

FROHLICH, C., and DAVIS, S. D. (1993), *Teleseismic b Values; or, Much Ado About 1.0*, J. Geophys. Res. *98*, 631–644.

GROSS, S., and RUNDLE, J. (1998), *A Systematic Test of Time-to-failure Analysis*, Geophys. J. Int. *133*, 57–64.

GUTENBERG, B., and RICHTER, C. F., *Seismicity of the Earth and Associated Phenomena* (Hafner, New York 1954).

HABERMANN, R. E. (1987), *Man-made Changes in Seismicity Rates*, Bull. Seismol. Soc. Am. *77*, 141–159.

HEINGARTNER, G. F., and SCHWARTZ, D. P. (1996), *Paleoseismic Evidence for Large Magnitude Earthquakes along the San Andreas Fault in the Southern Santa Cruz Mountains* (Abstract), EOS, Trans. AGU 1996 Fall Meeting Suppl. *77*, F462.

HOUSE, L. S., SYKES, L. R., DAVIES, J. N., and JACOB, K. H. (1981), *Identification of a possible seismic gap near Unalaska Island, eastern Aleutians, Alaska*. In *Earthquake Prediction, An International Review* (eds. Simpson, D. W., and Richards, P. G.) (AGU, Washington, D.C. 1981) pp. 81–92.

HUANG, Y., SALEUR, H., SAMMIS, C., and SORNETTE, D. (1998), *Precursors, Aftershocks, Criticality and Self-organized Criticality*, Europhys. Lett. *41*, 43–48.

IMAMURA, A., *Theoretical and Applied Seismology* (Maruzen, Tokyo 1937).

JAUMÉ, S. C. (1992), *Moment Release Rate Variations during the Seismic Cycle in the Alaska-Aleutians Subduction Zone* (Extended Abstract), Proc. Wadati Conf. on Great Subduction Earthquakes, University of Alaska, 123–128.

JAUMÉ, S. C., and ESTABROOK, C. H. (1992), *Accelerating Seismic Moment Release and Outer-rise Compression: Possible Precursors to the Next Great Earthquake in the Alaska Peninsula Region*, Geophys. Res. Lett. *19*, 345–348.

JAUMÉ, S. C., and SYKES, L. R. (1996), *Evolution of Moderate Seismicity in the San Francisco Bay Region, 1850 to 1993: Seismicity Changes Related to the Occurrence of Large and Great Earthquakes*, J. Geophys. Res. *101*, 765–789.

KELLEHER, J., and SAVINO, J. (1975), *Distribution of Seismicity before Large Strike-slip and Thrust-type Earthquakes*, J. Geophys. Res. *80*, 260–271.

KING, G. C. P., STEIN, R. S., and LIN, J. (1994), *Static Stress Changes and the Triggering of Earthquakes*, Bull. Seismol. Soc. Am. *84*, 935–953.

KISSLINGER, C., and KIKUCHI, M. (1997), *Aftershocks of the Andreanof Islands Earthquake of June 10, 1996, and Local Seismotectonics*, Geophys. Res. Lett. *24*, 1883–1886.

KNOPOFF, L., LEVSHINA, T., KEILIS-BOROK, V. I., and MATTONI, C. (1996), *Increased Long-range Intermediate-magnitude Earthquake Activity prior to Strong Earthquakes in California*, J. Geophys. Res. *101*, 5779–5796.

MOGI, K. (1969), *Some Features of Recent Seismic Activity in and near Japan (2). Activity before and after Great Earthquakes*, Bull. Earthquake Res. Inst., Univ. Tokyo *47*, 395–417.

MOGI, K., *Seismicity in western Japan and long-term earthquake forecasting*. In *Earthquake Prediction, An International Review* (eds. Simpson, D. W., and Richards, P. G.) (AGU, Washington, D.C. 1981) pp. 43–51.

MOGI, K., *Earthquake Prediction* (Tokyo, Academic Press 1985).

MOGI, K. (1979), *Two Kinds of Seismic Gaps*, Pure appl. geophys. *117*, 1172–1186.

MOGI, K. (1980), *Seismic Activity: Earthquake Prediction in and around the Tokyo Metropolitan Area*, Bull. Reg. Coord. Comm. Earthquake Prediction *2*, 20–21 (in Japanese).

NISHENKO, S. P., and JACOB, K. H. (1990), *Seismic Potential of the Queen Charlotte-Alaska-Aleutian Seismic Zone*, J. Geophys. Res. *95*, 2511–2532.

NUTTLI, O. W. (1979), *Seismicity in the Central United States*, Geol. Soc. Am. Rev. Eng. Geol. *4*, 67–93.

RUNDLE, J. B. (1988), *A Physical Model for Earthquakes 2. Application to Southern California*, J. Geophys. Res. *93*, 6255–6274.

SALEUR, H., SAMMIS, C. G., and SORNETTE, D. (1996), *Discrete Scale Invariance, Complex Fractal Dimension, and Log-periodic Fluctuations in Seismicity*, J. Geophys. Res. *101*, 17,661–17,677.

SAMMIS, C. G., SORNETTE, D., and SALEUR, H., *Complexity and earthquake forecasting*. In *Reduction and Predictability of Natural Disasters*, SFI Studies in the Sciences of Complexity (eds. Rundle, J. B., Klein, W., and Turcotte, D. L.) (Addison-Wesley, Reading, MA 1996) pp. 143–156.

SCHMITTBUHL, J., VILOTTE, J., and ROUX, S. (1996), *A Dissipation-based Analysis of an Earthquake Fault Model*, J. Geophys. Res. *101*, 27,741–27,764.

SCHWARTZ, S. Y. (1996), *Large Underthrusting Earthquakes in Subduction Zones with "Premature" Recurrence: Implications for the Seismic Gap Hypothesis* (Abstract), EOS, Trans. AGU 1996 Fall Meeting Suppl. *77*, F517.

SHAW, B. E., CARLSON, J. M., and LANGER, J. S. (1992), *Patterns of Seismic Activity Preceding Large Earthquakes*, J. Geophys. Res. *97*, 479–488.

SORNETTE, D., and SAMMIS, C. G. (1995), *Complex Critical Exponents from Renormalization Group Theory of Earthquakes: Implications for Earthquake Predictions*, J. Phys. I France *5*, 607–619.

SORNETTE, D., KNOPOFF, L., KAGAN, Y. Y., and VANNESTE, C. (1996), *Ranking-order Statistics of Extreme Events: Application to the Distribution of Large Earthquakes*, J. Geophys. Res. *101*, 13,883–13,893.

STEACY, S. J., and MCCLOSKEY, J. (1998), *What Controls an Earthquake Size? Results from a Heterogeneous Cellular Automaton*, Geophys. J. Int. *133*, F11–F14.

SYKES, L. R., and JAUMÉ, S. C. (1990), *Seismic Activity on Neighboring Faults as a Long-term Precursor to Large Earthquakes in the San Francisco Bay Region*, Nature *348*, 595–599.

SYKES, L. R., and NISHENKO, S. P. (1984), *Probabilities of Occurrence of Large Plate Rupturing Earthquakes for the San Andreas, San Jacinto, and Imperial Faults, 1983–2003*, J. Geophys. Res. *89*, 5905–5927.

SYKES, L. R., SCHOLZ, C. H., and SHAW, B. E. (1997), *Increased Rates of Moderate-size Events Preceding Large Earthquakes: The Prescence of a Self-organized Critical State May be Regarded as a Precursor Instead of an Impediment to Earthquake Prediction* (Abstract), EOS, Trans. AGU 1997 Fall Meeting Suppl. *78*, F465.

TOCHER, D. (1959), *Seismic History of the San Francisco Bay Region*, Calif. Div. Mines Spec. Rep. *57*, 39–48.

TRIEP, E. G., and SYKES, L. R. (1997), *Frequency of Occurrence of Moderate to Great Earthquakes in Intracontinental Regions: Implications for Changes in Stress, Earthquake Prediction, and Hazards Assessment*, J. Geophys. Res. *102*, 9923–9948.

VARNES, D. J. (1989), *Predicting Earthquakes by Analyzing Accelerating Precursory Seismic Activity*, Pure appl. geophys. *130*, 661–686.

VARNES, D. J., and BUFE, C. G. (1996), *The Cyclic and Fractal Seismic Series Preceding an m_b 4.8 Earthquake on 1980 February 14 near the Virgin Islands*, Geophys. J. Int. *124*, 149–158.

VOIGHT, B. (1989), *A Relation to Describe Rate-dependent Material Failure*, Science *243*, 200–203.

WARD, S. N. (1996), *A Synthetic Seismicity Model for Southern California: Cycles, Probabilities, and Hazard*, J. Geophys. Res. *101*, 22,393–22,418.

WESSEL, P., and SMITH, W. H. F. (1991), *Free Software Helps Map and Display Data*, EOS, Trans. AGU *72*, 445–446.

WILLIS, B. (1924), *Earthquake Risk in California 8. Earthquake Districts*, Bull. Seismol. Soc. Am. *14*, 9–25.

(Received July 6, 1998, revised December 7, 1998, accepted December 11, 1998)

Pure appl. geophys. 155 (1999) 307–334
0033–4553/99/040307–28 $ 1.50 + 0.20/0

Seismic Cycles and the Evolution of Stress Correlation in Cellular Automaton Models of Finite Fault Networks

CHARLES G. SAMMIS[1] and STEWART W. SMITH[2]

Abstract—A cellular automaton is used to study the relation between the structure of a regional fault network and the temporal and spatial patterns of regional seismicity. Automata in which the cell sizes form discrete fractal hierarchies are compared with those having a uniform cell size. Conservative models in which all the stress is transferred at each step of a cascade are compared with nonconservative ("lossy") models in which a specified fraction of the stress energy is lost from each step. Particular attention is given to the behavior of the system as it is driven toward the critical state by uniform external loading. All automata exhibit a scaling region at times close to the critical state in which the events become larger and energy release increases as a power-law of the time to the critical state. For the hierarchical fractal automata, this power-law behavior is often modulated by fluctuations that are periodic in the logarithm of the time to criticality. These fluctuations are enhanced in the nonconservative models, but are not robust. The degree to which they develop appears to depend on the particular distribution of stresses in the larger cells which varies from cycle to cycle. Once the critical state is reached, seismicity in the uniform conservative automaton remains random in time, space, and magnitude. Large events do not significantly perturb the stress distribution in the system. However, large events in the nonconservative uniform automaton and in the fractal systems produce large stress perturbations that move the system out of the critical state. The result is a seismic cycle in which a large event is followed by a shadow period of quiescence and then a new approach back toward the critical state. This seismic cycle does not depend on the fractal structure, but is a direct consequence of large-scale heterogeneity of these systems in which the size of the largest cell (or the size of the largest nonconservative event) is a significant fraction of the size of the network. In essence, seismic cycles in these models are boundary effects. The largest events tend to cluster in time and the rate of small events remains relatively constant throughout a cycle in agreement with observed seismicity.

Key words: Regional seismicity, seismic cycle, cellular automation, critical point, fractals.

Introduction

Regional seismicity has many characteristics of a critical system including a power-law Gutenberg–Richter magnitude frequency relation and a fractal spatial

[1] Department of Earth Sciences, University of Southern California, Los Angeles, CA 90089-0740, U.S.A.
[2] Geophysics Program, University of Washington, Seattle, WA, U.S.A.

distribution of hypocenters (HIRATA, 1989a,b; ROBERTSON *et al.*, 1995). This had led many investigators to explore the possibility of using formalisms from statistical physics to model the spatial, temporal and magnitude distributions (ALLÈGRE *et al.*, 1982; ALLÈGRE and LE MOUEL, 1994; RUNDLE, 1988a,b, 1989, 1993; RUNDLE *et al.*, 1995, 1996; SMALLEY *et al.*, 1985; SAHIMI *et al.*, 1993a,b; SORNETTE and SORNETTE, 1990; NEWMAN *et al.*, 1994; SORNETTE and SAMMIS, 1995; SALEUR *et al.*, 1996a,b; MOREIN *et al.*, 1997; FISHER *et al.*, 1997; DAHMEN *et al.*, 1998). In the course of analyzing seismicity as a critical phenomenon, an interesting controversy has arisen as to the interpretation of such models and their implications. The central issue is whether the crust is in a continuous state of self-organized criticality (SOC), or whether it repeatedly approaches and retreats from a critical state. The working hypothesis for this later view is that a large regional earthquake is the end result of a process in which the stress field becomes correlated over increasingly long scale-lengths. The largest event cannot occur until regional criticality has been achieved. This large event then destroys criticality on its network creating a period of relative quiescence after which the process repeats by rebuilding correlation lengths toward criticality and the next large event.

The reason this question of continuous vs. discontinuous criticality is important is that it bears directly on the question of whether earthquakes can be forecast. BAK and TANG (1989) argue that the crust is always in a state of SOC. Their analog is a uniform cellular automaton in which a ball dropped in a randomly chosen cell is equally likely to start an avalanche of any size at any time. The crustal equivalent is that all small earthquakes have the same probability of growing into a great event. This model has been used by GELLER *et al.* (1997) as a physical basis for their recent assertion that earthquake prediction is inherently impossible. If the earth is always in a state of SOC, they may be correct. However, if a large shock moves some associated region away from criticality, then the seismicity associated with the subsequent return to criticality (and the next large event) is expected to have a number of observable characteristics including an increase in event size with growing stress correlation length, and an energy release rate which increases as a power law of the time-to-failure.

The central question thus becomes: is the crust in a continuous state of SOC? Several recent observations suggest not. SYKES and JAUMÉ (1990) and TRIEP and SYKES (1997) have found that large earthquakes are preceded by a cluster of intermediate-sized events (within $2M_w$ units of the main shock) in a large surrounding region. KNOPOFF *et al.* (1996) found that all 11 earthquakes in California since 1941 with magnitudes greater than 6.8 were preceded by an increase in the rate of occurrence of earthquakes with magnitudes greater than 5.1 in the appropriate tectonic domain. There is also mounting evidence that the rate of intermediate-sized events decreases following a large earthquake—an observation generally interpreted as resulting from a regional stress shadow. Such shadows have been documented by SYKES and JAUMÉ (1990) following the 1989 Loma Prieta earthquake, by TRIEP and SYKES (1997) following the 1950 Assam shock, by HARRIS and SIMPSON (1996) following the 1857 Fort Tejon earthquake, and by Jones following the 1952 Kern County and 1992 Landers earthquakes. BOWMAN and SAMMIS (1997) and TRIEP and SYKES (1997) have suggested that these shadows represent the retreat of a region from a state of SOC. SORNETTE and SAMMIS (1995), SAMMIS et al. (1996), and SALEUR et al. (1996a,b) also argue that the observed power-law buildup of intermediate events before a great earthquake represents the approach of the appropriate region toward a state of SOC.

The power-law increase in regional seismicity before large events was first documented by BUFE and VARNES (1993) and BUFE et al. (1994) who found that the clustering of intermediate events before a large shock produces an increase in cumulative regional energy release (or in cumulative regional Benioff strain, $\varepsilon(t) = \sum_{j=1}^{N(t)} E_j^{1/2}$) that can be fit by a power-law time-to-failure relation of the form

$$\varepsilon(t) = A + B(t_c - t)^m \tag{1}$$

where t_c is the time of the large event, B is negative and m is usually about 0.3. SORNETTE and SAMMIS (1995) showed that the power law (1) is expected if the largest event is viewed as some sort of critical point for the region. In this case the mathematical techniques developed in statistical physics to describe critical phase transitions can be applied to seismicity. One technique, the renormalization group (RG), leads directly to (1) (see SALEUR et al., 1996a,b for a discussion of the application of the RG to regional seismicity). The renormalization group also explains the observed clustering of large events prior to the largest shock (SYKES and JAUMÉ, 1990) in terms of the growth of the spatial correlation length of the regional stress field. In this view, larger events are not possible until the stress field is correlated at a sufficient length to produce them. Before this time small events are unable to jump barriers and grow into big events. Only when criticality is reached is the stress field correlation on all scale lengths permitting events of all sizes up to the largest possible on the given fault network. The implication is that the smaller events are the agents by which longer stress correlation lengths are established — they effectively smooth the stress field at larger scale lengths (SAMMIS et al., 1995; BEN-ZION, 1996). It is important to note that the largest regional event need not occur at t_c in Equation (1). Rather, this is the time when the region reaches the critical state and a large event is possible.

An important observational question arises when fitting Equation (1) to seismicity data: how is the size of the "critical region" to be chosen for a given event? BOWMAN et al. (1999) showed that, for all 12 California earthquakes having magnitude m > 6.2, a critical region size could be found which optimized the fit to Equation (1). Moreover, they found that the logarithm of the radius R of these optimal critical regions scaled as the magnitude

$$\log R \propto 0.5 \, m \tag{2}$$

and that they could reject the null hypothesis that their fits were consistent with a random catalog with 99.9% confidence. A similar study by BREHM and BRAIL (1998) analyzed nineteen events from the New Madrid Seismic Zone and found log R $\propto$ 0.75 m. JAUMÉ and SYKES (1999) combined the two data sets and found the best fit is log R $\propto$ 0.36 m in better agreement with Equation (2). It is interesting that Equation (2) is compatible with the scaling of preparatory region with magnitude found by KEYLIS-BOROK and MALINOVSKAYA (1964) using a different method.

Another testable hypothesis that has emerged from the critical point model for seismicity is the possibility of log-periodic fluctuations in seismicity approaching criticality. SORNETTE and SAMMIS (1995) showed that if the spatial renormalization can only be made at a discrete fractal hierarchy of scale lengths, then the critical exponent is imaginary in time and Equation (1) becomes (to a first approximation, retaining only leading term in the periodicity)

$$\varepsilon(t) = A + B(t_c - t)^m \left[1 + C \cos\left(2\pi \frac{\log(t_c - t)}{\log \lambda} + \psi \right) \right]. \tag{3}$$

Such log-periodicity has been documented in several cases (SORNETTE and SAMMIS, 1995; VARNES and BUFE, 1996), but it has yet to be established as a universal precursor to large events. The modeling study presented below suggests that it may not be universal. However, if observed, log-periodicity allows a more precise estimate of t_c (SORNETTE and SAMMIS, 1995; SAMMIS et al., 1996).

The physical cause of log-periodicity is still an open question. One possibility is the existence of a discrete fractal hierarchy in the regional fault network. Such structures may be simulated by a discrete hierarchical fractal automaton (BARRIER and TURCOTTE, 1994; HUANG et al., 1998) which is explored further in this paper. Another possibility is that the earthquake sequence develops its own discrete hierarchy through stress shielding interactions in a highly heterogeneous region. This possibility has been demonstrated experimentally by ANIFRANI et al. (1995), numerically by SAHIMI and ARBABI (1996), and analyzed by HUANG et al. (1997). A third possibility is that log-periodicity may arise from random fluctuations in the power-law approach to criticality (HUANG, personal communication). Although not physically interesting, the observation of randomly generated log-periodic oscillations can be taken as evidence of power-law behavior and used to help determine t_c.

In this paper, we use cellular automaton models to explore conditions under which these simple systems alternately build then destroy stress correlation thereby simulating regional seismic cycles. We begin with the homogeneous conservative automaton originally formulated by BAK et al. (1987) and explore the power-law scaling and growth of stress correlation as the initial loading transient moves the system toward SOC. Once SOC is established in this system, large events do not destroy stress correlation and the system remains critical. Large events in the SOC state do not show the precursory increases in seismicity or post-event quiescence observed in real catalogs. We next explore the uniform nonconservative automaton discussed by OLAMI et al. (1992) in which a fraction of the energy is lost from each subevent in a cascade. We find that the largest cascades in this model destroy stress correlation thus moving the system out of the critical state and producing seismic cycles having many characteristics observed in real catalogs. Finally, we explore the discrete hierarchical fractal automaton previously studied by BARRIER and TUR-COTTE (1994) and HUANG et al. (1998). The largest events in this model also destroy stress correlation and produce seismic cycles for cases where the scale of the structural heterogeneity associated with the fractal structure is comparable to the scale of the network. We expand the previous work of HUANG et al. (1998) by exploring a range of fractal dimensions, by exploring the relationship between the scale of the heterogeneity and the size of the network, and by separating the effects of energy loss from effects of the discrete hierarchical structure. Our general result is that seismic cycles in the automaton model result from the finite size of the network and either heterogeneity on a scale comparable to the size of the network, a significant loss of energy from the network during the event, or a combination of both.

A Cellular Automaton Model for Regional Seismicity

We begin with a review of the basic elements of the cellular automaton, first exploring the homogeneous automaton developed by BAK et al. (1987) for conservative systems and extended by OLAMI et al. (1992) for nonconservative systems which map onto the BURRIDGE-KNOPOFF (1967) spring-block model of earthquakes. We then review the conservative fractal automaton as defined by BARRIER and TURCOTTE (1994) and extended to nonconservative cascades by HUANG et al. (1998).

The Homogeneous Cellular Automaton with no Losses

Since the homogeneous conservative automaton is known to exhibit SOC (BAK and TANG, 1989) we use it here to establish the conditions which characterize SOC as a basis for comparison with the nonconservative and fractal systems. The uniform cellular automaton defined by BAK et al. (1987) consist of a 2-D array of equal sized cells as in Figure 1a. The array is loaded by dropping "balls," one at a time, into randomly selected cells. A cell is considered full when it contains three balls. If a fourth ball is added to a full cell, the contents of that cell are distributed equally to the four nearest neighboring cells. If this redistribution results in four balls in a neighboring cell, then that cell also unloads to its four neighbors (one of which goes back into the original cell), and so on. Cells that are on the edge of the array dump one ball off the array while corner cells dump two balls off the array. When no cell contains four balls, the cascade ends and the random dropping of balls onto the array resumes. The term "conservative" is used here to denote the conservation of balls during each subevent in a cascade, even though balls are lost from the edges of the network.

The size of a cascade is measured by the number of cells that are involved. Cascades of all sizes occur from one cell up to the size of the array. Once the number of balls on the array reaches equilibrium, the distribution of avalanche sizes follows a power law and the system is in a state of self-organized criticality (SOC).

a) Uniform Grid

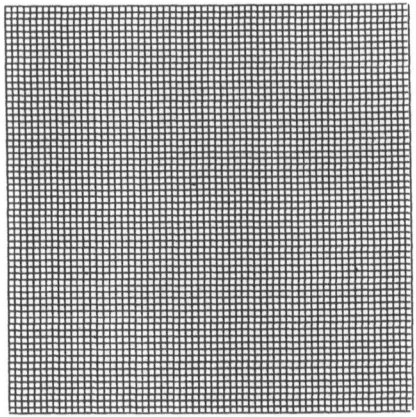

b) Discrete Fractal, R=2   c) Discrete Fractal, R=3

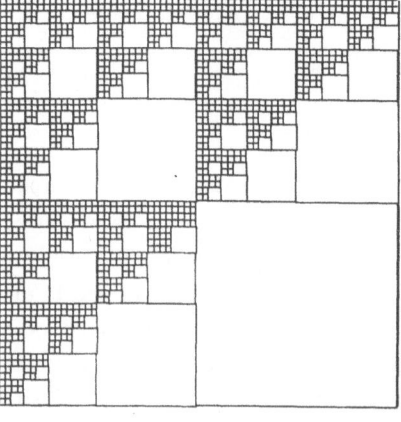 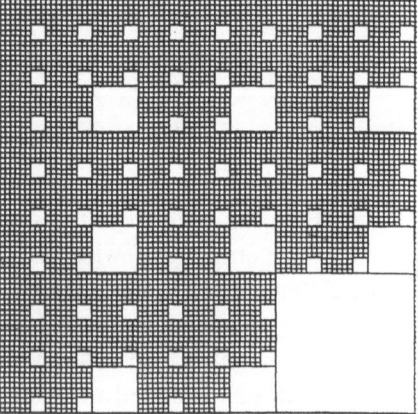

Figure 1
Grids used for (a) the uniform homogeneous automaton, (b) the discrete hierarchical fractal automaton with rescale factor $R = 2$, and (c) the discrete hierarchical fractal automaton with rescale factor $R = 3$.

A characteristic of SOC is that the individual events are not predictable in time, or space, or magnitude. The fact that regional seismicity follows the Gutenberg–Richter power-law relation between number and magnitude suggests the possibility that the crust is in a state of SOC, and that regional seismicity is a member of the class of phenomena described by the homogeneous automaton. If this is true, then the time place and magnitude of individual earthquakes is inherently unpredictable.

Interpreting regional seismicity in terms of the cellular automaton, the balls can be viewed as representing increments of stress, strain, or stored elastic energy—all are equivalent measures in an elastic solid. We will discuss the process in terms of stored energy. Each ball drop represents an equal increment of time. The random loading of the array corresponds to a slowly increasing tectonic stress while the unloading of a cell represents the stress redistribution associated with an earthquake. A larger cascade represents a spatially larger redistribution of stress and therefore corresponds to a larger earthquake. The cascades are assumed to take no time since the time-scale associated with an earthquake is short in comparison to the time-scale associated with the tectonic loading. Although the long-term average loading rate is constant for the array and may be viewed as due to the constant motion of tectonic plates, the random dropping of balls continuously introduces random fluctuations at the smallest scale. In the geological context, this corresponds to a continuous, small-scale, spatially random roughening of the stress field. A physical source for such small-scale roughening is not obvious. One might assert that it is the result of small earthquakes that produce stress redistribution at scale-lengths less than or equal to that of the smallest box. However, as will be discussed below, small events are expected to smooth the field at scale-lengths larger than themselves, not roughen it. The amplitude of this small-scale roughening in the model can be decreased by increasing the failure threshold from four balls to, say, 400 balls in which case 100 balls are distributed to each of the four neighbors. The roughening produced by each random ball drop is thereby proportionally reduced. We have found that such reduction in the amplitude of small-scale roughening has no observable effect on the temporal or spatial behavior of the automaton.

It is also possible to formulate the automaton such that the loading is equally distributed to all cells (OLAMI et al., 1992). In this case the energy in each cell is not an integral number of balls, but a real number between 0 and some arbitrary threshold which we take to be 1. The cell with the maximum energy s_{max} is identified and the energy in each cell is increased by $1 - s_{max}$. The energy in the maximal cell is then set to zero and its energy of 1 is distributed equally among each of its four nearest neighbors that receive 0.25 each. If this redistribution results in another cell having an energy $s \geq 1$, the energy in that cell is set to 0 and its energy, which may be greater than 1, is distributed in equal portions to its four nearest neighbors (which includes reloading the cell which failed initially). The cascade continues until no cell has $s \geq 1$. The time-scale in this case is set by the assumption of a uniform loading rate which requires that time increments be

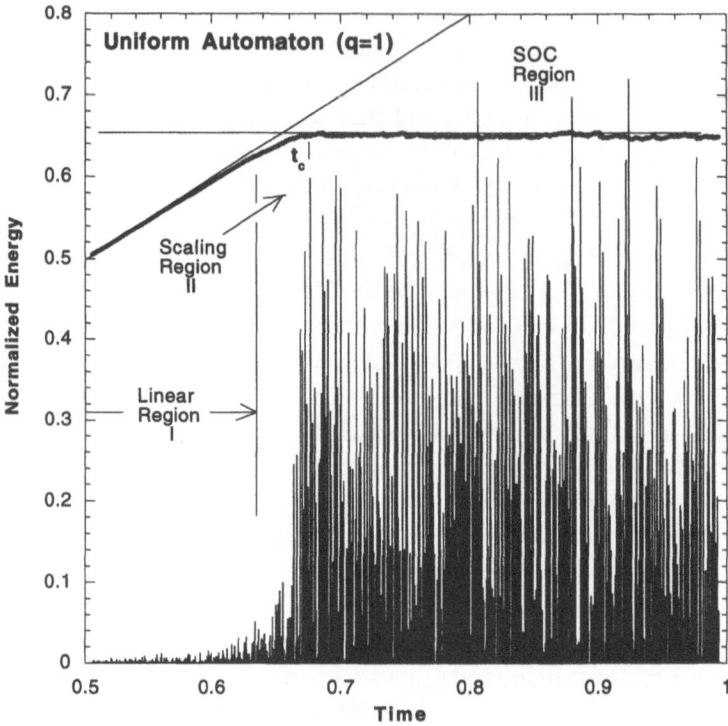

Figure 2

Normalized energy vs. normalized time for the uniform homogeneous automaton with no losses ($q = 1$). The upper heavy line shows energy on the grid while the catalog of events is shown below. The linear, scaling, and SOC regions are identified as discussed in the text.

directly proportional to increments in the energy. Since the constant of proportionality is arbitrary, we choose it to be unity, i.e., we take $\Delta t = \Delta s$. This means that a cell having an energy of 0.6 will fail in 0.4 time units if no other energy is added from neighboring cells. Again we assume that cascades take no time. In this case the automaton is initiated by assigning a random number in the interval (0,1) to each cell. This random initial distribution can be interpreted as the result of a combination of elastic heterogeneity and the prior history of events. This system evolves to an equilibrium state that is independent of the random starting distribution. We use uniform loading for all the simulations presented in this paper and discuss those results which are affected by this choice.

Figure 2 shows the catalog generated by a homogeneous 64×64 automaton with uniform loading and no losses. As in BAK and TANG (1989) the total number of cells which spill during a cascade is taken as the size of the event and as a measure of the energy released by the event. Also shown on this figure is the average energy density in the system. Initially, the average energy density increases linearly from its starting value of 0.5 (since each cell is given a random initial energy between 0 and 1). The initial slope is one since most of the energy added to the grid through the loading process remains on the grid. As the energy density grows, the size of the largest avalanche increases. As these larger avalanches begin to intersect the edges of the grid on a more regular basis, more energy is lost from the grid per unit time and the slope of energy-density vs. time curve in Figure 2 decreases. Eventually, the energy density on the grid becomes constant—energy is lost from the edges at the same rate it is added through the loading process.

We can thus identify three regions in the energy density curve labeled I, II, and III in Figure 2. Region I is the initial transient where most of the energy added to the grid remains on the grid. Region II is the transition region between region I and the steady-state region III that we identify as characterizing the SOC state. By fitting straight lines to regions I and III in Figure 2, the transition region II can be seen to extend from about $t = 0.64$ to $t = 0.67$. We will show below that region II is the "scaling region" close to SOC where the energy released by the cascades can be fit to the power law in Equation (1). Note that t_c does not correspond to the largest event in the sequence. Figure 3 shows the cumulative energy released by the events in Figure 2. The average energy density and three regions from Figure 2 are also plotted for reference.

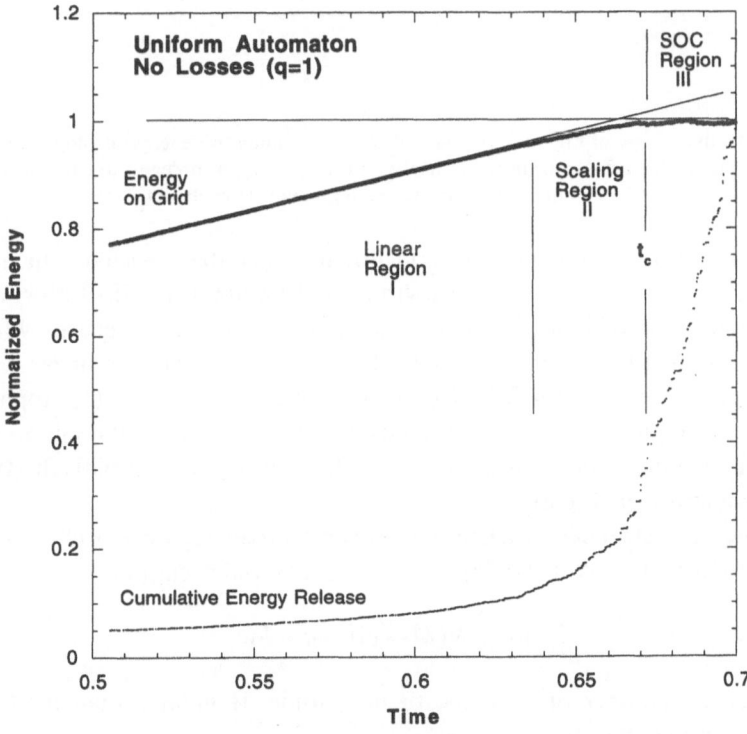

Figure 3
Cumulative release of scaled energy as a function of time for the uniform homogeneous automaton with no losses. The energy on the grid and regions I–III are replotted from Figure 2 for reference.

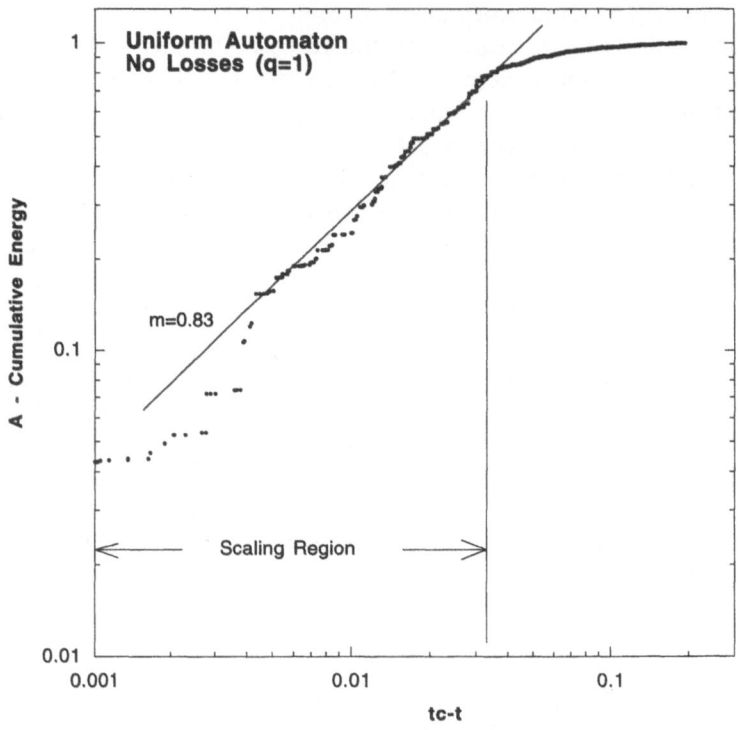

Figure 4

For the cumulative energy in Figure 3, the quantity $\log(A -$ cumulative energy) is plotted as a function of the logarithm of the distance from the critical point, $\log(t_c - t)$, in order to test the fit to Equation (1). The fit is linear in the scaling region II as expected.

To further investigate region II we fit Equation (1), the expected behavior in the scaling region close to criticality (SORNETTE and SAMMIS, 1995). Both t_c and A in this equation are known since A is the value of the cumulative energy when $t = t_c$ (see Fig. 3). Figure 4 is a plot of $\log(A - \Sigma E)$ vs. $\log(t_c - t)$. The linear portion of this curve indicates the scaling region II in which Equation (1) applies (with $m = 0.83$). Note that this region extends to about $t_c - t = 0.03$. Since we took $t_c = 0.67$ the scaling region covers $0.64 < t < 0.67$, in agreement with the transition region II identified in Figure 2.

The rate of occurrence of earthquake over a broad region is well described by the GUTENBERG–RICHTER (1956) frequency-magnitude relation

$$\log_{10} N(M > \mathrm{m}) = a - b\mathrm{m} \qquad (4)$$

where N is the number of events with magnitude M greater than m. Using the energy magnitude relation

$$\log_{10} E = c + d\mathrm{m} \qquad (5)$$

The Gutenberg–Richter relation becomes a power-law relation for the number of earthquakes having E_0 greater than E

$$N(E_0 > E) \sim E^{-b/d} = E^{-B} \tag{6}$$

with B commonly in the range 0.80–1.05. The observation that the cellular automaton produces this distribution of event sizes is commonly cited as evidence that the crust is in a state of SOC (BAK and TANG, 1989; OLAMI et al., 1992). We also find that events in the SOC region ($t > t_c = 0.67$) are consistent with Equation (6).

We end our discussion of the conservative homogeneous automaton by exploring the growth of stress correlation during the approach to criticality and the fluctuations in correlation in the SOC state. By stress correlation we mean the size of "critical clusters" which would cascade if one element were triggered. The simplest measure of stress correlation is the size distribution of the avalanches themselves. Since we are most interested in the largest critical cluster at any time, we can simply identify the largest event as a function of time. The only problem with this approach is that, at any given time, the largest cluster may not be triggered. We therefore plot the largest event in successive time intervals of $\Delta t = 0.01$ to minimize nucleation effects. Figure 5a shows the maximum event as a function of time for the case shown in Figures 2–4. The correlation length grows steadily in region I, at an accelerated rate in the scaling region II, and then remains relatively constant in the SOC region III. Figure 5b shows the rate of small events (events which have an area $< 5\%$ of the maximum event area observed over the entire time shown). Note that rate of small events remains relatively constant over the entire time interval in agreement with observations by ELLSWORTH et al. (1981).

The Homogeneous Cellular Automaton with Losses

OLAMI et al. (1992) explored the effect of a loss factor $0 < q < 1$ on the behavior of the uniformly loaded homogeneous automaton. In this case, if a cell contains an energy $s \geq 1$, its energy is set to zero and the energy qs is distributed equally to its four neighbors. The justification for this loss factor is clear if the automaton is being used to simulate the complexity of slip on a fault-plane. In this case q represents the fraction of energy lost to friction, to the formation of fault gouge, and to seismic radiation. However, the justification of q is not as clear when the automaton is being used to simulate regional seismicity. In this case, a cascade does not represent a growing slip patch, but the transfer of shear stress released by an earthquake. In an abstract sense, if s is viewed as the elastic energy stored in a cell, then the overall loss of energy from the system could be viewed as representing the losses to friction and other forms of nonelastic deformation, fragmentation, and radiation. Since this loss limits the size of the cascade, it could be argued that an earthquake with larger nonelastic losses transfers stress to a smaller region.

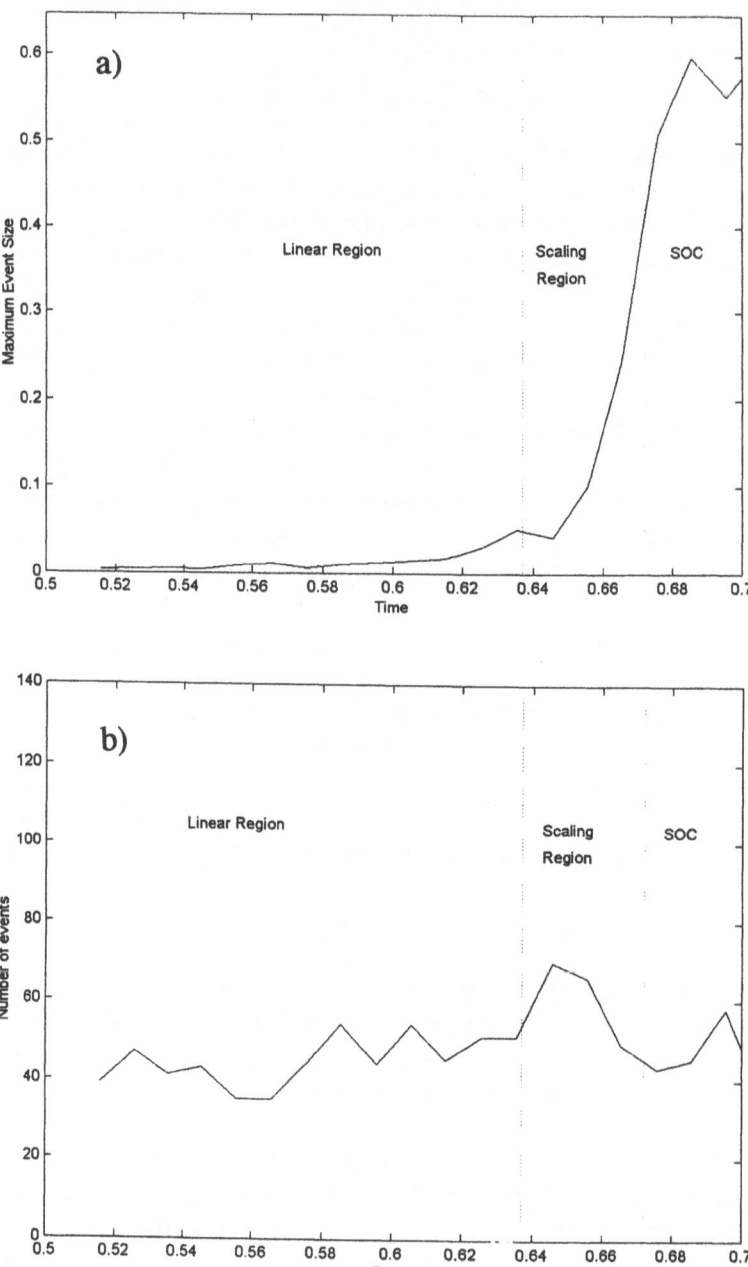

Figure 5
Temporal distribution of event sizes in the uniform automaton without losses. In panel (a) the maximum
event size is plotted as function of scaled time illustrating the growth of correlation length in the scaling
region leading up to the critical point and the relative constancy of correlation in the SOC region. Panel
(b) shows the temporal distribution of small events which remains approximately constant in agreement
with observed seismicity.

The q factor will be shown below to introduce memory into the automaton. The largest events leave "stress shadows" that move the system out of the critical state. This result seems to be at odds with OLAMI et al. (1992) who claim that their nonconservative automaton displays SOC based on its generation of power-law scaling of event sizes. However, KLEIN and RUNDLE (1993) point out that the scaling with system size found by OLAMI et al. is not consistent with their model being at the true scaling limit and their conclusions about critical behavior are therefore suspect. GRASSBERGER (1994) developed a fast computational algorithm that allowed him to explore the nonconservative model on larger grids and for longer times. He found that the scaling observed by OLAMI et al. was largely a boundary effect that is observed only in relatively small systems or in larger systems during the long transient period. Grassberger suggested that such boundary effects, or other forms of frozen heterogeneity, might be important in the earthquake problem and we agree. The significant factor appears to be the size of the largest event relative to the size of the network. Where the size of the largest event in the conservative automaton is limited only by the size of the system, we will show below that the size of the largest event in the nonconservative system is limited by the losses during the cascade. Significant retreats from the critical state only occur when the size of the largest event is a significant fraction of the size of the network.

Figure 6 shows the catalog generated by this model for $q = 0.9$ and the average energy density in the grid for selected values in the range of $0.5 \leq q \leq 1.0$. There are several striking differences between this figure and Figure 2. Most notably, the events here are much smaller. We found that the size of the events is proportional to $1/(\ln q)$. For $q < 1$, the size of the events is limited by the losses, not by the size of the grid. In fact, the size of the events is controlled by the clustering statistics of the array of random initial values assigned to the cells. A simple statistical analysis of an $n \times n$ array of random initial values in the range $(0,1)$ predicts that the maximum event size should vary approximately as $(\ln 4 - 2 \ln n)/\ln q$ which we have verified with simulations over a range of n and q.

Another contrast with Figure 2 is that the average energy on the grid in Figure 6 shows periodic structures in what should be the SOC region III. The period of this behavior, T, increases with decreasing q as $T = 1 - q$. This periodicity is also evident in the catalogs and appears to be due to stress shadows left behind by the large events. To understand why $T = 1 - q$, consider a cell that nucleates a large cascade. Initially, its load drops from 1 to 0. For a large cascade, it is likely that this will cause its four nearest neighbors to fire each of which returns $q/4$ units of stress to the initial cell. Hence, the nucleating cell ends up with a load of q and will therefore trigger another large cascade after $T = 1 - q$ time units have passed. Note that the relation $T = 1 - q$ holds at the extremes of $q = 1$ and $q = 0$. The case $q = 1$ corresponds to full stress transfer and SOC where there is no periodicity ($T = 0$). The case $q = 0$ corresponds to no stress transfer in which case the initial random distribution of loads is preserved and leads to an exact repetition of the catalog

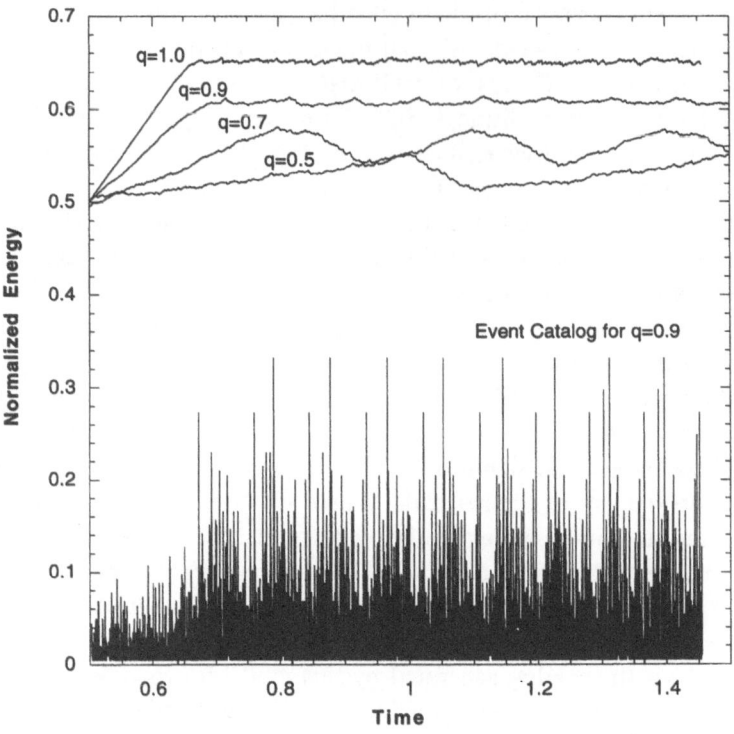

Figure 6
Normalized energy vs. normalized time for the uniform homogeneous automaton with losses ($q < 1$).
The upper heavy lines shows energy on the grid for a range of q while the catalog of events for $q = 0.9$
is shown below. Note the periodic fluctuations in energy on the grid and related periodicity in the
catalog.

after each full loading period $T = 1$ (recall that we defined the threshold of 1 to
correspond to one time unit). It is interesting that this periodicity does not appear
for the case of strong random loading (a four ball-threshold). In this case the
memory of stress shadows is destroyed by the random fluctuations introduced
through loading.

Figure 7 shows the cumulative energy corresponding to the catalog in Figure 6.
Note that the three regions can also be identified in the initial transient, but the
interpretation of region III as SOC is questionable because of the predictability
introduced by the periodic structure. The parameters A and t_c identified in Figure
7 are used in Figure 8 to test the suitability of Equation (1) to describe the scaling
region II. Note that an apparent scaling region can be identified extending to
$t_c - t = 0.05$ with a slope of $m = 0.8$. Figure 9a shows the largest event (our proxy
for maximum correlation length) as a function of time. Note that the losses produce
fluctuations in the maximum correlation length in region III which correspond to
the periodic structures in the energy on the grid and in the catalog (Fig. 6). Figure

9b shows that the rate of small events also remains relatively constant during the initial approach of lossy systems toward criticality. In Figure 7 we have only analyzed the initial transient for direct comparison with the conservative case. We have also analyzed the increasing portions of the cycles and found that they also can be fitted to Equation (1), although there is a bit more scatter.

We therefore interpret the periodic fluctuations in Figure 6 as reflecting the repeated approach and retreat of the system from the critical state. It is interesting that the retreat from criticality is not caused by a single large event. Rather, the decrease of energy on the grid is caused by a swarm of large events across the grid over the entire decreasing interval. Often the largest event occurs at the end of the swarm and marks the onset of the increase in grid energy leading to the next swarm. Note that as q decreases, the average energy on the grid in region III decreases while the amplitude of the fluctuation increases. Both effects are related to the decrease in the maximum size of the events with decreasing q. In general, as energy on the grid increases the correlation length increases. However, when the correlation length reaches the size of the maximum event for the given q, the events

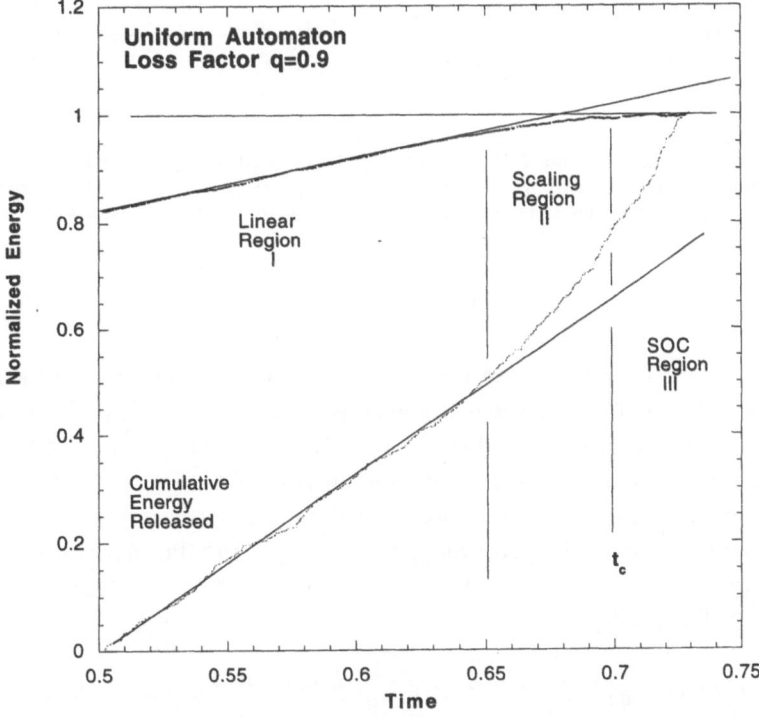

Figure 7

Normalized energy on the grid and cumulative energy released as a function of scaled time for the uniform automaton with loss factor $q = 0.9$. Note that regions I, II, and III can also be identified in this case.

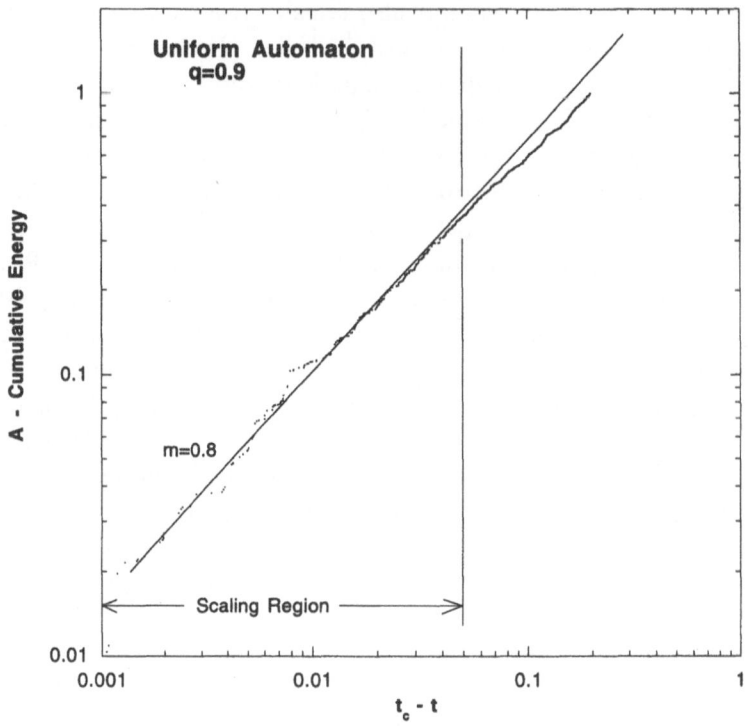

Figure 8
For the cumulative energy in Figure 7, the quantity log(A − cumulative energy) is plotted as a function of the logarithm of the distance from the critical point, log($t_c - t$), in order to test the fit to Equation (1). The fit is linear in the scaling region II for the loss factor $q = 0.9$.

of that size unload the grid ending the cycle. Since this correlation length is reached at a lower grid load for a lower q the average energy on the grid decreases with decreasing q. Also, for a given grid size n, there are statistically more limiting events at lower q since they are smaller. Since the energy ΔE lost by a limiting event is almost independent of q, there is more unloading and hence larger fluctuations at smaller q. To see why ΔE is almost independent of q, write the energy lost per event as the energy lost per cell, $(1 - q)$, times the size of the event and approximate $\ln q \approx (1 - q)$ for $q \approx 1$

$$\Delta E \approx (1-q) \frac{\ln 4 - 2 \ln n}{\ln q} \approx (1-q) \frac{2 \ln n - \ln 4}{(1-q)} = 2 \ln n - \ln 4. \qquad (7)$$

Hence, for $q < 1$ the system never reaches the critical state but is moved away by events at the largest allowable size.

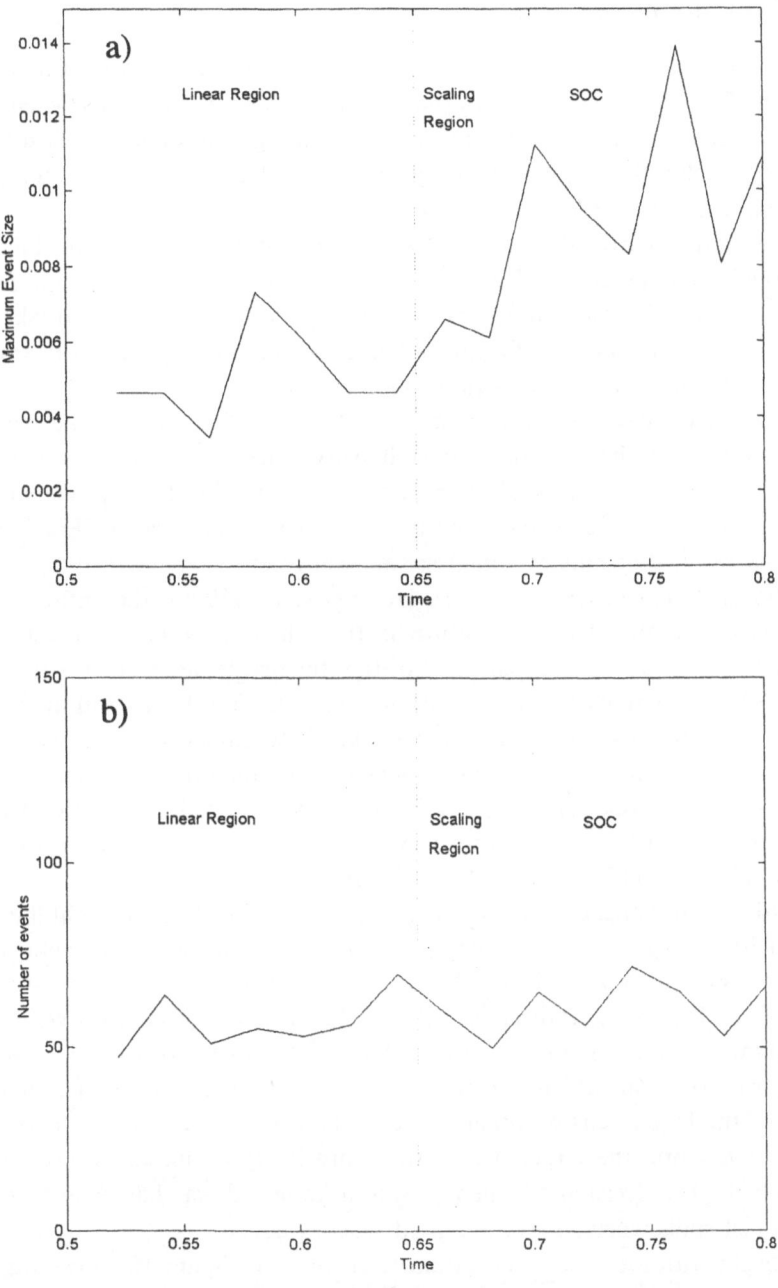

Figure 9
Temporal distribution of event sizes in the uniform automaton with loss factor $q = 0.9$. In panel (a) the maximum event size is plotted as function of scaled time. The correlation length again grows in the scaling region leading up to the critical point but the maximum event size fluctuates for $t > t_c$ in concert with the fluctuations in energy on the grid in Figure 6. Panel (b) shows the temporal distribution of small events which remains approximately constant in agreement with observed seismicity.

The Conservative Fractal Automaton

BARRIER and TURCOTTE (1994) explored the fractal automaton shown in Figure 1b. This fractal is discrete in that it is only strictly self-similar when rescaled by a factor of $R = 2$. For the case of uniform loading, the smallest cells still unload when they reach a threshold of 1, but progressively larger cells do not unload until they reach a larger load equal to their area.

Barrier and Turcotte also gave a slightly different interpretation to the meaning of a cascade. Unlike the uniform automaton where each cascade represents one earthquake, they view the failure of each cell as representing an earthquake. This is equivalent to viewing each cell as an individual fault and assuming a characteristic earthquake model in which the size of the earthquake is proportional to the size of the cell. For each cascade, they identify the largest cell to spill as the mainshock. Those members of the cascade that spill before the mainshock are identified as foreshocks while those that spill afterwards as aftershocks. Their principal result is that while all large events have many aftershocks, only about 28% have fore-shocks—a result that mimics actual seismic statistics.

In the modeling given here, as in HUANG *et al.* (1998), the entire cascade is viewed as one event. The assumption is that the time-scale of foreshocks and aftershocks is short in comparison to the time-scale associated with tectonic loading. The stress redistribution associated with the foreshocks and aftershocks is combined with that caused by the mainshock. These models capture two aspects of real seismicity missing from the homogeneous automaton: large events are more likely to occur on pre-existing large structures and regional fault networks tend to have a hierarchical fractal structure (AVILES *et al.*, 1987; OKUBO and AKI, 1987; HIRATA, 1989a; ROBERTSON *et al.*, 1995; OUILLON *et al.*, 1996).

In order to investigate the effect of spatial structure on the temporal fluctuations in seismicity, we also considered the discrete fractal having a rescale factor of $R = 3$ shown in Figure 1c. For both the $R = 2$ and $R = 3$ fractals we also investigated the effects of the loss factor q discussed above. The fractal automatons were initiated by assigning a random number in the interval $(0,1)$ to each of the 1×1 subcells of the uniform grid from which the fractal automaton is constructed. The initial load in each of the larger cells was then calculated as the average of the load in all its subcells. Therefore, the larger a cell, the more likely its initial average load was close to 0.50. The decision of when to spill a large cell was based on this average load—a cell spills when its average load reaches one.

We begin with the case of no attenuation ($q = 1$). Figure 10 shows the catalog of events and the energy on the grid during the initial transient approach to the critical state; Figure 10a is for the $R = 2$ grid in Figure 1b and Figure 10b is for the $R = 3$ grid in Figure 1c. The scaling region is again defined by the range of times where the energy on the grid has a slope of less than one and more than zero. Figure 11 shows the cumulative energy of the events for the two fractal models.

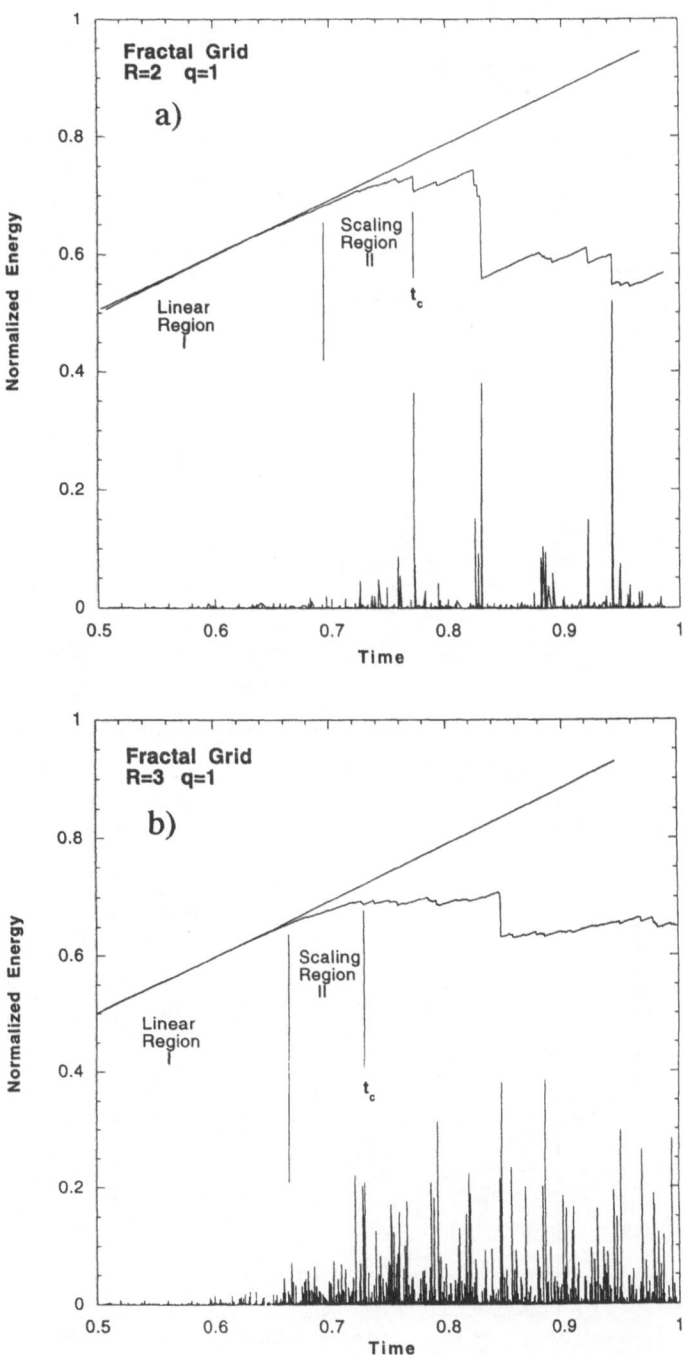

Figure 10
Total scaled energy on the grid and the catalog for the fractal grids in Figure 1 with no losses ($q = 1$).
Panel (a) is for a discrete rescale factor of $R = 2$ and panel (b) for $R = 3$.

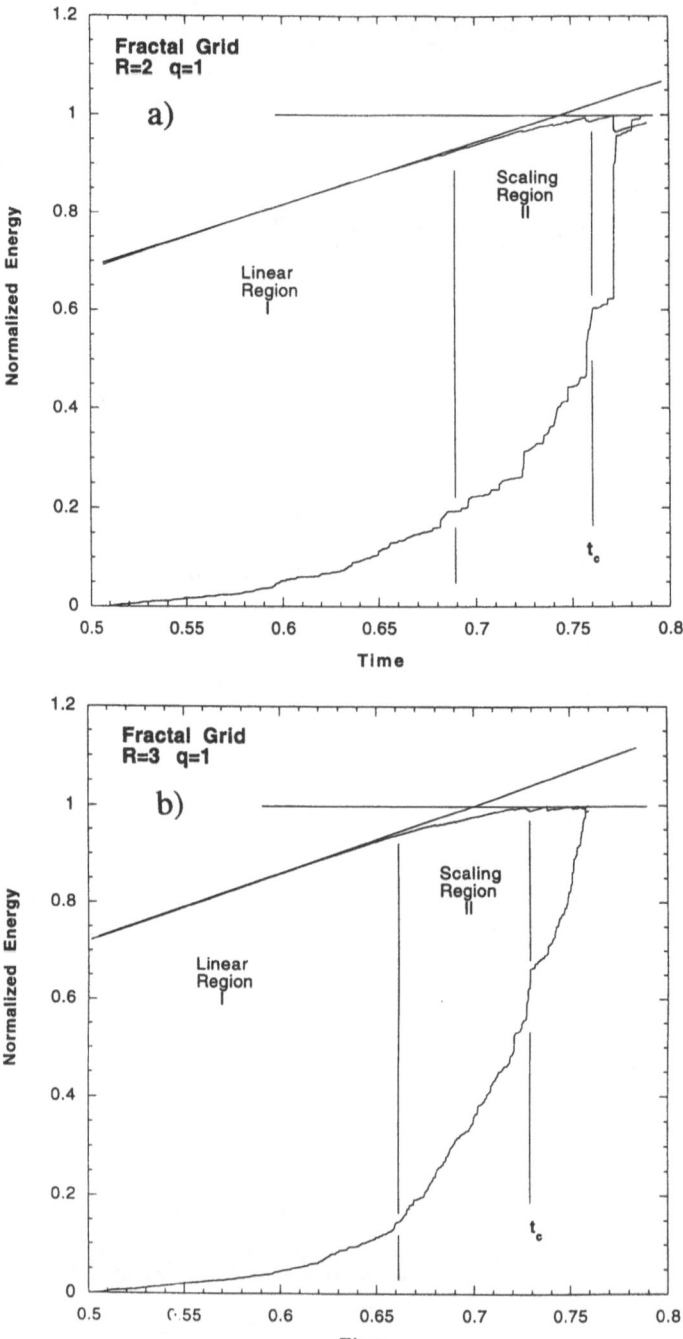

Figure 11
Cumulative scaled energy on the grid for the fractal grids in Figure 1 with no losses ($q = 1$). Panel (a) is for a discrete rescale factor of $R = 2$ and panel (b) for $R = 3$.

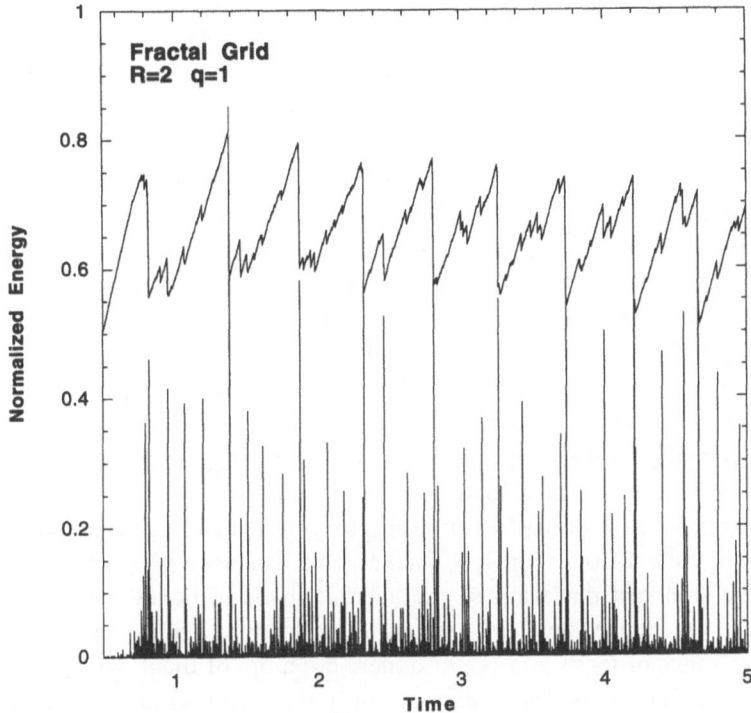

Figure 12
Energy on the grid and seismic catalog for several "cycles" of the $R = 2$ fractal automaton with no losses
$(q = 1)$.

Comparing these figures with the equivalent Figures 3 and 7 for the homogeneous model, note that the fractal models show larger fluctuations in the cumulative energy release. Figure 12 shows an extended catalog for the $R = 2$ grid and the energy on the grid. Note that the fractal structure introduces a periodicity into the energy on the grid. These "seismic cycles" are produced by big events in which the largest cell unloads and significant energy is lost from the grid. They do not require a fractal grid, only a grid in which the largest cell is a significant fraction of the grid size so its unloading produces an observable perturbation in the state of the system. In this view, each cycle represents the approach to criticality followed by a large event that moves the system away from the critical state. Note that the correlation and decorrelation of stress in the fractal model is forced by the structure since the larger structures are not allowed to unload until the average stress over the entire area becomes critical. When these large structures unload, the correlation of highly stressed regions is necessarily reduced.

As already demonstrated by HUANG et al. (1998), the discrete hierarchical structure produces log-periodic fluctuations in the power-law increase of cumulative energy release approaching the critical state of the form given by Equation (3).

Figure 13 shows the maximum event size as a function of time for the realization shown in Figure 11. The model used by HUANG *et al.* (1998) included a loss factor $q = 0.5$. We will see below that the introduction of loss into the fractal automaton enhances the amplitude of the log-periodic fluctuations.

It is important to note that while all realizations of this model produced a log-periodic sequence of discrete jumps in the maximum correlation length of the critical region as in Figure 13, this did not always produce log-periodic fluctuations in the cumulative energy release. A clear log-periodic progression of steps in cumulative energy was only observed when the intermediate events (one and two orders smaller than the maximum cell) tended to cluster in time. For those cases in which events of a given order are more evenly distributed in time, the jumps in scale only produce a subtle increase in slope of the cumulative energy producing the power-law increase before failure, but no evident log-periodic fluctuations.

The log-periodic sequence of jumps in correlation length is a direct consequence of the way in which the initial loads were assigned. Since we define the load in a box by the area average of loads in its subcells, a larger box is more likely to have an initial load close to 0.5 than is a smaller box. Since we are averaging random variables, the standard deviation of the average loads in boxes having N subcells decreases as $\sqrt{N}$. Hence, smaller boxes are more likely to trigger first because they have a larger range of loads and because there are more of them. Now, consider the $R = 2$ fractal in Figure 1b. Since there are only three second largest cells and nine third largest, it is more likely that the loads in the larger cells will cluster to produce log-periodic fluctuations in the cumulative energy. This may explain why only one or two log-periodic fluctuations are observed even though there are many more orders in the fractal hierarchy.

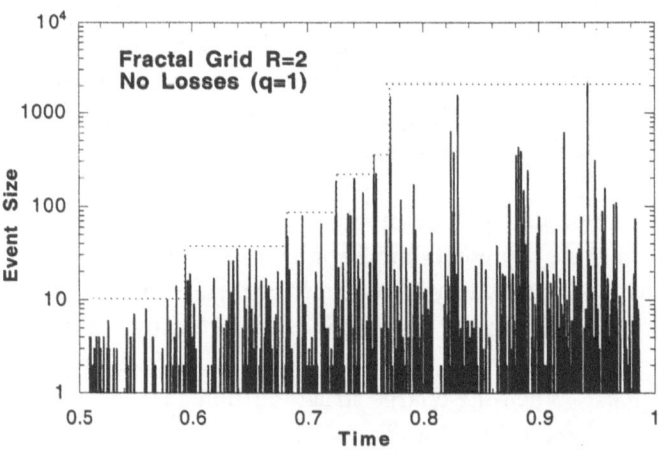

Figure 13
Seismic catalog for the $R = 2$ fractal grid with no losses illustrating the discrete jumps in event size leading up to the critical point.

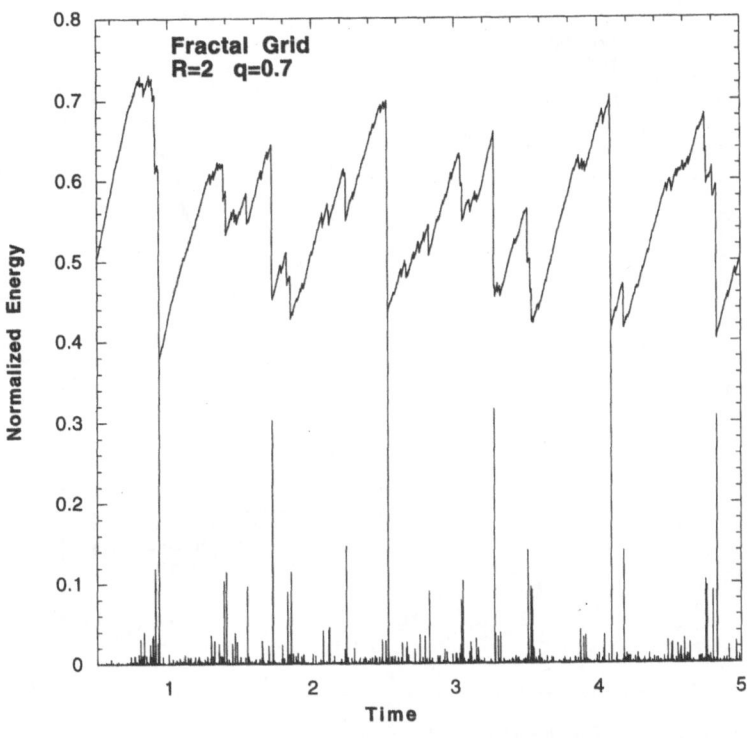

Figure 14
Energy on the grid and seismic catalog for several "cycles" of the $R = 2$ fractal automaton with $q = 0.7$.
Losses appear to produce longer and more irregular cycles than in Figure 12 where $q = 1$.

The Fractal Automaton with Losses

The calculations in the previous section were repeated with $q = 0.7$ and $q = 0.5$. Figure 14 shows the catalog and scaled energy on the grid for the case $R = 2$ and $q = 0.7$. Comparison with the $q = 1$ case (Fig. 10a) shows that the effect of losses in the fractal network is to make the cycles longer and more irregular. Figure 15 shows the cumulative Benioff strain (sum of the square root of the energy of each event through time) during an initial transient approach to SOC for the $R = 2$ and $R = 3$ grids. We plot Benioff strain here because it is the quantity usually analyzed for log-periodic fluctuations (SORNETTE and SAMMIS, 1995; HUANG et al., 1998). There is no theoretical reason for the choice of this parameter. Note that the wavelength of the log-periodic fluctuations is longer for the $R = 3$ case. This is the expected result based on the analysis of HUANG et al. (1998) since it takes longer for the correlation length to grow by a factor of 3 than by a factor of 2. However, a more definite quantitative analysis was frustrated by the large variations between different realizations of the model (probably due to the statistical nature of the log-periodic fluctuations themselves as discussed above) and by the sensitivity of the fitting procedure for the nonlinear equation (3).

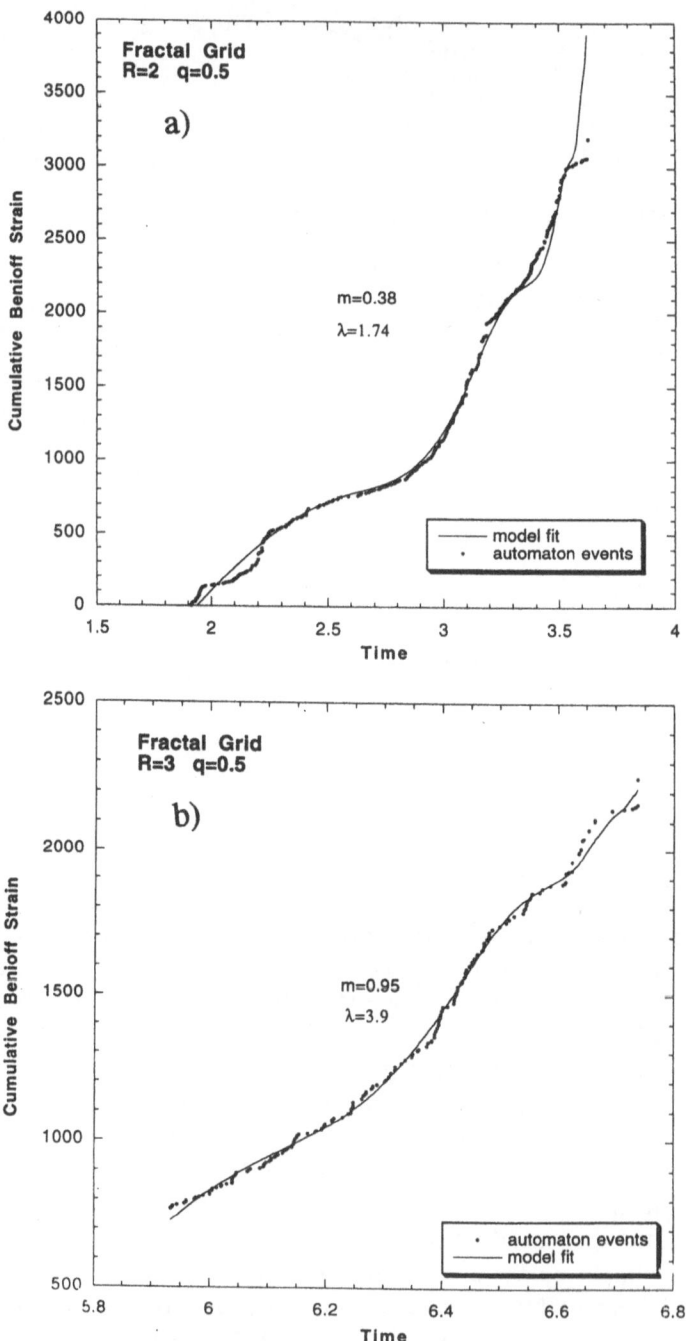

Figure 15
Cumulative Benioff strain for the fractal grids with loss factor $q = 0.5$. Note that the wavelength of the log-periodic oscillations increases with increasing R as expected.

Discussion and Conclusions

Self-organized criticality, as exhibited by the uniform cellular automaton, was proposed by BAK and TANG (1989) as a simple robust conceptual model for regional seismicity which offered an explanation for the ubiquitously observed Gutenberg–Richter power-law relation between the number of events and their energy. However, the uniform cellular automaton clearly oversimplifies the interaction between the various faults in a regional network in a number of ways. First, it considers only interactions between nearest neighbor cells while elastic interactions between faults are long-range. It also ignores time dependent effects that might arise from anelastic coupling between the crust and the mantle. Second, the cells are uniform whereas faults in natural networks are generally heterogeneous in both size and strength. Third, the only losses in the automaton are from the edges of the array whereas real earthquakes dissipate energy locally through friction, fragmentation, and elastic radiation. Fourth, a cellular automaton is usually large in comparison to the size of its individual cells so that boundary effects can be ignored. In most natural networks the size of the largest fault is a significant fraction of the size of the network.

There is mounting evidence that the spatial and temporal patterns of regional seismicity are more complicated than those predicted by the simple automaton. Other theoretical studies indicate, in contrast to the SOC paradigm, that the simulated seismicity patterns depend strongly on the assumed form of the interaction, heterogeneity, and rheology (BEN-ZION and RICE, 1995; BEN-ZION, 1996; FISHER et al., 1997; DAHMEN et al., 1998). In this paper we have shown that the inclusion of a local loss factor in the cascades leads to temporal cycles in the automaton seismicity which mimic many characteristics of observed seismic cycles. The largest events tend to be preceded by a power-law increase in seismicity associated with the clustering of intermediate-sized events and followed by a decrease leading to a period of relative quiescence. Even subtler characteristics of observed seismicity, such as the constant rate of small events through the cycle and the temporal clustering of larger events, are simulated by the lossy automaton. The seismic cycles associated with a local loss are a memory effect that is erased if strong random perturbations are introduced into the loading. Whether they are also destroyed by long-range interactions is an open question. We also found that the inclusion of structural heterogeneity produces similar cycles as long as the largest scale of the structure is comparable to the size of the network. Seismic cycles in lossy and structurally heterogeneous automatons are clearly seen to be associated with the repeated approach and retreat from the SOC state. The automaton allows us to document the growth in stress correlation and power-law energy increase in the scaling region near criticality and the destruction of correlation by large lossy events or by events associated with large structures and the subsequent movement of the system away from criticality.

The automatons provide other insights into spatial and temporal fluctuations in regional seismicity. For example, it is apparent that the largest event need not occur when the network reaches the critical point. At the critical point, the largest regional event is possible, but forecasting its actual occurrence becomes a nucleation problem that may ultimately limit the precision of an earthquake forecast (see also BOWMAN et al., 1999). Also, although we find that log-periodic fluctuations in the power-law increase of seismicity preceding large events can result from a discrete hierarchical fractal structure, such fluctuations are statistical in nature and need not always occur.

Probably the most significant and general insight gained by this study is the central importance of the finite size of a regional seismic network in relation to the size of its largest structural element. Virtually all the details of the seismic cycle in these models are the consequence of boundary effects. The largest events in such finite networks are fundamentally different than smaller events. The power-law buildup before the largest events is uncontaminated by the seismicity associated with neighboring events, an effect which makes power-law precursors difficult to observe for smaller events. These largest events, if they are associated with a large structural element and/or they result in a significant loss of energy, move the entire system away from its critical point to produce a regional seismic cycle.

Acknowledgements

We wish to thank Bill Klein for a helpful discussion of the issue of SOC in the nonconservative automaton. This work was supported by the National Science Foundation under grant EAR-9508040.

REFERENCES

ALLÈGRE, C. J., LE MOUEL, J. L., and PROVOST, A. (1982), *Scaling Rules in Rock Fracture and Possible Implications for Earthquake Predictions*, Nature *297*, 47–49.

ALLÈGRE, C. J., and LE MOUEL, J. L. (1994), *Introduction of Scaling Technique in Brittle Failure of Rocks*, Phys. Earth Planet Inter. *87*, 85–93.

ANIFRANI, J. C., LE FLOC'H, SORNETTE, D., and SOUILLARD, B. (1995), *Universal Log-periodic Corrections to Renormalization Group Scaling for Rupture Stress Prediction from Acoustic Emissions*, J. Phys. I. France *5*, 631–638.

AVILES, C. A., SCHOLZ, C. H., and BOATWRIGHT, J. (1987), *Fractal Analysis Applied to Characteristic Segments of the San Andreas Fault*, J. Geophys. Res. *92*, 331–344.

BAK, P., TANG, C., and WIESENFELD, K. (1987), *Self-organized Criticality: An Explanation of 1/f Noise*, Phys. Rev. Lett. *59*, 381–384.

BAK, P., and TANG, C. (1989), *Earthquakes as a Self-organized Critical Phenomenon*, J. Geophys. Res. *94*, 15,635–15,637.

BARRIER, B., and TURCOTTE, D. L. (1994), *Seismicity and Self-organized Criticality*, Phys. Rev. E *49*, 1151–1160.

BEN-ZION, Y. (1996), *Stress, Slip and Earthquakes in Models of Complex Single-fault Systems Incorporating Brittle and Creep Deformations*, J. Geophys. Res. *101*, 5677–5706.

BEN-ZION, Y., and RICE, J. R. (1995), *Slip Patterns and Earthquake Populations along Different Classes of Faults in Elastic Solids*, J. Geophys. Res. *100*, 12,959–12,983.

BOWMAN, D. D., and SAMMIS, C. G. (1997), *Observational Evidence for Temporal Clustering of Intermediate-magnitude Events before Strong Earthquakes in California* (Abst.), Seismol. Res. Lett. *68*, 324.

BOWMAN, D. D., OUILLON, G., SAMMIS, C. G., SORNETTE, A., and SORNETTE, D. (1999), *An Observational Test of the Critical Earthquake Concept*, J. Geophys. Res., still in press.

BREHM, D. J., and BRAILE, L. W. (1999), *Intermediate-term Earthquake Prediction Using the Modified Time-to-failure Method in Southern California*, Bull. Seismol. Soc. Am. *89*, 275–293.

BUFE, C. G., and VARNES, D. J. (1993), *Predictive Modeling of the Seismic Cycle of the Greater San Francisco Bay Region*, J. Geophys. Res. *98*, 9871–9883.

BUFE, C. G., NISHENKO, S. P., and VARNES, D. J. (1994), *Seismicity Trends and Potential for Large Earthquakes in the Alaska–Aleutian Region*, Pure appl. geophys. *142*, 83–99.

BURRIDGE, R., and KNOPOFF, L. (1967), *Model and Theoretical Seismology*, Seis. Soc. Am. Bull. *57*, 341–371.

DAHMEN, K., ERTAS, D., and BEN-ZION, Y. (1998), *Gutenberg–Richter and Characteristic Earthquake Behavior in Simple Mean-field Models of Heterogeneous Faults*, Phys. Rev. E *58*, 1494–1501.

ELLSWORTH, W. L., LINDH, A. G., PRESCOTT, W. H., and HERD, D. J. (1981), *The 1906 San Francisco Earthquake and the Seismic Cycle*, Maurice Ewing Monogr. *4*, 126–140, Am. Geophys. Union.

FISHER, D. S., DAHMEN, K., RAMANATHAN, S., and BEN-ZION, Y. (1997), *Statistics of Earthquakes in Simple Models of Heterogeneous Faults*, Phys. Rev. Lett. *78*, 4885–4888.

GELLER, R. J., JACKSON, D. D., KAGAN, Y. Y., and MULARGIA, F. (1997), *Earthquakes Cannot Be Predicted*, Science *275*, 1616–1617.

GRASSBERGER, P. (1994), *Efficient Large-scale Simulations of a Uniformly Driven System*, Phys. Rev. E *49*, 2436–2444.

GUTENBERG, B., and RICHTER, C. F. (1956), *Magnitude and Energy of Earthquakes*, Ann. di. Geofis. *9*, 1.

HARRIS, R. A., and SIMPSON, R. W. (1996), *In the Shadow of 1857—Effect of the Great Ft. Tejon Earthquake on the Subsequent Earthquakes in Southern California*, Geophysical Res. Lett. *23*, 229–232.

HIRATA, T. (1989a), *Fractal Dimension of Fault Systems in Japan: Fractal Structure in Rock Fracture at Various Scales*, Pure appl. geophys. *131*, 157–170.

HIRATA, T. (1989b), *A Correlation between the b Value and the Fractal Dimension of Earthquakes*, J. Geophys. Res. *94*, 7507–7514.

HUANG, Y., SALEUR, H., SAMMIS, C. G., and SORNETTE, D. (1998), *Precursors, Aftershocks, Criticality and Self-organized Criticality*, Europhys. Lett. *41*, 43–48.

HUANG, Y., OUILLON, G., SALEUR, H., and SORNETTE, D. (1997), *Spontaneous Generation of Discrete Scale-invariance in Growth-models*, Phys. Rev. E *55*, 6433–6447.

JAUMÉ, S. C., and SYKES, L. R. (1999), *Evolving Toward a Critical Point: A Review of Accelerating Seismic Moment/Energy Release Prior to Large and Great Earthquakes*, Pure appl. geophys., *155*, 279–306.

JONES, L. M., and HAUKSSON, E. (1997), *The Seismic Cycle in Southern California: Precursor or Response?* Geophys. Res. Lett. *24*, 469–472.

KEYLIS-BOROK, V. I., and MALINOVSKAYA, L. N. (1964), *One Regularity in the Occurrence of Strong Earthquakes*, J. Geophys. Res. *69*, 3019–3024.

KLEIN, W., and RUNDLE, J. (1993), *Comment on "Self-organized Criticality in a Continuous, Nonconservative Cellular Automaton Modeling Earthquakes*, Phys. Rev. Lett. *71*, 1288.

KNOPOFF, L., LEVSHINA, T., KEYLIS-BOROK, V. I., and MATTONI, C. (1996), *Increased Long-range Intermediate-magnitude Earthquake Activity Prior to Strong Earthquakes in California*, J. Geophys. Res. *101*, 5779–5796.

MOREIN, G., TURCOTTE, D. L., and GABRIELOV, A. (1997), *On the Statistical Mechanics of Distributed Seismicity*, Geophys. J. Int. *131*, 552–558.

NADEAU, R. M., FOXALL, W., and MCEVILLY, T. V. (1995), *Clustering and Periodic Recurrence of Microearthquakes on the San Andreas Fault at Parkfield, California*, Science *267*, 503–507.

NAKANISHI, H., SAHIMI, M., ROBERTSON, M. C., SAMMIS, C. G., and RINTOUL, M. D. (1993), *Fractal Properties of the Distribution of Earthquake Hypocenters*, J. Phys. I France *3*, 733–739.

NEWMAN, W., GABRIELOV, A., DURAND, T., PHOENIX, S. L., and TURCOTTE, D. L. (1994), *An Exact Renormalization Model for Earthquakes and Material Failure, Statics and Dynamics*, Physica D *77*, 200–216.

OKUBO, P. G., and AKI, K. (1987), *Fractal Geometry in the San Andreas Fault System*, J. Geophys. Res. *92*, 345–355.

OLAMI, Z., FEDER, H. J. S., and CHRISTENSEN, K. (1992), *Self-organized Criticality in a Continuous, Nonconservative Cellular Automaton Modeling Earthquakes*, Phys. Rev. Lett. *68*, 1244–1247.

OUILLON, G., SORNETTE, D., GENTER, A., and CASTAING, C. (1996), *The Imaginary Part of Rock Jointing*, J. Phys. I France *6*, 1127–1139.

ROBERTSON, M. C., SAMMIS, C. G., SAHIMI, M., and MARTIN, A. (1995), *The 3-D Spatial Distribution of Earthquakes in Southern California with a Percolation Theory Interpretation*, J. Geophys. Res. *100*, 609–620.

RUNDLE, J. B. (1988a), *A Physical Model for Earthquakes, I*. J. Geophys. Res. *93*, 6237–6254.

RUNDLE, J. B. (1988b), *A Physical Model for Earthquakes, II*. J. Geophys. Res. *93*, 6255–6274.

RUNDLE, J. B. (1989), *A Physical Model for Earthquakes, II*. J. Geophys. Res. *94*, 2839–2855.

RUNDLE, J. B. (1993), *Magnitude Frequency Relations for Earthquakes Using a Statistical Mechanical Approach*, J. Geophys. Res. *98*, 21,943–21,949.

RUNDLE, J. B., KLEIN, W., and GROSS, S. (1996), *Dynamics of a Traveling Density Wave Model for Earthquakes*, Phys. Rev. Lett. *76*, 4285–4288.

RUNDLE, J. B., KLEIN, W., GROSS, S., and TURCOTTE, D. L. (1995), *Boltzmann Fluctuations in Numerical Simulations of Nonequilibrium Lattice Threshold Systems*, Phys. Rev. Lett. *76*, 1658–1661.

SAHIMI, M., ROBERTSON, M. C., and SAMMIS, C. G. (1993a), *Fractal Distribution of Earthquake Hypocenters and its Relation to Fault Patterns and Percolation*, Phys. Rev. Lett. *70*, 2186–2198.

SAHIMI, M., ROBERTSON, M. C., and SAMMIS, C. G. (1993b), *Relation between the Earthquake Statistics and Fault Patterns, and Fractals, and Percolation*, Physica A *191*, 57–68.

SAHIMI, M., and ARBABI, S. (1996), *Scaling Laws for Fracture of Heterogeneous Materials and Rock*, Phys. Rev. Lett. *77*, 3689–3692.

SALEUR, H., SAMMIS, C. G., and SORNETTE, D. (1996a), *Discrete Scale Invariance, Complex Fractal Dimensions, and Log-periodic Fluctuations in Seismicity*, J. Geophys. Res. *101*, 17,661–17,677.

SALEUR, H., SAMMIS, C. G., and SORNETTE, D. (1996b), *Renormalization Group Theory of Earthquakes*, Nonlinear Processes in Geophysics *3*, 102–109.

SAMMIS, C. G., BOWMAN, D. D., SALEUR, H., HUANG, Y., SORNETTE, D., and JOHANSEN, A. (1995), *Log-periodic Fluctuations in Regional Seismicity before and after Large Earthquakes*, EOS Trans. Am. Geophys. Union, F405.

SAMMIS, C. G., SORNETTE, D., and SALEUR, H., *Complexity and Earthquake Forecasting, Reduction and Predictability of Natural Disasters, SFI Studies in the Sciences of Complexity*, vol. XXV (eds. J. B. Rundle, W. Klein, and D. L. Turcotte) (Addison-Wesley, Reading, Mass. 1996) pp. 143–156.

SORNETTE, D., and SAMMIS, C. G. (1995), *Complex Critical Exponents from Renormalization Group Theory of Earthquakes: Implications for Earthquake Predictions*, J. Phys. I. *5*, 607–619.

SORNETTE, A., and SORNETTE, D. (1990), *Earthquake Rupture as a Critical Point: Consequences for Telluric Precursors*, Tectonophysics *179*, 327–334.

SMALLEY, R. F., TURCOTTE, D. L., and SOLLA, S. A. (1985), *A Renormalization Group Approach to the Stick-slip Behavior of Faults*, J. Geophys. Res. *90*, 1894–1900.

SYKES, L. R., and JAUMÉ, S. (1990), *Seismic Activity on Neighboring Faults as a Long-term Precursor to Large Earthquakes in the San Francisco Bay Area*, Nature *348*, 595–599.

TRIEP, E. G., and SYKES, L. R. (1997), *Frequency of Occurrence of Moderate to Great Earthquakes in Intracontinental Regions: Implications for Change in Stress, Earthquake Prediction, and Hazards Assessments*, J. Geophys. Res. *102*, 9923–9948.

VARNES, D. J., and BUFE, C. G. (1996), *The Cyclic and Fractal Seismic Series Preceding an $M_b = 4.8$ Earthquake on 1980 February 14 near the Virgin Islands*, Geophys. J. Int. *124*, 149–158.

(Received October 4, 1998, revised January 21, 1999, accepted January 22, 1999)

Pure appl. geophys. 155 (1999) 335–353
0033–4553/99/040335–19 $ 1.50 + 0.20/0

Detailed Distribution of Accelerating Foreshocks before a M 5.1 Earthquake in Japan

Koshun Yamaoka,[1] Toru Ooida[1] and Yoshihiro Ueda[2]

Abstract—The M 5.1 event (May 23, 1993) which occurred in one of the most active swarm areas of Japan was preceded by foreshock activity. We obtained precise hypocenters of the foreshock-main-shock-aftershock sequences with a temporary seismic network installed just above the source region twenty days before the mainshock. The foreshocks are very unique in their accelerating activity; the acceleration in the number of foreshocks enabled us to estimate the time of the mainshock with time-to-failure analysis proposed by Voight (1988). Although substantial snow remained in the swarm area, we quickly installed the network because the time-to-failure analysis disclosed that the mainshock was impending. The temporary network provided detailed information on both the temporal and spatial distribution of the foreshock-mainshock-aftershock sequences. Foreshocks started fifty days before the mainshock and were distributed linearly at the base of the seismogenic layer with a length of 5 km and horizontal and vertical widths of about 1 km. The temporal change of the number of foreshocks is approximated by a power law, and the time of the mainshock can be estimated by extrapolating plots of the inverse of the daily number of events. An area of seismic quiescence appeared 40 hours before the mainshock and propagated with a rate of 20 m/hour. The mainshock occurred 2 km westward from the primary foreshock area. It was located at the base of the aftershock region. This process can be interpreted as source nucleation; preslip on the fault prior to the mainshock.

Key words: Time-to-failure analysis, seismicity, temporal variation.

Introduction

Detecting the generation process of earthquakes on faults is a key observation necessary for practical earthquake forecasting. The mechanism of earthquake generation has been intensively studied in terms of rock mechanics and friction physics, mostly based on laboratory experiments and numerical simulation (e.g., Dieterich, 1986, 1992; Ohnaka, 1992; Yamashita and Ohnaka, 1992). The phenomena which are expected from the experimental and theoretical studies, however, have been rarely observed in nature. This is mainly attributed to the lack of proper arrangement of sensors to detect slight changes of state in the vicinity of

---

[1] Research Center for Seismology and Volcanology, Graduate School of Science, Nagoya University, Chikusa-ku, Nagoya, 464-8602, Japan.
[2] Matsumoto Weather Station, Japan Meteorological Agency, Matsumoto, Japan. Present address: Fukui Weather Station, Japan Meteorological Agency, Fukui, Japan.

the source region. For example, the source nucleation (pre-slip) process is thought to initiate in the middle of the crust with a dimension of one tenth of the size of the main rupture. The strain changes for most inland earthquakes of M 7 class are so small that it is very hard to detect with strainmeters installed on the surface if they are not installed just above the source region.

On the other hand activity of microearthquakes sometimes exhibit the source process before major earthquakes. DODGE *et al.* (1996) analyzed immediate fore-shocks of earthquakes in California and found evidence of source nucleation. This means that microearthquakes sometimes act as a very sensitive sensor for stress or strain change. High sensitivity of microearthquake activity to such stress changes is also reported as induced seismicity caused by reservoir loading or earth tides. This is one reason why statistics of seismicity pattern are frequently used to monitor the state on earthquake faults.

Another reason for the lack of proper observations is a lack of successful estimation of the time of the mainshock. If the time of the mainshock is known beforehand precisely enough to install an appropriate and temporary observation network, even if it is too rough for practical earthquake prediction, we can obtain considerably more valuable data than we can do with the standard permanent observation network. For example, the Tokai region, Japan, is one such area where a large earthquake is thought to be impending. Fortunately, however, no large earthquakes have occurred there since the statement by ISHIBASHI (1981). Therefore, we have had very few chances to derive precise data for the earthquake generation process with the intended observation system.

In April of 1993 a M 5.1 earthquake occurred in the western region of Nagano Prefecture, Japan (Fig. 1). This earthquake was very unique because it was preceded by a very active sequence of foreshock activity. It started about fifty days prior to the mainshock and activated gradually and accelerated until the mainshock, which is commonly observed in pre-explosion seismicity of volcanoes (TOKAREV, 1973) and in some reservoir-induced earthquakes (PAPAZACHOS, 1973). The rate of increasing number of earthquakes enabled us to estimate the date of the mainshock based on the method proposed by VOIGHT (1988). The magnitude was also estimated from the dimension of the spatial distribution of foreshock activity.

Before establishing the temporary network we also submitted a statement to the Coordinating Committee for Earthquake Prediction of Japan that stated "A magnitude 5 earthquake is likely to occur in the southern area of Mt. Ontake at around April 10, 1993" (NAGOYA UNIVERSITY, 1993a). The general features of this activity were reported by NAGOYA UNIVERSITY (1993a,b). In this paper we thoroughly investigate the distribution of hypocenters of the foreshocks, referring to source nucleation models to clarify the process before the mainshock.

Observation

Having recognized the foreshock activity, we established a temporary network around the source region of the foreshocks. The distribution of the seismometer network is shown in Figure 2. There were three permanent stations in this region; two of them (MKO, TKN) were operated by Nagoya University, and the other one closest to the summit of Ontake Volcano is operated by the Japan Meteorological Agency (JMA) to monitor its volcanic activity. Four of the temporary stations (MIU, NGR, MKS, NAK) had been deployed when the foreshocks began, although the latter two were dead due to heavy snow. These stations were deployed in order to monitor the activity of Ontake Volcano which erupted two years before.

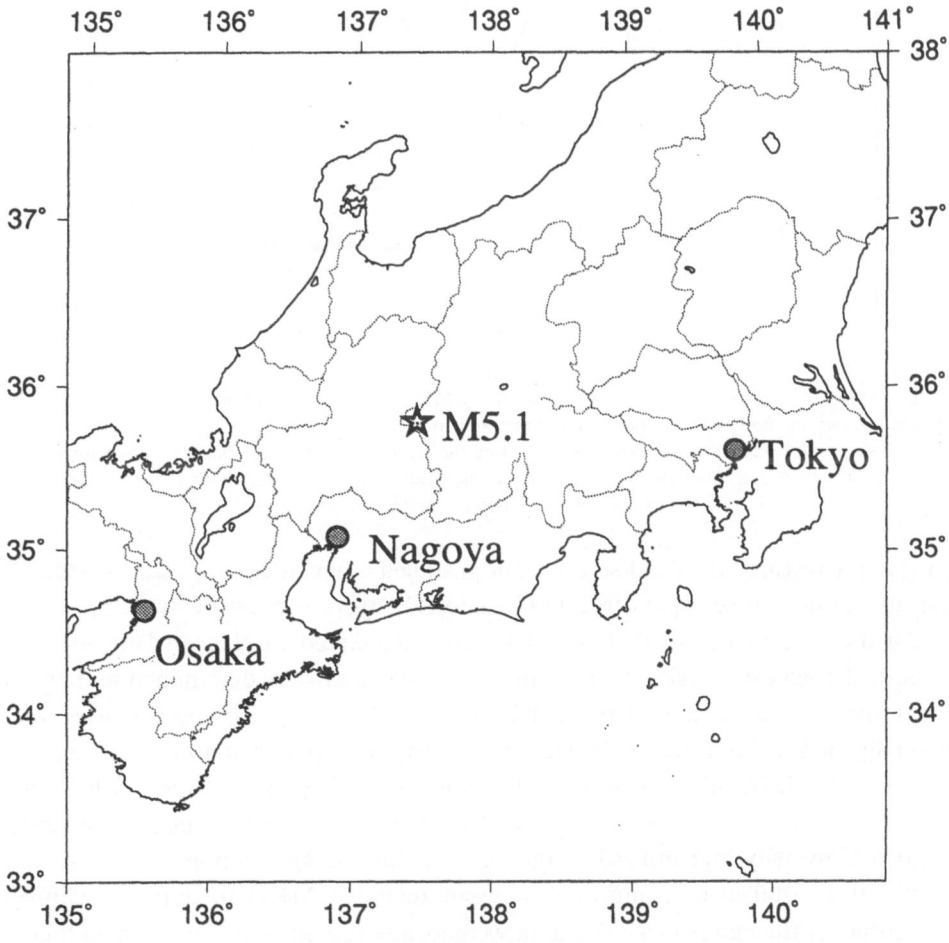

Figure 1

Location of the *M* 5.1 earthquake (April 23, 1993) studied in this paper. It occurred within one of the most active earthquake swarm regions in Japan.

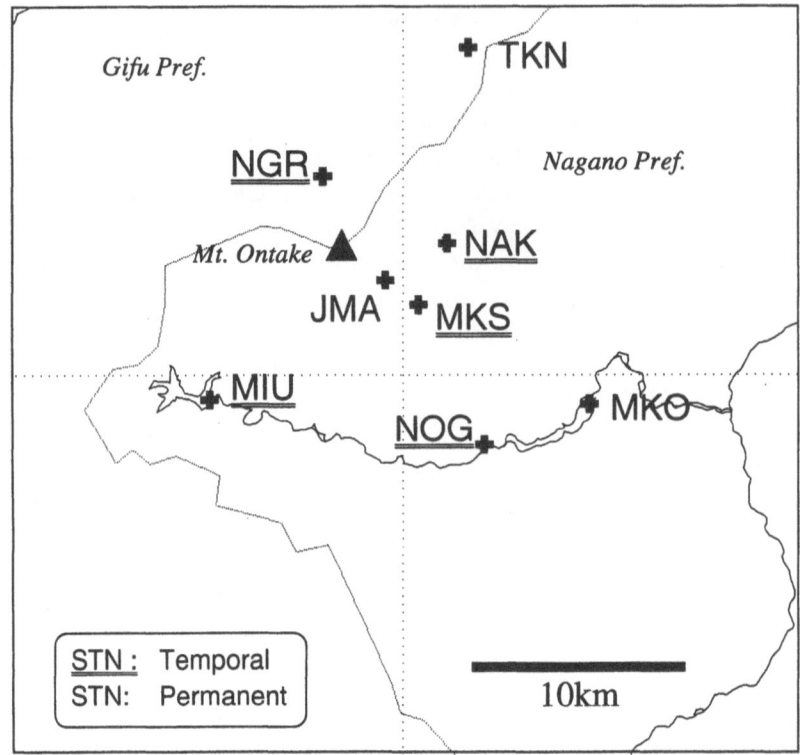

Figure 2
Station network used in this study. The stations labeled with the underlined name represent the
temporary stations, and those without underlines indicate the permanent ones. Solid line indicates the
bank of rivers and reservoirs. Dotted line indicates the border of Gifu and Nagano Prefectures. Mt.
Ontake in this map is a Quaternary volcano which is located about 100 km behind the volcanic front
of the Japanese islands.

We quickly restored the dead stations and installed a new temporary station (NOG)
just above the source region to attain a good depth resolution of hypocenters.

All the stations except that of JMA were telemetered to Nagoya University to
monitor the seismic activity. Hypocenters were automatically determined initially to
secure the up-to-date state of the activity as soon as possible. *P* and *S* phases were
manually picked later to be assured of the hypocenter distribution. The station
belonging to JMA was telemetered to Matsumoto Weather Station, where the
number of earthquakes were counted. The numbers were promptly reported to
Nagoya University and utilized in the decision for the next actions.

We also installed a strong motion seismometer at MKO to obtain unclipped
waveforms of the mainshock. The data were locally recorded in a 16-bit data logger
with a trigger. Based on the estimation of a magnitude of 5, we carefully chose the
gain of the amplifier and the trigger level. We succeeded in acquiring the entire

waveform of the mainshock without clipping. It enabled us to pick the S phase of the large event, which had been difficult with only the high-sensitivity seismic network.

Estimation of the Time of the Mainshock

Figure 3 shows the temporal variation of the daily number of earthquakes officially announced by JMA, based on the observation at the JMA station. The number of earthquakes increases day by day from the beginning of March 1993 until the mainshock. At JMA they counted the number of events whose $S-P$ times are less than 4 seconds in the routine monitoring. The station is located at the southern flank of Ontake Volcano, originally intended to monitor its volcanic activity. The volcano was quite dormant in this period and almost no seismic activity was observed related to its volcanic activity. On the other hand, the area at the southern foot of the volcano is one of the most active earthquake swarm areas in Japan. The earthquakes counted by JMA include the events in this region, but little variation is observed in any other area around the volcano during the activation period of the foreshocks. Therefore we can consider the increase of the

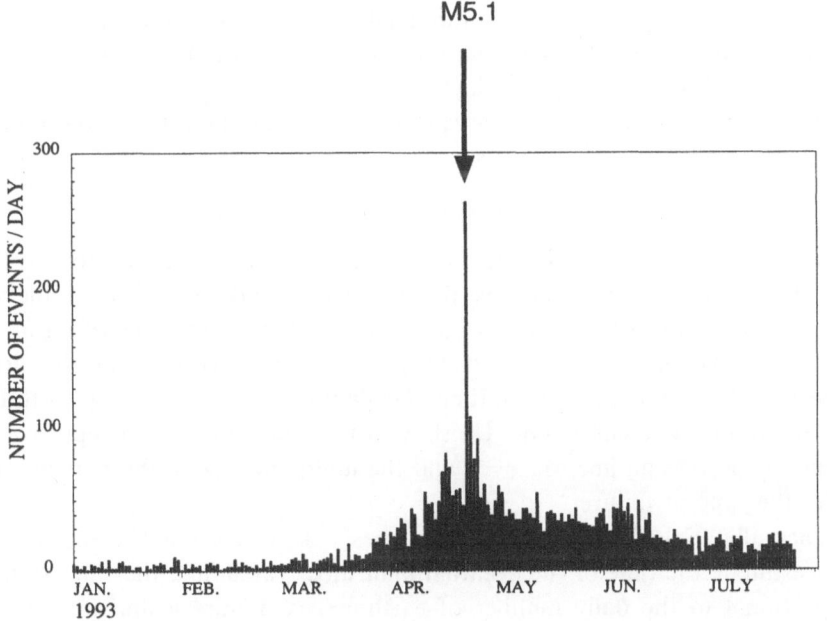

Figure 3
Daily number of earthquakes around Mt. Ontake counted by the Japan Meteorological Agency. The activity during the activation stage is attributed to the foreshocks.

earthquake frequency detected at JMA station to be the activation of the foreshocks.

Several authors have investigated activations of foreshocks before mainshocks. MOGI (1985) divided patterns of foreshock activity into two types: a discontinuous type and a continuous type. The discontinuous type is foreshock activity which is isolated in time from the mainshock-aftershock sequence. In the continuous type the activity increases until the mainshock occurs. There are very few reports of the continuous type of foreshock activity excepting earthquake sequences before explosive eruptions (TOKAREV, 1973) and reservoir-induced earthquakes (PAPAZACHOS, 1973).

JONES and MOLNAR (1979) investigated a global catalog of seismic activity and summarized the foreshock sequences. They demonstrated that the probability of foreshock occurrence increases with time empirically following a power law as:

$$n = a(\tau - t)^{-k}$$

rather than an exponential increase. In this equation, n denotes the number of earthquakes in unit time, t is the time and τ is the time of a mainshock. This is the same equation obtained by PAPAZACHOS (1973) and TOKAREV (1973) if we assume that the number of earthquakes is proportional to the strain rate in the foreshock area. The estimation of k varies from around 1 for JONES and MOLNAR (1979) to 2 for PAPAZACHOS (1973) and TOKAREV (1973). BUFE and VARNES (1993) and BUFE *et al.* (1994) applied time-to-failure analysis to major earthquakes in California and Alaska using the Benioff strain. They also formulated the accelerating seismic strain release with a power law.

These relationships are special cases of the empirical law for material failure as pointed out by VOIGHT (1988):

$$(d\varepsilon/dt)^{\alpha} \cdot (d^2\varepsilon/dt^2) = C$$

where ε is the strain of the material. For k of 1 to 2 α varies from 2 to 1.5. If α is greater than unity the strain increase follows a power law. In this case, the strain diverges to infinity at a finite time, which means failure of the material. For $\alpha = 1$ the strain increases exponentially and diverges to infinity with infinite time length.

VOIGHT (1988) showed that the time of failure can be estimated by plotting the inverse of strain rate against time. He showed that the strain rate is approximated by a linearly decreasing line for $\alpha = 2$ and the intersection with the horizontal axis provides the time of failure.

We actually applied his method to the foreshock sequence at the end of March 1993 to estimate the time of the eventual mainshock, assuming that the strain rate is proportional to the daily number of earthquakes. Figure 4 illustrates how we estimated the time of the mainshock with an inverse plot of the daily number of earthquakes. The estimated time of the mainshock is about ten days earlier than it actually occurred.

Time-to-Failure Analysis done before the Event.
(Done at the end of March, 1993)

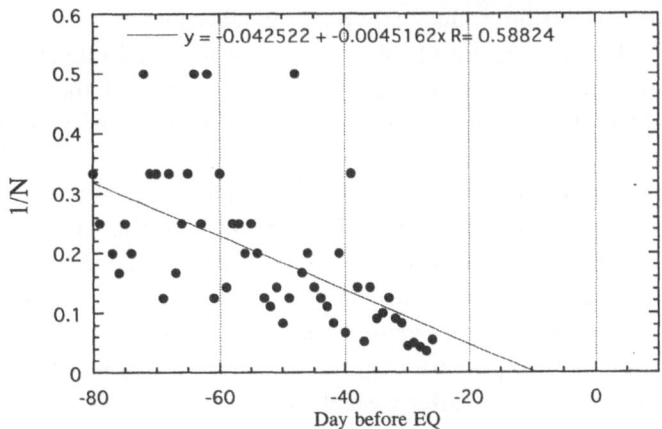

Figure 4

Time-to-failure analysis based on the VOIGHT (1988) method, assuming strain rate is proportional to the daily earthquake frequency (N). The time of failure is given by the time of the intersection of the t-axis with the extrapolation of the plot of $1/N$.

The value α is not always 2 as mentioned before, which appeared to be the case in the foreshock sequence. Linear extrapolation is still possible for $\alpha \neq 2$ ($\alpha > 1$) by adopting the transformation:

$$e = \varepsilon^{1-\alpha}.$$

Taking α appropriate for the foreshock increase and using e as the vertical axis, we can estimate the time of the failure with the same method as VOIGHT (1988). We adopted $\alpha = 1.5$, for example, for the linear extrapolation technique for the foreshock data. This value corresponds to the inverse of the square root of the frequency of occurrence, and yields a better estimation as shown in Figure 5.

In Figure 3 one may notice that the decay of aftershocks is apparently slower than that expected from Omori's law. This reflects an increase of seismic activity in the region where JMA officially counts the number of earthquakes, namely, the background swarm activity biased the count during the aftershock period. The actual decay of the aftershocks is shown in the later chapter and Figure 12.

Hypocenter Distribution

Figures 6 and 7 delineate the hypocenter distribution of foreshocks and after-shocks, respectively. The hypocenters plotted are the events after the NOG is installed just above the source region. The hypocenter locations are calculated with the picking of fixed five stations; MKO, NOG, TKN, NAK and MIU. This gives good resolution of relative hypocenter location. We used the HYPOMH (HIRATA and MATSU'URA, 1987) program for hypocenter determination with a one-dimensional layered velocity structure (Table 1) which is modeled referring to a seismic tomography analysis (HIRAHARA *et al.*, 1992). The errors of the hypocenters are estimated to be less than 150 m and 300 m in the horizontal and vertical directions, respectively.

The epicenters of the events during the last twenty days before the mainshock (Fig. 6) are linearly distributed from ENE to WSW with the length of 7 km. This direction is consistent with that of maximum shear in this region and also coincides with that of the fault of the *M* 6.8 earthquake which occurred in this area in 1984 (OOIDA *et al.*, 1989). Most of the events occurred in the eastern 5 km of the region. In contrast, a few tens of events occurred in the westernmost part, where the mainshock occurred later. Hereafter we call the eastern part of the foreshock region the primary foreshock region. The focal depths are around 9 km and deepening westward; 8 km at the eastern end of the region and 10 km at the western end. The spatial variation of the focal depth is consistent with that of the background swarm activity in this region (Fig. 8).

In Figure 8 the fore- and aftershocks are plotted on the depth contour of the lower limit of the usual earthquakes in this region. The lower limit is given by the average plus one standard deviation of the focal depth of the events in each 2-km grid. In this figure we used the hypocenter catalog of Nagoya University, based on its permanent station network. The depth of the lower limit is obtained more or less shallower than the focal depth of the earthquakes associated with the *M* 5.1 event.

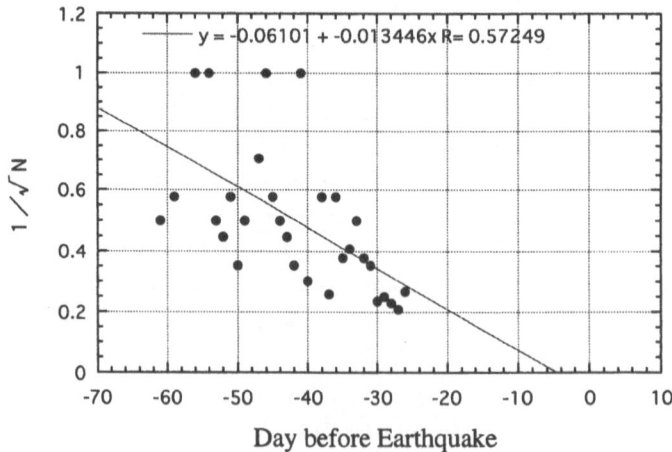

Figure 5

Time-to-failure analysis based on the VOIGHT (1988) method. In this figure we used the daily number of earthquakes from which the average number of background activity is subtracted, and the square root of $1/N$ is plotted against time.

Hypocenters of foreshocks
Apr. 11 - Apr 22, 1993

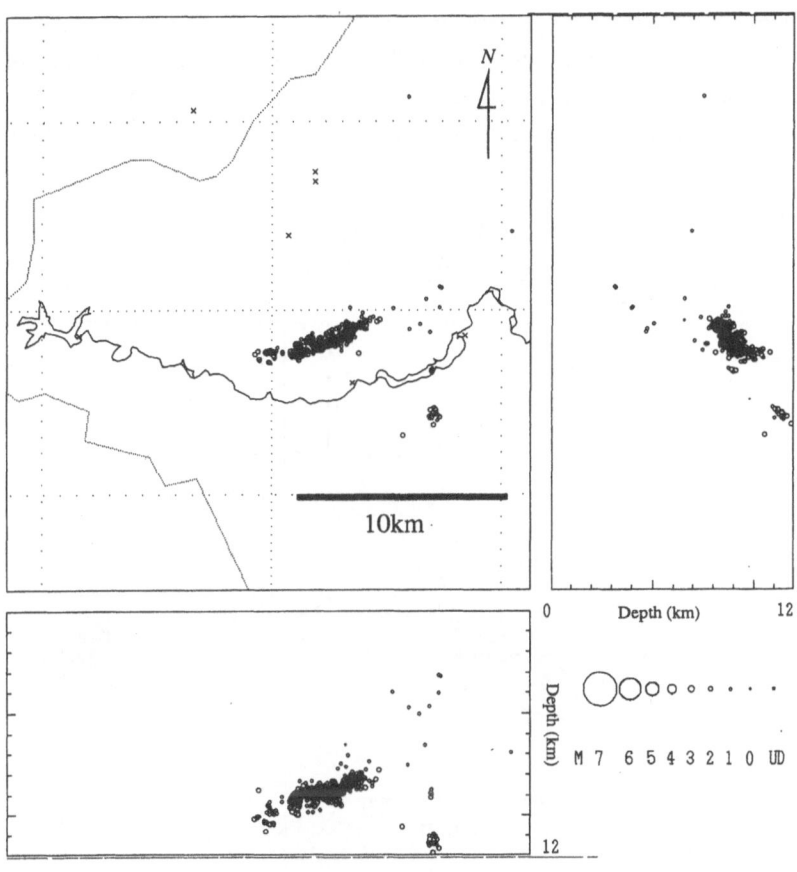

Figure 6
Hypocenter distribution of the foreshocks.

This may be primary attributed to the difference in the station distribution between the permanent and temporary network for this region.

Figure 7 shows the hypocenter distribution of the mainshock and the aftershocks. The mainshock occurred in the western part of the foreshock region, apart from the primary foreshock region. It is also located at the deepest point in the source area. In the standard hypocenter determination the depth of the mainshock is often questionable because it is sometimes determined without proper S-picking. Without S-picking focal depth is poorly determined due to the trade-off between depth and origin time. The signal of the high-sensitive seismometer is usually clipped for a large event and S phases are often impossible to pick. To avoid the difficulty we installed a strong-motion seismometer at the nearby station (MKO), and the S arrival is clearly picked from the unclipped data.

Temporal Variation of Hypocenter Distribution

In this section we examine the temporal variation of hypocenter distribution. The hypocenters are determined with five fixed stations to exclude the effect of the difference of station combination. Figure 9 shows the temporal variation of focal depth and epicenters along the linear trend of earthquakes. In the plot it is remarkable that the foreshocks are concentrated in the primary part in the eastern

Hypocenters of Aftershocks
Apr. 23 - May 27, 1993

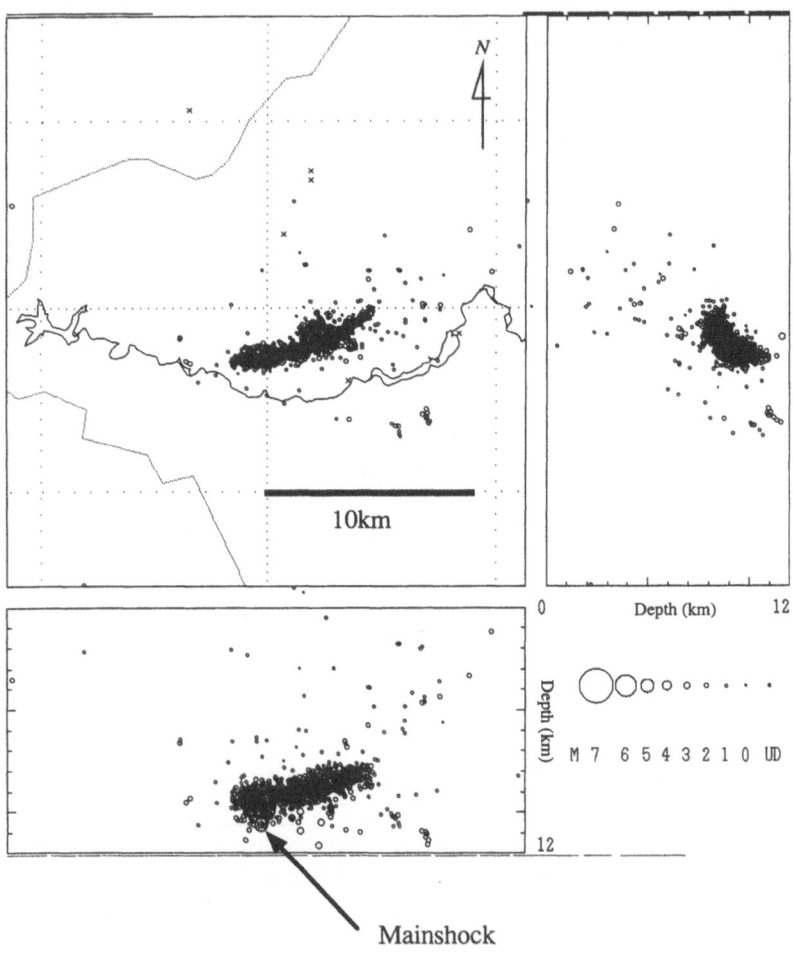

Figure 7
Hypocenter distribution of the mainshock and aftershocks. The mainshock occurred at the deepest point in the activity.

Table 1

P-wave velocity structure used in this study. S-wave velocity is obtained asuming V_p/V_s *to be 1.732*

Thickness of layer	P-wave velocity
2 km	4.7 km/s
20 km	5.95 km/s
5 km	6.7 km/s
semi-infinite	7.9 km/s

5 km, and the mainshock occurred 2 km apart from it. Just after the mainshock many events occurred around it, primarily in the western half of the foreshock area. The activity does not manifest a sudden increase in the easternmost part of the foreshock area, meaning that the rupture of the mainshock did not propagate across the entire foreshock area. In the depth section, it is very clear that the mainshock occurred at the deepest point and aftershocks distribute densely above the hypocenter of the mainshock extending to the top of the foreshock region.

Next we shall closely examine the time sequence of the foreshock activity. Figure 10 displays a close-up section of the temporal variations of hypocenters in the same region as in Figure 9. In the plot of epicenter variation, it is observed that the foreshocks propagated both eastward and westward. Westward propagation, which register a rate of about 0.04 km/day, is more obvious than the eastward propagation. Just before the mainshock a seismic quiescence is recognized in the western part of the primary foreshock region. This quiescence appeared two days

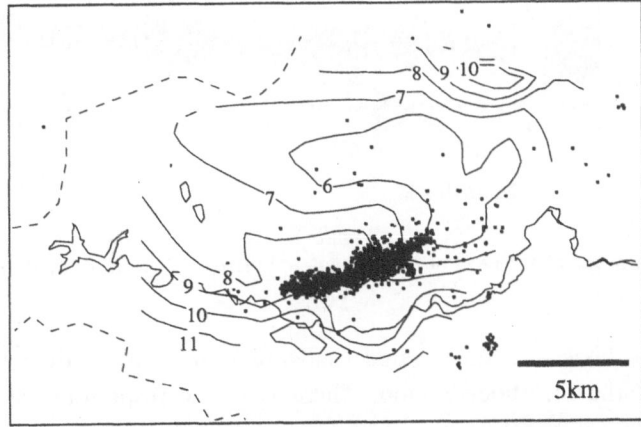

Figure 8

Epicenter distribution of the earthquakes associated with the *M* 5.1 event plotted on the depth contour of the lower limit of earthquake occurrence. The general trend of hypocenter depths studied in this paper coincides with the distribution of the lower limit of hypocenters in this region.

Time Sequence of earthquakes
(Apr. 10- May. 27, 1993)
Using Fixed 5 stations

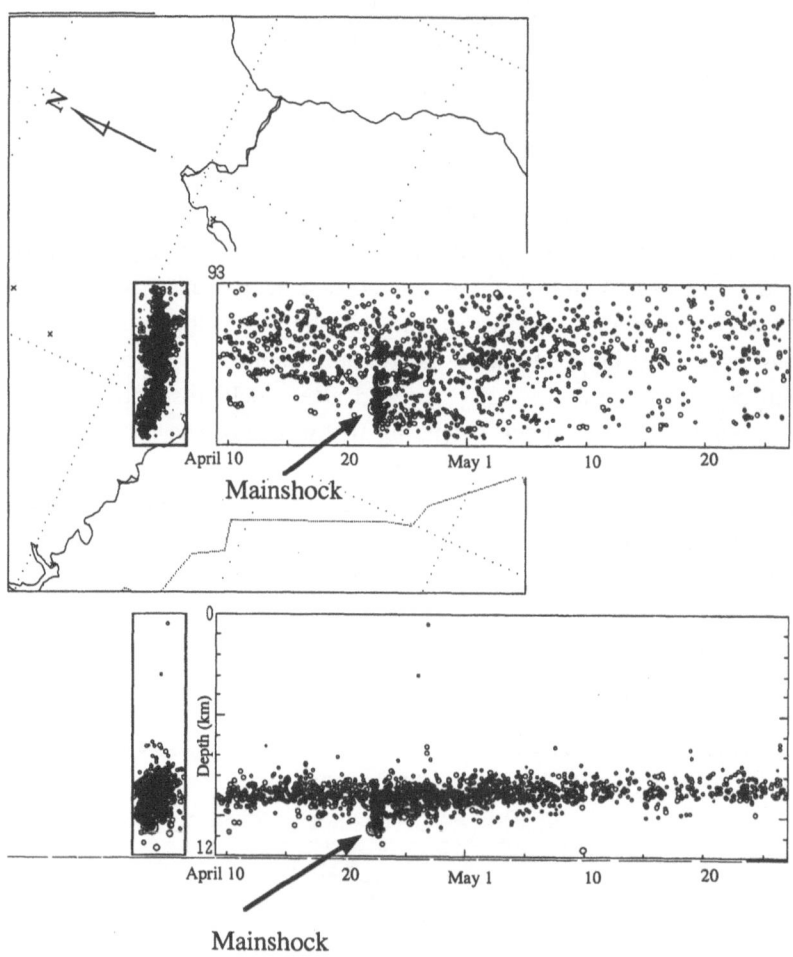

Figure 9
Temporal variation of the hypocenters of fore-, main- and aftershocks from April 10 through May 27, 1993.

before the mainshock and propagated eastward more rapidly than the westward propagation of the foreshock region. There is no corresponding size of seismic quiescence in the foreshock sequence, indicating that it did not appear by chance.

In the depth plot one may see that the hypocenters deepened until the eventual mainshock occurred. Since the hypocenters are deepening westward, it is difficult to recognize the depth change of the hypocenter at a certain point in the cross-sectional projection as shown in Figure 9. It is important to know whether or not the focal depth changed, because the temporal change of depth may represent that of the strain rate. In the ductile region, failure is governed by the creep law and the strength increases with the increase of the strain rate, resulting in the deepening of brittle-ductile transition depth.

In Figure 11 we depict the temporal variation of the focal depth of the events in the individual rectangles shown in the epicenter map. In all sections no obvious depth changes are seen in the foreshock sequence. In contrast, an abrupt increase of deeper events occurred at the time of the mainshock in most of the sections. This

Time Sequence of Foreshocks
(Apr. 10 - Apr, 24, 1993)

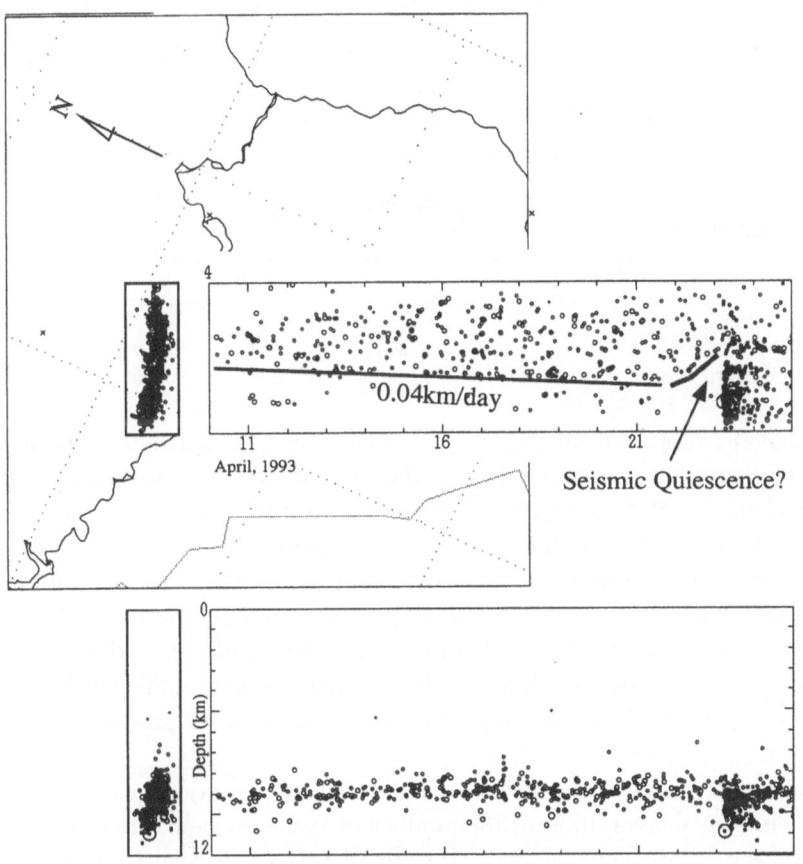

Figure 10

Temporal variation of the hypocenters of foreshocks from April 10 through April 24, 1993. Primary foreshock region spreads in both directions, particularly obvious is westward spreading with the rate of 0.04 km/day. A region of seismic quiescence is observed immediately before the mainshock.

Temporal Variation of Focal Depths

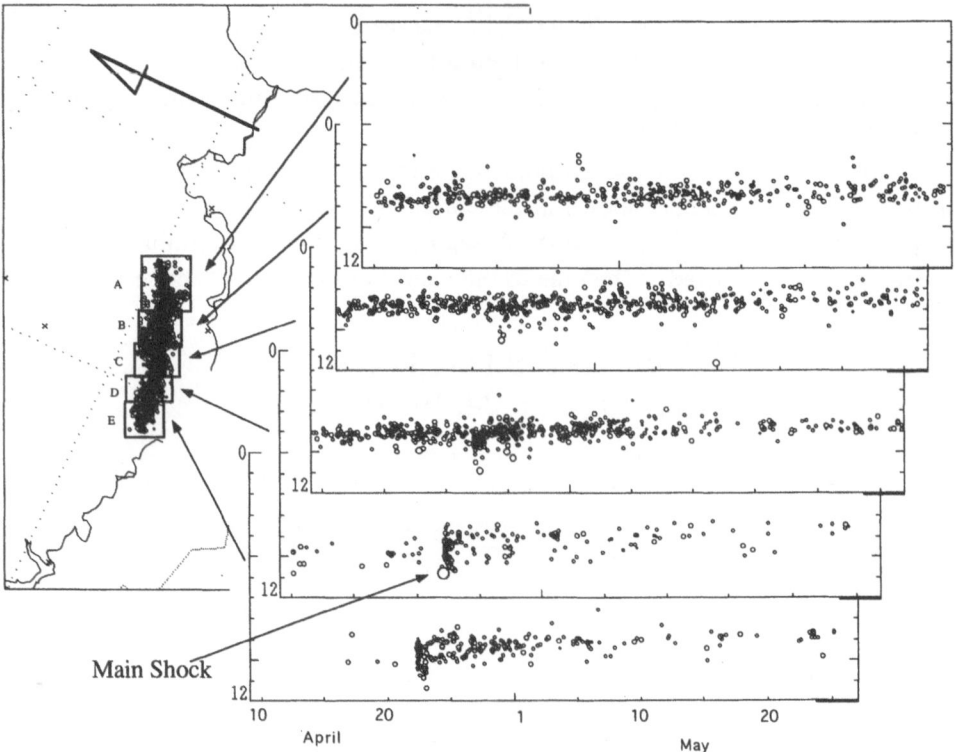

Figure 11
Temporal distribution of focal depths in the rectangles shown in the epicenter map on the left.

means that the rupture propagated in the area to include that below the foreshock region. Deeper events corresponding to the time of the mainshock are clearly seen in all sections except A, indicating that the main rupture did not propagate to this section. Aftershocks in the deeper part decrease with time in every section. In the entire period slight shallowing of hypocenter is seen in the section in which numerous foreshocks occurred. For example, in section A the mean depth is just above 9 km and slightly below 8 km at the start and the end of the period, respectively. Also, in sections B and C hypocenters became shallower by about 0.5 km in this period. This may represent the reduction of strain rate after the mainshock.

In order to manifest the temporal variation of activity in each region more clearly, we show the variation of the number of events in each one in Figure 12. In this figure the events whose magnitudes exceed 1.5 are counted in the same period in Figure 11. In regions C through E a sudden increase in the number of events is observed at the time of the mainshock, followed by a gradual decrease of after-shocks. In contrast, gradual increase and decrease of the activity which peaked at just after the mainshock are observed in regions A and B, where the main rupture may not have propagated. This is also supported by the fact that the decay rate is apparently slower in regions A and B than in the other regions.

Discussion

There are many possible interpretations for this foreshock activity. Here we attempt to explain the temporal change of the foreshock activity in terms of source nucleation of earthquakes. We try to apply a representative model of nucleation typically proposed by OHNAKA (1992). In his model nucleation starts where the resistance to rupture growth has a minimum value on the fault surface. Once nucleation starts the stress in the ruptured region is released, however accumulated in the surrounding region. If the nucleation starts in the brittle region of the crust, foreshocks are expected in the nucleation zone in the form of quasi-static rupture propagation. Sometimes a pronounced lull in foreshock activity appears just before the mainshock in the region in which the stress breakdown (or slip-weakening) has finished. The quasi-static rupture propagates both horizontally and vertically. When

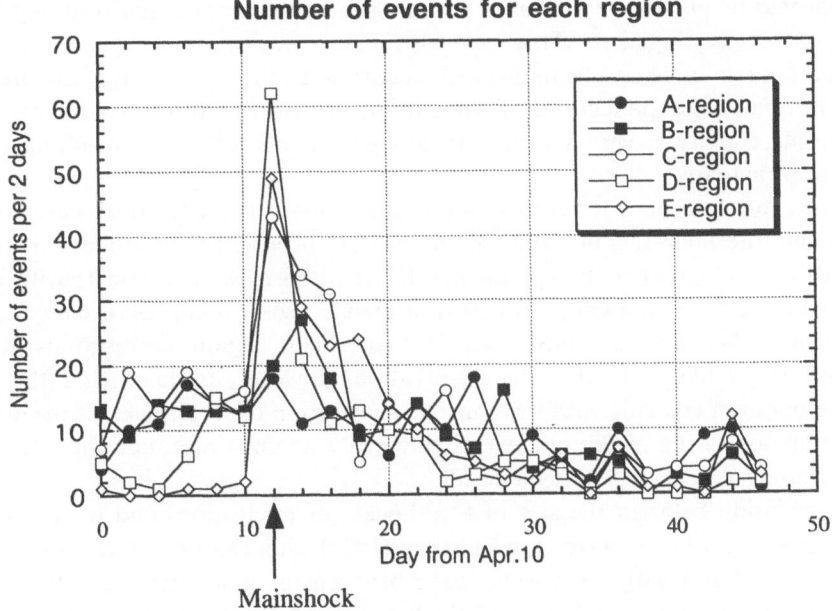

Figure 12
Temporal variation of the number of earthquakes in each region indicated in the epicenter map in Figure 11. The number of events above 1.5 is counted for every two days in the same period shown in Figure 11. The threshold of the magnitude 1.5 is chosen so that the Gutenberg-Richter relation holds above it in the frequency-magnitude distribution.

it propagates down to the base of the brittle seismogenic layer, where the friction is strongest, the main rupture starts.

The temporal and spatial distribution of the foreshocks revealed in this paper can be interpreted following the above model as shown in Figure 13. In the study region, nucleation starts somewhere above the hypocenter of the mainshock to the west of the primary foreshock region. Some events observed in the western part of the foreshock region are thought to be so-called immediate foreshocks which occur during the propagation of nucleation. Many small events in the primary foreshock region in the eastern 5 km are induced by minor stress change due to the nucleation; the activation is observed as an increase of the number and the area of the earthquakes. The nucleation progresses gradually and is finally intruded into the foreshock region, which results in the reduction of the stress level. This corresponds to the seismic quiescence observed just before the mainshock. Main rupture occurred when the nucleation reached the deepest part in which strength is maximum.

From the above discussion the activity may not be considered as a simple foreshock-mainshock-aftershock sequence but rather a combination of a foreshock-mainshock-aftershock sequences in the western part and a swarm activity in the eastern part. Both of them should influence each other by means of stress variation. The process can be modeled in two ways; the swarm causes the foreshock-mainshock-aftershock sequence or *vice versa*. In the former model the earthquake in the swarm are triggered by previous earthquakes, leading to stress increase and resulting in the activation of earthquakes. This can also cause nucleation to proceed in the neighboring region where the mainshock occurred. In the latter model, the progress of nucleation causes the increase of stress in the neighboring foreshock region where the seismic activity is very sensitive to a slight change of stress, resulting in the increase of activity.

Before the mainshock we estimated the magnitude of the mainshock based on the size of the foreshock region. Since the foreshocks occurred with a linear trend, we applied the analogy of rock experiments. In standard rock experiments with intact rock, AEs are becoming concentrated in a planar region obliquely oriented across a specimen. We first assumed that the foreshock region corresponds to the concentrated activity of AE. The observation, however, revealed that the main rupture does not coincide with the concentrated region of foreshocks. It means that our estimation of the magnitude was theoretically wrong—although the estimation of magnitude was successful as a result.

The relation between the size of foreshocks (or nucleation) and mainshocks is summarized by ELLSWORTH and BEROZA (1995) and DODGE *et al.* (1996). The former estimated it with the seismic wave between the first arrival and the abrupt increase in moment acceleration, and the latter estimated it with the distribution of foreshocks. Following their relation 5 km of foreshock region in this study corresponds to an earthquake of M 7 or larger, in contrast to the actual event of M 5.1. As mentioned before, it is interpreted that the nucleation proceeded beyond the primary foreshock region. Therefore the region in which accelerating activity is

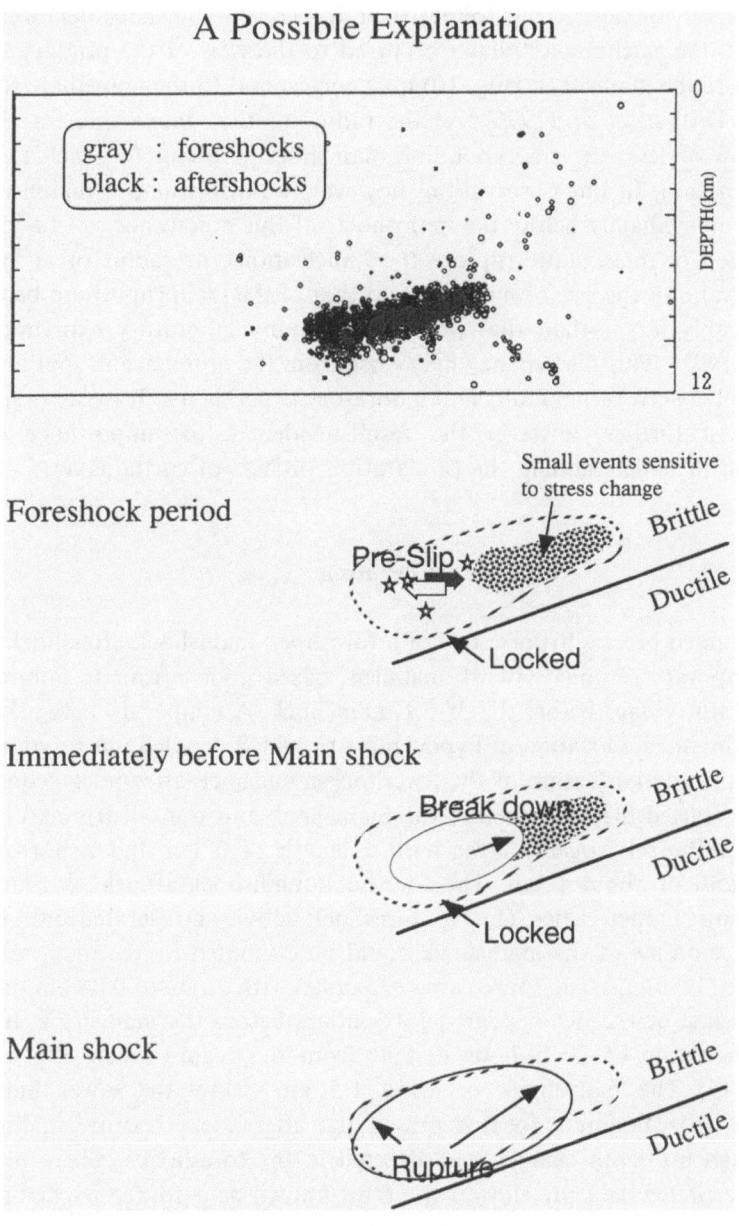

Figure 13
Schematic illustrations showing the interpretation of the activity in terms of the source nucleation process. Top figure delineates the vertical cross section of hypocenters for foreshocks and aftershocks to display the relative locations. Bottom three figures show an interpretation of the process during the foreshock activity. During the foreshock period pre-slip proceeds in the neighboring region of primary foreshocks, where earthquakes are triggered with a slight increase of stress due to pre-slip. In the pre-slip region several events occurred associated with the slip, though the bottom part is locked. Immediately before, the mainshock region of pre-slip (break-down) propagated into the primary foreshock region, reducing the activity due to stress release. Main rupture started at the bottom of the region where fault has been locked.

observed does not necessarily follow their foreshock-mainshock relation. On the other hand, the earthquakes which occurred to the west of the primary foreshock region before the mainshock (Fig. 10) may correspond to the immediate foreshocks defined in DODGE *et al.* (1996). As the radius of the "immediate foreshocks" is about 1 km or less, the corresponding mainshock is about 6, which results in a better estimation. In our observation, however, we also found a region of seismic quiescence immediately before the mainshock. If this quiescence can be interpreted as the region of quasi-static rupture (i.e., nucleation), its radius of at least 2 km makes the ratio of the size of nucleation to that of the main rupture to be 0.4 which is considerably larger than that observed in many laboratory experiments (e.g., OHNAKA, 1992). This discrepancy may arise from the difference in spatial variation of strength between actual faults and laboratory experiments. It is vastly speculative to discuss it further, however the result evidences the importance of actual observation in understanding the preparation process of earthquakes.

Conclusion

We obtained precise hypocenters of a foreshock-mainshock-aftershock sequence with a temporary seismic network installed, based upon a time-to-failure analysis performed thirty days before the M 5.1 mainshock. A temporary network intended to obtain the precise location of hypocenters provided detailed information on both temporal and spatial feature of the foreshock-mainshock-aftershock sequence. The foreshocks started fifty days before the mainshock and were distributed linearly at the base of the seismogenic layer with a length of 5 km and a horizontal and vertical width of about 1 km. The foreshock-mainshock-aftershock sequence has the following characteristics: (1) The foreshock activity accelerated until the mainshock and the time of the mainshock could be estimated by the method proposed by VOIGHT (1988). (2) The source area expanded with a rate of 0.04 km/day. (3) An area of seismic quiescence appeared forty hours before the mainshock and propagated with a rate of 20 m/hour distant from the region where the mainshock occurred. (4) The mainshock occurred 1.5 km below the lower limit of the foreshocks. (5) Maximum focal depth of the aftershocks became shallower with time, though no depth change was detected in the foreshocks. There observation can be interpreted as a pre-slip on the fault known as a source nucleation.

Acknowledgements

The staff of the Research Center for Seismology and Volcanology, Nagoya University supported the urgent installation of the temporary network and quickly responded to my statement of the estimation of earthquake occurrence. We also express appreciation to Max Wyss, Stephan McNutt, Stefan Wiemer and Naoyuki Fujii for their useful discussions. We likewise thank two anonymous reviewers for their critical comments.

REFERENCES

BUFE, C. G., and VARNES, D. J. (1993), *Predictive Modeling of the Seismic Cycle of the Greater San Francisco Bay Region*, J. Geophys. Res. *98*, 9871–9883.

BUFE, C. G., NISHENKO, S. P., and VARNES, D. J. (1994), *Seismic Trends and Potential for Large Earthquakes in the Alaska-Aleutian Region*, Pure appl. geophys. *142*, 83–99.

DIETERICH, J. H., *A model for the nucleation of earthquake slip*. In *Earthquake Source Mechanics*, M. Ewing Ser. 6 (eds. Das, S., Boatwright, J., and Scholz, C. H.) (American Geophysical union, Washington D.C. 1986) pp. 297–332.

DIETERICH, J. H. (1992), *Earthquake Nucleation on Faults with Rate- and State-dependent Strength*, Tectonophysics *211*, 115–134.

DODGE, D. A., BEROZA, G. C., and ELLSWORTH, W. L. (1996), *Detailed Observation of California Foreshock Sequences: Implications for the Earthquake Initiation Process*, J. Geophys. Res. *101*, 22,371–22,392.

ELLSWORTH, W. L., and BEROZA, G. C. (1995), *Seismic Evidence for Earthquake Nucleation Phase*, Science *268*, 851–855.

HIRAHARA, K., *et al.* (1992), *Three-dimensional P- and S-wave Velosity Structure in the Focal Region of the 1984 Western Nagano Prefecture Earthquake*, J. Phys. Earth *40*, 343–360.

HIRATA, N., and MATSU'URA, M. (1987), *Maximum-likelihood Estimation of Hypocenter with Origin Time Eliminated Using Nonlinear Inversion Technique*, Phys. Earth Planet. Int. *47*, 50–61.

ISHIBASHI, K., *Specification of a soon-to-occur seismic faulting in the Tokai district, central Japan, based upon seismotectoics*. In *Earthquake Prediction: An International Review*, M. Ewing Ser. 4 (eds. Simpson, D., and Richards, P.) (American Geophysical Union, Washington D.C. 1981) pp. 297–332.

JONES, L. M., and MOLNAR, P. (1979), *Some Characteristics of Foreshocks and their Possible Relationship to Earthquake Prediction and Premonitory Slip Faults*, J. Geophys. Res. *84*, 3596–3608.

MOGI, K., *Earthquake Prediction* (Academic Press 1985) 355 pp.

NAGOYA UNIVERSITY (1993a), *Prediction of M 5.1 Earthquake at Western Nagano Prefecture*, Rep. Coordinating Committee for Earthquake Prediction *50*, 141–145 (in Japanese).

NAGOYA UNIVERSITY (1993b), *The 1993 Seismic Activity in the Southeastern Foot of Mt. Ontake*, Rep. Coordinating Committee for Earthquake Prediction *50*, 132–140 (In Japanese).

OHNAKA, M. (1992), *Earthquake Source Nucleation: A Physical Model for Short-term Precursors*, Tectonophysics *211*, 149–178.

OOIDA, T., YAMAZAKI, F., FUJII, I., and AOKI, H. (1989), *Aftershock Activity of the 1984 Western Nagano Prefecture Earthquake, Central Japan, and its Relation to Earthquake Swarms*, J. Phys. Earth. *37*, 401–416.

PAPAZACHOS, B. C. (1973), *Foreshocks and Earthquake Prediction*, Tectonophysics *28*, 213–226.

TOKAREV, P. I. (1973), *Forecasting Volcanic Eruptions from Seismic Data*, Bull. Volcanology *35*, 243–250.

VOIGHT, B. (1988), *A Method for Prediction of Volcanic Eruptions*, Nature *332*, 125–130.

YAMASHITA, T., and OHNAKA, M. (1992), *Precursory Surface Deformation Expected from a Strike-slip Fault Model into which Rheological Properties of the Lithosphere are Incorporated*, Tectonophysics *211*, 179–199.

(Received July 6, 1998, revised December 6, 1998, accepted December 11, 1998)

Pure appl. geophys. 155 (1999) 355–379
0033–4553/99/040355–25 $ 1.50 + 0.20/0

❙Pure and Applied Geophysics

Foreshock Occurrence Rates before Large Earthquakes Worldwide

PAUL A. REASENBERG[1]

Abstract—Global rates of foreshock occurrence involving shallow $M \geq 6$ and $M \geq 7$ mainshocks and $M \geq 5$ foreshocks were measured, using earthquakes listed in the Harvard CMT catalog for the period 1978–1996. These rates are similar to rates ones measured in previous worldwide and regional studies when they are normalized for the ranges of magnitude difference they each span. The observed worldwide rates were compared to a generic model of earthquake clustering, which is based on patterns of small and moderate aftershocks in California, and were found to exceed the California model by a factor of approximately 2. Significant differences in foreshock rate were found among subsets of earthquakes defined by their focal mechanism and tectonic region, with the rate before thrust events higher and the rate before strike-slip events lower than the worldwide average. Among the thrust events a large majority, composed of events located in shallow subduction zones, registered a high foreshock rate, while a minority, located in continental thrust belts, measured a low rate. These differences may explain why previous surveys have revealed low foreshock rates among thrust events in California (especially southern California), while the worldwide observations suggest the opposite: California, lacking an active subduction zone in most of its territory, and including a region of mountain-building thrusts in the south, reflects the low rate apparently typical for continental thrusts, while the worldwide observations, dominated by shallow subduction zone events, are foreshock-rich.

Key words: Foreshock, foreshock rates, earthquake clustering.

Introduction

Short-term earthquake clustering, including the occurrence of foreshocks and aftershocks, is a widely observed phenomenon in shallow crustal seismicity. Nearly half of all earthquakes (most of them aftershocks) are included in short-term clusters (REASENBERG, 1985; DAVIS and FROHLICH, 1991; OGATA et al., 1995). Moderate or strong earthquakes are occasionally followed by stronger shocks nearby within a few days. This clustering makes possible short-term probabilistic earthquake forecasts after any earthquake based on triggered stochastic models (KAGAN and KNOPOFF, 1987) in which the occurrence times of the triggered events are Poissonian (with varying rate), while their rate, magnitude distribution and geographic distribution are empirically determined from the clustering behavior observed in the historic seismicity. After a moderate earthquake in an urbanized

[1] U.S. Geological Survey, 345 Middlefield Road, Menlo Park, California 94025, U.S.A.

area, the increased probability of an imminent, larger earthquake is of great concern. For example, after a moderate ($M = 5 \sim 6$) earthquake in either San Francisco or Los Angeles areas, where large earthquakes are considered likely in the long term, there follows a transient probability gain for a ($M \geq 7$) earthquake of $10^2 \sim 10^3$, relative to the respective regional long-term probabilities. In absolute terms, the conditional probability of such an earthquake ranges from a few tenths of a percent to a few percent, and has a half-life of about 1 day. The corresponding, long-term daily probabilities in these areas are approximately 0.01% and 0.004% per day, respectively.

Two approaches have been taken to develop triggered stochastic models of earthquake occurrence. The first is based directly on the observed frequency of foreshock-mainshock (*fs-ms*) pairs. Empirical estimates of the *fs-ms* pairing rate and the distribution of *fs-ms* magnitude differences have been made with numerous earthquake catalogs (JONES and MOLNAR, 1979; VON SEGGERN, 1981; JONES, 1984; BOWMAN and KISSLINGER, 1984; LINDH and LIM, 1995; ABERCROMBIE and MORI, 1996; MICHAEL and JONES, 1998). These measurements, which characterize the clusters' transient, pair-wise interactions, are analogous to the *a*-value and *b*-value in the Gutenberg-Richter (time-independent, noninteracting events) seismicity model. Such foreshock studies provided the basis for conditional probability estimates for the next Parkfield, California, earthquake (MICHAEL and JONES, 1998), and for characteristic earthquakes on selected fault segments in California (AGNEW and JONES, 1991).

In another approach, REASENBERG and JONES (1989, 1994) introduced a stochastic model of aftershock occurrence based on the observed rate (*a*-value), magnitude distribution (*b*-value) and temporal decay (*p*-value) of 62 aftershock sequences following $M \geq 5$ mainshocks in California. They assumed self-similarity of clustering in their "California generic model," and allowed the modeled "aftershocks" to take on magnitudes larger than their "mainshock"—in effect, applying an aftershock model to foreshock-mainshock pairing. This approach was advantageous in that the model and its uncertainty were readily determined (even in a limited region such as California) owing to the large number of recorded mainshock-aftershock sequences. To estimate the expected numbers of $M > 3$ aftershocks following the 1994 Northridge, California earthquake, the California generic model worked reasonably well (REASENBERG and JONES, 1994). However, in the type of application for which the interest is greatest—the case of potential $M > 5$ foreshocks prior to a larger earthquake in California—the model has not been directly tested through comparisons to observed earthquake occurrence, because insufficient numbers of such pairs have been recorded. Recently, the question was raised as to whether the California generic model accurately calculates the average, short-term conditional probability of large earthquakes in California (CALIFORNIA EARTHQUAKE PREDICTION EVALUATION COUNCIL, 1977).

Here, I use a worldwide earthquake catalog to estimate rates of *fs-ms* pairing and compare these rates to the corresponding probabilities calculated with the California generic model. Because the goal is to test a model derived from (and applied to) California seismicity, which is largely confined to the upper crust, I use only shallow earthquakes in the global catalog. The empirical foreshock rates are determined separately for mainshocks with $M \geq 6$ and $M \geq 7$, for mainshocks with thrust, normal and strike-slip mechanisms. The observed rates are compared to the California generic model probabilities and their ranges of uncertainty. Then, I examine the magnitude differences among the foreshock-mainshock pairs involving $M > 7$ mainshocks in the Harvard catalog, and attempt to use them to constrain a model for their distribution.

Data

Data for this study were taken from the Harvard catalog of centroid moment tensor solutions for large earthquakes (DZIEWONSKI *et al.*, 1987). For this study, I use M_s for magnitude when it is given (90% of the shallow events), and M_b when M_s is not given (10% of the shallow events), and refer to this magnitude simply as magnitude (M) (Fig. 1). The Harvard CMT catalog is complete from 1977–1996 for $M_w \geq 5.3$ or $M_w \geq 5.5$ events (KAGAN and KNOPOFF, 1980, and this study). Within the magnitude range $M = 5$ to $M = 5.5$, M and M_w are in close agreement, having a maximum cumulative difference of less than 0.1, so that the estimated levels of completeness for M_w may be assumed for M. For this study I selected earthquakes with $M \geq 5.0$, below the completeness levels of the catalog. This choice requires justification. The motivation for it stems from the ultimate aim of the study, which is to test the assumption that an aftershock model (REASENBERG and JONES, 1989) may be used to forecast larger earthquakes (i.e., may be used as a foreshock-mainshock model). Because such an application is currently carried out by the USGS for all $M \geq 5$ earthquakes in California, I tailored the study to this magnitude range. The expected effect of choosing the foreshock magnitude cutoff below the completeness level is to underestimate the actual foreshock rates. I estimated with a graphical method (straight-line fit to log cumulative counts between $M = 5.5$ and $M = 6.2$) that approximately 28% of the earthquakes ($M \geq 5.0$) are missing from the Harvard catalog. Hence, the retrospective foreshock rates presented below may be underestimated by this amount.

For the 5695 events ($M \geq 5.0$; depth ≤ 50 km) taken from the Harvard CMT catalog, I used the hypocentral locations listed as having originated from either NEIS or ISC. From 1977–1992 I used only ISC hypocenters (SMITH, 1995). Between 1993 and 1996, I used the combination of ISC and NEIC hypocenters listed in the Harvard catalog, which consists primarily of NEIS locations. Because many of the depths are fixed or poorly determined, depth information was used only in the selection of events to be included in the study; calculated distances between events are epicentral distances. The choice of a 50-km depth cutoff, while somewhat arbitrary, accommodates our desire to work with "shallow" events and includes the fixed depths of 10 and 33 km assigned to numerous events in the Harvard catalog for which more precise depth determination was impossible. Aside from the cut at 50 km, depths were not considered in this study.

Definition of Foreshocks

Foreshock-mainshock pairs used in this study were defined according to the following protocol:

1. The magnitudes of the first and second events in each pair (M_f and M_m, respectively) exceed fixed thresholds: $M_f \geq M_f^{\min}$ and $M_m \geq M_m^{\min}$.
2. The second event is not smaller than the first: $M_f \leq M_m$.

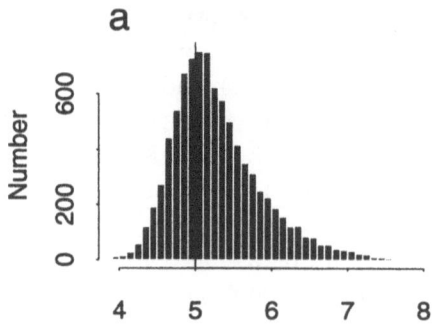

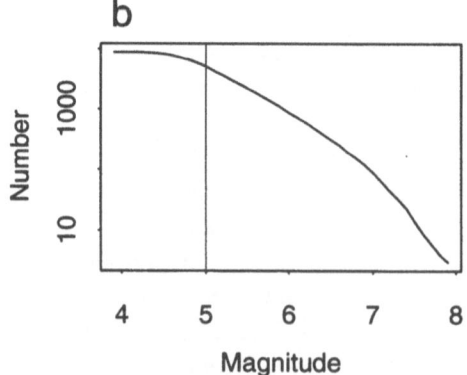

Figure 1
Magnitude distribution of earthquakes used in this study and taken from the Harvard CMT catalog, 1977–1996, 0–50 km depth. (a) Interval distribution; (b) cumulative distribution.

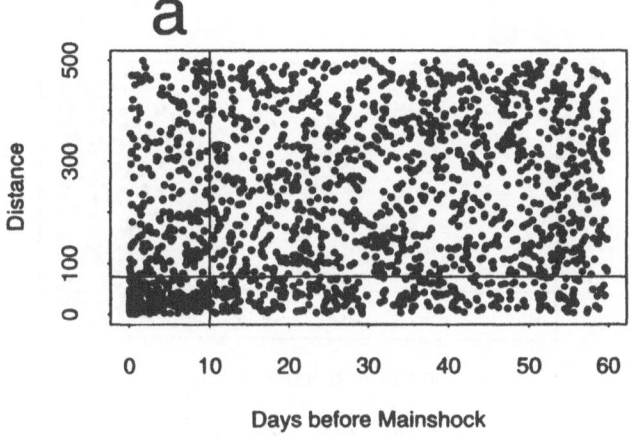

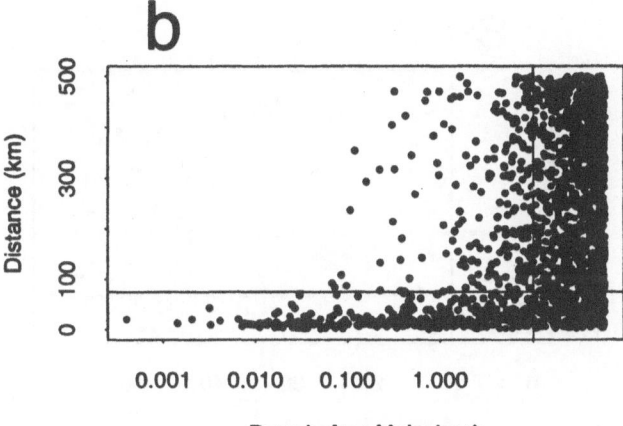

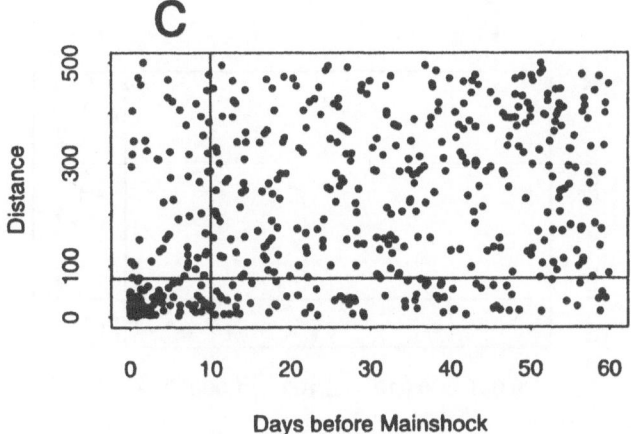

Fig. 2.

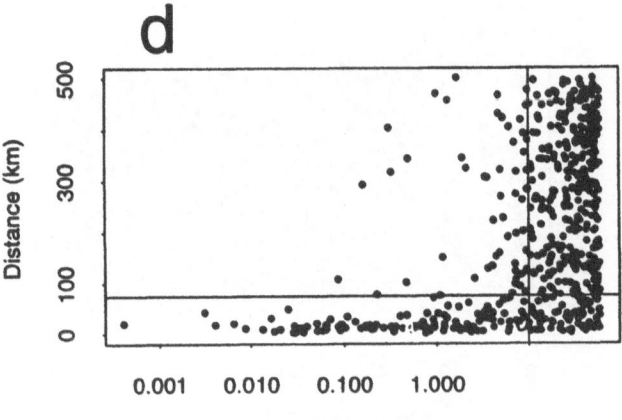

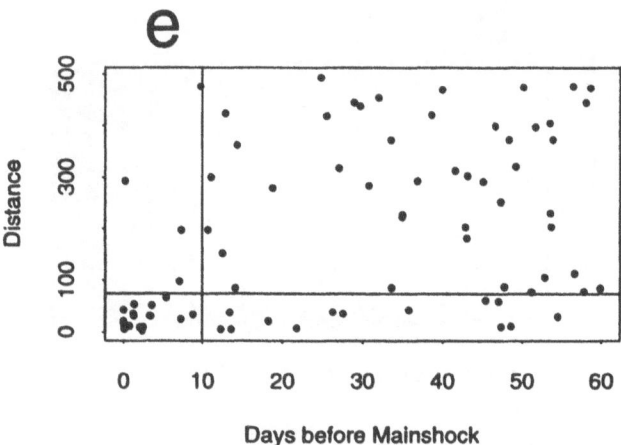

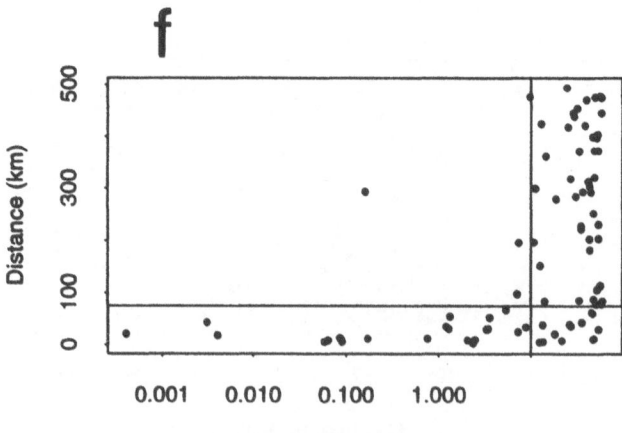

3. Interevent epicentral distance in a pair does not exceed dX and the interevent time difference does not exceed dT.

4. When two or more mainshocks are themselves clustered within dX and dT of each other, only the largest is considered a mainshock.

In the estimates of foreshock rates below, tied magnitudes are allowed in the above definition, in conformance with the definition in REASENBERG and JONES (1989, 1994), so that the present results should be directly comparable to their model. However, in the subsequent analysis of the distribution of magnitude differences among foreshock-mainshock pairs, I require that $M_f < M_m$, in conformance with the definition used by AGNEW and JONES (1991), LINDH and LIM (1995) and MICHAEL and JONES (1998), which facilitates comparison with those studies. For each qualifying mainshock, all qualifying events within the time dT and distance dX of it are identified as potential foreshocks. If more than one foreshock is thus associated with a given mainshock, only the largest foreshock is counted.

Choice of Spatial and Temporal Windows

The spatial and temporal windows used to define the foreshock-mainshock pairs were selected so as to capture the strongest and most obvious foreshock clustering activity in the Harvard CMT catalog. Figure 2 shows the earthquake pairing corresponding to three choices of M_f and M_m, with interevent separations reaching 500 km and 60 days. Figure 3 presents the corresponding cumulative distribution of interevent time and distance among the pairs. At interevent distances greater than approximately 75 km and interevent times greater than approximately 10 days, the density of pairs apparent in Figure 2 is approximately uniform with respect to both time and distance and the cumulative curves in Figure 3 are approximately linear, consistent with noninteracting earthquakes occurring uniformly in time and epicentrally distributed uniformly along fault zones (see, for example, KAGAN and KNOPOFF, 1980). At interevent distances and times less than 75 km and 10 days, the dense clustering of points in Figure 2 and steep assent of the cumulative distributions in Figure 3 reveal the foreshock clustering process. The apparent

Figure 2

Foreshock-mainshock pairs in the Harvard CMT catalog (1977–1994) for various mainshock and foreshock magnitude thresholds, shown as a function of the pair's interevent distance and interevent time. (a, c, e): linear time axis; (b, d, f): same data shown with log time axis. (a and b) Pairs with foreshock magnitude $M_f \geq 5$ and mainshock magnitude $M_m \geq 5$; $N = 1966$. (c and d) Pairs with $M_f \geq 5$ and $M_m \geq 6$; $N = 562$. (e and f) Pairs with $M_f \geq 5$ and $M_m \geq 7$; $N = 84$. The large ranges of interevent time and distance shown here reveal both the foreshock-related clustering and background or "chance" clustering. Pairs with interevent times less than 10 days and distances less than 75 km (the lower left quadrants formed by the solid lines) were used in the subsequent foreshock rate analysis.

range of foreshock-mainshock distance appears to be independent of the magnitude of the mainshock; the median and 90%-ile interevent distances are approximately 15 km and 30 km, respectively, for pairs involving $M > 5$, $M > 6$ and $M > 7$ main-shocks (Fig. 4). The consistency of these ranges suggests that they are essentially controlled by location errors and should be considered upper bounds. BOWMAN and KISSLINGER (1984) found an apparent interevent distance range (which includes an unstated location uncertainty) of 20–40 km for events located near Adak Island. OGATA *et al.* (1995) observed a range of apparent interevent distances

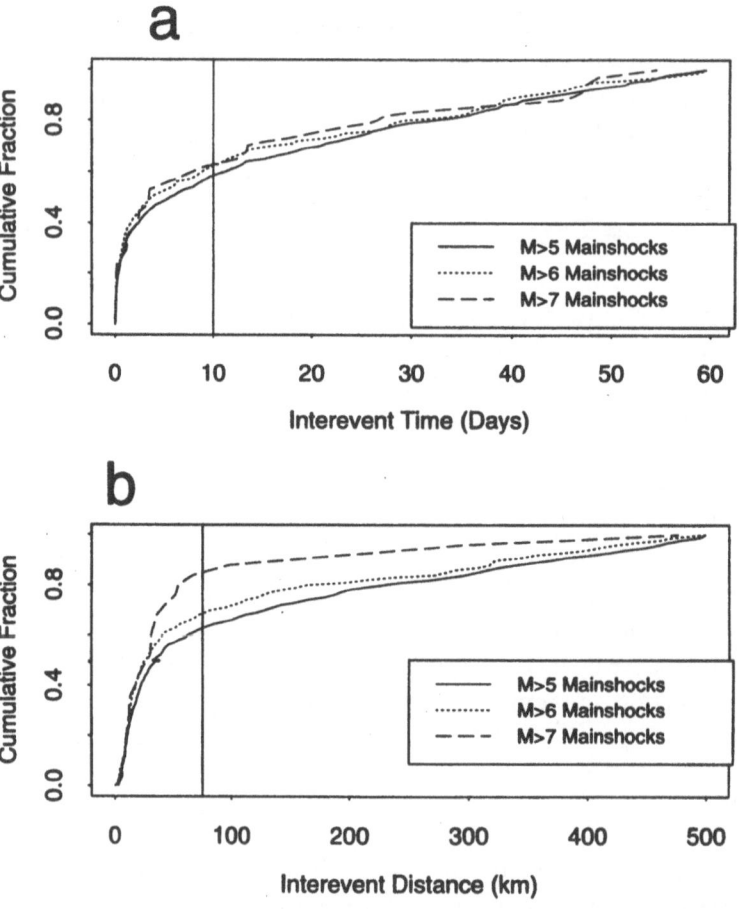

Figure 3
Interevent distances and times among foreshock-mainshock pairs found in the Harvard CMT catalog. (a) Cumulative distribution of interevent times among foreshock-mainshock pairs ·with interevent distances of 75 km or less. (b) Cumulative distribution of interevent distances among foreshock-main-shock pairs with interevent times of 10 days or less. Background seismicity, which contributes to the linear trends at large interevent times and distances, is not removed in this figure. Vertical lines indicate windows of 10 days and 75 km used in calculating foreshock rates.

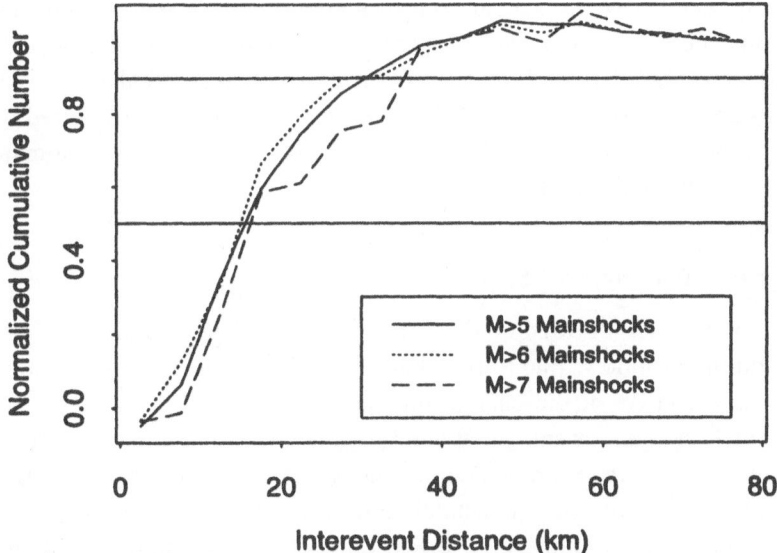

Figure 4

Normalized cumulative distributions of interevent distances among foreshock-mainshock pairs shown in Figure 3b, after correcting for "background rate." Background rate is defined as the average spatial density of pairs observed in each data set over the interevent distance range 200–500 km (and corresponds to the slopes of the flat portions of curves in Fig. 3b). Removal of these constant background rates accounts for the decreases in "cumulative" number seen in the upper tails of the distributions. Horizontal lines indicate the median and 90%-ile levels. For all three data sets (mainshocks with $M > 5$, $M > 6$ and $M > 7$), the median and 90%-ile interevent distances are approximately 15 km and 30 km, respectively. Numbers of pairs represented are 518, 161 and 26 for the $M > 5$, $M > 6$ and $M > 7$ mainshocks, respectively.

between $M \geq 4$ foreshocks and mainshocks in the JMA catalog of 10–20 km, but they warned that event location errors may be contributing to these distances. Among observations of well-located events, there is evidence that foreshocks often occur much closer to their mainshocks. Using standard catalog locations, JONES (1985) found that most foreshocks ($M > 3$) in southern California were located within approximately 1 km of the epicenters of their respective mainshocks. Using precise relative event locations (uncertainties of 0.1 to 0.3 km) obtained with a waveform correlation technique, DODGE et al. (1996) found that foreshocks before mainshocks ($4.7 < M < 7.3$) in California were located approximately 1 km or less from their mainshock's epicenter. Returning to the present study, the median epicentral error in the Harvard CMT catalog was estimated by SMITH and EKSTRÖM (1997) to be 25 km or greater for $M_w = 6.5$ events. Relative locations between foreshocks and mainshocks are probably better determined than this owing to common unmodeled velocity structure and station site effects. While we have neither relocated nor estimated uncertainty in the relative locations of these event pairs, it is reasonable to assume that at least some of the apparent interevent range of 15–30 km seen here reflects location errors. Clearly, the 75-km window used here is larger than necessary, and was adopted to insure that badly mislocated events would not be excluded from the analysis; it does not imply that the actual event separations are this large. To correct the resulting assays for this oversized window, a "background" rate, corresponding to chance pairings, must be estimated and removed from the pair counts.

Correction for "Background" Events

A certain number of earthquakes are expected to fall within our interevent distance and time windows due to the "background" seismicity in the regions of the mainshocks. The effect of the background seismicity is seen in the linear trends in Figure 3 at large interevent times and distances. To correct for this effect, background rates were estimated from the seismicity by counting earthquakes located within 75 km of each mainshock and occurring during the time period 300 to 100 days before each mainshock. Based on these rate estimates, the numbers of background earthquakes expected to fall into the actual foreshock windows were subtracted from the counts of possible foreshocks. This method may slightly overestimate the background rate because $M \geq 5$ aftershocks, which were not removed from the catalog, could be included in the background count, while foreshocks were limited to one per mainshock. However, inspection of the background events revealed none to be $M \geq 5$ aftershocks.

Focal Mechanisms

I categorized the mainshocks as either thrust, strike-slip or normal according to the plunge of the tension axis (p1) and plunge of the null axis (p2), as in TRIEP and SYKES (1997) (Table 1). These definitions include some oblique thrust events as "thrust," and some oblique normal events as "normal." The use of the plunges of principal axes to classify the focal mechanism leads to clear misclassifications. These arise because knowledge of which focal plane is the slip plane is not included in the CMT solution.

Table 1

Definition of focal mechanism types in Harvard CMT catalog

Focal mechanism	p1	p2
Thrust	>45°	—
Normal	<45°	<45°
Strike-slip	—	>45°

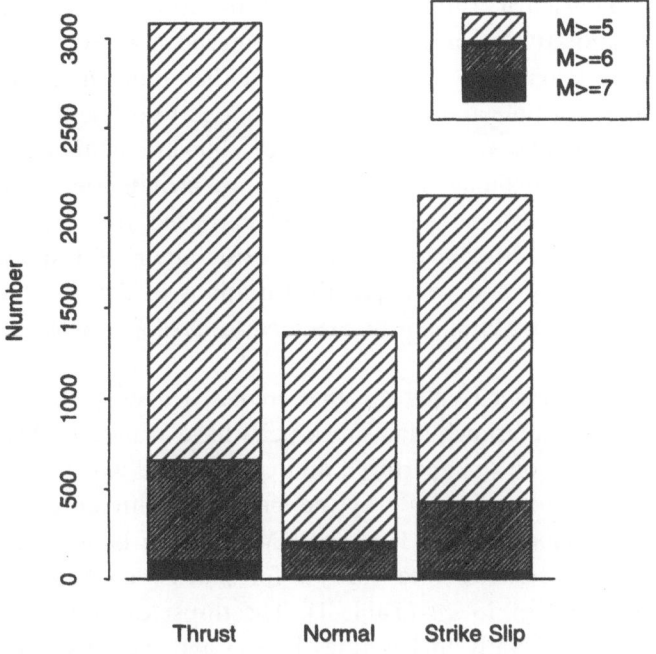

Figure 5
Earthquakes listed with depths of 50 km or less in the Harvard CMT catalog (1977–1996) and used in
this study, classified according to focal mechanism and magnitude.

Figure 5 shows the distribution of earthquake focal mechanisms among the
mainshocks used in this study. To characterize the *fs-ms* pairs, I used the mainshock's focal mechanism rather than the foreshock's, assuming it to be more
representative of the regional faulting style. Thus, in interpreting the results, this
classification may be applied to the region in which a potential foreshock is located,
but may not be used to condition the probability of a mainshock based on the
potential foreshock's focal mechanism.

Retrospective Foreshock Frequencies

The retrospective foreshock frequency associated with a set of mainshocks is
defined as the fraction of the mainshocks that are preceded by a foreshock. To
estimate this rate, I used time and distance windows of 0–10 days and 0–75 km, set
the foreshock magnitude threshold $M_f^{\min} = 5$, and considered mainshock magnitude
thresholds of $M_m^{\min} = 5$, $M_m^{\min} = 6$, and $M_m^{\min} = 7$. The resulting sets of foreshock-
mainshock pairs have magnitude differences distributed between zero and an upper
limit approximately given by $\Delta M = M_m^{\min} - M_f^{\min}$. I call ΔM the magnitude difference aperture. The larger the aperture is, the greater the number of foreshocks that

will be counted. This is true for any assumed distribution of magnitude difference. Therefore, to quantitatively compare different foreshock assays, it is necessary to normalize for the magnitude apertures used. Other differences among studies, including the choice of time and spatial windows used to define the foreshocks, are of less importance, because foreshock-mainshock pairs are tightly clustered in space and time, and thus self-defining. Oversized windows include too many background events, but this error can be approximately corrected for. However, no such clustering of magnitude difference apparently exists among foreshocks and main-shocks, as shown below, and consequently variations in the observable range of magnitude difference (aperture) can be expected to have a first-order effect on the resulting assay.

To compare the present results to other studies, I assume that the magnitude differences among the foreshock-mainshock pairs have a uniform distribution over their range of observation (or aperture), and calculate a "unit foreshock rate" by dividing the apparent foreshock rate by the magnitude difference aperture.

The $M \geq 6$ mainshocks in the Harvard CMT catalog between 1978 and 1996 ($N = 1108$) have a foreshock rate ($M \geq 5$ foreshocks, corrected for estimated background seismicity) of 13.2% (Table 1). The thrust earthquakes among these ($N = 533$) have a significantly higher rate (17.5%) and the strike-slip earthquakes ($N = 397$) have a significantly lower rate (8.0%). Among the $M \geq 7$ mainshocks in the Harvard CMT catalog ($N = 149$), both the overall frequency of foreshocks (16.5%) and the departures from this rate observed among subsets defined according to focal mechanism are similar to the rates in the $M \geq 6$ set. However, because of the smaller numbers of *fs-ms* pairs, the differences among the $M \geq 7$ subsets are not statistically significant.

The overall foreshock rate of 13.2% found for $M \geq 6$ mainshocks preceded by $M \geq 5$ foreshocks (13.2% per magnitude unit) is comparable to results obtained in other studies. JONES (1984) found that 7 out of 20 ($M \geq 5$) earthquakes which occurred in California between 1966 and 1980 were preceded by ($M \geq 2$) foreshocks (a foreshock rate density of 12% per magnitude unit). JONES and MOLNAR's (1979) study of worldwide ($M \geq 5$) foreshock activity before 161 $M \geq 7$ mainshocks (1914–1973) detected a rate of 24.8%, or approximately 12% per magnitude unit. More recently, ABERCROMBIE and MORI (1996) examined 59 ($M \geq 5$) mainshocks in California and Nevada, and found that 26 of them were preceded by ($M \geq 2$) foreshocks (15% per magnitude unit). MICHAEL and JONES (1998) looked at 33 $M \geq 5$ strike-slip earthquakes along the San Andreas fault physiographic province and found that 17 were preceded by ($M \geq 2$) foreshocks, corresponding to a rate density of 17% per magnitude unit. A similar result was also reported by LINDH and LIM (1995) for the same region, time period and magnitude ranges. BOWMAN and KISSLINGER (1984) determined that 15% of $m_b \geq 5.0$ earthquakes in the Adak thrust zone were preceded by foreshocks. While they estimated a completeness threshold of $M_f^{\min} = 2.5$ (m_b), inspection of their Figure 2 suggests an alternate threshold of, perhaps, $m_b = 3.8$, which would correspond to a foreshock rate density of about 12% per magnitude unit. Variations among these studies involving the region studied and the choice of spatial and temporal windows make it difficult to compare these results. Given these differences, the range of results is surprisingly narrow, from 12% to 17% per magnitude unit. This range coincides closely with the 95% confidence range estimated here for the foreshock rate among $M \geq 6$ main-shocks in the Harvard catalog) and may be considered a robust, worldwide meta-result.

The present results for $M_m^{\min} = 7$ do not follow this trend, however, and instead indicate a foreshock rate density of between 5.5% and 11.7% per magnitude unit (95% confidence range) (Table 2). However, because of the small number of $M \geq 7$ mainshocks ($N = 149$), these rates are not significantly different from the corre-sponding rates determined with the more numerous $M \geq 6$ events.

Comparison of the present results of those of VON SEGGERN's (1981) is made with the recognition that Von Seggern's analysis did not adequately correct for background seismicity. While Von Seggern's assay was known to include back-ground events, he did not attempt to quantify their presence. Instead, he interpreted the relative rates, but cautioned against assigning significance to the absolute rates. He concluded that "the rate . . . for shallow foreshocks is significant but still small," and that "the true rate of foreshock occurrence was less than 20%." Von Seggern's Table 4 reports a foreshock rate of 18% (9 out of 51) among $M_s \geq 7$ mainshocks of all kinds (mostly 0–100 km depth). This result may be compared most directly to the apparent rate of foreshocks before $M \geq 7$ mainshocks (19 out of 95, or 20%) observed in the present study before the estimated rate of background activity is removed (Table 2). The spatial and temporal distribution of foreshocks relative to their mainshocks observed by Von Seggern is similar to those seen in this study. Most of Von Seggern's foreshocks occurred within 10 days of their mainshocks, and most fell within 0.5 degree of the mainshocks.

Sensitivity to Parameter Choices

The above estimates of foreshocks rate depend on several choices made in the analysis, perhaps the most important being the minimum magnitude of foreshocks and the duration and spatial size of the foreshock window. I argued above that the results should be fairly insensitive to increases in the space and time window parameters as long as the window exceeds the natural scale of clustering. In addition, I suggested that the use of events with magnitudes below the completeness level of the catalog would produce foreshock assays approximately 28% low. I explore here the actual sensitivity of key results to these choices.

For both the $M \geq 6$ and $M \geq 7$ mainshock sets, I increased the time window to 20 days and increased the distance window to 100 km, separately. The apparent

Table 2

Summary of retrospective foreshock rates

Mainshock category	Number of mainshocks	Number of possible foreshocks (1)	Expected number of background events	Fraction of mainshocks preceded by a foreshock (2) (%)	95% confidence range (3) for foreshock frequency (%)	95% confidence range (3) for foreshock rate per unit magnitude difference (%)
Harvard CMT catalog ($M \geq 6$) 1978–1996						
All	1108	161	14.4	13.2	11.3–15.4	11.3–15.4
Thrust	533	103	9.7	17.5*	14.4–21.0	14.4–21.0
Normal	172	23	2.1	12.2	7.7–18.0	7.7–18.0
Strike-slip	397	34	2.4	8.0**	5.5–11.1	5.5–11.1
Harvard CMT catalog ($M \geq 7$) (1978–1996)						
All	149	26	1.5	16.5	10.9–23.4	5.5–11.7
Thrust	95	19	1.2	18.7	11.5–28.0	5.8–14.0
Normal	13	3	0.1	22.3	—	—
Strike-slip	41	4	0.3	9.0	—	—

* Foreshock rate among thrusts exceeds foreshock rate among all $M > 6$ earthquakes in the Harvard CMT catalog at the 95% confidence level.
** Foreshock rate among strike-slip events is lower than foreshock rate among all $M > 6$ earthquakes in the Harvard CMT catalog at the 99% confidence level.
(1) A spatial window of 75 km and a temporal window of 10 days were used.
(2) Corrected for estimated background rate.
(3) Range represents sampling uncertainty.

foreshock rates decreased in all cases (suggesting that I may be overcorrecting for the background). These changes are minor and fall within the given 95%-confidence ranges (reflecting sampling uncertainty) listed in Table 2.

Increasing the minimum magnitude from 5.0 to 5.5 resulted in higher estimates for (normalized) foreshock rates before $M \geq 6$ and $M \geq 7$ mainshocks. These increases correspond to apparent deficits in the ($M \geq 5.0$) catalog of 13% and 33%, respectively, which are consistent with the predicted deficit of 28%.

Dependence on Focal Mechanism

Foreshock rate varied significantly among subsets of mainshocks defined by their focal mechanisms. Among the ($M \geq 6$) thrust earthquakes in the Harvard CMT catalog, 17.5% were preceded by a ($M \geq 5$) foreshock, this rate being significantly higher (95% confidence) than the corresponding rate of 13.2% observed before all ($M \geq 6$) earthquakes. Also, strike-slip mainshocks were preceded by foreshocks only 8.0% of the time, a rate significantly lower than that for all ($M \geq 6$) events at the 99% confidence level. Thus, foreshocks occurred before shallow $M \geq 6$ thrust events at approximately twice the rate they did before shallow $M \geq 6$ strike-slip events.

The relatively high foreshock rates observed before thrust earthquakes appear to contradict a result of ABERCROMBIE and MORI (1996) (their Fig. 2b), who reported a lower frequency of foreshocks before reverse mainshocks than before strike-slip mainshocks ($M \geq 5$) in California and western Nevada. There are at least two possible ways to resolve these apparent differences. First, ABERCROMBIE and MORI (1996) considered small numbers of earthquakes: 14 reverse mainshocks (2 with foreshocks) were compared to 38 strike-slip mainshocks (19 with foreshocks). Statistical error in these samples may partially mask the underlying frequencies. In addition, ABERCROMBIE and MORI (1996) were able to use the well-determined depths in the California and Nevada catalogs to establish a clear inverse dependence of foreshock rate on the depth of the mainshock, over a depth range from 0 to about 20 km. Because reverse events tend to be deeper than strike-slip events in California, at least some of the apparent dependence on focal mechanism they report might actually be a coupled depth (via normal stress) effect, as suggested by ABERCROMBIE and MORI (1996).

The present result appears to contradict JONES (1984), who found no foreshocks preceding four ($M_L \geq 5$) thrust mainshocks in southern California, and who also suggested that the foreshock rate for thrust earthquakes in California might be lower than the rate for strike-slip events (15% per magnitude unit, based on 16 mainshocks), although they cautioned that their data set was too small to definitively distinguish between these rates.

Also, my result appears at first to conflict with JONES and MOLNAR (1979), who concluded that there was a higher foreshock rate before large earthquakes in non-subduction zones than in subduction zones. They reported a foreshock rate of 20% (10% per magnitude unit) among 120 $M \geq 7$ earthquakes in subduction zones, and a rate of 29.3% (15% per magnitude unit) among 41 $M \geq 7$ events in non-subduction zones. (Here, I assumed a magnitude aperture of 2 in Jones and Molnar's study.) However, there are too few data in this study to infer a significant rate difference with confidence; the 90% confidence ranges are, respectively, 7%–14% per magnitude unit (subduction zone events) and 9%–22% per magnitude unit (non-subduction zone events).

In spite of the difficulties in comparing these studies, the apparent discrepancies between them and the present worldwide result suggests that California foreshocks may be a special case, owing to the particular regional tectonics there, and are not representative of worldwide foreshock rates. This possibility is explored below.

Subduction Zones and Continental Thrusts

In the Harvard ($M \geq 6$) assay, 17.5% of the thrust-type mainshocks were preceded by foreshocks, and most of these are located in the shallow portions of the major circum-Pacific subduction zones (Fig. 7a). These events involve the subduction of water-saturated oceanic crust in the uppermost portion of descending slabs and have a median depth of 30 km. In contrast, only 3 foreshocks were found among the 35 thrust earthquakes in the Himalayan collision belt involving the Indio-Australia and Eurasian plates, and none were found before the 7 thrust (and oblique thrust) earthquakes in California. These events are mountain-building earthquakes involving the transpression of continental, crystalline rock and have a median depth of 15 km. The contrast suggests that the foreshock rate may be higher in shallow subduction zones than in continental thrust belts, although the number of events in these regional subsets are generally too small to prove it. This could explain both the high foreshock rate for thrust earthquakes observed in this study and the low rates observed among California thrust earthquakes (JONES, 1984; ABERCROMBIE and MORI, 1996).

Foreshocks preceding normal mechanism earthquakes occurred along the east African rift, on the spreading centers in the Indian Ocean, and along the western Pacific subduction zones from New Zealand to Japan (Fig. 7b).

Most of the foreshocks that preceded ($M \geq 6$) strike-slip mainshocks lie along the convergent boundary between the Pacific and Indio-Australian plates, from the New Guinea Thrust to the New Hebrides Thrust (Fig. 7c). Three were located in California. In other regions, few strike-slip earthquakes were preceded by foreshocks.

Prospective Foreshock Frequencies

From the perspective of real-time, short-term earthquake hazard assessment, it is the probability of earthquakes following earthquakes, not preceding them, that is

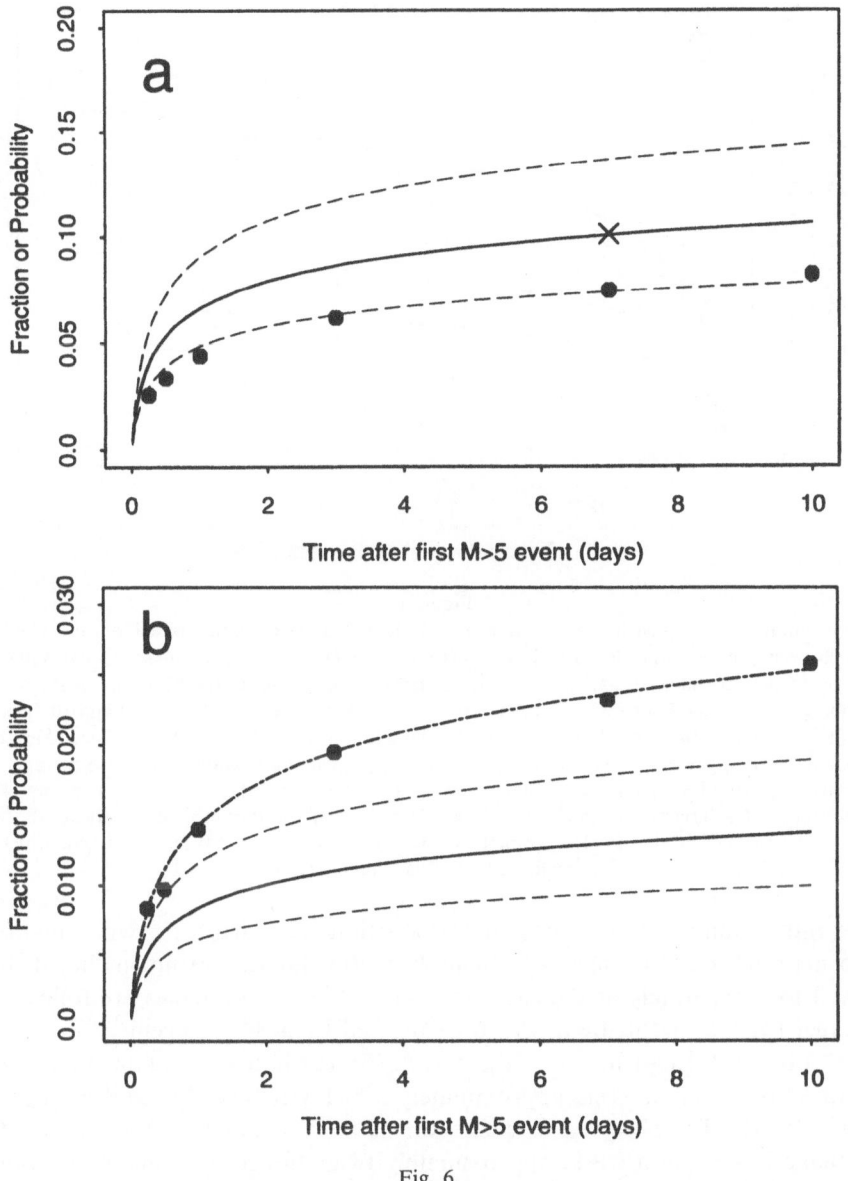

Fig. 6.

of interest. Here I use the retrospective foreshock frequencies, such as those given in Table 1, to calculate "prospective foreshock frequencies." I calculate the fraction of shallow $M \geq 5$ earthquakes in the Harvard catalog that were followed within time periods up to 10 days and within distances up to 75 km by a larger earthquake, and compare these frequencies to the corresponding prospective probabilities calculated with a generic model of aftershock activity in California

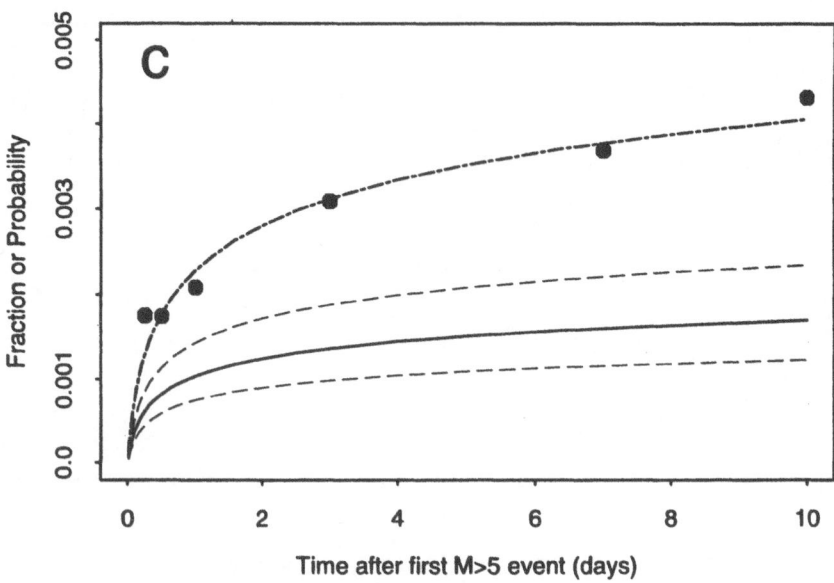

Figure 6

Observed cumulative frequencies (solid circles) and predicted probabilities based on the California generic clustering model (solid lines) of larger earthquakes which occur after an $M \geq 5$ earthquake, as a function of the time after the $M \geq 5$ earthquake. Broken lines indicate the 95% confidence range of California generic model. Observed frequencies are calculated at times 0.25, 0.5, 1, 3, 7 and 10 days after the first ($M \geq 5$) event. (a) Case of any larger earthquake following a $M \geq 5$ event. (b) Case of an $M \geq 6$ earthquake following an $M \geq 5$ event. (c) Case of an $M \geq 7$ earthquake following an $M \geq 5$ event. " $\times$ " in (a) marks a generic forecast often announced by the USGS after significant California earthquakes of "a 10% chance of a larger event in the next week." In (b) and (c), dash-dot lines indicate the world generic model (defined by the model parameters $a = -1.5$, $b = 0.8$, $p = 1.0$ and $c = 0.05$), which was foward-fitted to the data points shown.

(REASENBERG and JONES, 1989, 1994). Earthquakes that otherwise qualify as $M \geq 5$ mainshocks, but which are themselves aftershocks, are not included in the survey. I look separately at the cases in which $M \geq 5$ earthquakes are followed by any larger ($M \geq 5$) event, by a $M \geq 6$ event, and by a $M \geq 7$ event.

The observed frequency of larger ($M \geq 5$) earthquakes following a $M \geq 5$ earthquake is 7.5% in a week, approximately 25% lower than the California generic model (Fig. 6a). The observed frequency of $M \geq 6$ earthquakes following a $M \geq 5$ earthquake is 2.3% in a week, approximately twice the generic model and outside is 95% confidence range (Fig. 6b). The observed frequency of $M \geq 7$ earthquakes following a $M \geq 5$ earthquake is approximately 0.4%, again approximately twice the generic model and outside its 95% confidence range (Fig. 6c). An explanation for this highly non-self-similar result involves the catalog incompleteness, which has caused the observed frequencies of $M > 5$ mainshocks to be artificially low. Each observation point in Figure 6 represents a ratio between an observed number of foreshock-mainshock pairs and an observed number of potential ($M \geq 5$) fore-shocks. In the cases of pairs involving $M \geq 6$ or $M \geq 7$ earthquakes, the catalog deficiency equally affects both quantities in the ratio, since both rely directly on the ability to observe $M \geq 5$ earthquakes; the combined effect of the missing events cancels out in the ratio. But in the case of pairs involving two $M \geq 5$ events, the error associated with incompleteness enters twice (and is, hence, squared), while the error in the number of potential foreshocks is still only directly affected by the incompleteness. Consequently, the ratio (and points in Fig. 6a) will be artificially lowered. From this line of reasoning, I conclude that the observed frequencies of $M \geq 6$ and $M \geq 7$ earthquakes following $M \geq 5$ events (points in Figs. 5b and 5c) are unbiased estimates of the true frequencies, while the $M \geq 5$ observations (points in Fig. 6a) are artificially low. In Figure 6a, the " $\times$ " indicates the probability usually stated in USGS forecasts in California after a $M \geq 5$ mainshock ("10% probability of an equal or larger event in the next 7 days"). While the corresponding frequency observed in the Harvard catalog is 7.5%, extrapolation from the $M > 6$ and $M > 7$ results suggests that the actual worldwide rate may be as high as 15%.

The observed frequencies of $M > 6$ and $M > 7$ earthquakes in the Harvard catalog were used to forward fit a new model (which I call the "world generic model") having the same form as in REASENBERG and JONES (1989) but defined by new parameter values ($a = -1.5$, $b = 0.8$, $p = 1.0$ and $c = 0.05$). Numerically, the biggest difference between these models is in the a-value. But there are other important differences as well. The world generic model may be better determined than the California generic model because it is based on 171 foreshock-mainshock pairs, compared to only 62 aftershock sequences in the REASENBERG and JONES (1989) compilation. In addition, the world generic model directly reflects the behavior of large earthquakes involved in foreshock-mainshock sequences, while the California generic model is an extrapolation from a smaller aftershock activity.

Summary and Discussion

The overall foreshock rate of 13.2% per magnitude unit found here for all $M > 6$ earthquakes in the Harvard CMT catalog (1978–1996) lies within the range of estimates of foreshock rate density (12%–17% per magnitude unit) obtained by other studies using regional and worldwide catalogs. The higher rate estimates within this range (for example, those reported in MICHAEL and JONES, 1998 and LINDH and LIM, 1995) were obtained with relatively small numbers of $M \geq 5$ mainshocks (33 and 30, respectively) and were limited to strike-slip earthquakes in the San Adreas Fault zone. Larger and broader studies (JONES and MOLNAR, 1979; JONES, 1984; AGNEW and JONES, 1991; and the present one) find slightly lower overall rates in the range 12%–15%.

This study found with high statistical confidence that the foreshock rate before $M \geq 6$ thrust mainshocks in the Harvard catalog is about twice the corresponding

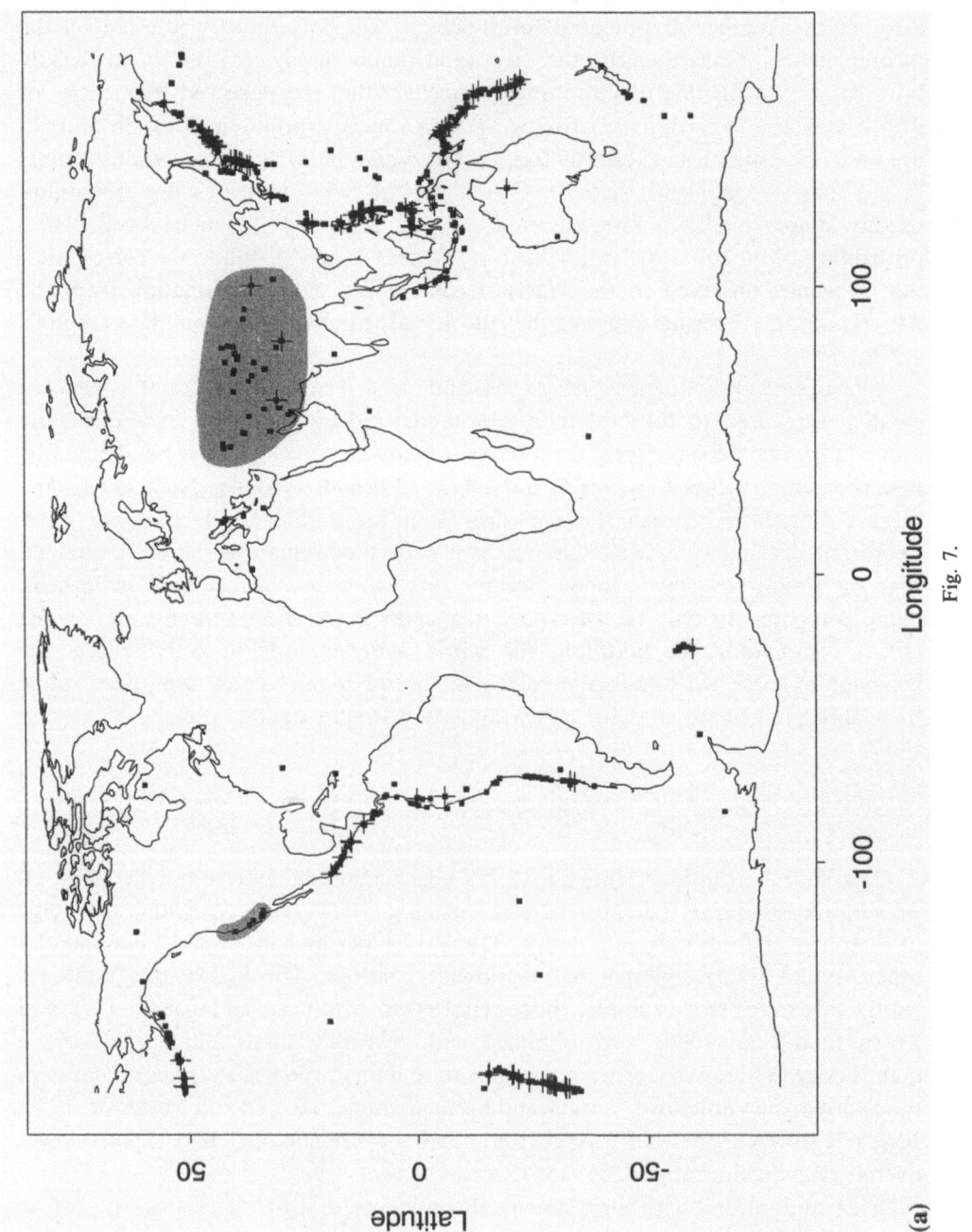

Fig. 7.

(a)

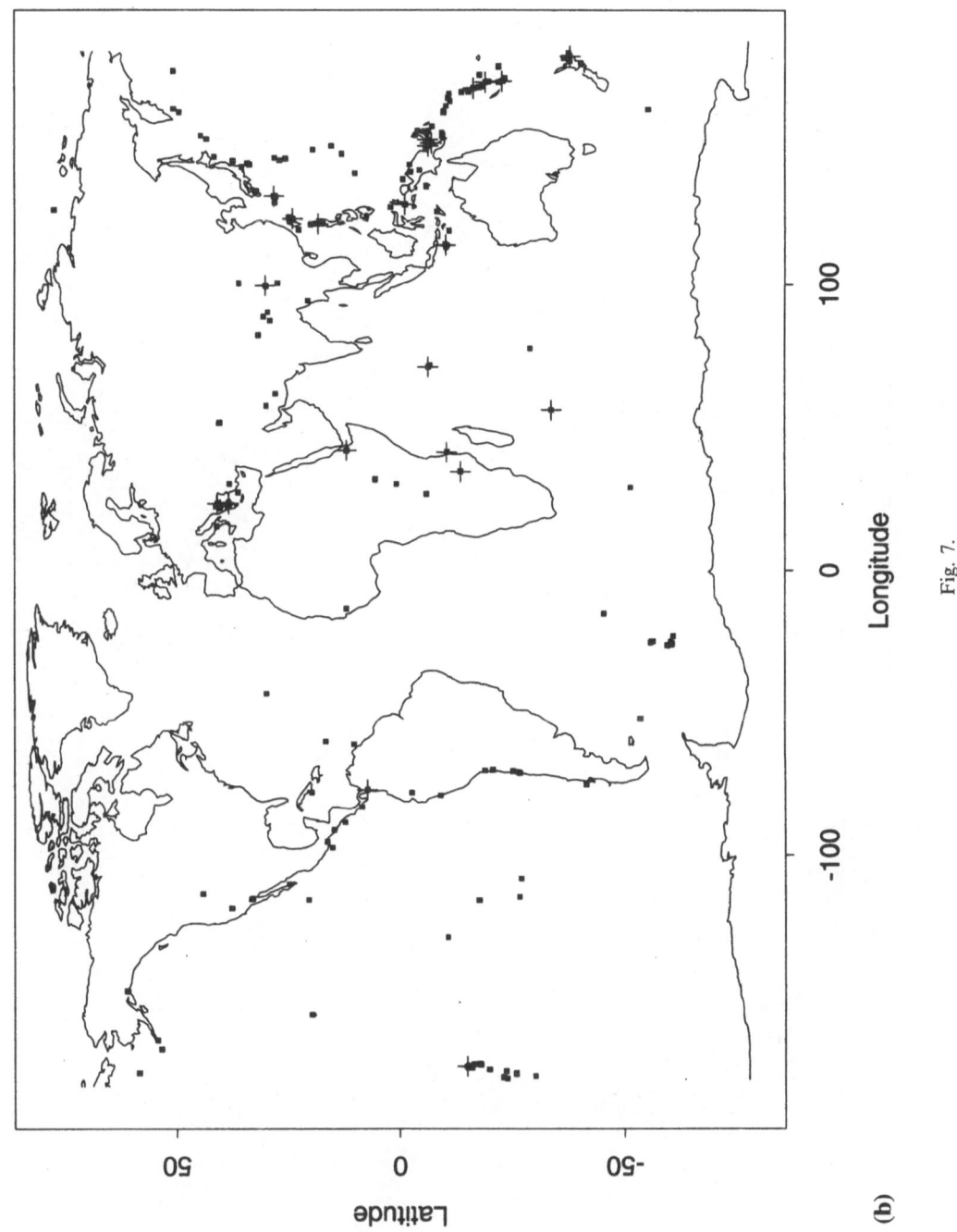

Fig. 7.

(b)

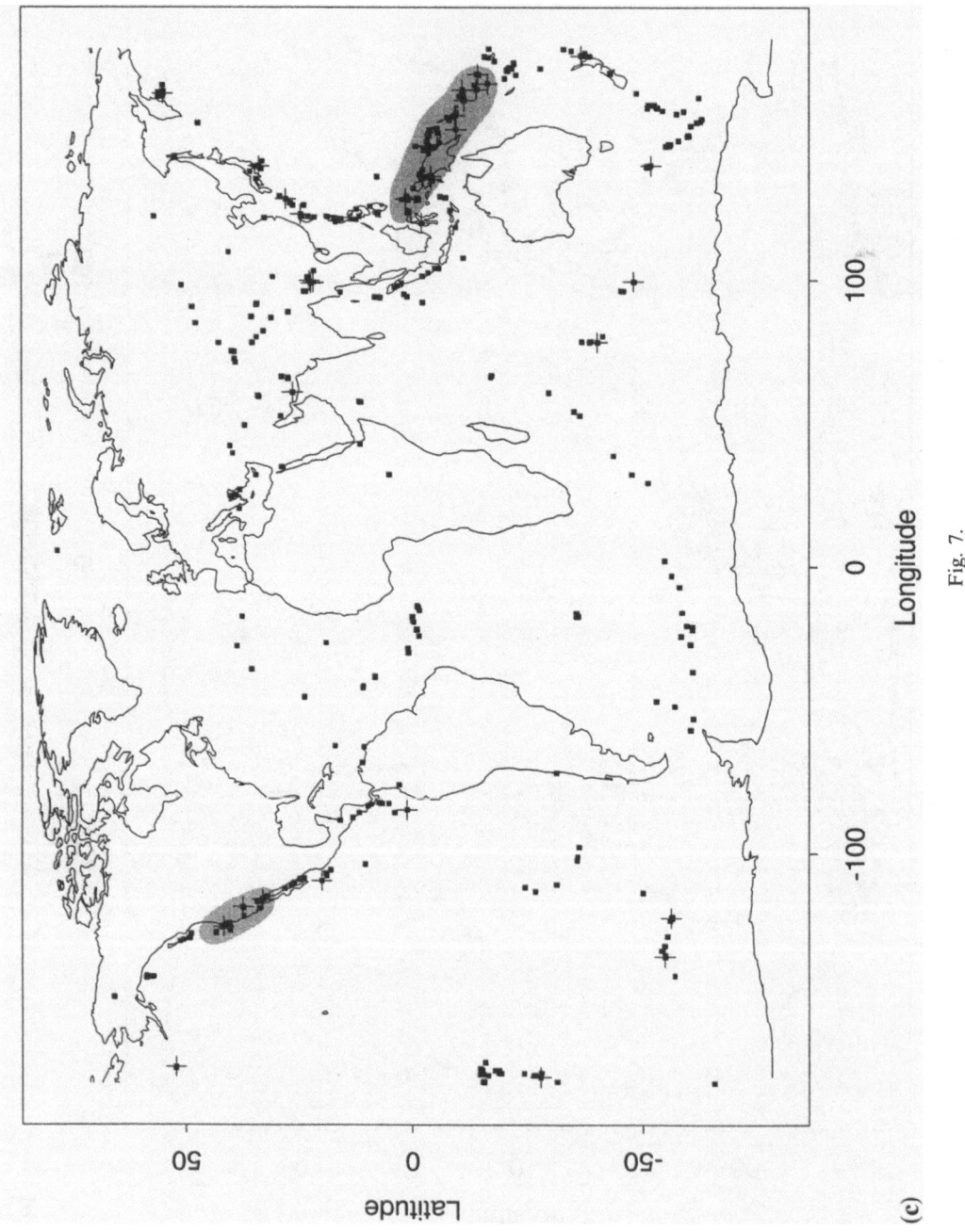

Fig. 7.

(c)

rate before strike-slip mainshocks. The apparent contradiction between the resolvable trend in the worldwide data of higher rates among thrusts than strike-slips and the opposite trend inferred in studies of California $M \geq 5$ earthquakes (JONES, 1984; ABERCROMBIE and MORI, 1996) may be resolved by hypothesising that most California thrusts (with the exception of off-shore events associated with the subduction of the Juan de Fuca plate) are typical of continental thrust belts which involve unsaturated, crystalline continental rock (such as the Himalayan collision belt) but not typical of subduction zone thrust events which occur in young, water-saturated ocean sediments. This would imply that the generic California model may be representative of most earthquakes in California, but may significantly underestimate the conditional probabilities following potential foreshocks in Cascadia, a region whose clusters may be more typical of shallow subduction zones.

The worldwide occurrence of $M \geq 6$ and $M \geq 7$ earthquakes in the 10-day periods following (and 75-km distance range from) $M > 5$ earthquakes in the Harvard catalog exceeds by a factor of about 2 the rate predicted by the California generic model (Figs. 5b and 5c). This difference may be understood, in part, by the dominance of shallow subduction thrusts in the Harvard data set (which were found to have a relatively high foreshock rate), the presence of an active strike-slip plate boundary in California (associated with a low foreshock rate worldwide), and the absence of an active subduction zone in most of California (an important exception being the Cascadia megathrust). In this sense, the California generic model can be considered "correct" for most of California, while the world generic model would better represent worldwide foreshock-related conditional probabilities. In certain other regions where sufficient seismological data are available, regionally determined generic clustering models (see, for example, REASENBERG et al., 1990; JONES et al., 1995) may provide a better basis than the worldwide generic model for estimating short-term earthquake probabilities.

Figure 7

Locations of Harvard catalog $M > 6$ mainshocks (1978–1996) and their ($M > 5$) foreshocks, separated according to the focal mechanism of the mainshock. Solid squares indicate mainshocks; plus signs indicate foreshocks. (a) Thrust events. Shaded areas, which mark continental thrust zones, have a lower foreshock rate (3 out of 35 mainshocks) than is found along the shallow portions of the major circum-Pacific subduction zones. (b) Normal events. (c) Strike-slip events. Shaded areas along the New Hebrides and New Guinea thrusts (presumably including strike-slip events on transform faults associated with the subduction zone) and the San Andreas Fault zone have average or higher foreshock rates, compared to that for strike-slip events in the rest of the world.

Acknowledgements

I am grateful for valuable discussions and suggestions provided by many of my colleagues, including Rachel Abercrombie, Bill Bakun, Jim Davis, Rich Eisner, Bill Ellsworth, Lucy Jones, Allan Lindh, Mark Matthews, Andy Michael, Jim Mori and David Oppenheimer. I thank Lucy Jones, Andy Michael and Francesco Mulargia for their helpful and critical reviews of the manuscript.

REFERENCES

ABERCROMBIE, R. E., and MORI, J. (1996), *Occurrence Patterns of Foreshocks to Large Earthquakes in the Western United States*, Nature *381*, 303–307.

AGNEW, D. C., and JONES, L. M. (1991), *Prediction Probabilities from Foreshocks*, J. Geophys. Res. *96*, 11,959–11,971.

BOWMAN, J. R., and KISSLINGER, C. (1984), *A Test of Foreshock Occurrence in the Central Aleutian Island Arc*, Bull. Seismol. Soc. Am. *74*, 181–197.

CALIFORNIA EARTHQUAKE PREDICTION EVALUATION COUNCIL (1977), Minutes, February 28, 1997 Meeting, Pasadena, California.

DAVIS, S. D., and FROHLICH, C. (1991), *Single-link Cluster Analysis of Earthquake Aftershocks: Decay Laws and Regional Variations*, J. Geophys. Res. *96*, 6335–6350.

DODGE, D. A., BEROZA, G. C., and ELLSWORTH, W. L. (1996), *Detailed Observations of California Foreshock Sequences: Implications for the Earthquake Initiation Process*, J. Geophys. Res. *101*, 22,371–22,392.

DZIEWONSKI, A. M., EKSTRÖM, G., FRANZEN, J. E., and WOODHOUSE, J. H. (1987), *Global Seismicity of 1977: Centroid-moment Tensor Solutions for 471 Earthquakes*, Phys. Earth Planet. Inter. *45*, 11–36.

JONES, L. M. (1984), *Foreshocks (1966–1980) in the San Andreas System California*, Bull. Seismol. Soc. Am. *74*, 1361–1380.

JONES, L. M. (1985), *Foreshocks and Time-dependent Earthquake Hazard Assessment in Southern California*, Bull. Seismol. Soc. Am. *75*, 1669–1679.

JONES, L. M., and MOLNAR, P. (1979), *Some Characteristics of Foreshocks and their Possible Relationship to Earthquake Prediction and Premonitory Slip on Faults*, J. Geophys. Res. *84*, 3596–3608.

JONES, L. M., CONSOLE, R., DILUCCIO, F., and MURRA, M. (1995), *Are Foreshocks Mainshocks Whose Aftershocks Happen to be Big?* (Abstract), EOS Trans. AGU *76*, 388.

KAGAN, Y. Y., and KNOPOFF, L. (1980), *Spatial Distribution of Earthquakes: The Two-point Correlation Function*, Geophys. J. R. Astr. Soc. *62*, 303–320.

KAGAN, Y. Y., and KNOPOFF, L. (1987), *Statistical Short-term Earthquake Prediction*, Science *236*, 1563–1567.

LINDH, A. G., and LIM, M. R. (1995), *A Clarification, Correction and Updating of Parkfield, California, Earthquake Prediction Scenarios and Response Plans*, U.S.G.S. Open-File Report 95–695.

MICHAEL, A. J., and JONES, L. M. (1998), *Seismicity Alert Probabilities at Parkfield, California, Revisited*, Bull. Seismol. Soc. Am. *88*, 117–130.

OGATA, Y., UTSU, T., and KATSURA, L. (1995), *Statistical Features of Foreshocks in Comparison with Other Earthquake Clusters*, Geophys. J. Int. *121*, 233–254.

REASENBERG, P. A. (1985), *Second-order Moment of Central California Seismicity, 1969–1982*, J. Geophys. Res. *90*, 5479–5496.

REASENBERG, P. A., and JONES, L. M. (1989), *Earthquake Hazard after a Mainshock in California*, Science *243*, 1173–1176.

REASENBERG, P. A., and JONES, L. M. (1994), *Earthquake Aftershocks: Update*, Science *265*, 1251–1252.

REASENBERG, P. A., OKADA, Y., and YAMAMIZU, F. (1990), *Earthquake Hazard after a Mainshock in the Kanto-Tokai Districts, Japan* (Abstract), EOS Trans. AGU, *71*, 908.

SMITH, G. (1995), *CMT associated with ISC*, URL http://tempo.harvard.edu/ ~ smith/CMT-ISC.html.

SMITH, G. P., and EKSTRÖM, G. (1997), *Interpretation of Earthquake Epicenter and CMT Centroid Locations, in Terms of Rupture Length and Direction*, Phys. Earth Planet. Inter. *102*, 123–132.

TRIEP, E. G., and SYKES, L. R. (1997), *Frequency of Occurrence of Moderate to Great Earthquakes in Intracontinental Regions: Implications for Changes in Stress, Earthquake Prediction and Hazards Assessments*, J. Geophys. Res. *102*, 9923–9948.

VON SEGGERN, D. (1981), *Seismicity Parameters Preceding Moderate to Major Earthquakes*, J. Geophys. Res. *86*, 9325–9351.

(Received July 31, 1998, revised/accepted December 18, 1998)

 To access this journal online:
http://www.birkhauser.ch

Pure appl. geophys. 155 (1999) 381–394
0033–4553/99/040381–14 $ 1.50 + 0.20/0

| **Pure and Applied Geophysics** |

Time Distribution of Immediate Foreshocks Obtained by a Stacking Method

KENJI MAEDA[1]

Abstract—We apply a stacking method to investigate the time distribution of foreshock activity immediately before a mainshock. The foreshocks are searched for events with $M \geq 3.0$ within a distance of 50 km and two days from each mainshock with $M \geq 5.0$, in the JMA catalog from 1977 through 1997/9/30. About 33% of $M \geq 5.0$ earthquakes are preceded by foreshocks, and 50–70% in some areas. The relative location and time of three types of representative foreshocks, that is, the largest one, the nearest one to the mainshock in distance, and the nearest one in time, are stacked in reference to each mainshock. The statistical test for stacked time distribution of foreshocks within 30 km from and two days before mainshocks shows that the inverse power-law type of a probability density time function is a significantly better fit than the exponential one for all three types of representative foreshocks. Two explanations possibly interpret the results. One is that foreshocks occur as a result of a stress change in the region, and the other one is that a foreshock is the cause of a stress change in the region and it triggers a mainshock. The second explanation is compatible with the relationship between a mainshock and aftershocks, when an aftershock happens to become larger than the mainshock. However the values of exponent of the power law obtained for stacked foreshocks are significantly smaller than those for similarly stacked aftershocks. Therefore the foreshock–mainshock relation should not be explained as a normal aftershock activity. Probably an increase of stress during foreshock activity results in apparently smaller values of the exponent, if the second explanation is the case.

Key words: Foreshocks, accelerating seismicity, power-law, stacking method, aftershocks.

1. Introduction

Foreshocks have expectantly become one of the most promising phenomena to predict earthquakes. However it is very difficult to actually predict earthquakes on the basis of foreshock activity. This is because we have not yet developed a technique to distinguish foreshock activity from background seismicity, and we have insufficient knowledge of the physical mechanism of foreshock occurrence. Generally the total number of observed foreshocks immediately (several days) before a mainshock is not very large, but often no foreshock is observed before a

[1] Meteorological Research Institute, 1-1, Nagamine, Tsukuba, Ibaraki, 305-0052, Japan. Tel.: + 81-298-53-8682, Fax: + 81-298-51-3730, E-mail: mkenji@bea.hi-ho.ne.jp

mainshock. This makes it difficult to obtain the generic features of foreshock activity by which we could infer the physical process of initiation of a mainshock. In order to overcome the shortage of data, we apply a stacking method and obtain a statistical space-time distribution of immediate foreshocks in reference to mainshocks. The stacked distribution will be meaningful if one assumes that every mainshock has a similar process of failure.

As introduced in the next section, there are examples in which the seismicity is activated acceleratively in a time range of several tens of years or several months before the occurrence of a mainshock. It is also known from rock fracturing experiments in the laboratory that the rate of acoustic emission (AE) increases rapidly before the final main failure of the rock under some conditions (for example, MOGI, 1962; SCHOLZ, 1968, 1990). The characteristic of these phenomena is that the increasing rate of seismicity before a mainshock can be approximated by the form of an inverse power law of the remaining time to the occurrence of the mainshock. This feature is very useful for the prediction of earthquakes because the form of the function has a singular point in time and implies the possibility of occurrence time prediction of mainshocks.

In this paper we focus on the foreshock activity immediately (two days) before the mainshock by stacking many foreshock–mainshock sequences in Japan. Thereafter we will show that the possibility of occurrence of immediate foreshocks also increases acceleratively in the form of an inverse power of remaining time to a mainshock. We also discuss the difference between the values of the exponent for the foreshock activity and those for aftershocks known as Omori's law. In this paper, we refer to an 'inverse power' function as simply a 'power' function.

2. Examples of Accelerated Seismic Activities in a Wide Range of Time Scales

BUFE and VARNES (1993) pointed out that the long-term (more than 200 years) acceleration of seismic release (seismic moment, Benioff strain, or event count) in the seismic cycle can be synthesized for the San Francisco Bay region before the 1906 San Francisco earthquake ($M \sim 8$). They also stated that the shorter-term (about 50 years) acceleration of seismic release of earthquakes with $M \geq 5$ in this region can be observed before the 1989 Loma Prieta earthquake (M 7.1). According to their analysis, the rate of seismic release is proportional to an inverse power of the remaining time to failure, and they concluded that the Loma Prieta earthquake should have been predictable with an uncertainty of two years in time and 0.5 in magnitude (Fig. 1). BUFE et al. (1994) analyzed the seismicity trends in the Alaska–Aleutian region, and found that the seismic release is accelerated within the recent decade in four zones along trenches. They applied the time-to-failure analysis, that is, fitting a power-law type of time function with the data, and estimated the expected occurrence times and magnitude of future large earthquakes

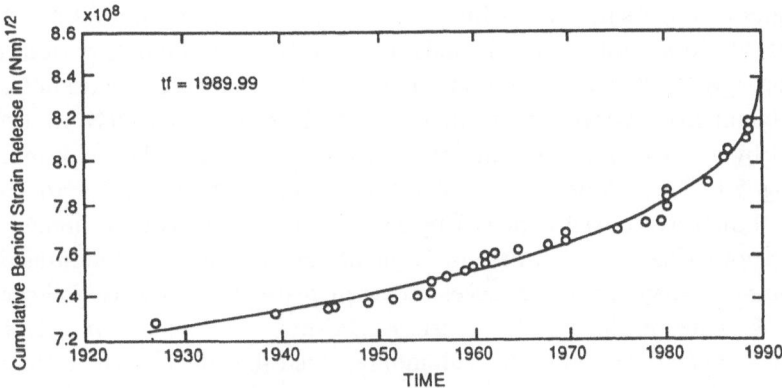

Figure 1
Accelerative increase of the cumulative Benioff strain of earthquakes with $M \geq 5$ within about 100-km wide and 400-km long area in northern California observed before the 1989 Loma Prieta earthquake (M 7.1) (BUFE and VARNES, 1993). t_f represents the expected time of a large earthquake obtained by fitting the power-law function to the data. The figure is slightly modified from the original one.

for four regions. In one of four regions, the Adak earthquake (M 7.6) actually occurred in 1996 (expected time was 1994.1 and magnitude 7.8), while in the rest of the regions no large earthquake has occurred yet. As for the Loma Prieta and Adak earthquakes, YAMAMOTO (1994) also tried to estimate the expected times and

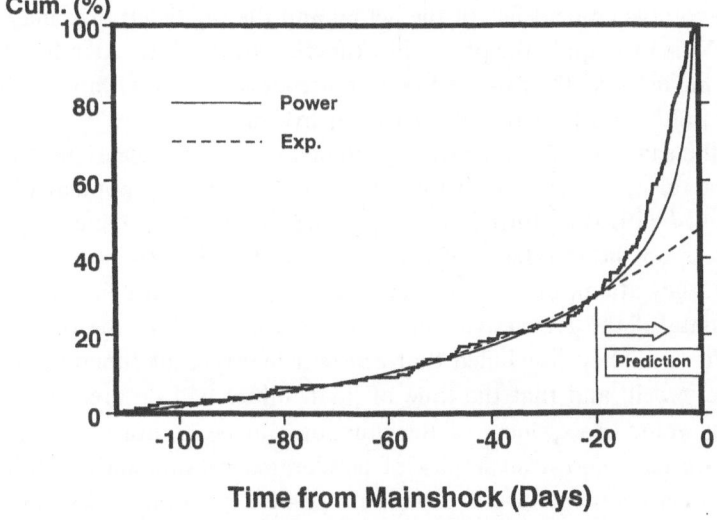

Figure 2
Accelerative increase of foreshocks observed before and within 10 km from the 1993 western Nagano earthquake, Japan (M 5.1). Solid and broken lines represent the predicted activity by fitting the power law and exponential type of functions, respectively, to the data up to 20 days before the mainshock.

magnitudes of them similarly to Bufe's method, although using a modified power-law type of fitting function originally derived from a laboratory test of rock fracturing. KNOPOFF et al. (1996) reported that all eleven earthquakes in California with nominal magnitudes greater than or equal to 6.8 from 1941 to 1993 were preceded by an increase in the rate of occurrence of earthquakes with magnitudes exceeding 5.1. These long-term accelerating seismic activities prior to large and great earthquakes are well reviewed by JAUMÉ and SYKES (this volume).

On the other hand, within the time scale of one year or several months there are few examples which show an accelerative increase of seismicity followed by a mainshock. One of them is the western Nagano earthquake (M 5.1) in 1993. YAMAOKA et al. (1993 and in this volume) reported that the occurrence time of this event could be predictable by applying the formula of VOIGHT (1988) to the seismicity data. It is known that applying the formula of Voight is equivalent to applying the power-law type of function (BUFE and VARNES, 1993). Here we present our original results of prediction made by applying the power-law type and exponential type of functions to the JMA (Japan Meteorological Agency) earth-quake catalogue data of all magnitudes, from 120 to 20 days before and within 10 km from the 1993 western Nagano mainshock (Fig. 2). Obviously, the power-law type of function better represents the rapid increase of activity in the prediction period before the mainshock than does the exponential type. The time difference between the actual occurrence of the mainshock and the expected time obtained by fitting the power-law function is less than one hour. However, the error in estimating the occurrence time of the mainshock is as sizable as 50 days in this case, thus, the favorable coincidence of the actual and the predicted time may be a result of chance. When we apply the power-law function to the data extending to five days before the mainshock, the expected occurrence time is also within one hour of the actual time, although the error decreases to five days.

As for the example in a considerably smaller space-time scale, the AE activity in the rock fracturing experiment in the laboratory is the most popular phenomenon. YOSHIDA et al. (1994) performed the experiment to produce a stick slip by loading a shear stress on the interfaces of a rock placed between two other rocks. In the experiment, they attempt to predict the occurrence time of the final failure by using the count data of the AE activity up to five seconds before the time of the main fracture (Fig. 3). They concluded that a power-law type of function approximates the data very well, and that the time of final failure will be predicted with minor errors if the value of exponent of the function can be accurately estimated.

So far we have shown examples of accelerated seismic activity followed by a mainshock or a main fracture in a wide range of space-time scales. The character-istic feature of this phenomenon is that a power-law type of time function can well approximate the increasing rate of the events. However, the characteristics of the foreshock activity immediately before the mainshock are not well discussed because of the shortage of data. In the following section we focus on the foreshock activity

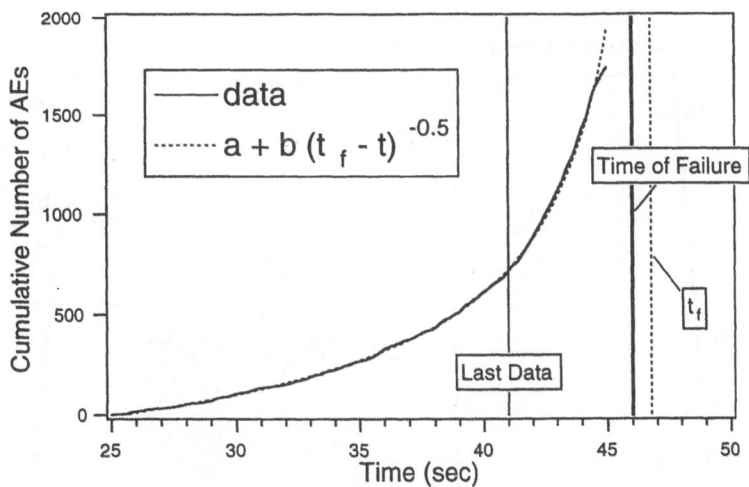

Figure 3

Accelerative increase of microfractures before the main rupture in a laboratory rock fracturing experiment. The broken line represents the predicted activity by fitting a power-law type of function to the data up to five seconds before the main rupture. The parameters a, b, and t_f are estimated by fitting the power-law function to the data, and t_f represents the expected time of failure (after YOSHIDA et al., 1994).

just two days before a mainshock, and investigate the statistical features of the time distribution of immediate foreshocks using a stacking method.

3. Stacking Method for Space-time Distribution of Immediate Foreshocks

Our investigation is based on the JMA earthquake catalogue data from 1977 through 1997/9/30. Taking into account the magnitude completeness of the observation system during this period (ISHIKAWA, 1987), we use only the earthquakes with $M \geq 3.0$. Firstly, we select mainshocks from the catalogue. Mainshocks are defined as earthquakes with $M \geq 5.0$, depth ≤ 60 km except for aftershocks. The definition of an aftershock is as follows:

distance: $\log L \leq 0.5 M_m - 1.8$

time: $t \leq 10^{[0.17 + 0.85(M_m - 4.0)]/1.3} - 0.3$

magnitude: $M_a < M_m$,

where L, t, M_m, and M_a represent an epicentral distance from a mainshock, time in days from occurrence of a mainshock, magnitude of a mainshock, and that of an aftershock, respectively. This criterion of the aftershock definition is basically derived from the empirical formula obtained by UTSU (1970). Next, we carry out a

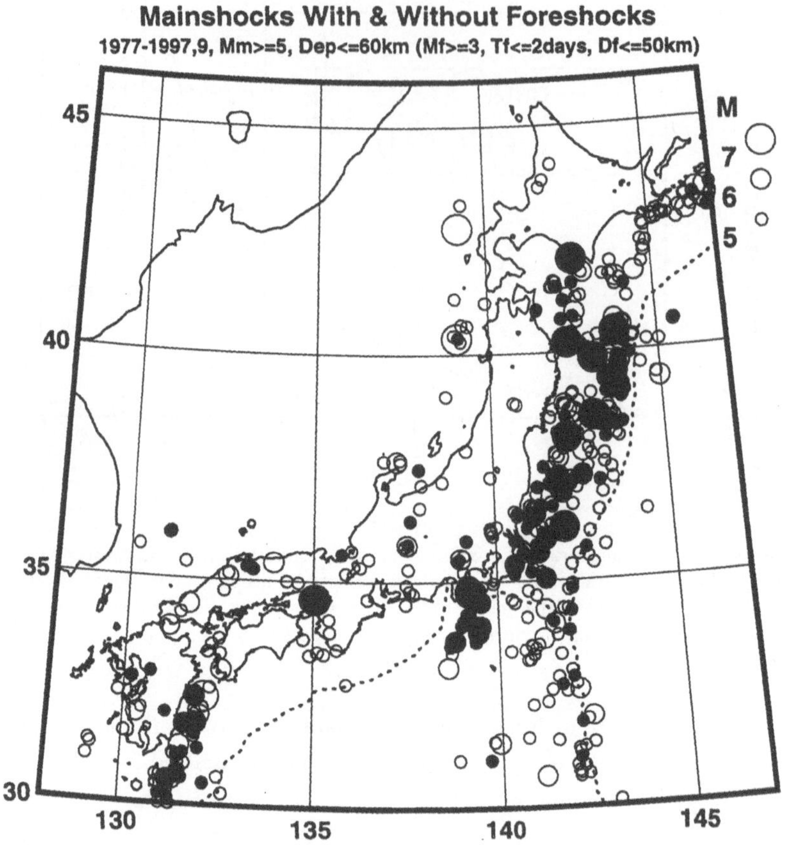

Figure 4

Epicentral distribution of investigated mainshocks with $M \geq 5$. Solid and open circles represent the mainshocks, preceded by foreshock(s) and no foreshock, respectively.

search for foreshocks with $M \geq 3.0$ and depth ≤ 100 km within a distance of 50 km and a time of two days from each mainshock. The total number of mainshocks selected is 589, and 192 (33%) of them are preceded by foreshock(s) as defined above.

The epicentral distribution of mainshocks preceded by foreshock(s) and no foreshocks is shown in Figure 4. The rate of mainshocks preceded by foreshock(s) is as high as about 50–70% in the regions of Sanriku and Off-Ibaraki (northeastern region along the trench), Hyuganada (southwestern region near the coast) and the Izu volcanic area (south–central region). On the contrary, the rate is very low and is less than about 20% in the inland area. Some characteristics associated with foreshock activities along the Japan trench and the Izu region are reported by MAEDA (1993, 1996).

As, generally, only a few foreshocks precede each mainshock, we stack many foreshock–mainshock sequences to obtain the statistical features of immediate foreshocks. This stacking method implies that we assume that the probabilistic distribution of immediate foreshocks is similar for different mainshocks. In order to avoid overestimating a specific foreshock activity, we select one representative foreshock for each mainshock and stack the relative location and time of the representative foreshock in reference to each mainshock. Three types of the representative foreshock are proposed: the first type is the largest foreshock, the second is the foreshock nearest in distance to the mainshock, and the last is the foreshock nearest in time to the mainshock. All of these foreshocks are searched for events within a distance of 50 km and a time of two days from each mainshock. However, as little difference can be found between the distributions for the three types of representative foreshocks, we mainly discuss the distribution of the largest foreshocks in the following sections.

4. Results

The stacked space-time distribution of the largest foreshocks is shown in Figure 5. The origin of the axis corresponds to the space-time location of each mainshock. From this figure we find that the probability of occurrence of immediate foreshocks increases as the time and the distance approaches a mainshock. Particularly, the rate of increase as a function of time is marked within a distance of about 30 km from a mainshock. The pattern of foreshock distribution obtained in this study is very similar to the results obtained by stacking all the foreshocks in Japan (OGATA et al., 1995), in California (JONES, 1985), and in Italy (CONSOLE et al., 1993). The time distribution of immediate foreshocks is mainly discussed in this paper, as the mainshocks preceded by foreshocks occur mostly in the sea area (Fig. 4) where the hypocenter is not so accurately determined.

Next, we investigated which type of function, power or exponential, can better approximate the accelerated distribution of immediate foreshocks. The data within 30 km distant from each mainshock are used in the following analysis, as the accelerative increase within that distance is remarkable. We adopt the function form of $A1 + A2 \cdot (A3 - t)^{-A4}$ as a power-law type, and $B1 + B2 \cdot \exp(B3 \cdot t)$ as an exponential type. The parameter values of $A1 \sim A4$ and $B1 \sim B3$ are estimated by using the computer program coded by UTSU (1997), which was originally developed to analyze aftershock activity on the basis of the maximum likelihood method. Figure 6 exhibits the cumulative data of the time distribution of the largest foreshocks together with the best-fitting curves of the power and exponential types. It can be seen from this figure that the power-law type of function is a better fit than the exponential one as a whole. To investigate more objectively, we first apply the Kolmogorov test, which is useful to determine the standard goodness of the model fitness on the basis of the maximum difference between the data and a model. The result of this test also suggests that the power-law type is a better fit, although both types of the models cannot be rejected with a five percent level of confidence except one case. One exception is that the exponential model for the nearest foreshocks in time is rejected by the data. Then we calculate the AIC values. The estimated values of the parameters and the AIC values for the two types of time functions for the three types of representative foreshocks are listed in Table 1. The parameter $A1$ is fixed to be 0, as the AIC value for this case is smaller (that is, this fits the data better) than that for the case of $A1$ being treated as a free parameter. This implies that the effect of the background seismicity on the stacked time distribution is very small. The table shows that the AIC values calculated for the power function are considerably smaller than those for the exponential one for all three types of foreshocks, which also means that the power type fits significantly better than the exponential one. JONES and MOLNAR (1979), and OGATA et al. (1995) also obtained a similar result, therefore this seems to be a fairly universal characteristic of immediate foreshocks. The value of exponent ($A4$) obtained in this study is significantly smaller than that (1.120) by OGATA et al. (1995). This distinction

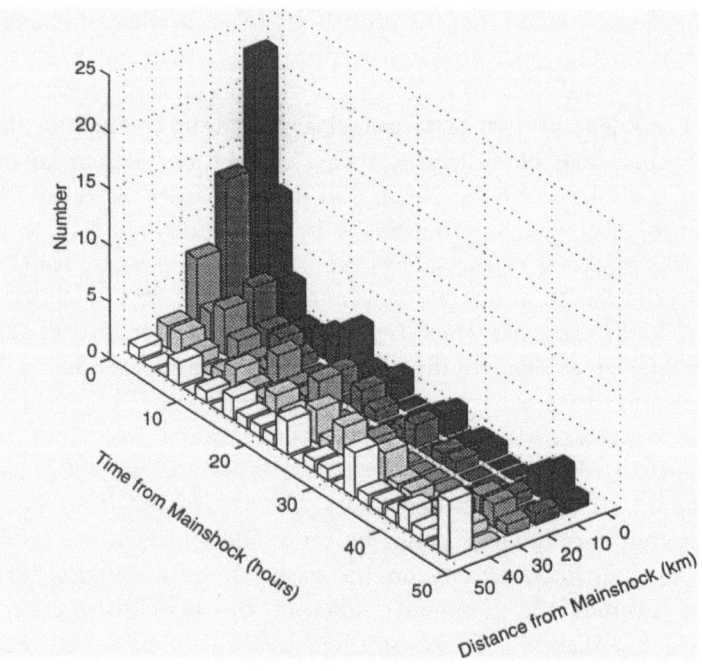

Figure 5
Space-time distribution of the largest foreshocks obtained by a stacking method. The origin of the axis represents the space-time location of each mainshock.

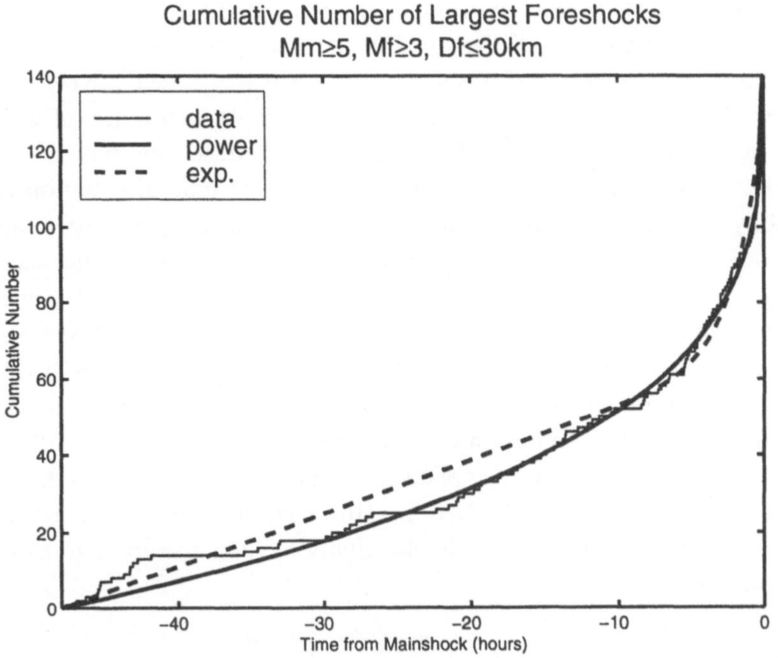

Cumulative Number of Largest Foreshocks
Mm≥5, Mf≥3, Df≤30km

Figure 6

Cumulative number of the largest foreshocks within 30 km from and two days before mainshocks. Solid and broken lines represent the fitted curves of power law and exponential functions, respectively.

probably derives from the differentiation of the stacking methods, that is, they stack all the foreshock activities, whereas we stack only the representative foreshocks.

5. Discussion

Two different explanations are possible to interpret the accelerative increase of the stacked time distribution of immediate foreshocks. The first is that foreshocks are caused by the change of stress field near a mainshock, and the second is that a foreshock causes the stress change and a mainshock eventually.

In the first explanation, the instability is caused by, for example, the accelerative increase of pre-slip in the region near the subsequent mainshock area, and foreshocks are the result of a stress change in the region. YAMAMOTO (1998), for example, tries to explain the accelerative increase of microfracture in intact rocks as well as a long-term increase of seismicity, by assuming a constant rate of stress increase and the power-law type of strength distribution. Qualitatively, this concept interprets the accelerative increase of foreshocks as follows. A fracture or a pre-slip begins to occur in the weaker portion of the area as the applied stress increases. This accelerates the increase of the stress in the rest of the area. Consequently the probability of occurrence of foreshocks also increases acceleratively. However, foreshocks may not necessarily occur, as the actual occurrence of foreshocks depends on the distribution of asperities that produce the seismic waves as foreshocks. In this explanation, the power-law type of time distribution could be obtained by assuming an appropriate stressing rate and strength distribution.

The other possible explanation for the power-law type of the accelerating seismicity categorized in the first explanation is made by some researchers (e.g., SORNETTE and SAMMIS, 1995; SALEUR et al., 1996; SAMMIS and SMITH, in this volume). They have developed a model of seismicity on the basis of statistical physics in a state of self-organized criticality (SOC), and demonstrated that the power law is expected if the largest event is viewed as a type of "critical" point for the region. This model is mainly applied to the increase of the long-term seismicity as introduced in the preceding section. However, it is not clear whether this hypothesis is applicable to the foreshock activity in such a short term as two days before a mainshock.

Table 1

Values of the parameters and the AIC for the power and exponential types of the time functions, estimated by the maximum likelihood method for the three types of representative foreshocks with $M \geq 3.0$ within 30 km from and two days before mainshocks with $M \geq 5.0$. The parameter A1 is fixed to be 0, as this fits the data better than A1 being treated as a free parameter. N represents the number of data

LARGEST FORESHOCKS ($N = 133$)				
POWER TYPE	A1	A2	A3	A4	AIC
	0	14.01	0.000	0.7263	−191.0
EXP. TYPE	B1	B2	B3		AIC
	1.381	33.53	0.5133		−163.3

NEAREST FORESHOCKS IN DISTANCE ($N = 149$)				
POWER TYPE	A1	A2	A3	A4	AIC
	0	18.41	0.06416	0.7901	−238.1
EXP. TYPE	B1	B2	B3		AIC
	1.551	43.91	0.6072		−228.6

NEAREST FORESHOCKS IN TIME ($N = 136$)				
POWER TYPE	A1	A2	A3	A4	AIC
	0	19.28	0.02864	1.029	−442.9
EXP TYPE	B1	B2	B3		AIC
	0.8076	78.74	0.8240		−388.6

The second explanation for the immediate foreshock distribution is that a foreshock is the cause of a stress change in the region and that it triggers a mainshock. This is compatible with the relationship between a mainshock and aftershocks, for the case when an aftershock happens to become larger than the mainshock (JONES et al., 1995). Consistent with this idea, the increase of stacked foreshocks is interpreted as an apparent feature obtained by stacking representative foreshocks in reference to mainshocks. If we stack the occurrence times of mainshocks in reference to representative foreshocks, we will obtain the decaying distribution with the power law. This power law of the stacked time distribution of mainshocks can be interpreted similarly as aftershocks are distributed in the form of the power law known as Omori's formula.

If the second explanation is the more realistic, it is worth investigating in order to clarify the difference of the stacked time distributions of foreshocks and aftershocks. Therefore, we compare the values of the exponent of the power law for both distributions. When we calculate the values of the exponent p for aftershocks, we adopt the same method as used to obtain those for foreshocks. That is, we select three types of representative aftershocks for each mainshock, and stack their relative location and time respectively in reference to each mainshock. Three types of the representative aftershocks are the largest aftershock, the aftershock nearest in distance to the mainshock, and the aftershock nearest in time to the mainshock. All of these aftershocks are searched for events within a distance of 50 km and two days from each mainshock with $M \geq 5.0$. To avoid the effect of swarm type activities, we use only the cases in which the magnitude difference (M_d) between a mainshock and the largest aftershock is larger than 1.0.

The results are as follows: the values of the exponent for the largest foreshocks and the largest aftershocks are 0.73 ± 0.06 and 0.92 ± 0.06; those for foreshocks and aftershocks nearest in distance are 0.79 ± 0.07 and 1.02 ± 0.08; and those for foreshocks and aftershocks nearest in time are 1.03 ± 0.07 and 1.24 ± 0.07. The difference of the exponent for the largest foreshocks and aftershocks is illustrated in Figure 7. All of these results show that the value of the exponent for foreshocks is significantly smaller than that for aftershocks. However, we should note that the difference for the type of events nearest in time might be artificially derived from the magnitude difference between foreshocks and mainshocks. One additional important matter to be noted is that the difference of the exponent between foreshocks and aftershocks becomes small and cannot be considered significant, if the value of the exponent for aftershocks is calculated for all the activities including the swarm type (that is the case for $M_d > 0$). Regardless, the results mean that the relationship between foreshocks and mainshocks is distinctive from that for mainshocks and generic aftershocks.

DIETERICH (1994) proposed a theoretical model to explain earthquake clustering, especially aftershock activity on the basis of the nucleation process. According to his model, the decay rate of aftershocks depends on the stressing rate applied in

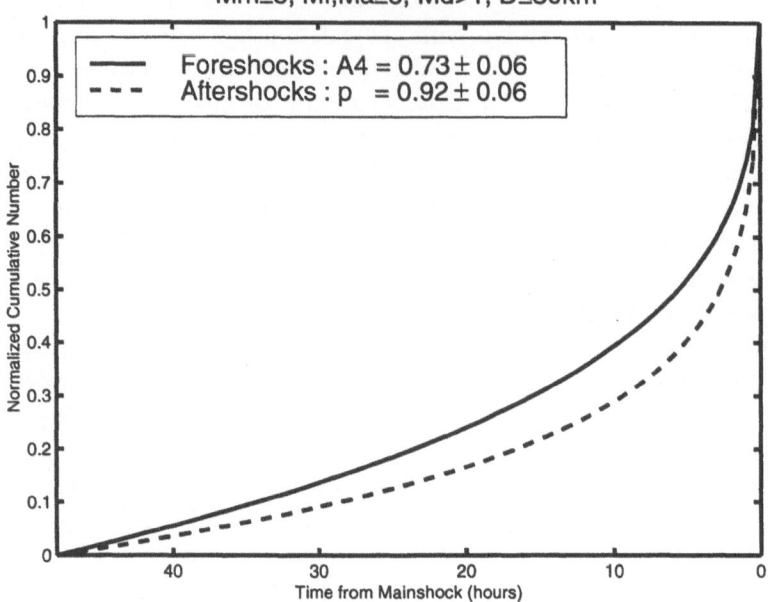

Figure 7
Comparison between the increasing rate of the stacked largest foreshocks and the decay rate of the stacked largest aftershocks. Solid and broken lines represent the power-law curves fitted to the largest foreshocks and the largest aftershocks, respectively.

the region. Thus one possible interpretation for the smaller value of the exponent for foreshocks is as follows: The stressing rate increases acceleratively because of a pre-slip in some weaker portion of the stressed area, which derives a slower decay rate of aftershock activity, and one of the aftershocks becomes larger than the foreshock by chance.

As we have discussed above, both explanations possibly interpret the power-law type of stacked time distribution of immediate foreshocks. However, we cannot determine in this study which explanation of the two is the more realistic one. This is mainly because the stacking method applied here obscures the details of each activity and does not represent the entire activities of foreshocks or aftershocks. Further investigation will be necessary to answer this question.

6. Conclusions

We apply a stacking method to investigate the time distribution of immediate foreshocks on the assumption that every mainshock has a similar process to failure, and summarize our conclusions as follows.

(1) The statistical test for stacked time distribution of foreshocks reveals that the inverse power-law type of a probability density time function is a significantly better fit than the exponential one for all three types of representative foreshocks.

(2) Two different explanations possibly interpret the result. The first is that foreshocks occur as a result of a stress increase in the region, and the second one is that a foreshock is the cause of a stress change that triggers a mainshock.

(3) The values of the exponent of the power law obtained for stacked fore-shocks are significantly smaller than those for similarly stacked aftershocks. This implies that the foreshock–mainshock relationship should be different from a normal aftershock activity, even if a foreshock triggers a mainshock. Probably an increase of stress during foreshock activity results in the smaller values of the exponent.

In the actual activity of immediate foreshocks, the number of foreshocks is usually very small. Therefore, it seems very difficult to predict the orrurrence time of a mainshock on the basis of the number of events, even if the stacked time distribution of foreshocks represents an accelerative stress increase before a main-shock. However, further investigation of the immediate foreshock activity could be the key to understanding the physical process which leads to a mainshock.

Acknowledgements

The author thanks Paul A. Reasenberg and an anonymous reviewer for helpful comments and suggestions which have enhanced the manuscript. Special gratitude is due to Max Wyss who provided the author the opportunity of discussions at the Seismicity Patterns Conference in Nikko, Japan, and has made useful comments regarding the manuscript.

References

BUFE, C. G., NISHENKO, S. P., and VARNES, D. J. (1994), *Seismicity Trends and Potential for Large Earthquakes in the Alaska–Aleutian Region*, Pure appl. geophys. *142*, 83–99.

BUFE, C. G., and VARNES, D. J. (1993), *Predictive Modeling of the Seismic Cycle of the Greater San Francisco Bay Region*, J. Geophys. Res. *98*, 9871–9883.

CONSOLE, R., MURRU, M., and ALESSANDRINI, B. (1993), *Foreshock Statistics and their Possible Relationship to Earthquake Prediction in the Italian Region*, Bull. Seismol. Soc. Am. *83*, 1248–1263.

DIETERICH, J. (1994), *A Constitutive Law for Rate of Earthquake Production and its Application to Earthquake Clustering*, J. Geophys. Res. *99*, 2601–2618.

ISHIKAWA, Y. (1987), *Change of JMA Hypocenter Data and Some Problems*, Quarterly J. Seismol. *51*, 47–56 (in Japanese).

JAUMÉ, S. C., and SYKES, L. R. (1999), *Evolving Towards a Critical Point: A Review of Accelerating Seismic Moment/Energy Release Prior to Large and Great Earthquakes*, Pure appl. geophys., this volume.

JONES, L. M. (1985), *Foreshocks and Time-dependent Earthquake Hazard Assessment in Southern California*, Bull. Seismol. Soc. Am. *75*, 1669–1679.

JONES, L. M., CONSOLE, R., LUCCIO, F. D., and MURRU, M. (1995), *Are Foreshocks Mainshocks whose Aftershocks Happen to be Big? Evidence from California and Italy* (Abstract), EOS, Trans. AGU *76*, 388.

JONES, L. M., and MOLNAR, P. (1979), *Some Characteristics of Foreshocks and their Possible Relationship to Earthquake Prediction and Premonitory Slip on Faults*, J. Geophys. Res. *84*, 3596–3608.

KNOPOFF, L., LEVSHINA, T., KEILIS-BOROK, V. I., and MATTONI, C. (1996), *Increased Long-range Intermediate-magnitude Earthquake Activity Prior to Strong Earthquakes in California*, J. Geophys. Res. *101*, 5779–5796.

MAEDA, K. (1993), *An Empirical Alarm Criterion Based on Immediate Foreshocks—A Case Study for the Izu Region*, J. Seism. Soc. Japan *2*, 45, 373–383 (in Japanese).

MAEDA, K. (1996), *The Use of Foreshocks in Probabilistic Prediction Along the Japan and Kuril Trenches*, Bull. Seismol. Soc. Am. *86*(1A), 242–254.

MOGI, K. (1962), *Study of Elastic Shocks Caused by the Fracture of Heterogeneous Materials and its Relations to Earthquake Phenomena*, Bull. Earthquake Res. Inst. Univ. Tokyo *40*, 125–173.

OGATA, Y., UTSU, T., and KATSURA, K. (1995), *Statistical Features of Foreshocks in Comparison with Other Earthquake Clusters*, Geophys. J. Int. *121*, 233–254.

SALEUR, H., SAMMIS, C. G., and SORNETTE, D. (1996), *Discrete Scale Invariance, Complex Fractal Dimensions, and Log-periodic Fluctuations in Seismicity*, J. Geophys. Res. *101*, 17661–17677.

SAMMIS, C. G., and SMITH, S. W. (1999), *Seismic Cycles and the Evolution of Stress Correlation in Cellular Automaton Models of Finite Fault Networks*, Pure appl. geophys., this volume.

SCHOLZ, C. H. (1968), *Microfracturing and the Inelastic Deformation of Rock in Compression*, J. Geophys. Res. *73*, 1417–1432.

SCHOLZ, C. H., *The Mechanics of Earthquake and Faulting* (Cambridge University Press 1990) pp. 1–41.

SORNETTE, D., and SAMMIS, C. G. (1995), *Complex Critical Exponents from Renormalization Group Theory of Earthquakes: Implications for Earthquake Predictions*, J. Phys. I. *5*, 607–619.

UTSU, T. (1970), *Aftershocks and Earthquake Statistics (2)—Further Investigation of Aftershocks and Other Earthquake Sequences Based on a New Classification of Earthquake Sequences*, J. Fac. Sci. Hokkaido Univ. Ser. 7(3), 197–266.

UTSU, T., *IASPEI Software Library* (IASPEI 1997).

VOIGHT, B. (1988), *A Method for Prediction of Volcanic Eruptions*, Nature *332*, 125–130.

YAMAMOTO, K. (1998), *Estimation of Fracture Stress for Intact Rocks and Possibility of Long-term Earthquake Prediction*, J. Seismol. Soc. Japan *50*, 169–180 (in Japanese).

YAMAOKA, K., OOIDA, T., and UEDA, Y. (1999), *Detailed Distribution of Foreshocks Acceleration before M 5.1 Earthquake in Japan*, Pure appl. geophys., this volume.

YAMAOKA, K., UEDA, Y., OOIDA, T., and YAMAZAKI, F. (1993), *Earthquake Swarm Predicting the M 5.1 Event near Mt. Ontake, Central Japan*, Abstract for Seismol. Soc. Japan Fall Meeting, 314 (in Japanese).

YOSHIDA, S., OHNAKA, M., and SHEN, L. F. (1994), *Prediction of Failure Time in Slip Failure Experiments*, Abstract for Seismol. Soc. Japan Fall Meeting, 23 (in Japanese).

(Received October 1998, revised January 16, 1999, accepted January 22, 1999)

 To access this journal online:
http://www.birkhauser.ch

Pure appl. geophys. 155 (1999) 395–408
0033–4553/99/040395–14 $ 1.50 + 0.20/0

⎮ **Pure and Applied Geophysics**

Pattern Characteristics of Foreshock Sequences

Yong Chen,[1] Jie Liu,[2] and Hongkui Ge[3]

Abstract—Earthquake clusterings in both space and time have various forms, in particular, two typical examples are the foreshock sequences and earthquake swarms. Based on the analysis of 8 foreshock sequences in mainland China during 1966–1996, this study concentrates on the pattern characteristics of foreshock sequences. The following pattern characteristics of foreshock sequences have been found: (1) the epicenters of foreshock sequences were densely concentrated in space; (2) the focal mechanisms of foreshocks were similar to that of the main shock. Such consistency of focal mechanisms with main shocks did not exist in aftershock series as well as in several earthquake swarms; (3) we found no case in mainland China during the past thirty years that a main shock is preceded by an earthquake clustering with inconsistent focal mechanisms. Finally, we found 5% of the main shocks in mainland China are preceded by foreshock sequences.

Key words: Foreshock sequence, seismic pattern, consistency of focal mechanism.

1. Introduction

Foreshocks are the precursors that undoubtedly are physically related to the occurrence of main shocks, and the usefulness of foreshocks for prediction, particularly for imminent earthquake prediction, is essentially important.

If an earthquake or a few earthquakes occurred before a main shock at the same place of the main shock, obviously, this or these few earthquakes could be considered as the foreshocks of the main shock. However, it is very difficult to distinguish this kind of foreshock from the background earthquake activity. For this reason we turned our interests toward the foreshock sequences. A foreshock sequence is an earthquake clustering which consists of many foreshocks. Among the various definitions of foreshocks (Wyss, 1991; Seggern, 1981), we found four common points existing:

---

[1] Institute of Geophysics, Chinese Academy of Sciences, Beijing 100101, China.

[2] Center for Analysis and Prediction, China Seismological Bureau, Beijing 100036, China.

[3] Department of Petroleum Engineering, University of Petroleum, Dongying, Shandong 257062, China.

1. foreshock sequence is one of high-level seismic activities, i.e., the earthquake occurrence rate in foreshock sequences is considerably higher than that of the background level;
2. foreshock sequences occur before the main shock (a few hours to a few days);
3. foreshock sequences occur at the same place as main shocks, therefore, foreshock sequences can indicate the place of a forthcoming main shock;
4. the magnitude of the main shock is greater than that of any earthquake foreshock sequences.

A causal relationship between the foreshocks and main shocks has been recognized by scientists for decades (SEGGERN, 1981; WYSS, 1991). In the present paper our focus concentrates on the discrimination of foreshock sequences and earthquake swarms, i.e., on the recognition of the differences of the seismicity pattern between foreshock sequences and other sequences not followed by a main shock. We start with the case studies.

2. Case Studies: Foreshock Sequences

Normally, seismic activity was very low around the epicentral region of the Haicheng earthquake (m = 7.3, February 4, 1975. CHEN, 1979). However, from

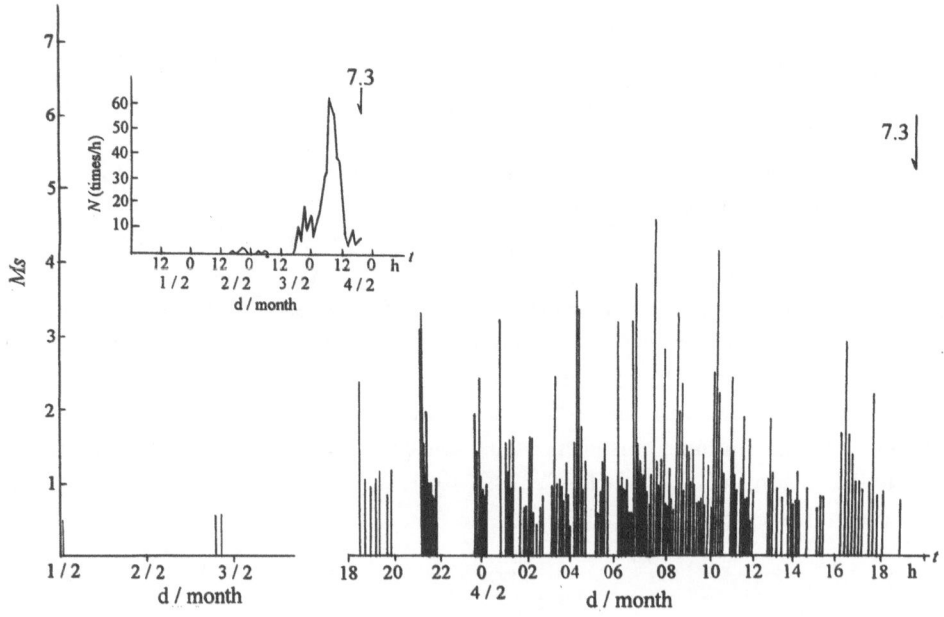

Figure 1

Magnitude of foreshocks which occurred near Haicheng during the period 1 to 4 February, 1975. The main shock of magnitude 7.3 struck at 7:36 p.m., 4 February. The insert diagram illustrates the average number of foreshocks per hour from 1 until 4 February (CHEN *et al.*, 1988).

February 1, 1975, 521 foreshocks were recorded at Shipengyu seismic station located about 20 km from the epicenters (CHEN, 1979). Several days before the main shock and apparently near the epicenter of them, the earthquake activity began to increase—a common feature of foreshocks (Fig. 1). Statistical studies of earthquake catalogs have clearly shown that the rate of foreshock occurrence is well above the background level (BOWMAN and KISSLINGER, 1984).

The distributions of main shock, foreshocks and aftershocks in the Haicheng earthquake were obtained based on the regional seismographic network of Liaoning Province (Fig. 2). It can be seen from Figure 2 that the epicenters of foreshocks were spatially, densely concentrated. Using arrival times from six local seismic stations, Jones et al. found that the foreshock activity of the Haicheng earthquake was located within a small, approximately equidimensional volume with a diameter of a few km (JONES et al., 1983).

Because the accuracy of determination of earthquake epicenters by a regional network is not high enough (about ± 5 km), we use the differences of t_s (arrival time of S wave) and t_p (arrival time of P wave) obtained from the nearest station to study the spatial clustering of foreshocks. Figure 3 shows the distances from earthquake epicenters to the station according to the values of $t_s - t_p$. It can be found in Figure 3 that the $t_s - t_p$ of foreshocks were similar ($t_s - t_p = 2.5$ sec). Considering all the foreshocks located in the same direction of the station (Fig. 2), we concluded that the epicenters of fore-shocks were spatially, densely concentrated. On the other hand, the epicenters of aftershocks were vastly scattered ($t_s - t_p$ ranging from 1 sec to 5.5 sec, see Fig. 3).

The fault plane solutions of foreshocks, main shock and aftershocks of Haicheng earthquake are given in Figure 4. Figures 4(a), 4(b) show the fault plane solutions of foreshocks which coincide with that of the main shock (Fig. 4(c)). The focal mechanisms of eight major aftershocks were shown in Figures 4(d)–4(k). It is obvious that the mechanisms of aftershocks vary from one quake to another. Only 2 focal plane solutions of foreshocks were given in Figure 4 because of the limited monitoring capacity of the network in the 1970s during the time the focal plane solutions could only be obtained for those earthquakes with magnitude 4 and above. Similarly, we can find the same feature from the foreshock sequence of Xingtai earthquake of 1966 (Fig. 5).

In the present study, our interest is not the exact focal mechanism of each foreshock (in fact, it is very difficult to collect enough data to determine the focal mechanism of each event, because most foreshocks are not large enough), rather our interest is in the consistency of focal mechanisms of many foreshocks in foreshock sequences. Therefore we must search for additional methods to monitor the changes of consistency of focal mechanisms.

Yong Chen *et al.*

Table 1

Foreshock sequences of mainland China during 1966–1996

Earthquake	m	Latitude	Longitude	Foreshock sequence start from	main shock	total no.	$m_{mainshock} - m_{foreshock}$	Distance from Main shock (km)
Xingtai	6.8	37.18	114.54	1966-03-06	03-08-05h	23	1.5	10
Wushi	6.1	41.19	79.22	1971-03-21	03-24-04h	29	0.1	10
Haicheng	7.3	40.63	122.81	1975-02-01	02-04-19h	521	2.6	15
Menla	5.7	21.18	101.30	1975-10-23	10-28-07h	33	1.0	20
Lunglin	7.3	24.22	98.38	1976-05-29	05-29-23h	16	2.0	20
Ninglang	6.7	27.30	101.05	1976-11-04	11-07-02h	16	3.2	15
Datong	5.8	39.57	113.50	1991-03-21	03-25-02h	13	4.6	10
Menlian	7.3	22.10	99.60	1995-07-10	07-12-08h	125	1.8	15

$m_{mainshock} - m_{foreshock}$ means the magnitude differences between the main shocks and the maximum foreshocks.

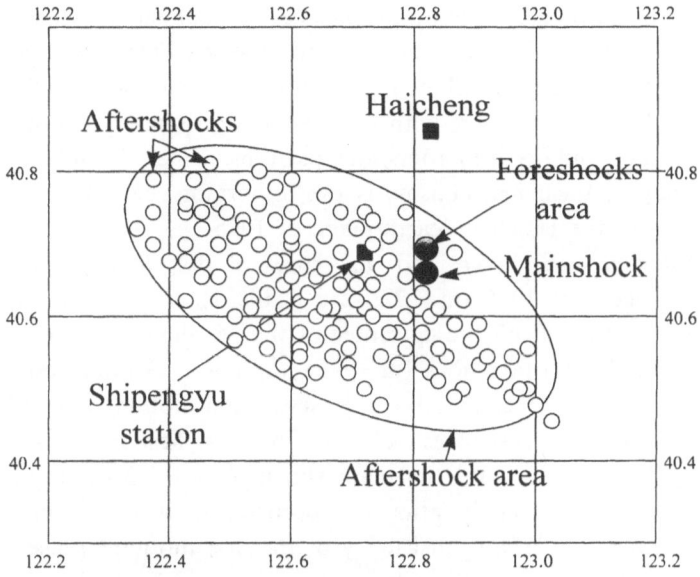

Figure 2
The distribution of foreshocks, main shock and aftershocks of the Haicheng earthquake of 1975.

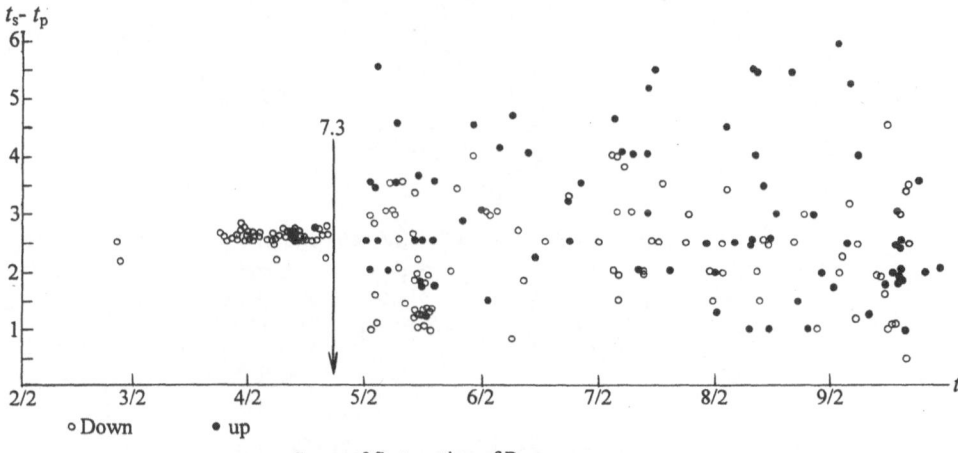

Sense of first motion of P-waves

Figure 3
$t_s - t_p$ and senses of first motions of P waves recorded at Shipengyu station during 1–9, February, 1975, arrow represents the occurrence time of the main shock of Haicheng, m = 7.3.

For a given station and for a given epicenter area, we can monitor the changes of focal mechanisms by using the senses of first motions recorded in the station. It also can be seen from Figure 3 that the senses of the *P*-wave first motions at Shipengyu station (the station nearest the epicenter for foreshocks) were all of the same senses ("down") before the occurrence of the main shock which indicated no remarkable changes of focal mechanisms of foreshocks. There were 79 first P-wave motions identified clearly at Shipengyu station among the 521 foreshocks; 78 foreshocks of the 79 events were with the first motion sense of "down." Contrastingly, great changes did occur in aftershock patterns, some "down" and some "up." The consistency of focal mechanisms is the important characteristic of the fore-shock sequence of the Haicheng earthquake of 1975.

Figure 6 displays the $t_s - t_p$ and the senses of first P-waves motions of the Menlian earthquake of 1995 (m = 7.2, see Table 1), which were recorded at a seismic station 120 km from the epicenter of the Menlian earthquake. Before the occurrence of the largest foreshock (m = 6.2), the $t_s - t_p$ of the foreshock sequence ranged from 15 sec to 17 sec, and all the senses of first motions were "down." After the occurrence of the largest foreshock, the senses of first P-waves motions kept "down" for all the events until the main shock of m = 7.3 occurred. The senses started to change immediately after the occurrence of the main shock: some "down" and some "up." The consistency of the first motion's senses was used to predict whether the largest earthquake past over or not in the Menlian case.

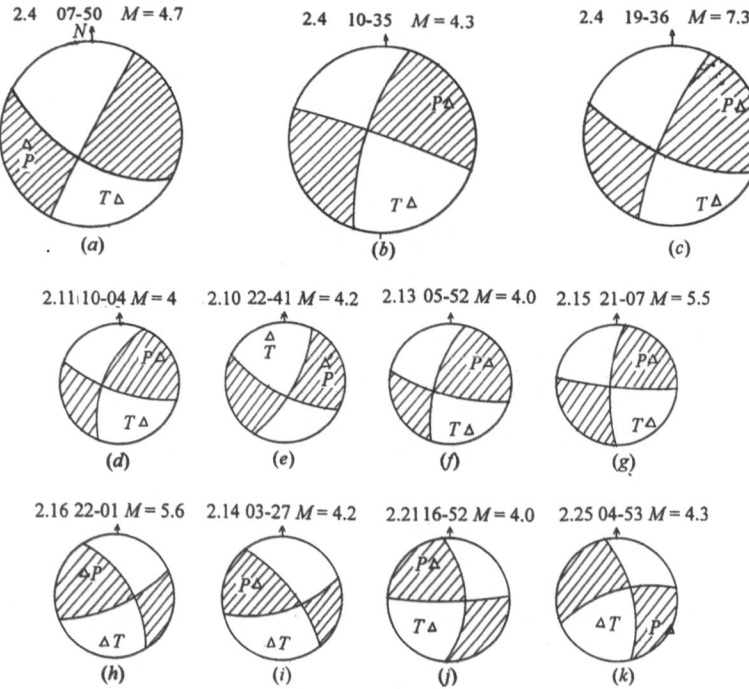

Figure 4
The focal mechanisms of Haicheng earthquake of 1975. (a), (b) foreshocks; (c) main shock; (d)–(k) aftershocks.

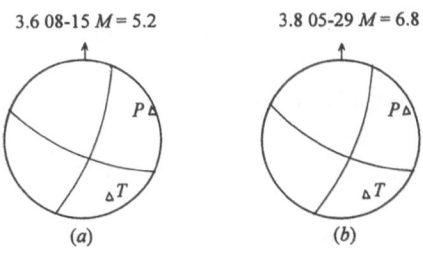

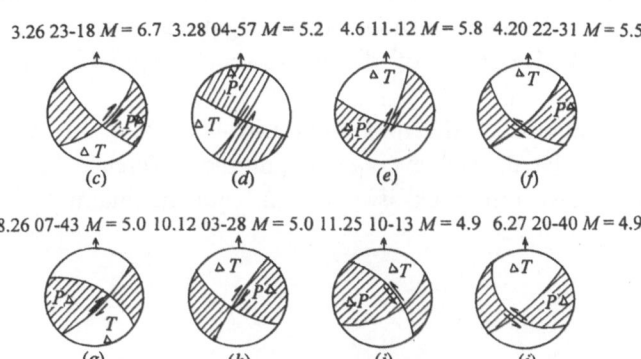

Figure 5
The focal mechanisms of Xingtai earthquakes of 1966 (after CHEN, 1979). (a) Foreshocks; (b) main shock; (c)–(j) aftershocks.

Another characteristic of the Haicheng foreshock sequences is that the temporal process of foreshock sequences can be divided into four steps (Fig. 7(b)):

I. very low seismic activity (long-time average level);
II. very high seismic activity level (reaching several tens of events per day before main shock);
III. a few hours to a day of seismicity "quiescence";
IV. main shock occurrence.

Two other foreshock sequences, before the Xingtai and Datong earthquakes, respectively, were also given in Figures 7(a) and 7(c). The four steps of the temporal process in foreshock sequences, i.e., low activity—very high activity (few days)—quiescence (few hours to a day)—main shock occurrence, existed in all the above three foreshock sequences, however the time intervals of each step varied in different cases. Generally speaking, foreshock series are separated by a few hours of quiescence from the main shock. From the above three case studies it can be seen that the sudden transition of foreshock activities from intensive occurrence to temporary quiescence may be an important indicator to large earthquake occurrence.

Assume we define foreshock sequences according to the following conditions (LIN *et al.*, 1994):
- within 5 days before the main shock;
- within 20-km distance to the main-shock epicenter;
- earthquake occurrence rate exceeds ten events with magnitude 1.5 and above per day.

We found that there were 159 major earthquakes with magnitude m > 5.5 in the mainland China during 1966–1996, based on a preliminary determination of the earthquake catalog of China (provided by the Center of Analysis and Prediction, China Seismological Bureau). Among them eight earthquakes have foreshock sequences (Table 1). The ratio of the main shocks preceded by foreshocks is about 5%.

The two magnitudes, of the maximum foreshock and the main shock, are not systematically related. The average magnitude difference between the main shock and the largest foreshock is about 2.1 for mainland China, based on Table 1.

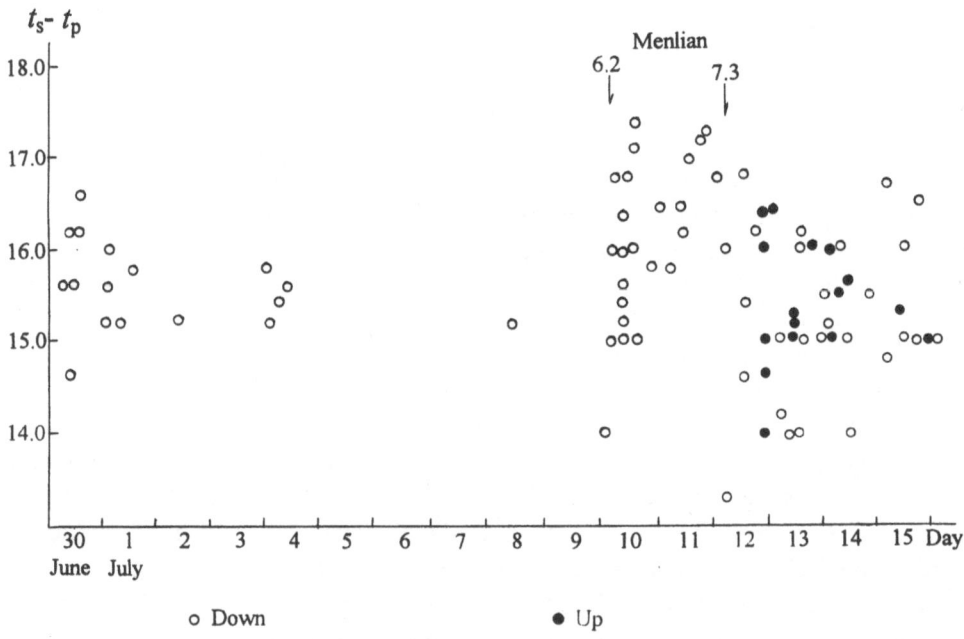

Figure 6

$t_s - t_p$ and senses of first motions of P waves recorded at a seismic station 120 km from the Menlian earthquake epicenter during 1–15 of July. Two arrows represent the occurrence times of the largest foreshock and main shock, respectively.

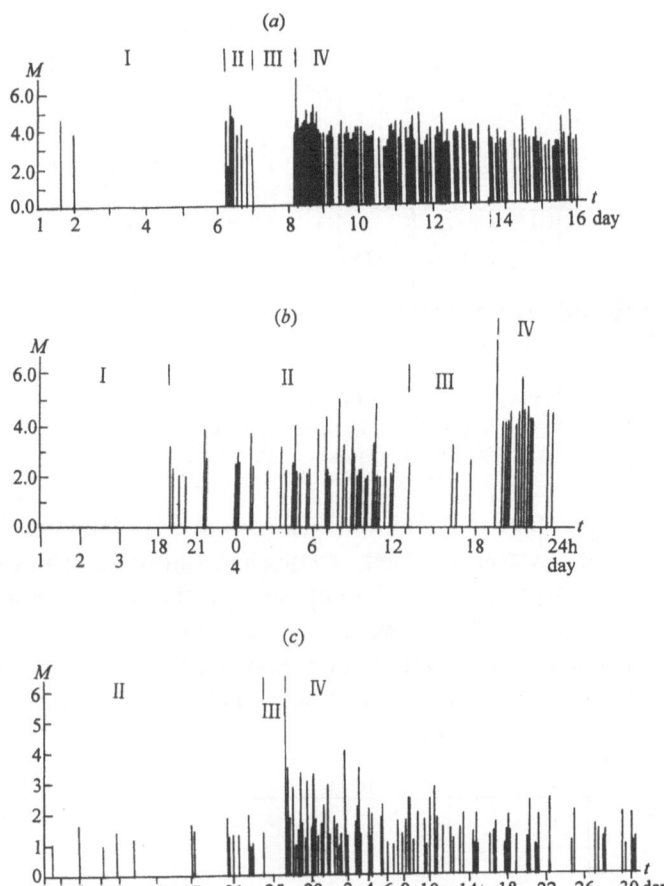

Figure 7

Foreshock sequences. (a) Xingtai earthquake (March 8, 1966, m = 6.8); (b) Haicheng earthquake (February 4, 1975, m = 7.3); (c) Datong earthquake (March 1, 1991, m = 5.8); I, II, III, IV represent different steps in the process of foreshock sequences (see text).

3. Case Studies: Earthquake Swarms

Another form of earthquake clustering, many earthquakes occuring in the same place but no larger earthquake following them, was called an earthquake swarm.

During 4–24 March, 1973, an earthquake swarm occurred in the Huoshan region (31.37°–31.48°N, 116.15°–116.21°E, see Table 2). 562 earthquakes were recorded at Huoshan seismic station, including three earthquakes with magnitude ≥ 4. The epicenters of Huoshan earthquake swarms concentrated in a small region within an area of 48 km². When many small tremors were felt by local people, the question arose as to whether these increasing activities were an earthquake swarm or a foreshock sequence (that means will a larger earthquake follow them)?

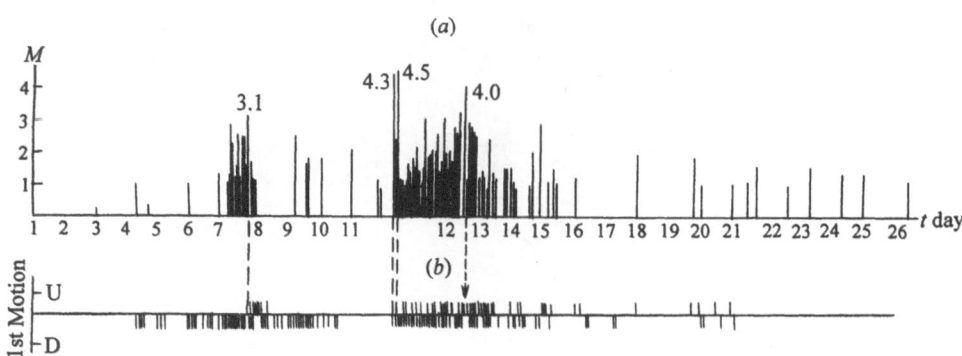

Figure 8
Huoshan earthquake swarm (1973, March 4–24). (a) Earthquake sequence: $m - t$ plot; (b) the senses of
the first motion of P waves as the function of time.

The magnitude vs. time plot and the senses of the first motion of P waves of the
Huoshan earthquake swarm recorded at Huoshan station are shown in Figure 8.
During 1972, one year before the occurrence of the swarm, Huoshan station
recorded 53.4% of "up" of the senses of the first motion and 46.6% of "down" for
background earthquakes in the Huoshan region. However, since 3 March of 1973
all of the senses of the first motion of P wave (47 events) became "down," after

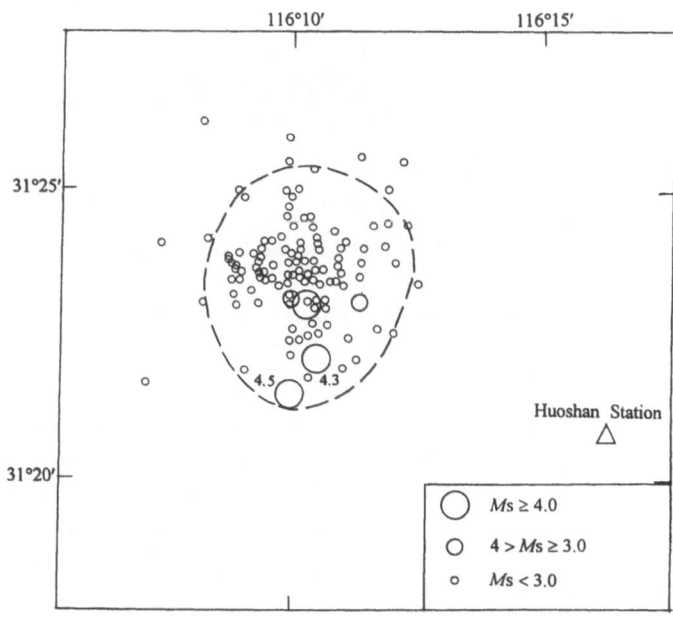

Figure 9
Huoshan earthquake swarm spatial distribution.

Table 2

Earthquake swarms in north China during 1966–1996

Swarms	m_{max}	Lat.	Long.	Time duration	$m_{max} - m_2$	Total number
Ningyang	5.0	35.70	116.88	1970.08.10-09.04	1.7	68
Heshun	5.2	37.38	113.50	1971.04.24-12.01	0.4	315
Wuzhong	5.2	37.84	106.25	1971.06.10-09.20	0.3	85
Pingyao	4.7	37.17	112.23	1971.11.26-01.26	0.9	132
Heshun	4.7	37.38	113.49	1972.01.14-03.26	1.1	120
Huoshan	4.5	31.37	116.17	1973.03.04-05.11	0.1	562
Zechuan	5.0	38.02	106.30	1973.11.29-01.08	0.2	44
Xinghe	4.9	37.58	115.08	1974.05.21-08.10	0.6	48
Shenwo	5.2	41.21	123.62	1974.11.18-01.31	1.1	227
Baodi	4.9	39.13	117.4	1976.08.02-09.02	0.5	31
Liangcheng	5.2	40.64	112.53	1976.10.14-11.14	0.5	39
Liangcheng	5.2	40.50	112.50	1977.03.02-06.24	0.9	57
Fouxi	5.1	41.95	121.30	1977.06.05-06.11	1.5	21
Weihai	4.6	37.52	122.05	1980.05.14-05.22	0.5	38
Linxian	5.1	36.05	113.93	1980.06.04-11.13	1.7	195
Liaoyang	5.1	40.68	122.67	1981.08.12-08.14	0.8	13
Baotou	5.1	40.62	109.39	1982.11.21-04.10	0.0	142
Huairou	4.9	40.47	116.56	1982.12.10-12.18	1.2	39
Xiongyue	4.8	40.45	122.10	1982.11.04-05.12	0.4	114
Dachun	4.7	40.10	115.85	1985.11.21-11.23	1.2	15
Sizilan	4.8	37.72	115.24	1986.02.15-02.27	0.3	32
Ningwu	4.9	38.00	106.03	1987.08.10-10.30	0.0	170
Gaixian	4.8	40.43	122.39	1989.01.12-03.17	1.8	53
Yanqing	5.0	40.60	115.81	1990.07.15-08.23	2.2	136
Sheyang	5.0	33.84	120.43	1992.10.22-11.09	1.0	18
Luanxian	5.4	39.81	118.48	1995.10.06-10.23	2.4	154
Shungyi	4.5	40.16	116.50	1996.12.16-02.28	0.5	264

which an earthquake of m = 3.1 occurred. After a brief period of first motion scattering the senses became "down" again until a larger earthquake with m = 4.5 occurred. Figure 9 pictures the seismic station site and the earthquake epicenters' distribution of the Huoshan earthquake swarm. Obviously, if the earthquakes occurred in random directions from the station, the ratio of "up" and "down" could provide no information about the mechanisms. In the Huoshan region (Fig. 9), however, all of the earthquakes were located in the same west-north direction to the station. This seems to indicate that the consistency of the focal mechanism is the indicator which shows whether the biggest earthquake past over or not.

The earthquake swarms with the largest magnitudes of 4.5 and above which occurred in North China during 1966–1996, were listed in Table 2. Comparing Table 2 to Table 1, it can be seen that the number of earthquake swarms is greater than that of foreshock sequences.

Most earthquake swarms usually are small. It is difficult to observe clearly the senses of first *P*-wave motion in earthquake swarms with a very small magnitude.

New pattern characteristics might be found if new information of earthquake swarms could be received and applied. In many cases we could use the ratio of maximum amplitudes of P waves to that of S waves to monitor the changes of the focal mechanism of earthquakes. According to the double-couple model of the earthquake mechanism, the ratio of A_p/A_s is a function of the azimuth angle. An earthquake swarm did occur in Baodi county (117.2°E, 39.7°N, see Table 2) of Hebei province during July–September 1976. The nearest Nanshan seismic station was located 65 km from the swarm. The focal mechanism changes of the earthquake swarm could be monitored by the A_p/A_s recorded at the Nanshan station. Figure 10 indicates the changes of A_p/A_s of the Baodi earthquake swarm during August 1976. It is evident that the changes were small before the major earthquake (m = 4.9) occurrence of 21 August, and with the changes expanding (from 21 August to 1 September), no major earthquakes occurred. JING *et al.* (1976) studied the ratio of the maximum vertical amplitudes of P and S waves of the Haicheng earthquake of 1975 recorded at 12 stations, and they found that the ratio appeared quite stable. In comparison with five earthquake swarms which occurred elsewhere in China, Jing *et al.* said "this may be helpful in distinguishing foreshock sequences from earthquake swarms." JONES and MOLNAR (1979) considered foreshocks before three groups of main shocks and they found that the ratio of the amplitudes of the P and S waves was approximately the same; suggesting that the faulting mechanisms are the same for events in each sequence. The observation results before the occurrence of earthquake m = 4.9 in the Baodi region are in agreement with the results of Jones and Molnar.

It should be emphasized that the determination of focal mechanisms requires a denser seismographic network (OGATA *et al.*, 1995, 1996). However there are many methods available to monitor the changes of the focal mechanism of earthquakes; such as the changes of the senses of the first motion of P waves (CHEN, 1979) and

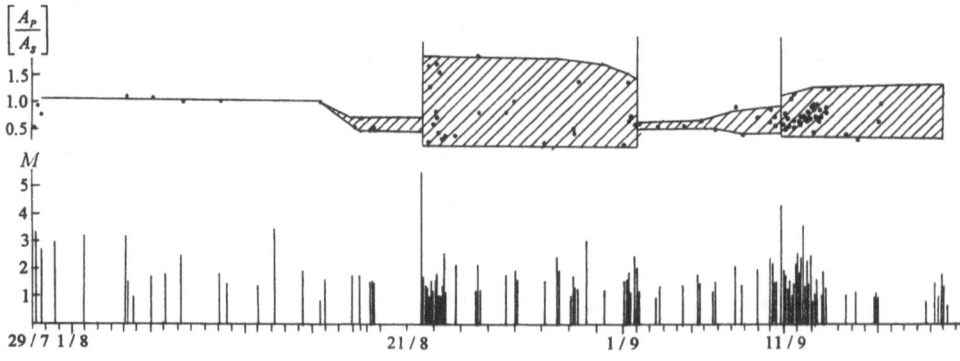

Figure 10
Baodi earthquake swarm (117.2°E, 39.7°N, July to September, 1976). $m - t$ plot of Baodi earthquake swarm; A_p/A_s ratio vs. time.

the ratio of the amplitudes of the P and S waves, etc. The monitoring capacities of these methods, especially the amplitude ratio method, are very limited for monitoring the changes of the focal mechanism of earthquakes.

4. Discussion and Conclusion

Some case studies of foreshock sequences and earthquakes swarms were performed in mainland China, which evidence that the number of earthquake swarms is greater than that of foreshock sequences. Therefore, the statistical discrimination between foreshocks and other earthquake clusters is a very important task of earthquake prediction.

It was commonly believed that the occurrence of an earthquake increases the occurrence probability of another larger earthquake in the same place because the first earthquake may be the foreshock (JONES, 1985). This paper points out that this is not always true.

The consistency of the focal mechanism is an important characteristic for a group of earthquakes, which could be used to discriminate foreshock sequences and several earthquake swarms. For the former, the focal mechanisms are consistent and similar to that of the main shock, and for the latter, they perhaps are inconsistent.

Studies exist to demonstrate similar consistent focal mechanisms in earthquake swarms (JONES, 1985). However, the most difficult problem in the study of foreshocks and earthquake swarms is the lack of data. With the installation of dense networks after the occurrence of the main shock, substantial aftershock data are available while data of foreshocks and earthquake swarms are relatively limited. Under such conditions, we need more data to demonstrate the consistency or inconsistency of earthquake swarms.

A very important point to make is that we found no case in mainland China during the past 30 years in which the main shock is preceded by an earthquake clustering with inconsistent focal mechanisms. That is, the consistency of focal mechanisms of an earthquake clustering (both for foreshock sequence and earthquake swarm) can be used to be an indicator showing if the largest earthquake in the clustering has past over or not.

The percentage of larger earthquakes which are preceded by foreshock sequences (within 5 days and 20 km) is about 5% in the mainland China during the past 30 years. These case analyses provide us with the possibility of predicting those earthquakes with foreshock sequences.

JONES (1985) found the probability that an earthquake will be followed by an earthquake with larger magnitude within 5 days (and 10 km is 6 ± 0.5 percent in southern California.) SEGGERN (1981) analyzed 510 "main shocks" based on LASA or NORSAR data of 1970–1977 with the lower cutoff of m = 5.8. A search for definitive foreshocks, based on a significantly short-time delay to the main shock, revealed that the true rate of foreshock occurrence was less than 20%. These results are in good agreement with our result. Indeed, discriminating foreshocks from other earthquake clusters require the study of many foreshock sequences and background activity to demonstrate that the same characteristic is unique or more common in foreshocks. Therefore more case analysis and data accumulation are needed because the comparisons between foreshock sequences and sequences not followed by a main shock should be performed. This will allow us to estimate the probability that a sequence which exhibits these characteristics would in turn be followed by a large event.

Acknowledgments

This research is sponsored by the NFSC (59574203) and by the State Seismological Foundation of China (95–070437).

REFERENCES

BOWMAN, J., and KISSLINGER, C. (1984), *A Test of Foreshock Occurrence in the Central Aleutian Arc*, Bull. Seismol. Soc. Am. *74*, 181–198.

CHEN, Y. (1979), *Consistency of Focal Mechanism as a New Parameter in Describing Seismic Activity*, Acta Geophisica Sinica *22*(2), 142–159.

CHEN, Y., TSOI, K. L., CHEN, F. B., GAO, Z. H., and ZHOU, Q. J., *The Great Tangshan Earthquake of 1976—An Anatomy of Disaster* (Pergamon Press 1988).

JIN, Y., ZHAO, Y., CHEN, Y., YAN, J. Q., and ZHOU, Y. R. (1976), *A Characteristic Feature of the Dislocation Model of Foreshocks of the Haicheng Earthquake, Liaoning Province*, Acta Geophysica Sinica *19*(3), 156–164.

JONES, L. M. (1985), *Foreshocks and Time-dependent Earthquake Hazard Assessment in Southern California*, Bull. Seismol. Soc. Am. *75*, 1669–1680.

JONES, L. M., and MOLNAR, P. (1979), *Some Characteristics of Foreshocks and their Possible Relationship to Earthquake Prediction and Premonitory Slip on Faults*, J. Geophys. Res. *84*(B7), 3596–3608.

JONES, L. M., WANG, B. Q., XU, S. X., and FITCH, T. J. (1983), *The Foreshock Sequence of the February, 4, 1975, Haicheng Earthquake (m = 7.3)*, Acta Seismologica Sinica *5*(1), 1–14.

LIN BANGHUI, LI DAPENG, LIU JIE, and WU ANXU (1994), *Study on Foreshocks and Foreshock Sequences*, Acta Seismologica Sinica 16 (Supp.), 24–38.

OGATA, Y., UTSU, T., and KATSURA, K. (1995), *Statistical Features of Foreshocks in Comparison with Other Earthquake Cluster*, Geophys. J. Int. *121*, 233–254.

OGATA, Y., UTSU, T., and KATSURA, K. (1996), *Statistical Discrimination of Foreshocks from other Earthquake Cluster*, Geophys. J. Int. *127*, 17–30.

SEGGERN, D. (1981), *Seismicity Pattern Preceding Moderate to Major Earthquakes*, J. Geophys. Res. *86*(B10), 9325–9351.

WYSS, M. (ed.) (1991), *Evaluation of Proposed Earthquake Precursors*, American Geophysical Union.

(Received June 20, 1998, revised/accepted December 17, 1998)

Pure appl. geophys. 155 (1999) 409–423
0033–4553/99/040409–15 $ 1.50 + 0.20/0

┃Pure and Applied Geophysics

Precursory Activation of Seismicity in Advance of the Kobe, 1995, $M = 7.2$ Earthquake

V. G. Kossobokov,[1] K. Maeda[2] and S. Uyeda[3]

Abstract—A succession of precursory changes of seismicity characteristic to earthquakes of magnitude 7.0–7.5 occurred in advance of the Kobe 1995, $M = 7.2$, earthquake. Using the Japan Meteorological Agency (JMA) regional catalog of earthquakes, the *M8* prediction algorithm (Keilis-Borok and Kossobokov, 1987) recognizes the *time of increased probability*, TIP, for an earthquake with magnitude 7.0–7.5 from July 1991 through June 1996. The prediction is limited to a circle of 280-km radius centered at 33.5°N, 133.75°E. The broad area of intermediate-term precursory rise of activity encompasses a 175 by 175-km square, where the sequence of earthquakes exhibited a specific intermittent behavior. The square is outlined as the second-approximation reduced area of alarm by the "Mendocino Scenario" algorithm, *MSc* (Kossobokov *et al.*, 1990). Moreover, since the *M8* alarm starts, there were no swarms recorded except the one on 9–26 Nov. 1994, located at 34.9°N, 135.4°E. Time, location, and magnitude of the 1995 Kobe earthquake fulfill the *M8-MSc* predictions. Its aftershock zone ruptured the 54-km segment of the fault zone marked by the swarm, directly in the corner of the reduced alarm area. The Kobe 1995 epicenter is less than 50 km from the swarm and it coincides with the epicenter of the *M* 3.5 foreshock which took place 11 hours in advance.

Key words: Earthquake prediction, algorithms *M8* and *MSc*, seismicity, Japan.

Introduction

Each major earthquake raises the question as to whether or not it was preceded by some phenomena that could be considered as precursory. The "precursors" defined in advance (see e.g., Wyss, 1991) are of particular interest, since the earthquake verifies their significance and reliability. Here we examine this question as applied to the recent January 16, 1995, $M = 7.2$, Kobe (Hyogo-ken Nanbu) earthquake, Southern Japan (Japan Meteorological Agency, 1997; Hashimoto *et al.*, 1996), and two completely reproducible intermediate-term earthquake prediction algorithms known as *M8* and *MSc* (Keilis-Borok and Kossobokov, 1990a; Kossobokov *et al.*, 1990). The destructive Kobe earthquake case-history poses an

---

[1] International Institute of Earthquake Prediction Theory and Mathematical Geophysics, Russian Academy of Sciences, Moscow, Russia.

[2] Meteorological Research Institute, Japan Meteorological Agency, Tsukuba, Japan.

[3] RIKEN, Earthquake Prediction Research Center, Tokai University, Japan.

additional problem in monitoring seismic patterns, since major contributions to actual seismic hazard may result in our life-time, not from the largest but comparatively moderate-size earthquakes. Thus, a multitude of magnitude, as well as spatial and temporal, ranges should be considered simultaneously in a hierarchy of predictions that facilitate earthquake hazard mitigation.

Prediction Algorithms M8 and MSc

As has been shown in previous publications (KEILIS-BOROK, 1996; PEREZ and SCHOLZ, 1997) most large earthquakes are preceded by a set of rather simple universal symptoms of instability. These symptoms, known from studies on nonlinear dynamics, include the rise of intensity of background perturbations in a system and their concentration, as well as burst-like reaction to an excitation. The integral estimates of these characteristics form the basis of the *M8* algorithm designed for intermediate-term prediction of earthquakes (KEILIS-BOROK and KOSSOBOKOV, 1987, 1990a). The first testing of the algorithm dates to 1984 when it was applied retroactively to diagnose *times of increased probability*, TIPs for the world's largest (magnitude 8 or above) earthquakes, hence its name. Subsequently the algorithm is the subject of many studies on global and regional scales (KEILIS-BOROK, 1996).

In most applications, *M8* diagnoses TIPs in circles with a radius determined by the magnitude threshold M_0 which defines what large earthquake we intend to predict. The energy-space-time scaling is the essential part of the algorithm. Therefore, the magnitude scale should reflect the size of earthquake sources (e.g., for many catalogs this is equivalent to maximal magnitude reported). The algorithm analyses normalized integral characteristics of seismicity in each circle of investigation, CI. It issues a TIP if the values of these characteristics are abnormally high compared to the ranges observed in the circle over a longer background period. A TIP is declared and usually lasts for five years. In some cases, seismic changes may re-establish the limits of norm and anomaly, and therefore may cancel or extend the TIP. The ultimate definition of the algorithm is given by its source code and prefixed profiles (HEALY *et al.*, 1992, 1997).

Algorithm M8. For the reader's convenience, we describe the scheme of the *M8* algorithm in brief:

Prediction is aimed at the earthquakes of magnitude M_0 and above. We consider different values of M_0 with a step 0.5. The seismic territory is scanned by overlapping circles with the diameter $D(M_0)$. Within each circle the sequence of earthquakes is considered with aftershocks removed $\{t_i, m_i, h_i, b_i(e)\}$, $i = 1, 2 \ldots$. Here t_i is the origin time, $t_i \leq t_{i+1}$; m_i is the magnitude, h_i—focal depth, and $b_i(e)$—the number of aftershocks during the first e days. The sequence is normalized by the lower magnitude cutoff $M_{\min}(\tilde{N})$, $\tilde{N}$ being the standard value of average annual number of earthquakes in the sequence. As mentioned above, the magnitude scale should reflect the size of earthquake sources. Accordingly, if reported the M_s-type magnitude is taken for larger events, while the commonly determined m_b magnitude is used for smaller ones.

Several running averages are computed for this sequence in the sliding time windows $(t - s, t)$ and magnitude range $M_0 > M_i \geq M_{\min}(\tilde{N})$. They measure intensity of earthquake flow, its deviation from the long-term trend, and clustering of earthquakes. The averages include: $N(t)$—the number of the main shocks. $L(t)$—the deviation of $N(t)$ from its long-term trend, $L(t) = N(t) - N_{\mathrm{cum}}(t - s) \cdot (t - s) / (t - t_0 - s)$, $N_{\mathrm{cum}}(t)$ being the cumulative number of the main shocks with $M \geq M_{\min}(\tilde{N})$ from the beginning of the sequence t_0 to t. $Z(t)$—linear concentration of the main shocks estimated as ratio of the average diameter of the source l to the average distance between them r. $B(t) = \max_i \{b_i\}$—the maximal number of aftershocks (a measure of earthquake clustering); the earthquake sequence $\{i\}$ is considered in the time window $(t - s', t)$ and in the magnitude range $(M_0 - p, M_0 - q)$.

Each of the functions N, L, Z is calculated for $\tilde{N} = 20$ and $\tilde{N} = 10$. As a result, the earthquake sequence is given a robust description by seven functions: N, L, Z (twice each), and B.

"Very large" values are identified for each function, using the condition that they exceed Q percentiles (i.e., they are higher than $Q\%$ of the encountered values).

An alarm or TIP, "time of increased probability," is declared for five years, when at least 6 of 7 functions, including B, become "very large" within a narrow time window $(t - u, t)$. To stabilize predictions, this condition is required for two consecutive moments, t and $t + 0.5$ years.

The following standard values of parameters indicated above are prefixed in the original version of the $M8$ algorithm: $D(M_0) = 111.111 \cdot (\exp(M_0 - 5.6) + 1)$ km, where $\exp(x) \approx 10^{0.43x}$ is the natural exponent of x (this gives 384 km, 560 km, 854 km and 1333 km for $M_0 = 6.5$, 7.0, 7.5 and 8 respectively), $s = 6$ years, $s' = 1$ year, $g = 0.5$, $p = 2$, $q = 0.2$, $u = 3$ years, $Q = 75\%$ for B and 90% for the other six functions.

The territorial uncertainty of $M8$ predictions can be reduced significantly, from 4 to 14 times, using the second-approximation prediction algorithm MSc (KOS-SOBOKOV et al., 1990), known also as "Mendocino Scenario." The second approximation is achieved after additional analysis of the low-level seismicity in the area of a TIP. MSc searches for an episode of "anomalous quiescence" when a part of the area of a TIP, which was steadily active in its formation, exposes a sudden and rather short (a few months) quiescence. That is, the MSc algorithm outlines such an area of the territory of alarm in which the activity is high and has been interrupted for a short time (the interruption must have a sufficient temporal and/or spatial span). In many cases such an intermittent episode in seismic regime occurs in a narrow vicinity of the expected large earthquake. It may start long after a TIP beginning, thus, reducing the temporal span of the alarm. An application of MSc

requires a catalog which systematically reports the earthquakes of magnitude lower than the minimal threshold used by *M8*, i.e., $M_{\min}(20)$; therefore, in certain cases the second approximation could not be achieved with the existing data sources.

Algorithm MSc was designed by retroactive analysis of seismicity prior to the Eureka earthquake (1980, $M = 7.2$) near Cape Mendocino in California, hence its name. For reader's convenience, we describe here the scheme of the *MSc* algorithm in brief:

Given a TIP diagnosed for certain territory **U** at the moment **T**, the algorithm is aimed to find within **U** a *smaller* area **V** in which the predicted earthquake has to be expected. An application of the algorithm requires a reasonably complete catalog of earthquakes with magnitudes $M \geq (M_0 - 4)$ which is usually lower than a minimal threshold used by *M8*.

The essence of *MSc* can be summarized as follows. Territory **U** is coarse-grained into small squares of $s \times s$ size. Let (i, j) be the coordinates of the centers of the squares. Within each square (i, j) the number of earthquakes $n_{ij}(k)$, aftershocks included, is calculated for consecutive short time windows u months long, starting from the $(T - 6$ years$)$ onward, to allow for the earthquakes which contributed to the TIPs diagnosis; k is the sequence number of a time window. In this manner the time-space considered is divided into small boxes (i, j, k) of the size $(s \times s \times u)$. "Quiet" boxes are singled out for each small square (i, j); they are defined by the condition that $n_{ij}(k)$ is below the Q percentile n_{ij}. The clusters of q or more quiet boxes connected in space or in time are identified. Area **V** is the territorial projection of these clusters. The *standard values of parameters* adjusted for the case of the Eureka earthquake are the following: $u = 2$ months $Q = 10\%$, $q = 4$, and $s = 3D/16$, D being the diameter of the circle used in algorithm *M8*.

Application of the Algorithms Using the NEIC Data

Since 1990 the algorithms are applied systematically for a research real-time intermediate-term prediction in those regions worldwide where seismic catalogs are available and complete enough for the analyses. In particular, since 1992 in collaboration with the United States Geologic Survey, we carry out the experimental prediction of earthquakes with magnitude 7.5 and above in the Circum-Pacific as a rigid Test of *M8* (HEALY *et al.*, 1992). The later prediction is based on analysis of the NEIC GLOBAL HYPOCENTERS DATA BASE CD-ROM (1989) and its updates to the present. It also includes the territory of Japan.

The territory of Japan and adjacent regions were among the first regions in which the algorithms were tested although by retroactive application (KEILIS-BOROK and KOSSOBOKOV, 1990b). The prediction using NEIC GHDB aims at M 8.0 + and M 7.5 + events. In fact, the prediction results, both of retroactive as well as of forward testing (HEALY *et al.*, 1992; KOSSOBOKOV, 1994; KOSSOBOKOV *et al.*, 1996a), prove the efficiency of the algorithms here.

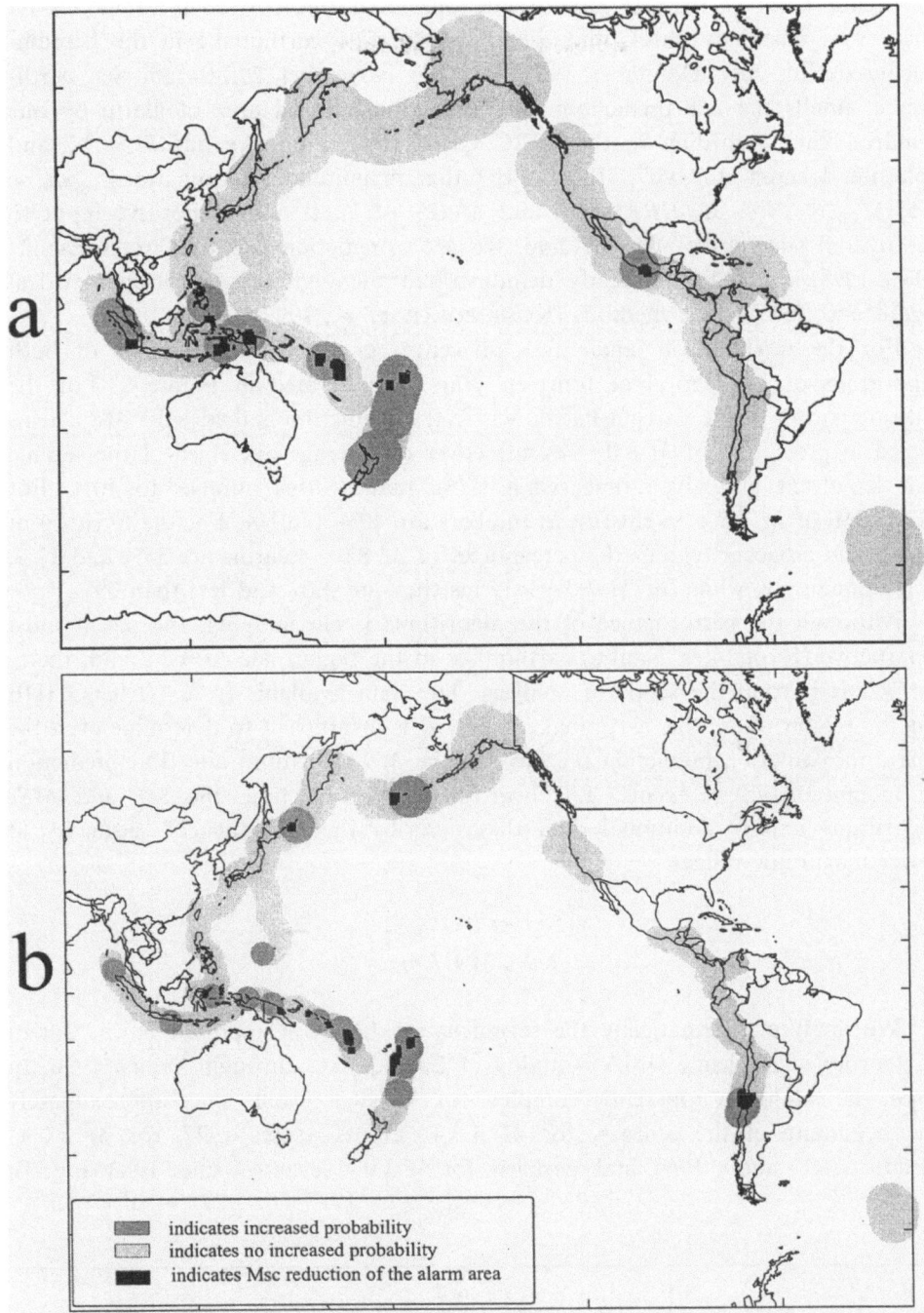

Figure 1
The regions of increased probability in the Circum-Pacific, July 1998 to December 1998 determined by the *M8* and *MSc* algorithms aimed to predict magnitude 8.0 + (a) and 7.5 + (b) events.

Figure 1 shows the *M8-MSc* predictions (KOSSOBOKOV *et al.*, 1999) as on July 1998. One may check that the only M 7.5 + earthquake in the Circum-Pacific during July–December 1998, i.e., the November 29, Ceram Sea earthquake, fulfills the *M8* prediction and misses the reduced area of alarm by one hundred km. According to the NEIC QED, the earthquake has $M_s = 7.7$ and epicenter located at 2.051°S 124.925°E (other magnitude determinations – $m_b =$ 6.5, $M_W GS = 7.8$, $M_W HRV = 7.7$, and $MeGS = 8.1$). It adds a positive input to the overall statistics of the *M8* and *M8-MSc* predictions in the Circum-Pacific, 1985–1998, which have already demonstrated the high (above 99%) statistical significance level of the methods (KOSSOBOKOV *et al.*, 1999).

For the territory of Japan and adjacent regions the performance of both algorithms over a period of fourteen years is illustrated in Figure 2. For the entire territory of the Circum-Pacific where the prediction is made, the *M8* alarms aimed at prediction of M 8.0 + events cover on average one third of the seismic belt length at any given time, while *MSc* reduces this number to 10%. For prediction of M 7.5 + events these numbers are 40% and 6%. For the territory of Japan and adjacent regions the percentages of M 8.0 + alarms are 35% and 12%, correspondingly, while for M 7.5 + alarms they are 33% and less than 3%.

Although the performance of the algorithms is encouraging, the recent most destructive (Hyogo-ken Nanbu) earthquake in the region has $M = 7.2$ and, therefore, falls beyond the scope of analysis. The data available from NEIC GHDB for the region where this earthquake occurred is insufficient to determine even the first approximation prediction (i.e., to run the *M8* algorithm) aimed at prediction of magnitude 7.0 + events. To cover them by prediction, the *M8* and *MSc* algorithms require additional data that describe the dynamics of seismicity at lower magnitude ranges.

The JMA Data

We analyze systematically the seismicity of Japan as reported in the Japan Meteorological Agency (JMA) Catalog of Earthquakes through August 1996. In total, the catalog is apparently complete for M 6.0 + events since approximately the beginning of the century, for M 4.5 + events—since 1927, for M 4.0 + events—since about 1964, and, perhaps, for M 3.0 + events—since 1983 (Fig. 3).

Figure 2

The space-time distribution of the *M8* and *MSc* alarms aimed to predict M 8.0 + (a) and M 7.5 + (b) earthquakes in Japan and adjacent territories, 1985–1997. The territory considered is on the left. The space-time distribution of real time alarms and the great earthquakes (stars) are given on the right. Space coordinate is given as the distance along the belt.

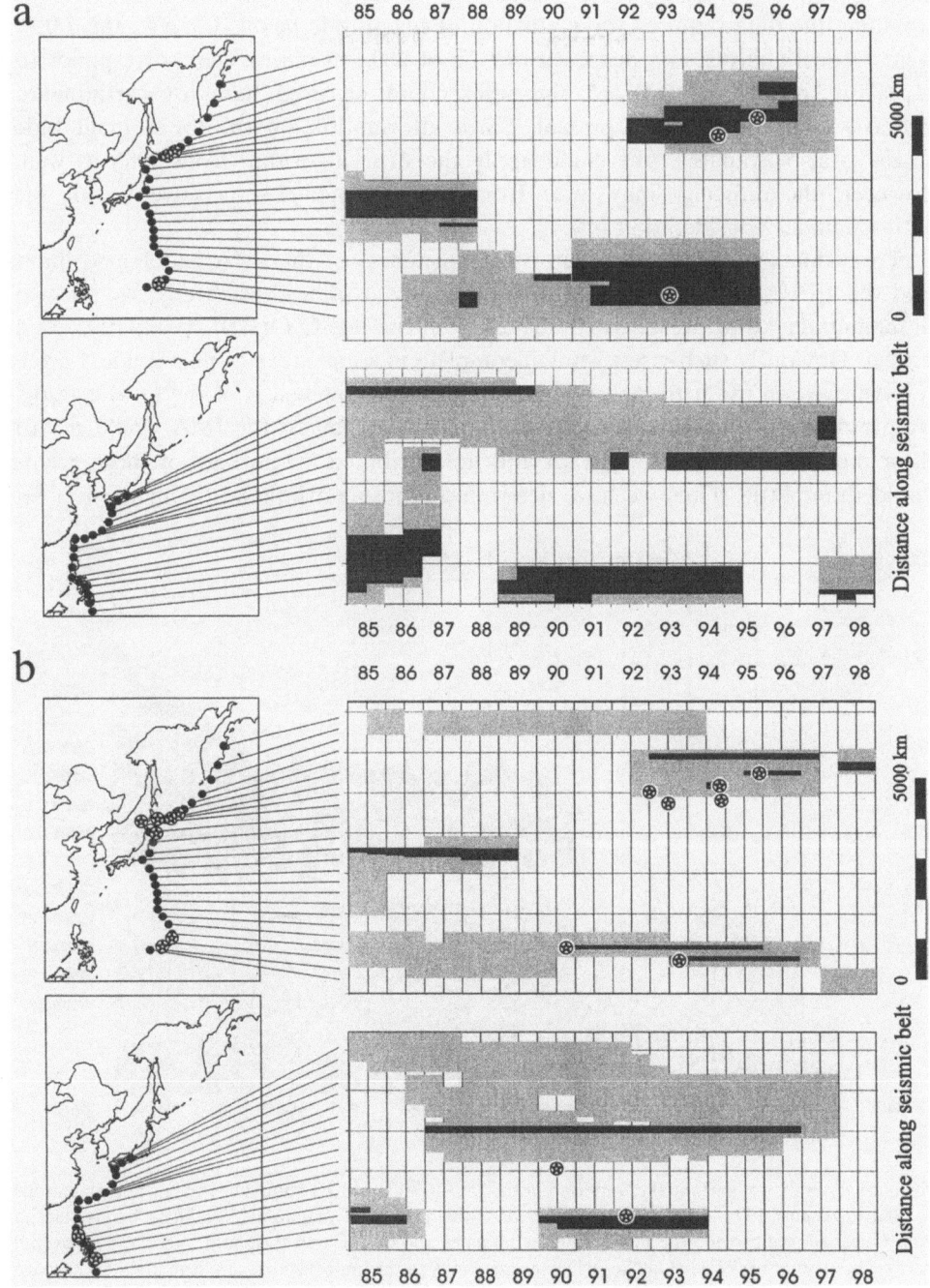

Thus, locally the JMA Catalog is indeed more complete than the NEIC GHDB providing the data required for prediction of magnitude below 7.5, e.g., the 7.0 + events. Local analysis in a dense set of CI's of 280-km radius, that corresponds to $M_0 = 7.0$, shows that the *M8* algorithm could be used here for earthquakes forecasts from 1985 to the present. Since information on the lower magnitude ranges is also available, one could apply the *MSc* algorithm from 1985 as well. However, the difficulty may arise from inhomogeneous data coverage of the territory under consideration.

To evaluate the territorial span of completeness of the JMA Catalog we have used the NEIC GHDB as an independent source of data. Specifically, we selected all magnitude 4.0 or greater earthquakes from the NEIC GHDB, 1963–1995, as a test set. Obviously such a test set is incomplete in some areas at magnitude 4 level, however we can use it to measure the territorial completeness of the local catalog. We presume that in the areas of its territorial completeness the JMA would record all or most events from the test set and the attributed magnitudes of those events would differ little. Thus we have determined those earthquakes from the test set

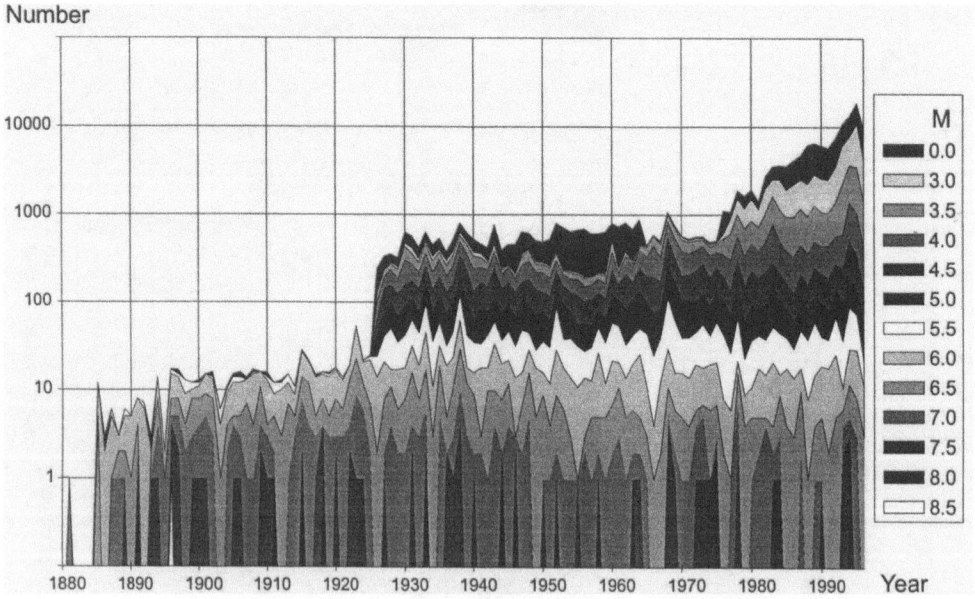

Figure 3
The annual number of earthquakes reported in the JMA Catalog of Earthquakes by time and magnitude in 1880 to August 1996. Each band corresponds to half-a-unit of magnitude *M*. These bands stacked from higher ranges provide the number of earthquakes above a certain threshold. Note: (1) an overall increase in the number of earthquakes which corresponds to the improvement of seismographs; (2) sharp changes at 1885, 1896, 1926, 1959–1964, and 1976–1983; (3) certain stability of the number of the magnitude 6.0 and larger events from 1896; (4) rather uniform width of the bands since 1964 in agreement with the Gutenberg-Richter relationship down to magnitude 4.0 and since 1983 down to magnitude 3.0.

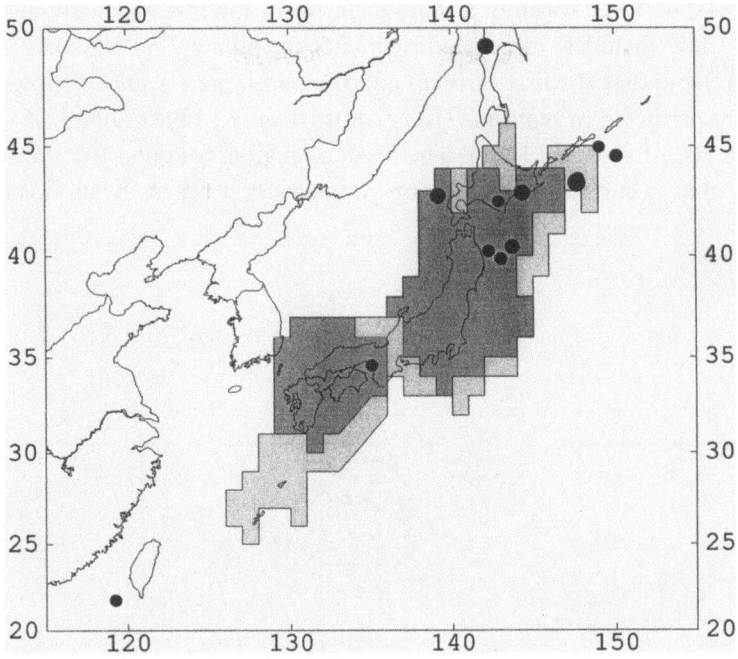

Figure 4

Spatial completeness of the JMA Catalog of Earthquakes and recent shallow earthquakes of magnitude 7.0 and larger (1985 to the present).

which have an equivalent among all magnitude 3.0 or greater earthquakes from the JMA Catalog. In our definition the equivalent events differ by less than one minute in time, 0.5° in latitude, 0.7° in longitude, and 33 km in depth. No additional limitation on magnitudes was set. We coarse-grain the territory into one by one degree cells between 20–50°N and 120–150°E. For each cell we count the total number of earthquakes from the test set in it (N) and those from the total that have an equivalent in the JMA Catalog (n). The ratio (n/N) characterizes the completeness of the JMA Catalog in a given cell. The spatial distribution of this ratio displays a high level of completeness (above 75%) for most of the Japan Islands. However, the completeness at Northern Hokkaido and along the Kuril, Ryukyu, and Izu trenches is considerably lower. Figure 4 shows the contours of 75 and 50% completeness along with the epicenters of all magnitude 7.0 and above earthquakes from the JMA Catalog, 1985–1996. The contours partially split the whole territory into two regions of a high level of completeness separated along 136°E, where the Japan subduction zone seismicity expires to the west. Note that the outlined areas differ slightly when other time interval or higher magnitude ranges (up to 5.0 and above) are considered.

The epicenters of magnitude 7.0 or greater events in 1985–1996 mark the northeastern edges of the two regions (Fig. 4). In the eastern region most of them occurred in 1993–1995, forming a unique cluster in the history of the instrumental seismology that includes five magnitude 7.5 or greater earthquakes and their aftershocks. Note that the first of them and the two largest events of magnitude 8.1 and 8.0 were predicted in real time (KOSSOBOKOV *et al.*, 1994, 1996b), in course the Test of *M8* (HEALY *et al.*, 1992). The western region contains the only *M* 7.0 + earthquake that occurred near Kobe on January 16, 1995 at 20:46 GMT.

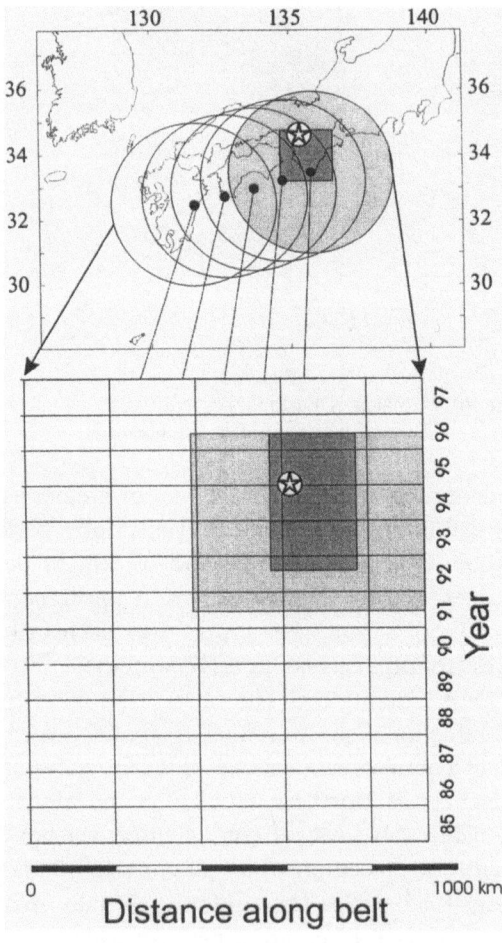

Figure 5

Space-time distribution of alarms: The territory of the five circles of investigation of 280-km radius (upper part) and temporal evolution of alarms (lower part). The *M8* times of increased probability (shaded light) are limited to one circle and 5-year interval; the *MSc* second approximation (shaded dark) narrows down the prediction to 175 × 175 km square and four years. The epicenter of the 1995 Kobe earthquake indicated with a star fulfills the *M8-MSc* predictions.

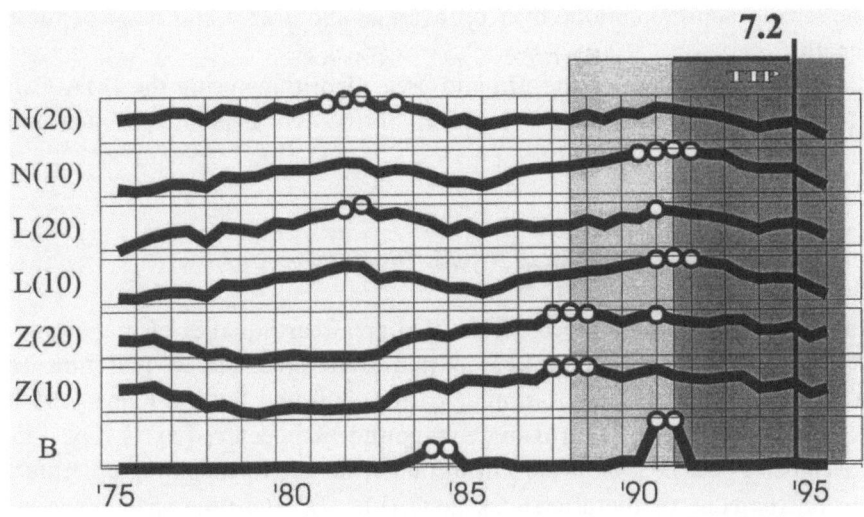

Figure 6

Functions of the *M8* algorithm in the circle centered at 33.5°N, 135.75°E with radius of 280 km for magnitude 7.0+ prediction. The abnormally high values of each function marked with circles concentrate in the 3-year interval (shaded light) ending on July 1, 1991. The observed coincidence of the abnormally high values, by the *M8* definition, starts the 5-year TIP (shaded dark). The time of the 1995 Kobe earthquake (vertical line) falls into this period.

Application of the Algorithms Using the JMA Catalog

To check whether the dynamics of seismicity prior to the 1995 Kobe earthquake followed the behavior suggested by the *M8* and *MSc* algorithms as precursory, we set $M_0 = 7.0$ and apply the algorithms to the territory of the western region. The upper part of Figure 5 shows the five CI's whose centers are evenly distributed along the line connecting 32.5°N, 131.6°E and 33.5°N, 135.75°E. They cover most of the western region including Western Honshu, Sikoku, and Ryukyu Islands. The retroactive monitoring of seismicity from 1975 by application of the *M8* algorithm aimed at M 7.0+ earthquakes supplies the only TIP in the most eastern CI spanning the period July 1991 through June 1996. The values of the *M8* algorithm functions counted in this CI are shown in Figure 6. We see that all but one measure of activity, namely $N(20)$, raised to their highest values, forming a TIP by the middle of 1991.

As on July 1991 the *MSc* algorithm gives no reduction of the *M8* prediction (the lower part of Fig. 5). The second approximation emerges one year later in July 1992 when *MSc* recognizes the precursory behavior for the 175 by 175-km square shown in Figures 5 and 7. The same area remains through each of the six half-year updates prior to January 16, 1995. In total, the space-time volume of alarm measured by distance along the belt times year occupies 2800 and 800 km × year in the first and the second approximation, correspondingly. In other words, the alarms of the first and the second approximation cover on average about 23.0 and 6.6% of the total belt length.

Thus, the combination of the *M8* and *MSc* algorithms using the JMA Catalog data through 1994 pinpoints the location of the Kobe 1995 epicenter and its aftershocks.

A Shorter-term Observation

Unusual swarms briefly preceded the two great earthquakes of magnitudes 8.1 and 8.0 to the northeast of Hokkaido that were predicted in real time (KOS-SOBOKOV *et al.*, 1994, 1996a). A similar pattern is found for the Kobe 1995 main shock. Ten earthquakes of approximate magnitude 3.5 occurred at 34.9°N, 135.4°E on November 9–26, i.e., about 50 km distant from the forthcoming epicenter and verging on the edge of the aftershock zone (Fig. 7). The time span between this "swarm" and the main shock is about six weeks. Although the seismicity of Japan produces numerous swarm activities that are not followed by a large event, the swarm of November 9–26 is remarkable. Indeed, it is the only cluster of such kind in the 280-km radius circle of the *M8* TIP ever recorded by JMA; similar space-time density of earthquakes above magnitude 3.0 were previously observed in clear

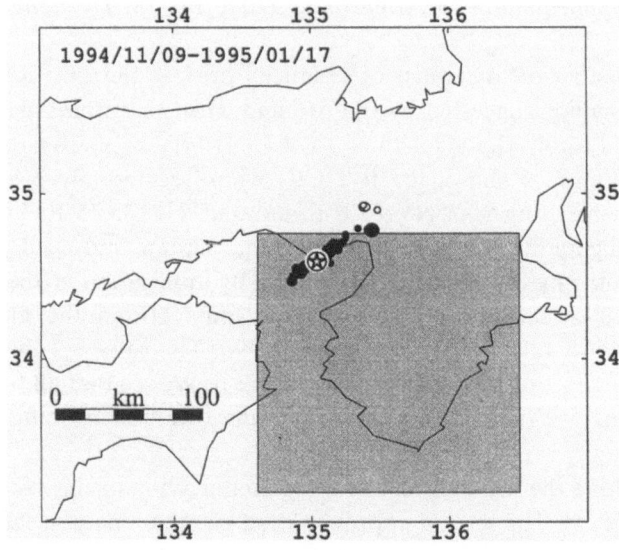

Figure 7

The localization of the *M8* alarm by the *MSc* algorithm (shaded square). Note a remarkable swarm (open circles) of November 9–26, the only one within the *M8* alarm limits, followed by the quake of January 9, 1995 on the fault then ruptured with the 1995 Kobe earthquake epicenter (star) and its aftershocks (solid circles). The epicenter of the 11-hour foreshock coincides with that of the main event.

aftershock sequences of only markedly larger events. It should be mentioned that this, apparently short-term, activation progresses: On January 9 a magnitude 3.1 event struck the fault of the future main rupture after which the epicentral area of the Kobe earthquake effected a magnitude 3.5 foreshock at 9:28 GMT on January 16.

Discussion and Conclusion

A novel understanding of the dynamics of seismicity has emerged over the last decade. The lithosphere, its earthquake-prone part in particular, is regarded as a hierarchical nonlinear system with dissipation of energy (e.g., NEWMAN et al., 1994). Among sources of nonlinearity of the system is abrupt triggering of an earthquake when stress exceeds strength on a segment of a fault. Powerful sources of nonlinearity also may arise from the multitude of processes (such as filtration of fluids and stress corrosion, buckling and microfracturing, phase transition of the minerals, etc.) that control distribution of strength within the hierarchy of blocks and faults. Except for unique cases, none of the sources of nonlinearity predominates, as a result the others can be neglected. Therefore, from a position of intermediate-term earthquake prediction, the lithosphere appears to be a chaotic system. Predictability of chaos could be achieved, up to a certain limit, after averaging and/or after recognizing the beginning of a certain scenario that often surfaces in the process considered (MONIN, 1994).

The methods we used in this paper account for these conclusions on predictability of nonlinear systems by describing the preparation stage of the system in terms of robust averages. However, our analysis of the 1995 Kobe earthquake indicates rather than proves emerging possibilities of earthquake monitoring supported by intermediate-term prediction algorithms. The observed progression of seismic dynamics suggests that certain phenomena that are usually disregarded as short-term premonitory may become such on the background of an intermediate-term alarm. Our results for the 1995 Kobe earthquakes together with those achieved in a real-time testing Circum-Pacific (HEALY et al., 1992; KOSSOBOKOV, 1994; KOSSOBOKOV et al., 1994, 1996a,b, 1999) favor a hierarchical, step-by-step prediction technique which accounts for a multi-scale escalation of seismic activity to the main rupture. It starts with recognition of earthquake-prone zones for a number of magnitude ranges, then follows with determination of long- or intermediate-term areas and times of increased probability, and, finally, may come out with a short-term alert. The prediction of the Haicheng earthquake of February 4, 1975 remains the unique example of how this approach could lead to actual success in saving lives and reducing damage, although at that time some of the steps were lucky guesses, rather than statistically justified conclusions (ZHANG-LI et al., 1984).

There is a prevailing opinion that earthquake hazard mitigation is more valuable than prediction. However, if one addresses the problem from a hierarchical viewpoint, it is clear that seismic hazard expectation is actually based on a zero-approximation prediction and its value strongly depends on how accurately the assumptions are used to derive seismic potential. Moreover, the practice of evaluating hazard potential by using exclusively historic data might be seriously biased due to a rather short period of observation. In most cases, the areas that have already exposed themselves as dangerous might be quasi relaxing while others are ready to go in the next strike. From an economical point of view, the issue of costs-and-benefits is still not clear for the expenditures on earthquake hazard mitigation in most seismic regions. Thus it is very possible that new strategies, based on reliable, statistically justified prediction methods, could outperform the existing ones (MOLCHAN, 1997). That is why testing of prediction algorithms is crucially important.

The ultimate test of any prediction method is the advance prediction. Inevitably each advance prediction experiment requires many years of a tedious, book-keeping investigation due to the infrequent occurrence of significant earthquakes. The procedures of such book-keeping should be rather transparent, so that other interested parties can repeat and/or revise *a posteriori* the results of the prediction experiment (HEALY *et al.*, 1992). All this may explain why we know of few studies on testing predictions. Nonetheless we repeat there is not other way achieve statistical justification of a prediction method except for actual prediction of earthquakes. The more predictions, fulfilled or not fulfilled, the stronger becomes our confidence for accepting or rejecting the underlying hypothesis we obtain. The results of this paper suggest extending to lower-magnitude ranges the testing of the *M8* and *MSc* algorithms in those segments of Circum-Pacific where regional data are complete enough for an adequate application.

The accumulated results of prediction by the *M8* and *MSc* algorithms (i.e., growing set of success and errors) provide a *pied à terre* for further development of prediction methods. The significance level achieved by these algorithms (KOSSOBOKOV *et al.*, 1997) might already be enough to address the question of development of civil-defense and/or economic instruments, among them activation of the existing low key safety measures that provide prevention of a significant part of the damage.

Acknowledgements

We thank the anonymous reviewers for their suggestions that helped to improve the paper. The study was supported by the grants ISTC 415-96, NSF EAR-9804859, and INTAS 93-809-ext.

REFERENCES

HASHIMOTO, M., SAGIYA, T., TSUJI, H., HATANAKA, Y., and TADA, T. (1996), *Coseismic Displacements of the 1995 Hyogo-ken Nanbu Earthquake*, J. Phys. Earth *44*, 255–279.

HEALY, J. H., KEILIS-BOROK, V. I., and LEE, W. H. K. (eds.) (1997), *Algorithms for Earthquake Statistics and Prediction*, IASPEI Software Library, Volume 6.

HEALY, J. H., KOSSOBOKOV, V. G., and DEWEY, J. W. (1992), *A Test to Evaluate the Earthquake Prediction Algorithm, M 8*, U.S. Geological Survey Open-File Report 92–401, 23 pp. with 6 Appendices.

GLOBAL HYPOCENTERS DATA BASE CD-ROM (1989), NEIC/USGS, Denver, CO (computer file).

JAPAN METEOROLOGICAL AGENCY (1997), The Seismological Bulletin of the Japan Meteorological Agency (1926.1.1–1996.6.30)

KEILIS-BOROK, V. I. (1996), *Intermediate-term Earthquake Prediction*, Proc. Natl. Acad. Sci. USA *93*, 3748–3755.

KEILIS-BOROK, V. I., and KOSSOBOKOV, V. G. (1987), *Periods of high probability of occurrence of the World's strongest earthquakes*, Computational Seismology 19, Allerton Press Inc., pp. 45–53.

KEILIS-BOROK, V. I., and KOSSOBOKOV, V. G. (1990a), *Premonitory Activation of Seismic Flow: Algorithm M 8*, Phys. Earth Planet. Inter. *61*, 73–83.

KEILIS-BOROK, V. I., and KOSSOBOKOV, V. G. (1990b), *Times of Increased Probability of Strong Earthquakes (M = 7.5) Diagnosed by Algorithm M 8 in Japan and Adjacent Territories*, J. Geophys. Res. *95*, 12,413–12,422.

KOSSOBOKOV, V. G. (1994), *Intermediate-term Changes of Seismicity in Advance of the Guam Earthquake on August 8, 1993*, EOS Transactions *75*, No. 25, AGU 1994 Western Pacific Geophysics Meeting, Additional Abstracts, SE22A-10.

KOSSOBOKOV, V. G., HEALY, J. H., DEWEY, J. W., and TIKHONOV, I. N. (1994), *Precursory Changes of Seismicity before the October 4, 1994 Southern Kuril Islands Earthquake*, EOS Transactions *75*, No. 44, 1994 AGU Fall Meeting Addendum, S51F-11.

KOSSOBOKOV, V. G., HEALY, J. H., DEWEY, J. W., SHEBALIN, P. N., and TIKHONOV, I. N. (1996a), *A Real-time Intermediate-term Prediction of the October 4, 1994 and December 3, 1995 Southern-Kuril Islands Earthquakes*, Computational Seismology 28, Nauka, Moscow, 46–55.

KOSSOBOKOV, V. G., HEALY, J. H., DEWEY, J. W., and ROMASHKOVA, L. L. (1996b), *Intermediate-term Changes of Seismicity in Advance of the 10 June 1996 Delaroff Islands Earthquake*, EOS Transactions *77*, 1996 AGU Fall Meeting, S31A-08.

KOSSOBOKOV, V. G., KEILIS-BOROK, V. I., and SMITH, S. W. (1990), *Localization of Intermediate-term Earthquake Prediction*, J. Geophys. Res. *95* (B12), 19,763–19,772.

KOSSOBOKOV, V. G., HEALY, J. H., and, DEWEY, J. W. (1997), *Testing an Earthquake Prediction Algorithm: A Global Increase of Seismic Activity?* IASPEI 29th General Assembly (August 18–28, 1997, Thessaloniki, Greece), Abstracts, 346.

KOSSOBOKOV, V. G., ROMASHKOVA, L. L., KEILIS-BOROK, V. I., and HEALY, J. H. (1999), *Testing Earthquake Prediction Algorithms: Statistically Significant Advance Prediction of the Largest Earthquakes in the Circum-Pacific, 1992–1997*, Phys. Earth Planet. Inter. *111*, 187–196.

MOLCHAN, G. M. (1997), *Earthquake Prediction as a Decision-making Problem*, Pure appl. geophys. *149*, 233–247.

MONIN, A. S. (1994), *Predictability of Chaotic Phenomena*, Russian Journal of Computational Mechanics *1*(3), 3–16.

NEWMAN, W. I., GABRIELOV, A., and TURCOTTE, D. L. (eds.) (1994), *Nonlinear Dynamics and Predictability of Geophysical Phenomena*, Geophysical Monograph Series, IUGG-AGU, Washington, D.C.

PEREZ, O. J., and SCHOLZ, C. H. (1997), *Long-term Seismic Behavior of the Focal and Adjacent Regions of Great Earthquakes during the Time between two Successive Shocks*, J. Geophys. Res. *102*, 8203–8216.

WYSS, M. (ed.) (1991), *Evaluation of Proposed Earthquake Precursors*, AGU, Washington, D.C.

ZHANG-LI, C., PU-XIONG, L., DE-YU, H., DA-LIN, Z., FENG, X., and ZHI-DONG, W. (1984), *Characteristics of Regional Seismicity before Major Earthquakes*, Earthquake Prediction (UNESCO, Paris), 505–521.

(Received July 28, 1998, revised January 4, 1999, accepted January 22, 1999)

Pure appl. geophys. 155 (1999) 425–442
0033–4553/99/040425–18 $ 1.50 + 0.20/0

Pure and Applied Geophysics

The Variation of Stresses due to Aseismic Sliding and its Effect on Seismic Activity

NAOYUKI KATO[1,2] and TOMOWO HIRASAWA[3]

Abstract—Numerical simulation of recurring large interplate earthquakes in a subduction zone is conducted to explore the effects of aseismic sliding on the variation of stresses and the activity of small earthquakes. The frictional force obeying a rate- and state-dependent friction law is assumed to act on the plate interface in a 2-D model of uniform elastic half-space. The simulation results show that large earthquakes repeatedly occur at a constant time interval on a shallow part of the plate interface and that aseismic sliding migrates from the upper aseismic zone as well as from the lower aseismic zone into the central part of the seismogenic zone before the occurrence of a large interplate earthquake. This spatiotemporal variation of aseismic sliding significantly perturbs the stresses in the overriding plate and in the subducting oceanic plate, leading to the precursory seismic quiescence in the overriding plate and the activation of the intermediate-depth earthquakes of down-dip tension type. After the occurrence of a large interplate earthquake, the activity of the intermediate-depth earthquakes of down-dip compression type in the subducting slab is expected to increase and migrate downward. This is because the downward propagation of postseismic sliding causes the downward migration of compressional-stress increase in the down-dip direction of the plate interface. The simulation result further indicates that episodic events of aseismic sliding may occur when the spatial distributions of friction parameters are significantly nonuniform. The variation of stresses due to episodic sliding is expected to cause seismicity changes.

Key words: A rate- and state-dependent friction law, seismic cycle, precursory seismic quiescence, postseismic sliding, intermediate-depth earthquake, episodic sliding.

Introduction

It is known that seismic activity is not necessarily stationary in time or in space. Seismic activity is thought to be affected by the variation of regional stress field and, therefore, it is related to large earthquakes. The activity of intraplate earthquakes in southwestern Japan shows some periodicity, and seems to be controlled by large interplate earthquakes along the Nankai trough (e.g., SHIMAZAKI, 1976; HORI and OIKE, 1996). The change in regional seismic activity following a large

[1] Geological Survey of Japan, 1-1-3 Higashi, Tsukuba 305-8567, Japan.
[2] Present address: Department of Geological Sciences, Brown University, Providence, RI 02912-1846, USA. Fax: +1-401-863-2058, E-mail: Naoyuki-Kato@brown.edu
[3] Graduate School of Science, Tohoku University, Aoba-ku, Sendai 980-8578, Japan.

earthquake is well correlated to the coseismic change in the Coulomb failure function, which represents the effective shear stress that promotes shear rupture (e.g., REASENBERG and SIMPSON, 1992; KING et al., 1994; TAYLOR et al., 1998). The change in regional stresses due to large earthquakes in the above cases is considerably smaller than the stress drop of earthquakes, suggesting a small change in stress significantly affects seismic activity. Conversely, seismic activity may be an indicator of changes in stress field.

WYSS et al. (1981) and DMOWSKA et al. (1988) indicated that the stress relaxation due to preseismic sliding may affect seismic activity, leading to precursory seismic quiescence observed for many large earthquakes (e.g., OHTAKE et al., 1977; KISSLINGER, 1988; WYSS and HABERMANN, 1988). Recently, KATO et al. (1997) performed numerical simulations of seismic cycles in a subduction zone by adopting a laboratory-derived friction law to demonstrate that the regional stresses might be varied in an interseismic period due to aseismic sliding on a plate interface, where the regional stresses mean the stresses around the source area of a large interplate earthquake. They suggested that the regional stress relaxation due to aseismic sliding prior to a large interplate earthquake may rationalize the precursory seismic quiescence. In the present paper, developing the model of a subduction zone by KATO et al. (1997), we indicate a possibility that the stress variation due to aseismic sliding explains a variety of seismic activity both in the overriding continental plate and in the subducting oceanic plate.

Model

We perform a numerical simulation of recurring large interplate earthquakes in a subduction zone to evaluate the spatiotemporal variation in regional stress field, using the same model adopted by KATO et al. (1997). We review the simulation method below. See KATO and HIRASAWA (1997a), in addition to KATO et al. (1997), for more detail of the method of simulation including model parameter selection.

DIETERICH (1979) and RUINA (1983) developed a rate- and state-dependent friction law, in which the frictional force was assumed to depend on the rate and the history of sliding, from laboratory experiments of frictional sliding on prepared rock surfaces. It has been found from numerous laboratory studies that the friction law explains the frictional behavior of rocks at hydrothermal conditions of source regions of earthquakes (e.g., CHESTER and HIGGS, 1992; BLANPIED et al., 1995) as well as at room temperature. Further, the friction law has been successfully applied to the modeling of seismic cycle on a plate interface in the earth to simulate the sliding behavior on it (TSE and RICE, 1986; STUART, 1988; RICE, 1993; STUART and TULLIS, 1995).

In the friction law, the friction coefficient μ is described as a function of a sliding rate V and a state variable θ, which depends on the history of sliding. There are several versions of the rate- and state-dependent friction law, although they are related to one another (e.g., MARONE, 1998). In the present study, we use the "slowness" version of the friction law. According to, for instance, BEELER et al. (1994), the friction coefficient μ obeying the slowness-version law is written as follows:

$$\mu = \mu_0 + a \ln(V/V_*) + b \ln(\theta V_*/L), \tag{1}$$

$$d\theta/dt = 1 - \theta V/L, \tag{2}$$

where V_* is a reference speed given arbitrarily, and μ_0, a, b, and L are constant. When $d\theta/dt = 0$, the friction coefficient takes a steady-state value μ_{SS}, which is written by $\mu_0 + (a - b) \ln(V/V_*)$.

RUINA (1983) showed that rate-weakening $\partial \mu_{SS}/\partial V < 0$, that is $(b - a) > 0$, is a necessary condition for the occurrence of unstable (seismic) slip. He further found that slip tends to be more unstable as $(b - a)$ becomes larger or L becomes smaller. A large positive $(b - a)$ and a small L indicate a large and rapid decrease in friction coefficient with an increase in slip rate, leading to unstable slip. On the other hand, stable (aseismic) sliding occurs when $(b - a) < 0$. Thus, the values of a, b, and L control the sliding behavior of a fault.

It should be remarked that KATO and HIRASAWA (1997a) and KATO et al. (1997) used a slightly different version of the friction law called the "slip" version (e.g., BEELER et al., 1994), where the evolution of state variable θ is written by

$$d\theta/dt = -(\theta V/L) \ln(\theta V/L) \tag{3}$$

instead of (2). Since the slowness version better describes the healing process of faults than the slip version as reported by BEELER et al. (1994) from their experimental studies, the slowness version is considered to be more appropriate for application to the modeling of seismic cycles. However, the simulation results obtained for the two versions of the friction law are not appreciably different from each other as reported by KATO and HIRASAWA (1997b) except that the amplitude of preseismic sliding immediately before an earthquake is significantly larger in the case of the slowness version than the slip version. This difference in the amplitude of preseismic sliding between the two versions was pointed out by DIETERICH (1992) and ROY and MARONE (1996) in their models of a straight fault in a 2-D infinite medium and a slider-block system.

We consider a thrust fault with a dip angle ϕ in a 2-D uniform elastic half-space (Fig. 1), in which pure thrust faulting is assumed to take place. We regard the fault of an infinite width along strike as the boundary between a continental plate and a subducting oceanic plate. A Cartesian coordinate system (x, y) is taken as shown in Figure 1, where ξ is the distance along the plate interface measured from the trench. The frictional force, $\mu\sigma_n^{\text{eff}}$ where μ is the friction coeffcient obeying the rate- and state-dependent friction law and σ_n^{eff} is the effective normal stress, is assumed to act only on the region of $0 \le \xi \le \xi_N$. We further assume that stable sliding at a sliding rate equal to the average rate of relative plate motion, V_{pl}, always takes place on the remaining region ($\xi > \xi_N$) of the plate interface. Under the condition of static equilibrium we can obtain the analytical expressions of stresses caused by quasi-static sliding on the plate interface from elasticity theory (e.g., RANI and SINGH, 1992). When an earthquake occurs, the condition of static equilibrium is violated. We therefore approximately evaluate the dynamic shear-stress acting on the plate interface by introducing the effect of strain energy dissipation due to elastic wave radiation (RICE, 1993). By doing this we may numerically compute the evolution of slip on the plate interface and the stresses in its vicinity.

We assume the distributions with respect to depth of a and $(a-b)$ as shown in Figure 2, taking account of the temperature dependence of friction parameters (e.g, BLANPIED et al., 1995) and the results of previous simulation studies (e.g., STUART, 1988; KATO and HIRASAWA, 1997a). The depth range of 12.93 km $\le y \le$ 52.22 km is the seismogenic zone because of negative $(a-b)$. The minimum value of $(a-b)$ is -1.6×10^{-4} in the depth range of 15 km $\le y \le$ 50 km. In the present study, we examine two cases of the distribution of L with depth as shown in Table 1. In Case 1, $L = 2$ cm independent of depth. In Case 2, a region of $L = 5$ cm (30 km $< y < 40$ km) is put between two regions of $L = 1$ cm. Since unstable slip is difficult to achieve in a region of large L, we expect that the region of $L = 5$ cm in Case 2 will behave as a barrier to rupture propagation and that a significantly nonuniform and

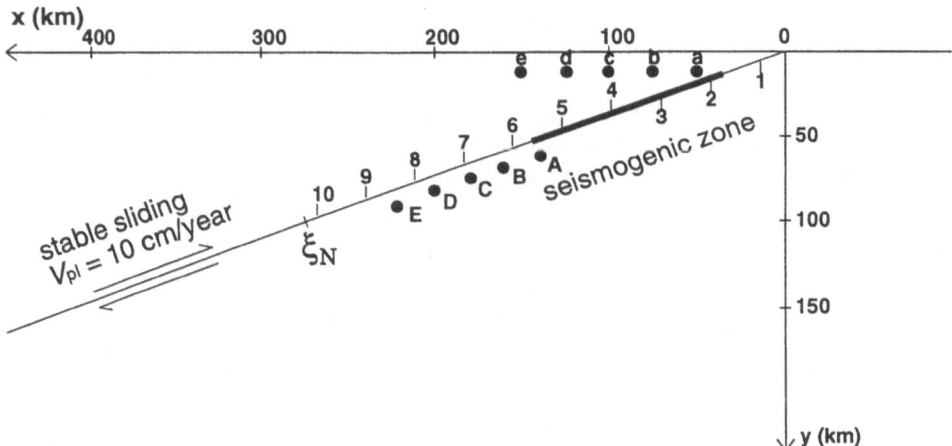

Figure 1
A 2-D model of subduction zone. The thick line on the plate interface indicates the seismogenic zone with negative values of $(a-b)$. Numerals 1 to 10 on the plate interface are points at which simulated slip histories are displayed in Figures 3 and 8. Symbols a to e and A to E are points at which histories of compression stresses are displayed in Figures 5 and 9, and Figure 7, respectively.

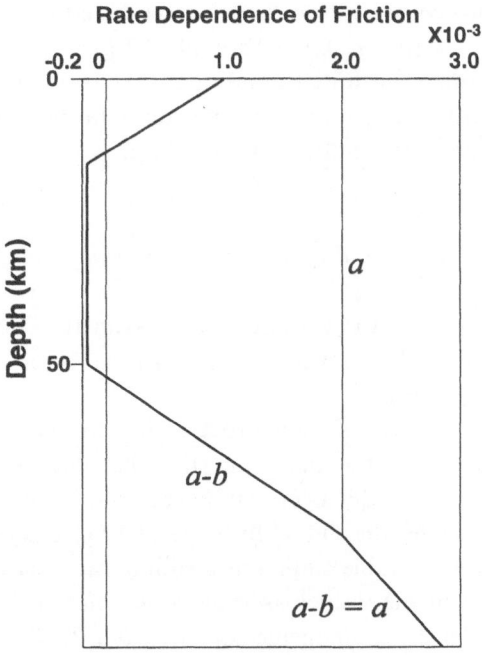

Figure 2
The variation with depth of friction parameters.

unsteady sliding process will occur (KATO and HIRASAWA, 1999). We take these values of friction parameters so as to satisfy observed characteristics of large interplate earthquakes in subduction zones such as the recurrence interval and the coseismic slip. It should be remarked that the values of L assumed in the present study are generally smaller than those in the preceding study (KATO *et al.*, 1997) based on the slip version of friction law; otherwise the recurrence interval of simulated large interplate earthquakes would be very long.

The values of remaining parameters are assumed as follows; dip angle $\phi = 20°$, rigidity $= 30$ GPa, Poisson's ratio $= 0.25$, and steady relative plate motion $(V_{pl}) =$

Table 1

Assumed values of L and simulation results for Cases 1 and 2

Case	L	T_r	u_{seis}
1	2.0 cm	93 years	3.2 m
2	5.0 cm (30 km $<y<$ 40 km)	111 years	5.1 m
	1.0 cm (elsewhere)		

L, characteristic slip distance.
T_r, recurrence interval of large interplate earthquakes.
u_{seis}, average value of coseismic slip on the seismogenic zone.

10 cm/year. Taking into consideration the lithostatic load and the hydrostatic pore pressure, we assume $\sigma_n^{\text{eff}} = (\rho - \rho_w)gy$, where $\rho = 2.7$ g/cm^3, $\rho_w = 1.0$ g/cm^3, $g = 9.8$ m/s^2, and y is the depth. The initial condition is that a uniform sliding rate of $0.01 V_{\text{pl}}$ exists on the entire plate interface. The computation is initiated by the load due to V_{pl} on the deeper part of the plate interface.

Possible Mechanism of Precursory Seismic Quiescence

We explore in this section the variation of stress in the overriding plate before a large interplate earthquake in Case 1 and discuss a possible mechanism of precursory seismic quiescence.

After a period of transient sliding process due to the initial condition, the simulated sliding process on the plate interface becomes periodic, where large interplate earthquakes repeatedly occur at a constant time interval of 93 years (Table 1). Figure 3 delineates the sliding histories at 10 points on the plate interface (Fig. 1) for a seismic cycle. Stable sliding at a sliding rate nearly equal to V_{pl} takes place at Points 8 to 10 in the deeper aseismic zone, and significant coseismic slip occurs at Points 2 to 5 in the seismogenic zone $(a - b < 0)$. The average seismic slip u_{seis} of 3.2 m as tabulated in Table 1 is the average value of seismic slip over the seismogenic zone, where the seismic slip is defined as slip at a slip rate equal to or greater than 1 cm/s following KATO and HIRASAWA (1997a). At Points 1, 6, and 7 in the aseismic zones adjacent to the seismogenic zone, a significant event of aseismic sliding occurs immediately after the occurrence of a large earthquake.

Figure 4 displays the spatiotemporal variation of slip distance on the plate interface before a large earthquake, demonstrating that aseismic sliding also takes place in the seismogenic zone. The aseismic sliding propagates both from the upper and lower aseismic zones into the central part of the seismogenic zone, where preslip occurs just before the earthquake. This spatiotemporal change in slip affects the regional stress field. Figure 5 shows the histories of compression stress, σ_{xx}, perpendicular to the trench axis at 5 points at a depth of 10 km in the overriding plate (a to e in Fig. 1) before the earthquake. Due to the aseismic sliding mostly in the deeper part of the plate interface, σ_{xx} increases nearly at a constant rate at first. Subsequently the rate of increase in σ_{xx} becomes small and finally σ_{xx} decreases because of the propagation of aseismic sliding into the seismogenic zone. This decrease in compression stress σ_{xx} is expected to lower the activity of earthquakes of reverse-fault type in the overriding plate to result in the seismic quiescence. The significant change in stresses immediately before the earthquake (Fig. 5) is caused by the preslip in the central part of the seismogenic zone (Fig. 4). This stress change may be related to foreshock activity.

It should be noted that the present model is 2-D and accordingly the region of preseismic sliding is regarded as infinitely long in the direction perpendicular to the

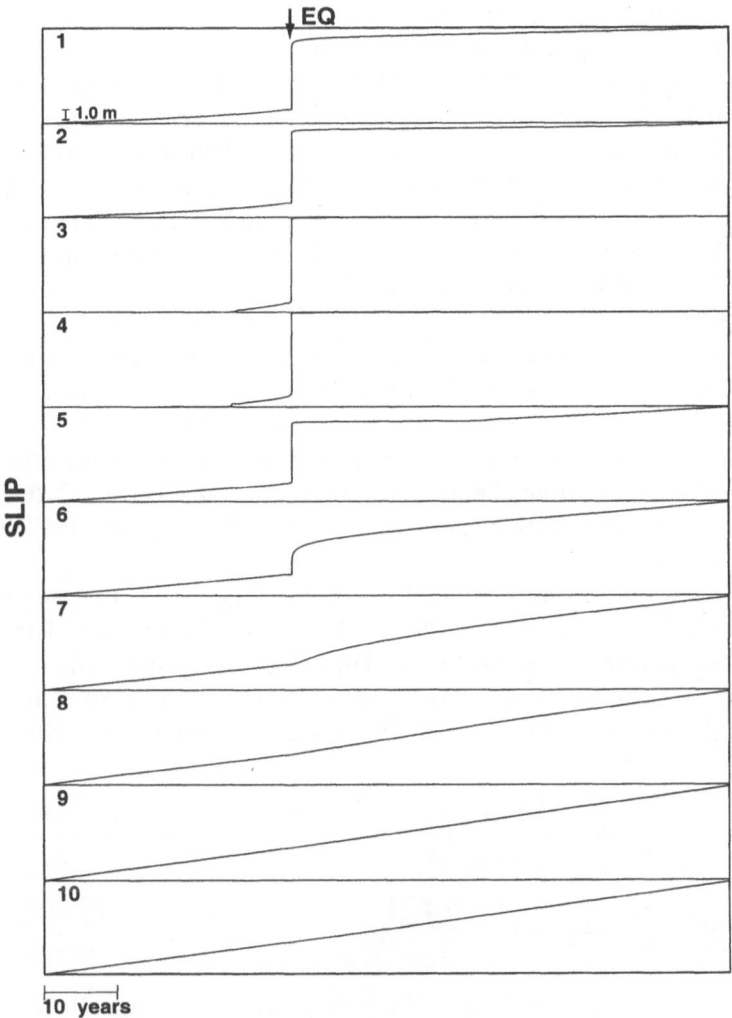

Figure 3

Slip histories at 10 points on the plate interface in Case 1 for a seismic cycle. See Figure 1 for the locations of points 1 to 10.

x-axis. SAGIYA and TADA (1994) determined the spatial distribution of slip on the plate interface for pre-, co-, post-, and inter-seismic periods in the source region of the 1946 Nankai earthquake ($M = 8.1$), southwestern Japan, from geodetic data. Their results indicate that the variations of preseismic sliding in the direction along the trench axis are not greatly large. Although the resolution in space and time is limited due to sparseness of data, the results of SAGIYA and TADA (1994) suggest a 2-D model may well approximate preseismic sliding of a relatively long term.

To evaluate the expected time interval of precursory seismic quiescence, we read the onset time of decrease in σ_{xx} at $y = 10$ km as denoted by tic marks in Figure 5.

The time interval, T_q, from the stress decrease to the occurrence of a large earthquake is plotted against the distance x from the trench axis in Figure 6. If most earthquakes in the overriding plate are generated by the compression stress perpendicular to the trench axis, the present result reveals that the seismic quiescence should appear both just above the source area of the coming large interplate earthquake and in an inner region of an island-arc trench system. The values of T_q are different for different locations just above the source area of main shock due to the migration of maximum slip rate of preslip in the seismogenic zone, while T_q in the region of $x > 100$ km is about 1 year.

This stress decrease due to the propagation of aseismic sliding may quantitatively explain the mechanism of precursory seismic quiescence as suggested by KATO et al. (1997). The present model may also reconcile the following empirical knowledge of precursory seismic quiescence reported by OHTAKE (1980) and MOGI (1985): (i) The precursory seismic quiescence often appears in a wider area than the source area of a large earthquake. (ii) The duration of quiescence is a few years to a few decades for great earthquakes of $M \approx 8$. See KATO et al. (1997) for more detailed discussion.

There are few observational examples of precursory seismic quiescence appearing in the overriding plates, because the depth control of hypocenter determination of offshore earthquakes is generally poor. The precursory seismic quiescences of the 1978 Miyagi-Oki, Japan, earthquake of $M = 7.4$ (TAKAGI, 1980) and the 1986 Adak, central Aleutian, earthquake of $M = 8.0$ (KISSLINGER, 1988) seem to occur

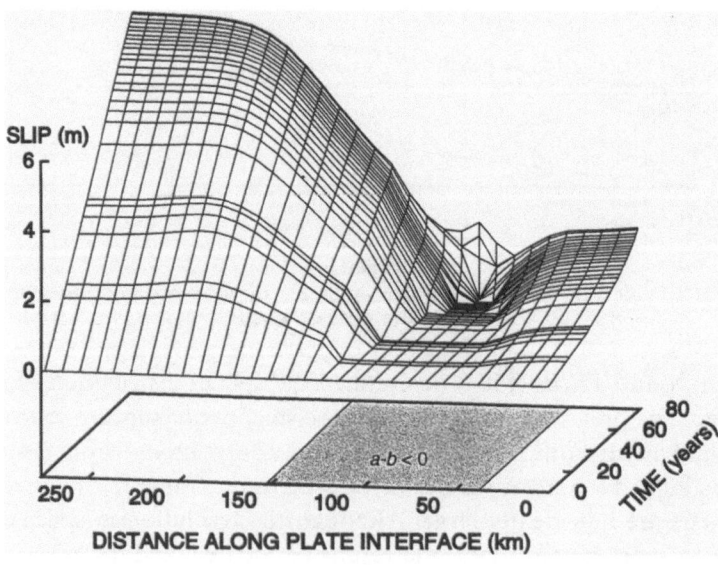

Figure 4
The spatiotemporal variation of slip on the plate interface before a large interplate earthquake in Case 1.

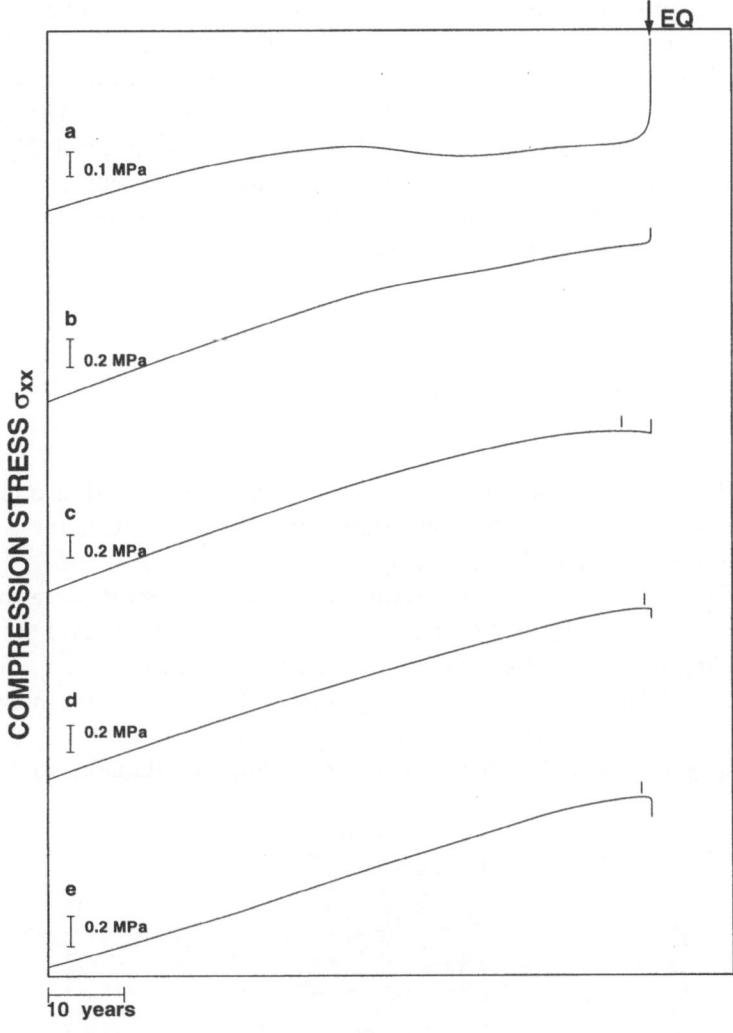

Figure 5

The histories of horizontal compression stress σ_{xx} before a large interplate earthquake at 5 points in Case 1. Points a to e are located at $x = 50$, 75, 100, 125, and 150 km and their depths are all 10 km (Fig. 1). Tic marks indicate the times of peak σ_{xx}.

in the overriding plates. KATO and WIEMER (1998) quantitatively compared the observed seismicity data prior to the 1986 Adak earthquake with a numerical model similar to the present one.

The most important difference between the present model and the model of KATO *et al.* (1997) is in the version of the rate- and state-dependent friction law. The present model uses the slowness version while KATO *et al.* (1997) used the slip version. We find that the spatiotemporal variation of σ_{xx} and the time interval, T_q, from the onset of stress decrease to the occurrence of a large interplate earthquake expected for the present model are similar to those of KATO *et al.* (1997). We performed simulations of many cases by changing the values of model parameters for both versions of the rate- and state-dependent friction law. The recurrence interval, T_r, of large interplate earthquakes and the time interval T_q generally increase with the characteristic slip distance L. T_q seems to be roughly 1% to 2% of T_r in all the cases. It is concluded that the differences in the two rate- and state-dependent friction laws do not appreciably affect the characteristics of precursory seismic quiescence.

Effects of Aseismic Sliding on Intermediate-depth Earthquakes

The variation of aseismic sliding on the plate interface should also change the stresses in the subducting slab. This stress variation may affect the activity of intermediate-depth earthquakes in the slab.

Figure 7 shows the histories for a seismic cycle of compression stress, $\sigma_{\xi\xi}$, in the down-dip direction of the plate interface at 5 points (A to E in Fig. 1). Since aseismic sliding propagates from the lower aseismic zone to the upper seismogenic zone before the large earthquake, $\sigma_{\xi\xi}$ generally decreases in the slab. At the occurrence of a large interplate earthquake, $\sigma_{\xi\xi}$ suddenly increases. Postseismic sliding propagates from the source region of a large earthquake to the deeper

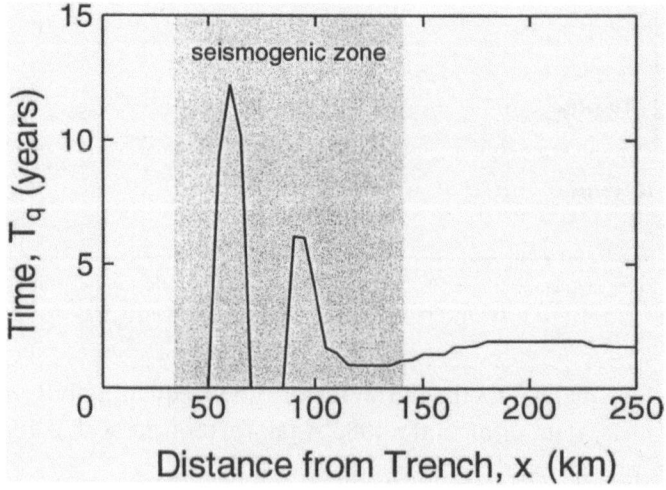

Figure 6

Time interval T_q from the maximum stress σ_{xx} at the depth of 10 km to the occurrence of a large interplate earthquake with respect to the horizontal distance, x, from the trench in Case 1. The shaded zone denotes the region just above the seismogenic zone $(a - b < 0)$ of the plate interface.

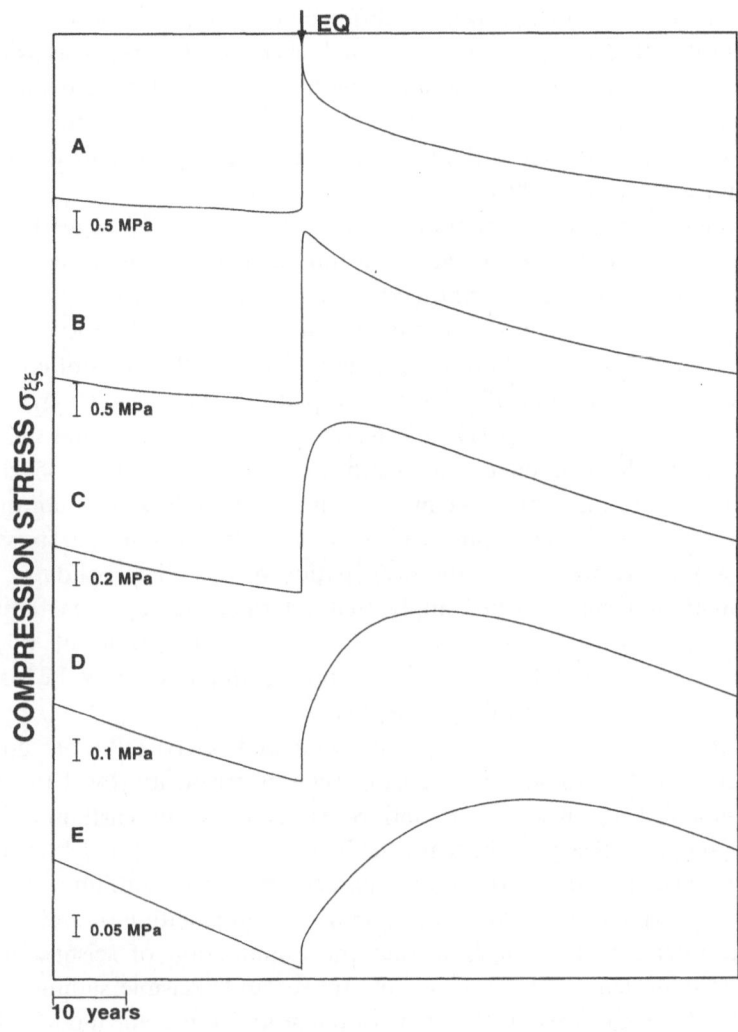

Figure 7

Histories of compression stress $\sigma_{\xi\xi}$ in the down-dip direction of the plate interface at 5 points in the subducting slab in Case 1. Points A to E are located at $x = 140$, 160, 180, 200, and 220 km, and their depths are 10 km deeper than the plate interface (Fig. 1).

aseismic zone as shown in Figure 3, and accordingly the increase of $\sigma_{\xi\xi}$ in a deeper part lasts for a longer time compared with that in a shallower part.

It should be remarked that the occurrence of postseismic sliding was reported by KAWASAKI *et al.* (1995), HEKI *et al.* (1997) and NISHIMURA *et al.* (1997) following large interplate earthquakes along the Japan trench from the observations of strainmeters and GPS. From GPS observations for the 1994 Far Off Sanriku earthquake of $M_w = 7.8$, NISHIMURA *et al.* (1997) found that the source region of postseismic sliding was not appreciably different from that of the main shock at first although it with time spread over the shallower and the deeper aseismic zone. The variation of cumulative slip amount with time is well approximated by a logarithmic function (HEKI et al., 1997). These observed characteristics of postseismic sliding can be reproduced in the present model (see Fig. 3). This was discussed by KATO and HIRASAWA (1997c).

The simulated temporal variation of slab stress (Fig. 7) suggests that the tension-type earthquakes would be active at intermediate depths in the subducting slab before a large interplate earthquake and the compression-type earthquakes would be active after it. This is consistent with the observation by LAY et al. (1989) who compiled the seismicity data of intermediate-depth earthquakes of compression and tension types before and after large interplate earthquakes in circum-Pacific subduction zones. WYSS et al. (1997) and IGARASHI et al. (1998) found that activity of intermediate-depth earthquakes was significantly changed after the occurrence of the 1994 Far Off Sanriku earthquake in a region considerably wider than the source area of the main shock. This can reasonably be explained if the stresses were changed in a wider region due to the propagation of postseismic sliding.

The simulation results further imply that intermediate-depth earthquakes of compression type may migrate downward due to propagation of postseismic sliding. The examination of intermediate-depth earthquakes may be useful for detecting postseismic sliding at deeper depths.

MATSUMURA (1997, 1998) and MATSUMURA and KATO (1999) examined the temporal variation of seismic characteristics such as seismicity, focal mechanisms, and predominant frequencies of P and S waves of small earthquakes in the subducting Philippine Sea plate beneath the Tokai district, central Japan, for more than 15 years. They found that the above characteristics vary with time and that the variation can be explained by stress variation due to propagation of aseismic sliding on the plate interface. This suggests that the examination of seismic activity is useful for detecting temporal variations of stresses and aseismic sliding.

We neglect viscoelasticity in the asthenosphere in the present model. However, viscoelastic relaxation in the asthenosphere considerably contributes the variations of stresses and seismicity in subducting oceanic plates (e.g., TAYLOR et al., 1996). We should develop a more comprehensive model including viscoelastic effects in future studies to precisely explain the variation of seismicity.

Effects of Episodic Sliding on Seismic Activity

Only large characteristic earthquakes at a constant recurrence interval are reproduced in Case 1, where the values of friction parameters are uniform over the seismogenic zone. When nonuniformity in stresses or friction parameters is introduced, more realistic and complicated sliding behavior can be simulated as shown by BEN-ZION and RICE (1995) and KATO and HIRASAWA (1999). We introduce nonuniformity in the characteristic slip distance L in Case 2 (Table 1) to discuss the effect of a slightly more complicated stress variation on seismic activity due to episodic events of aseismic sliding. Such nonuniformity in L as assumed in Case 2 is considered possible for the following reason: From laboratory studies, L is known to be dependent on the characteristics of a gouge materials between sliding surfaces such as the thickness of gouge layer (e.g., BIEGEL et al., 1989; MARONE and KILGORE, 1993). Surface geometry of plate interface is nonuniform and, accordingly, the thickness of gouge layer involved in a plate interface may be nonuniform (e.g., TANIOKA et al., 1996).

Figure 8 shows the histories of frictional sliding on the plate interface before the occurrence of a large interplate earthquake. At least three episodic events occur in the regions of $L = 1$ cm, and they are arrested at the region of $L = 5$ cm. The durations of these three events are longer than several days. Thus they are regarded as aseismic events. Episodic events of various durations can be found in related simulations (KATO and HIRASAWA, 1999).

Episodic events cause complicated variations of stresses in the overriding plate. Figure 9 shows the histories of compression stress σ_{xx} at 5 points at 10 km depth before a large interplate earthquake. We find significant decreases or increases in compression stress associated with the occurrence of episodic events. These episodic stress variations are not directly related to the occurrence of a large interplate earthquake, though they might increase or decrease the seismic activity in the overriding plate. When the barrier (the region of $L = 5$ cm) is broken, a large earthquake occurs over the entire seismogenic zone. The compression stress σ_{xx} starts to decrease several months before the large interplate earthquake in Case 2, indicating that a precursory seismic quiescence is also expected in Case 2, though its time interval is slightly shorter than Case 1.

It may be difficult to detect aseismic sliding from only the observations of seismic activity. A large event of episodic sliding should be accompanied by abnormal crustal deformation. Accordingly, the observation of crustal deformation combined with that of seismicity change is important to detect episodic sliding. There are some reports that abnormal crustal deformation and significant changes in seismicity were simultaneously observed. BEAVAN et al. (1984) found coherent tilt signals in a leveling survey in the Shumagin Islands, Alaska, where the Pacific plate subducts beneath the North American plate. They found a significant increase in microseismicity at shallow depths contemporaneous with the tilt signals, indicating that an event of episodic sliding on the plate interface can explain both the tilt signals and the stress increase that results in the seismicity change. SACKS et al. (1981) found strain events with three borehole strainmeters installed in the Izu Peninsula, Japan, following the 1978 Izu-Oshima earthquake ($M_{\text{JMA}} = 7.0$) to show that these strain events may be elucidated by aseismic sliding events on the fault plane where aftershocks of the Izu-Oshima earthquake occurred. They further suggested that the stress change due to aseismic sliding caused the two largest aftershocks. KIMATA (1992) reported a strain event in 1985–1987 in the Tokai district, central Japan, detected by electro-optical distance measurements. He pointed out that the decrease in seismic activity in this region was contemporaneous with the strain event. LINDE *et al.* (1996) found an abnormal strain signal with the duration of about a week in two borehole strainmeters corresponding to a creep event on the San Andreas fault in central California. They reported that small earthquakes occurred at similar times and places as the creep event.

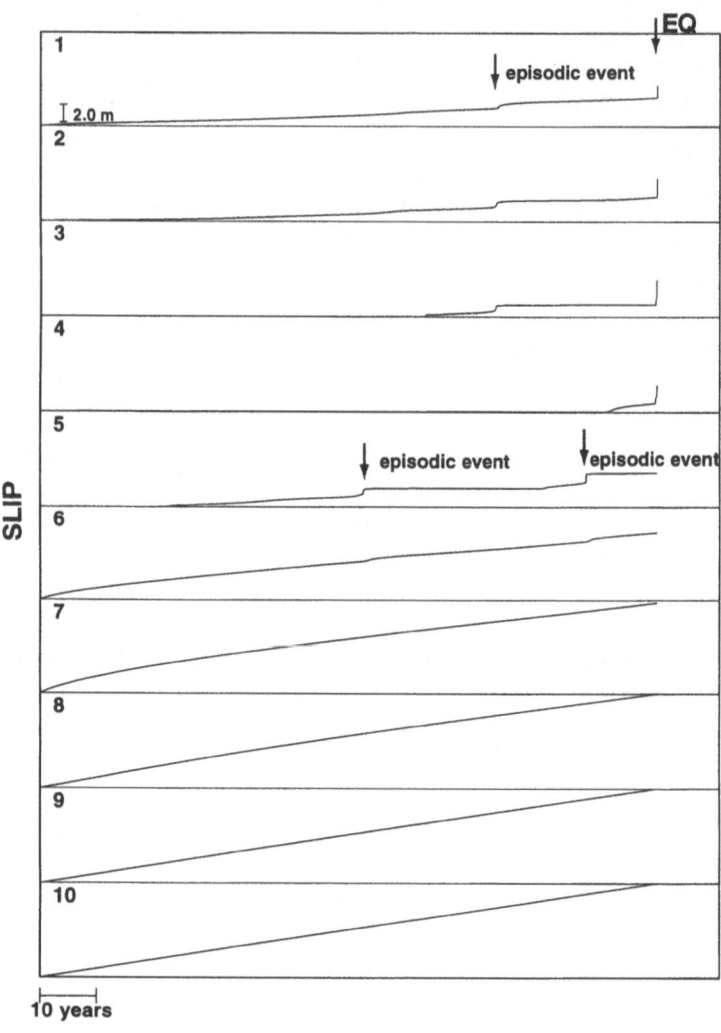

Figure 8
Slip histories at 10 points on the plate interface before a large interplate earthquake in Case 2. See Figure 1 for the locations of points 1 to 10.

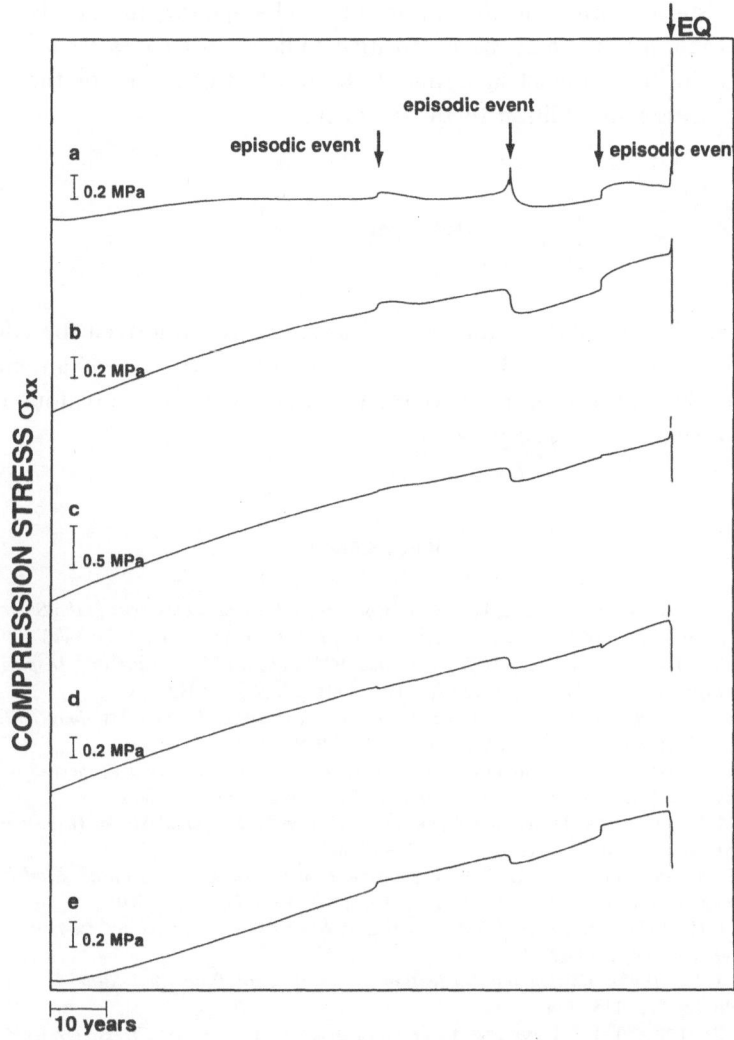

Figure 9
Histories of horizontal compression stress σ_{xx} before a large interplate earthquake at 5 points in Case 2. Points a to e are located at $x = 50, 75, 100, 125,$ and 150 km and their depths are all 10 km (Fig. 1). Tic marks indicate the times of peak σ_{xx}.

Conclusion

From the present numerical simulation, we show that aseismic sliding on a plate interface significantly varies the stress field and may affect the seismic activity. The present model can explain the precursory seismic quiescence and the temporal variation in activity of intermediate depth earthquakes. Since it may be difficult to detect the occurrence of aseismic sliding from only seismic observations, observations of crustal deformation are important to characterize the events of aseismic sliding. Seismic activity may be a sensitive indicator of stress variation. Seismic activity should be examined in terms of the focal mechanisms or the earthquake generating stresses in addition to the seismicity.

Acknowledgements

We thank M. Ohtake for stimulative discussion, R. Dmowska for informing us of valuable references, and M. Wyss for permitting the use of an unpublished manuscript. We are also grateful to W. D. Stuart and an anonymous referee for critical comments and suggestions.

REFERENCES

BEAVAN, J., BILHAM, R., and HURST, K. (1984), *Coherent Tilt Signals Observed in the Shumagin Seismic Gap: Detection of Time-dependent Subduction at Depth?*, J. Geophys. Res. *89*, 4478–4492.

BEELER, N. M., TULLIS, T. E., and WEEKS, J. D. (1994), *The Roles of Time and Displacement in the Evolution Effect in Rock Friction*, Geophys. Res. Lett. *21*, 1987–1990.

BEN-ZION, Y., and RICE, J. R. (1995), *Slip Patterns and Earthquake Populations along Different Classes of Faults in Elastic Solids*, J. Geophys. Res. *100*, 12,959–12,983.

BIEGEL, R. L., SAMMIS, C. G., and DIETERICH, J. H. (1989), *The Frictional Properties of a Simulated Gouge Having a Fractal Particle Distribution*, J. Struct. Geol. *11*, 827–846.

BLANPIED, M. L., LOCKNER, D. A., and BYERLEE, J. D. (1995), *Frictional Slip of Granite at Hydrothermal Conditions*, J. Geophys. Res. *100*, 13,045–13,064.

CHESTER, F. M., and HIGGS, N. G. (1992), *Multimechanism Friction Constitutive Model for Ultrafine Quartz Gouge at Hydrothermal Conditions*, J. Geophys. Res. *97*, 1859–1870.

DIETERICH, J. H. (1979), *Modeling of Rock Friction 1. Experimental Results and Constitutive Equations*, J. Geophys. Res. *84*, 2161–2168.

DIETERICH, J. H. (1992), *Earthquake Nucleation on Faults with Rate- and State-dependent Strength*, Tectonophysics *211*, 115–134.

DMOWSKA, R., RICE, J. R., LOVISON, L. C., and JOSELL, D. (1988), *Stress Transfer and Seismic Phenomena in Coupled Subduction Zones During the Earthquake Cycle*, J. Geophys. Res. *93*, 7869–7884.

HEKI, K., MIYAZAKI, S., and TSUJI, H. (1997), *Silent Fault Slip Following an Interplate Thrust Earthquake at the Japan Trench*, Nature *386*, 595–598.

HORI, T., and OIKE, K. (1996), *A Statistical Model of Temporal Variation of Seismicity in the Inner Zone of Southwest Japan Related to the Great Interplate Earthquakes along the Nankai Trough*, J. Phys. Earth *44*, 349–356.

IGARASHI, T., MATSUZAWA, T., UMINO, N., and HASEGAWA, A. (1998), *Spatial and Temporal Stress Changes in and around the Subducting Pacific Plate beneath the Northeastern Japan Arc*, Abstracts Japan Earth and Planetary Science Joint Meeting, 30 (in Japanese).

KATO, N., and HIRASAWA, T. (1997a), *A Numerical Study on Seismic Coupling along Subduction Zones Using a Laboratory-derived Friction Law*, Phys. Earth Planet. Inter. *102*, 51–68.

KATO, N., and HIRASAWA, T. (1997b), *Probable Crustal Deformation Prior to the Hypothetical Tokai Earthquake: A Numerical Experiment*, EOS Trans. Am. Geophys. Union *78*(46), Fall Meet. Suppl., F452.

KATO, N., and HIRASAWA, T. (1997c), *A Numerical Simulation of Postseismic Sliding on a Plate Boundary*, J. Seism. Soc. Jpn. *50*, 241–250 (in Japanese with English abstract).

KATO, N., and HIRASAWA, T. (1999), *Nonuniform and Unsteady Sliding of a Plate Boundary in a Great Earthquake: A Numerical Simulation Using a Laboratory-derived Friction Law*, Pure appl. geophys. *155*, 93–118.

KATO, N., and WIEMER, S. (1998), *Regional Stress Relaxation due to Preseismic Sliding as an Explanation of the Seismicity Rate Changes Observed Prior to the 1986 $M_w8.0$ Adak, Alaska, Earthquake*, EOS Trans. Am. Geophys. Union *79*(45), Fall Meet. Suppl., F643.

KATO, N., OHTAKE, M., and HIRASAWA, T. (1997), *Possible Mechanism of Precursory Seismic Quiescence: Regional Stress Relaxation Due to Preseismic Sliding*, Pure appl. geophys. *150*, 249–267.

KAWASAKI, I., ASAI, Y., TAMURA, T., SAGIYA, T., MIKAMI, N., OKADA, Y., SAKATA, M., and KASAHARA, M. (1995), *The 1992 Sanriku-Oki, Japan, Ultra-slow Earthquake*, J. Phys. Earth *43*, 105–116.

KIMATA, F. (1992), *Strain Event in 1985–1987 in the Tokai Region, Central Japan*, J. Phys. Earth *40*, 585–599.

KING, G. C. P., STEIN, R. S., and LIN, J. (1994), *Static Stress Changes and the Triggering of Earthquakes*, Bull. Seismol. Soc. Am. *84*, 935–953.

KISSLINGER, C. (1988), *An Experiment in Earthquake Prediction and the 7 May 1986 Andreanof Islands Earthquake*, Bull. Seismol. Soc. Am. *78*, 218–229.

LAY, T., ASTIZ, L., KANAMORI, H., and CHRISTENSEN, D. H. (1989), *Temporal Variation of Large Interplate Earthquakes in Coupled Subduction Zones*, Phys. Earth Planet. Inter. *54*, 258–312.

LINDE, A. T., GLADWIN, M. T., JOHNSTON, M. J. S., GWYTHER, R. L., and BILHAM, R. G. (1996), *A Slow Earthquake Sequence Near San Juan Bautista, California in December 1992*, Nature *383*, 65–68.

MARONE, C. (1998), *Laboratory-derived Friction Laws and Their Application to Seismic Faulting*, Annu. Rev. Earth Planet. Sci. *26*, 643–696.

MARONE, C., and KILGORE, B. (1993), *Scaling of the Critical Slip Distance for Seismic Faulting with Shear Strain in Fault Zone*, Nature *362*, 618–621.

MATSUMURA, S. (1997), *Focal Zone of a Future Tokai Earthquake Inferred from the Seismicity Pattern around the Plate Interface*, Tectonophysics *273*, 271–291.

MATSUMURA, S. (1998), *Possible Change of Locking State Between the Plates in the Tokai Area Indicated by the Central Shizuaka Earthquake of Oct. 5, 1996*, J. Seism. Soc. Jpn. *50*, Suppl. 251–261 (in Japanese with English abstract).

MATSUMURA, S., and KATO, N. (1999), *Recognition of a Locked State in Plate Subduction from Microearthquake Seismicity*, Pure appl. geophys. *155*, 669–682.

MOGI, K., *Earthquake Prediction* (Academic Press, Tokyo 1985).

NISHIMURA, T., MIURA, S., TACHIBANA, K., HASHIMOTO, K., SATO, T., HORI, S., MURAKAMI, E., NIDA, K., MISHINA, M., HIRASAWA, T., and MIYAZAKI, S. (1997), *Cohesive Region on the Plate Boundary East Off Tohoku Inferred from GPS Observation*, Abstracts Japan Earth and Planetary Science Joint Meeting, 690 (in Japanese).

OHTAKE, M. (1980), *Earthquake Prediction Based on the Seismic Gap with Special Reference to the 1978 Oaxaca, Mexico Earthquake*, Rept. Natl. Cent. Disaster Prev. *23*, 65–110 (in Japanese with English abstrct).

OHTAKE, M., MATUMOTO, T., and LATHAM, G. V. (1977), *Seismicity Gap near Oaxaca, Southern Mexico as a Probable Precursor to a Large Earthquake*, Pure appl. geophys. *115*, 375–385.

RANI, S., and SINGH, S. J. (1992), *Static Deformation of a Uniform Half-space to a Long Dip-slip Fault*, Geophys. J. Int. *109*, 469–476.

REASENBERG, P. A., and SIMPSON, R. W. (1992), *Response of Regional Seismicity to the Static Stress Change Produced by the Loma Prieta Earthquake*, Science *255*, 1687–1690.

RICE, J. R. (1993), *Spatio-temporal Complexity of Slip on a Fault*, J. Geophys. Res. *98*, 9885–9907.

ROY, M., and MARONE, C. (1996), *Earthquake Nucleation on Model Faults with Rate- and State-dependent Friction: Effects of Inertia*, J. Geophys. Res. *101*, 13,919–13,932.

RUINA, A. L. (1983), *Slip Instability and State Variable Friction Laws*, J. Geophys. Res. *88*, 10,359–10,370.

SACKS, I. S., LINDE, A. T., SNOKE, A. T., and SUYEHIRO, S., *A slow earthquake sequence following the Izu-Oshima earthquake of 1978*, In *Earthquake Prediction: An International Review* (eds. Simpson, D. W., and Richards, P. G.) (American Geophysical Union, Washington, D.C. 1981) pp. 617–628.

SAGIYA, T., and TADA, T. (1994), *Estimation of Interplate Coupling Strength beneath the Shikoku Island Deduced from Crustal Movements Data*, Abstracts Seism. Soc. Japan Meeting, No. 2, 50 (in Japanese).

SHIMAZAKI, K. (1976), *Intra-plate Seismicity and Inter-plate Earthquakes: Historical Activity in Southwest Japan*, Tectonophysics *33*, 33–42.

STUART, W. D. (1988), *Forecast Model for Great Earthquakes at the Nankai Trough Subduction Zone*, Pure appl. geophys. *126*, 619–641.

STUART, W. D., and TULLIS, T. E. (1995), *Fault Model for Preseismic Deformation at Parkfield, California*, J. Geophys. Res. *100*, 24,079–24,099.

TAKAGI, A. (1980), *Concluding Remarks and Precursory Seismic Activity of the 1978 Miyagi-Oki Earthquake*, Proceedings of Earthquake Prediction Research Symposium, Seismological Society of Japan and Subcomittee of Earthquake Prediction, National Committee of Geophysics, Science Council of Japan, 231–241 (in Japanese).

TANIOKA, Y., RUFF, L., and SATAKE, K. (1996), *What Control the Lateral Variation of Large Earthquake Occurrence along the Japan Trench*, Island Arc *6*, 261–266.

TAYLOR, M. A. J., DMOWSKA, R., and RICE, J. R. (1998), *Upper Plate Stressing and Back Arc Seismicity in the Subduction Earthquake Cycle*, J. Geophys. Res. *103*, 24,523–24,542.

TAYLOR, M. A. J., ZHENG, G., RICE, J. R., STUART, W. D., and DMOWSKA, R. (1996), *Cyclic Stressing and Seismicity at Strongly Coupled Zones*, J. Geophys. Res. *101*, 8363–8381.

TSE, S. T., and RICE, J. R. (1986), *Crustal Earthquake Instability in Relation to the Depth Variation of Frictional Slip Properties*, J. Geophys. Res. *91*, 9452–9472.

WYSS, M., and HABERMANN, R. E. (1988), *Precursory Seismic Quiescence*, Pure appl. geophys. *126*, 319–332.

WYSS, M., HASEGAWA, A., and UMINO, N. (1997), *Mapping the Extent of After-creep Following the 1994 (M_w7.8) Off-Sanriku Earthquake by Seismic Quiescence*, Internal Rept., 10 pp.

WYSS, M., KLEIN, F. W., and JOHNSTON, A. C. (1981), *Precursors to the Kalapana M = 7.2 Earthquake*, J. Geophys. Res. *86*, 3881–3900.

(Received July 22, 1998, revised December 10, 1998, accepted December 11, 1998)

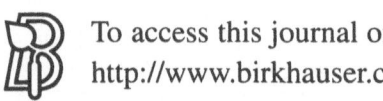

To access this journal online:
http://www.birkhauser.ch

Pure appl. geophys. 155 (1999) 443–470
0033–4553/99/040443–28 $ 1.50 + 0.20/0

▌**Pure and Applied Geophysics**

Precursory Seismic Quiescence before the 1994 Kurile Earthquake ($M_w = 8.3$) Revealed by Three Independent Seismic Catalogs

KEI KATSUMATA[1] and MINORU KASAHARA[1]

Abstract—We have found that the $M_w = 8.3$ Kurile earthquake on October 4, 1994 followed an outstanding seismic quiescence starting 5–6 years before the mainshock near the ruptured area. We have analyzed three independent seismic catalogs: Institute of Seismology and Volcanology, Hokkaido University (ISV), Japan Meteorological Agency (JMA) and International Seismology Center (ISC). In spite of selecting different magnitude bands and time windows all three catalogs presented the common feature of the seismic quiescence. This fact strongly suggests that the seismic quiescence should not be a man-made change but actually occurred. Moreover we have confirmed that the seismic quiescence was the most significant and the earthquake was the largest in the past twenty-five years in this region. Therefore we confidently interpret this seismic quiescence as an indication of a preparation process for the $M_w = 8.3$ Kurile earthquake.

Key words: Seismicity pattern, seismic quiescence, Kurile, Hokkaido Toho-Oki, earthquake prediction.

1. Introduction

The Kurile Islands (Hokkaido Toho-Oki) earthquake ($M_w = 8.3$) occurred on October 4, 1994 in the southern part of the Kurile Islands (Fig. 1). The focal mechanism was not a low-angle thrust-type, the centroid depth was large and the stress drop was high (KIKUCHI and KANAMORI, 1995; TANIOKA *et al.*, 1995). The azimuth and the dip of the aftershock area were most probably parallel to the trench axis and near vertical, respectively (KATSUMATA *et al.*, 1995; FURUKAWA, 1995). Coseismic crustal deformations were clearly consistent with the vertical fault plane (TSUJI *et al.*, 1995; OZAWA, 1996). These facts strongly suggest that this event is a lithospheric earthquake: an intra-plate event that ruptures through a substantial part of the subducting oceanic lithosphere (KIKUCHI and KANAMORI, 1995).

Many authors have reported that precursory seismic quiescences occurred in and around focal areas several years before earthquakes: Tonga-Kermadec (WYSS *et al.*, 1984), Tokachi-Oki (MOGI, 1969; HABERMANN, 1981b), Oaxaca (OHTAKE *et*

---

[1] Institute of Seismology and Volcanology, Graduate School of Science, Hokkaido University, Sapporo, Japan, 0600810. E-mail: katsu@eos.hokudai.ac.jp

al., 1977; HABERMANN, 1981b; McNALLY, 1981), Aleutians (HABERMANN, 1981a; KISSLINGER, 1988), Lima (HABERMANN, 1981a), Colima (HABERMANN, 1981b; McNALLY, 1981), Hawaii (WYSS *et al.*, 1981; WYSS, 1986), Hokkaido (TAYLOR *et al.*, 1991), Kuriles (HABERMANN, 1981b), Morgan Hill (HABERMANN and WYSS, 1984), San Andreas (WYSS and BURFORD, 1985, 1987; WYSS and HABERMANN, 1988), Landers (WIEMER and WYSS, 1994) and Izu-Oshima (WYSS *et al.*, 1996). TAKANAMI *et al.* (1996) have found a seismic quiescence starting three years before the $M_w = 8.3$ Kurile mainshock in 1994. Their analysis was based on an earthquake catalog produced by ISV.

To confirm the results of TAKANAMI *et al.* (1996), we have analyzed three different catalogs compiled by not only ISV but also both JMA and ISC. These institutions independently locate hypocenters and estimate magnitudes. If seismic quiescences have actually occurred in an area, they should be detected by different seismic catalogs. An apparent change in seismicity rate is easily brought on by artificial reasons: deployment of new seismic stations, closing of old seismic stations, changes in seismograph, waveform recording system and magnitude estimation algorithm (HABERMANN, 1987, 1991). Therefore, to compare the results from the three seismic catalogs provides us with evidence to verify whether or not a seismic quiescence detected in a target area is a fact.

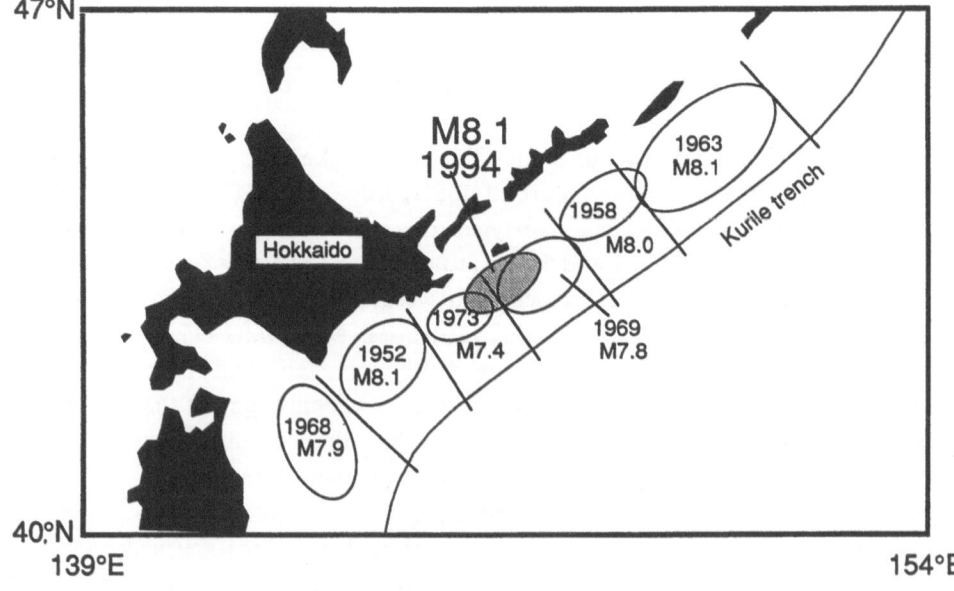

Figure 1
Great earthquakes off the coast of Hokkaido and off the southern Kurile Islands. JMA determined magnitudes. Ellipses roughly show each aftershock area.

Table 1

Characteristic parameters used in this study for the ISV, the JMA and the ISC seismic catalog

	ISV	JMA	ISC
Target area	144–149°E	144–149°E	144–149°E
	42–44.5°N	42–44.5°N	42–44.5°N
Time window	1 March, 1985–	1 January, 1977–	1 January, 1970–
	3 October, 1994	3 October, 1994	3 October, 1994
Length of time window (days)	3504	6485	9042
Magnitude	$M \geq 3.0$	$M \geq 4.3$	$M \geq 5.0$
Depth (km)	0–150	0–150	0–150
Number of earthquakes (original)	1445	946	526
Number of earthquakes (declustered)	1390	491	336

2. Data

Table 1 summarizes key parameters of the data used in this study. Details of each seismic catalog are described below.

2.1 ISV

ISV operates a regional seismic network called Hokkaido Seismic Network which consists of about thirty stations in Hokkaido and the northern part of the Honshu Islands. A typical station consists of three seismometers in a vault, i.e., one vertical and two horizontal (north-south and east-west) components, with a natural frequency of 1 Hz, amplifiers with magnifications of 30 to 72 dB and 10 or 16 bits analogue-to-digital converters. All waveform data are telemetered continuously to ISV in Sapporo by dedicated telephone lines of Nippon Telephone and Telegram Company. ISV started the field installation of the network in 1976 and finished it in 1985. Moreover thirty-one short-period seismic stations deployed by Japan Meteorological Agency (JMA) were added to the Hokkaido Seismic Network in 1997. A drastic change in the data processing system of ISV occurred on 20 May, 1993. Before the day, arrival times of *P*- and *S*-waves were read on seismograms from a 24-channel pen recorder with a paper speed of 1 cm/s and magnitudes were determined at each station using the following equation,

$$M_{F-P} = 2.75 \log T_{F-P} - 2.24, \tag{1}$$

where M_{F-P} is magnitude and T_{F-P} is time in seconds measured from the arrival time of *P*-waves to the time when the amplitudes of the coda of *S*-waves return to the ground noise level. M_{F-P}s at some stations were averaged to obtain the representative magnitude of an event.

In 1993 the WIN system running on a UNIX workstation (URABE and TSUKADA, 1992) was installed at ISV. WIN is a powerful tool to show waveform data on a computer display, to read manually arrival times and maximum amplitudes, to calculate the hypocenter using HYPOMH (HIRATA and MATSU'URA, 1987) and to show a seismicity map. Magnitudes are determined at each station using the following equation (WATANABE, 1971),

$$0.85M_A - 2.50 = \log Av + 1.73 \log r \quad (r < 200 \text{ km}), \quad (2)$$

where M_A is a magnitude, Av is the maximum velocity amplitude in cm/s on the vertical component and r is the epicentral distance in km. Therefore, the ISV catalog includes two different magnitudes estimated using (1) and (2). That is the reason why a magnitude shift and stretch occurred in 1993.

From the original seismic catalog of ISV, hypocenters with more than five P-wave and one S-wave readings were selected for relocation in a rectangular area (144–149°E, 42–44.5°N). First, we calculated hypocenters using HYPOMH (HIRATA and MATSU'URA, 1987) without station corrections and plotted residuals vs. epicentral distances for each station. Since the residuals were found to shift linearly as a function of epicentral distance, a straight line was fitted using the least-squares method. We obtained station corrections as a function of epicentral distance for each seismic station. Then the station corrections were added to arrival times and hypocenters relocated using HYPOMH.

After the relocation of hypocenters we made a correction for the magnitude change in 1993 in the target area of this study. The b-values were estimated to be 1.4 and 0.8 for a period between 1 March, 1985 and 20 May, 1993 (Period of old system) and a period between 20 May, 1993 and 3 October, 1994 (Period of new system), using events larger than $M = 3.5$ and 3.0, respectively. In this magnitude band the seismicity rate in Period of new system decreased to 0.4 compared to that in Period of old system. This is caused by the drastic shift and stretch of magnitude. Thereafter we assumed that the background seismicity rate and b-value in Period of old system was the same as that in Period of new system. We, derived the equation:

$$M_{\text{new}} = 1.83M_{\text{old}} - 3.45,$$

where M_{old} is a magnitude in Period of old system and M_{new} is the corresponding magnitude in Period of new system. Note that this equation is available for events with $M_{\text{old}} = 3.5$ and larger.

After the correction, we plotted the number of earthquakes vs. magnitude in the study area to estimate the magnitude of completeness, M_c. The algorithm we use interprets the point of the peak in the (non-cumulative) frequency-magnitude distribution as M_c. We concluded that all events larger than $M = 3.0$ were reported homogeneously between 1 March, 1985 and 3 October, 1994, that is, $M_c = 3.0$. 1445 earthquakes larger than $M = 3.0$ which occurred from 1 March, 1985 to 3 October, 1994 were selected in the area between 144.0–149.0°E and 42.0–44.5°N. Earth-quakes were selected with depths between 0 and 150 km beyond the depth of 60–70 km at which the seismic fault of the mainshock extended, because the seismicity change should be estimated not only on the seismic fault but also in its surrounding portion. Figure 2a shows epicenters including clustered events.

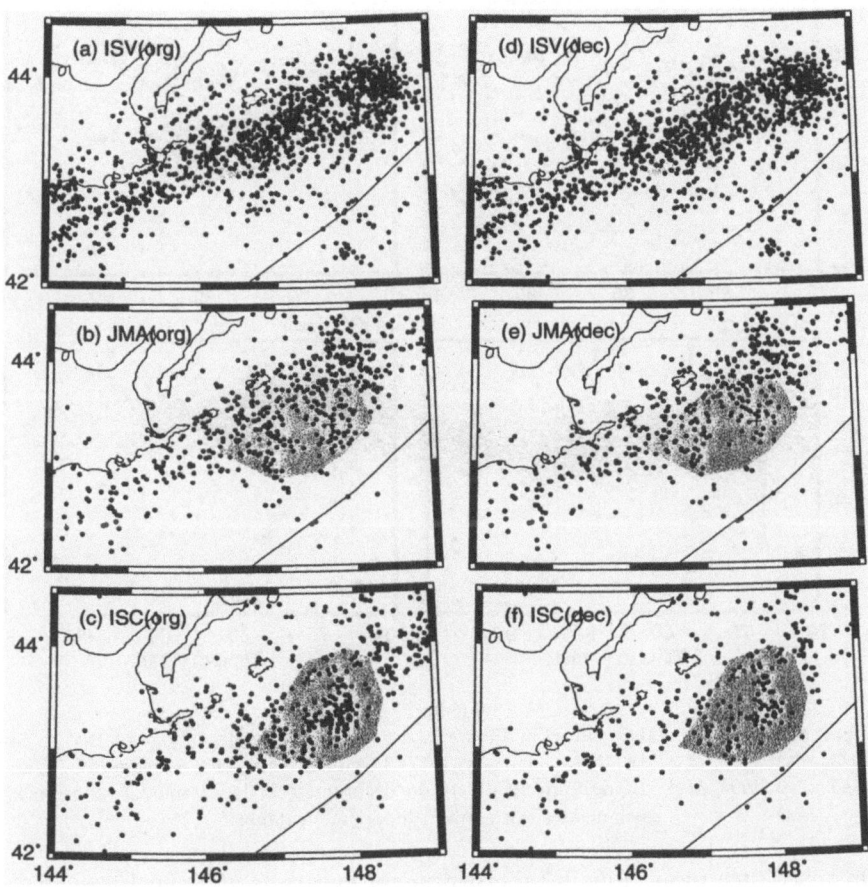

Figure 2

Epicenters of earthquakes used in this study. Hatched areas show the aftershock area of the 1994 Kurile earthquake, which are estimated using each catalog. (a)–(c) are original catalogs of ISV (1 March, 1985–3 October, 1994, $M \geq 3.0$), JMA (1 January, 1977–3 October, 1994, $M \geq 4.3$) and ISC (1 January, 1970–3 October, 1994, $M \geq 5.0$), respectively. (d)–(f) are declustered catalogs, excluding aftershocks and earthquake swarms from the original catalogs.

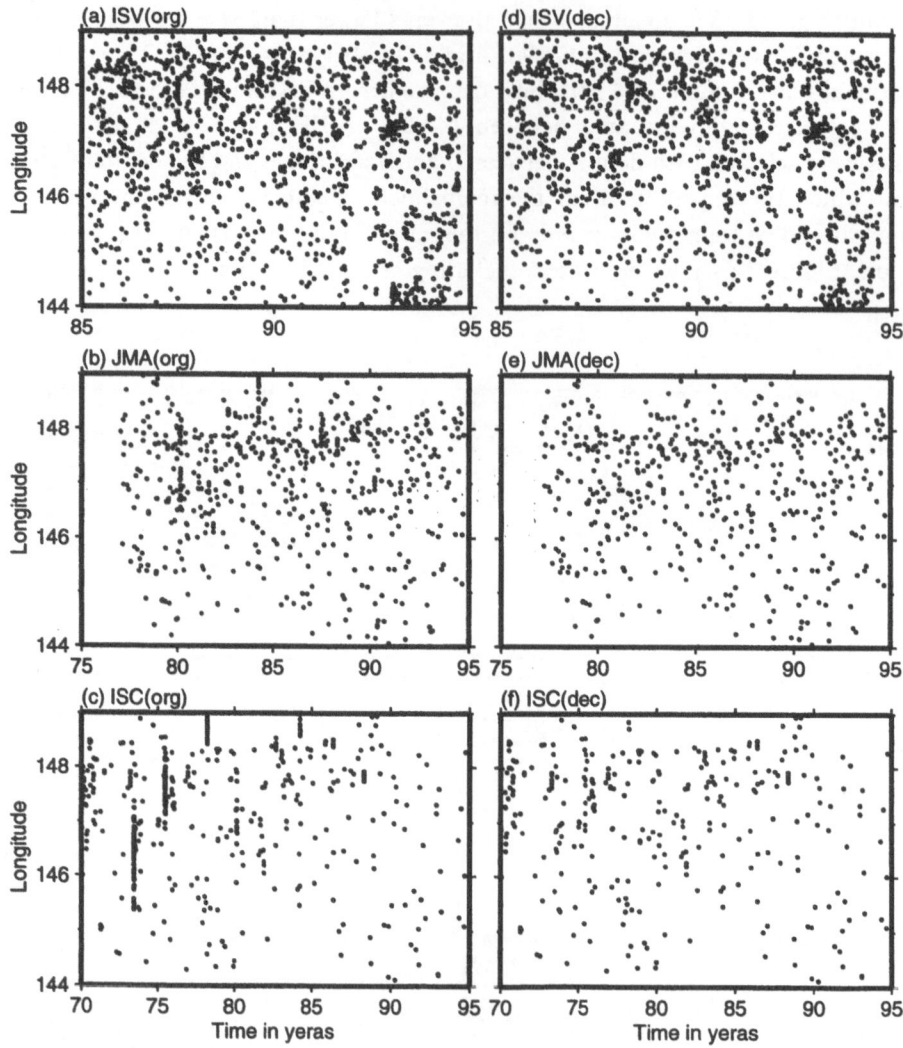

Figure 3

Space-time plots of earthquakes shown in Figure 2. (a)–(c) are original catalogs of ISV (1 March, 1985–3 October, 1994, $M \geq 3.0$), JMA (1 January, 1977–3 October, 1994, $M \geq 4.3$) and ISC (1 January, 1970–3 October, 1994, $M \geq 5.0$), respectively. (d)–(f) are declustered catalogs, excluding aftershocks and earthquake swarms from the original catalogs.

On space-time plots (Fig. 3a) the dependent events as aftershocks and swarms were carefully distinguished from background seismicity by hand; aftershocks were removed and outstanding swarms were substituted with the largest event in the sequence. This is a declustered catalog, including 1390 earthquakes, used in analysis (see Figs. 2d and 3d). There is scant difference: only 3% of the events were removed. As mentioned below, we have investigated that the process of declustering did not affect our results since this method is rather subjective.

2.2 JMA

JMA has its own seismic network in Japan and it has located hypocenters of earthquakes since 1926. The JMA hypocenter catalog has been generally used for research on seismicity in Japan. This catalog, however, has changed in quality because of the renewal of observation systems and the application of new techniques for data analysis (ICHIKAWA, 1987). In 1961 the manual calculation of the hypocenter was substituted by a computer system. JMA introduced a new travel timetable for the Kurile Islands region in 1978 (ICHIKAWA, 1978), and EMT equation in 1977 to estimate earthquake magnitudes (KANBAYASHI and ICHIKAWA, 1977; TAKEUCHI, 1983). Using the same method as applied to the ISV catalog to estimate magnitude completeness we obtained $M_c = 4.3$ for the JMA catalog, that is, earthquakes larger than $M = 4.3$ were uniformly reported in the target area and in the period between 1977 and 1994. Figures 2b and 3b show the epicenter map and a space-time plot including clustered events, respectively. After declustering (see Figs. 2e and 3e) the number of earthquakes was reduced from 946 to 491, with depths selected ranging from 0 to 150 km.

2.3 ISC

ISC publishes a monthly seismic bulletin, which is based on arrival times reported by regional seismic stations worldwide. ISC calculates hypocenters and magnitudes using these data. Thus the resulting catalog is independent of the ISV and the JMA catalogs (Figs. 2c and 3c). For the ISC catalog we estimated M_c as 5.0, using the same method as applied to the ISV and JMA catalogs. The declustered catalog includes 336 earthquakes larger than $M = 5.0$ located in the study area between 1 January, 1970 and 3 October, 1994 and depths selected from 0 to 150 km (Figs. 2f and 3f).

Figure 4 clearly shows the extent of time length and M_c of each catalog. The ISV catalog includes smaller events, therefore the number of declustered events is larger than that of the ISC catalog, though the time length is shorter. The JMA catalog contains features between those of the ISV and ISC catalogs.

3. Method

We have applied the ZMAP method (WIEMER and WYSS, 1994) to the three catalogs to image areas exhibiting a seismic quiescence. WIEMER and WYSS (1994) described details of the method. Therefore we only provide a brief summary in this paper. The study area was divided into grids spacing 0.1° in latitude and longitude (Table 2). A circle was drawn around each grid point and its radius was increased until it included a number N of earthquakes with magnitude larger than M_c. Such circle defined the resolution circle for a given N. For ISV, JMA and ISC, the couples (M_c, N) were (3.0, 100), (4.3, 100) and (5.0, 50), respectively. In areas with higher seismicity the radius of the resolution circle is markedly smaller. For each circle we can plot the cumulative number of events vs. time. Only grid points with a resolution circle smaller than r km were selected, and their number is named N_{grid}. For ISV, JMA and ISC, the couples (r, N_{grid}) were (60 km, 470), (80 km, 414), and (80 km, 573), respectively. For each grid point, at any time t $(t_0 < t < t_e - T_w)$, for a given time window T_w (Fig. 5), the Z-value is calculated by the equation,

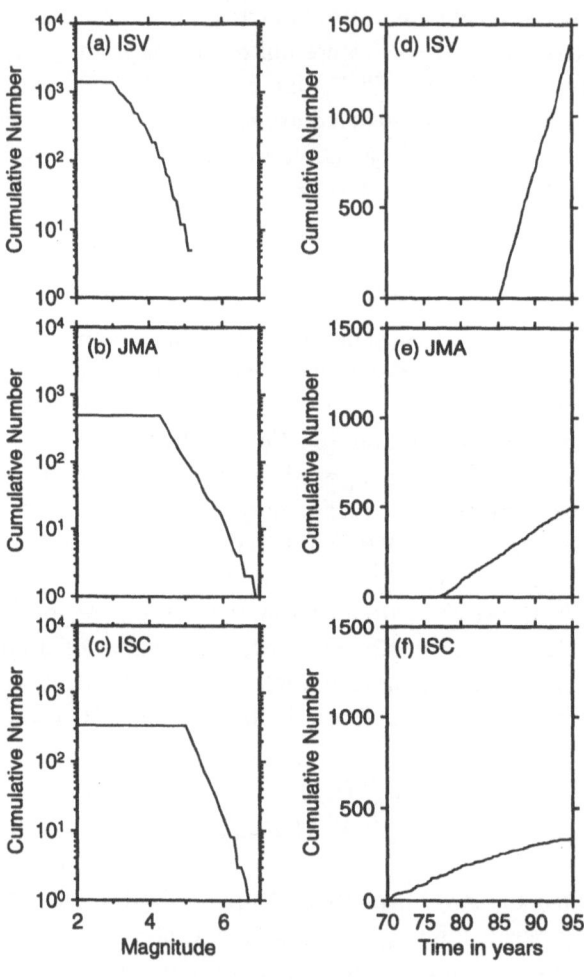

Figure 4

(a)–(c) are plots of the cumulative number of earthquakes shown in Figure 2 vs. magnitudes for the ISV, the JMA and the ISC catalogs, respectively. (d)–(f) are plots of the cumulative number vs. time for each catalog.

Table 2

Parameters used in the computer program ZMAP

Grid size	ISV 0.1°×0.1°	JMA 0.1°×0.1°	ISC 0.1°×0.1°
Number of grids	470	414	573
Number of earthquakes in a circle	100	100	50
T_w (years)	5.0	5.0	5.0
Resolution radius (km)	60	80	80
Bin length (days)	12	22	30
Samples in a cumulative number plot	292	294	301

$$z(t) = \frac{(R_{\mathrm{all}} - R_{wl})}{\sqrt{\dfrac{\sigma^2_{\mathrm{all}}}{n_{\mathrm{all}}} + \dfrac{\sigma^2_{wl}}{n_{wl}}}},$$

where R_{all} is the mean rate in the overall period including T_w (from t_0 to t_e), R_{wl} the mean rate in the considered time window (from t to $t + T_w$). σ_{all} and σ_{wl} are the standard deviations in these periods, and n_{all} and n_{wl} the number of samples. The Z-value, calculated for all times t between t_0 and $t_e - T_w$, by the equation is statistically appropriate for estimating seismicity rate change in a time window T_w in contrast with background seismicity.

When we detect the Z-value anomaly, we should also estimate how strong or how significant it is. For this purpose we used the alarm-cube method (Wyss and

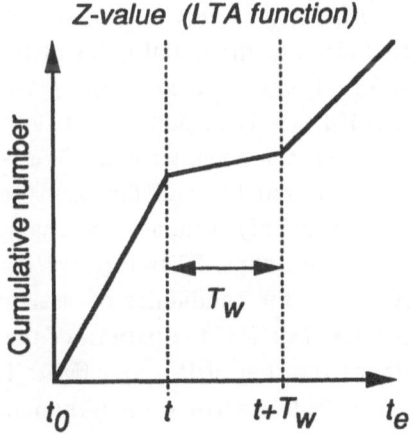

Figure 5

Schematic explanation of how to calculate Z-values. R_{all} is the mean rate in the overall period (from t_0 to t_e), R_{wl} the mean rate in the time window (from t to $t + T_w$), where t is the "current" time ($t_0 < t < t_e$) and T_w is the length of the time window in year. σ_{all} and σ_{wl} are the standard deviations in these periods, and n_{all} and n_{wl} the number of samples. The Z-value was calculated for all times t between t_0 and $t_e - T_w$ (see the text).

MARTIROSYAN, 1999): space-time search for anomalies. The space-time search illustrates in a space-time representation how often anomalies of the same kind as the one prior to the large event occurred in the study area. If similar anomalies appear frequently without any following large event, the reliability of the anomaly is judged very low. If the most significant seismic quiescence is followed by the largest event in the target area in all the time period, we can plausibly interpret the anomaly as a preparation process to that event.

4. Analysis

4.1 ZMAP

The Z-maps shown in Figure 6 present time slices for the ISV catalog every six month between 1986 and 1989.5 based on the declustered catalog. In Figures 7 and 8 the Z-maps for JMA and ISC catalogs were shown every year between 1982.5 and 1989.5. The time window, T_w, in which the mean rate is compared to the mean background rate, is 5 years. Though from all three catalogs a decrease of seismicity has been obviously detected in the target area, the shape of portions with the high Z-value anomaly (red areas) appears to be different. For ISV, there are two areas with the high Z-value (red areas) larger than $+4.0$. One is located west of 146°E on the maps from 1986.0 to 1987.5 (Anomaly ISV-1). Another one appears to be close to the initial point of the rupture of the 1994 Kurile mainshock on the maps from 1988.5 to 1989.0 (Anomaly ISV-2). Only one anomaly with high Z-value (a red area) was detected on JMA maps from 1987.5 to 1989.5 (Anomaly JMA-1). This anomaly is also located close to the initial point of the rupture. No anomaly corresponding to Anomaly ISV-1 was detected on the JMA maps. Anomaly ISV-2 is consistent with Anomaly JMA-1: both Anomaly ISV-2 and Anomaly JMA-1 started at a similar time and existed at a similar place. We believe that the common feature from the two catalogs is reliable and far from being due to man-made changes. Anomaly ISV-1 was probably a man-made change because no anomaly appeared west of 146°E on JMA maps. Therefore we concluded that Anamoly ISV-2 and Anomaly JMA-1 were the candidates of seismic quiescence associated with the 1994 Kurile mainshock. For ISC the pattern of the high Z-value anomaly (a red area) is rather different from that of ISV and JMA. The anomaly appears to surround the initial point of the rupture like a ring (Anomaly ISC-1). No change in

Figure 6

Time slices of Z-value distribution every six months between 1986 and 1989.5 using the ISV declustered catalog. The length of time window T_w is 5 years. Only grid points with a radius of resolution circle smaller than 60 km were selected. Their number is 470. The epicenter of the mainshock ($+$) and the aftershock area (a polygon bounded by a thin line) is indicated. Red color (positive Z-value) represents a decrease in seismicity rate.

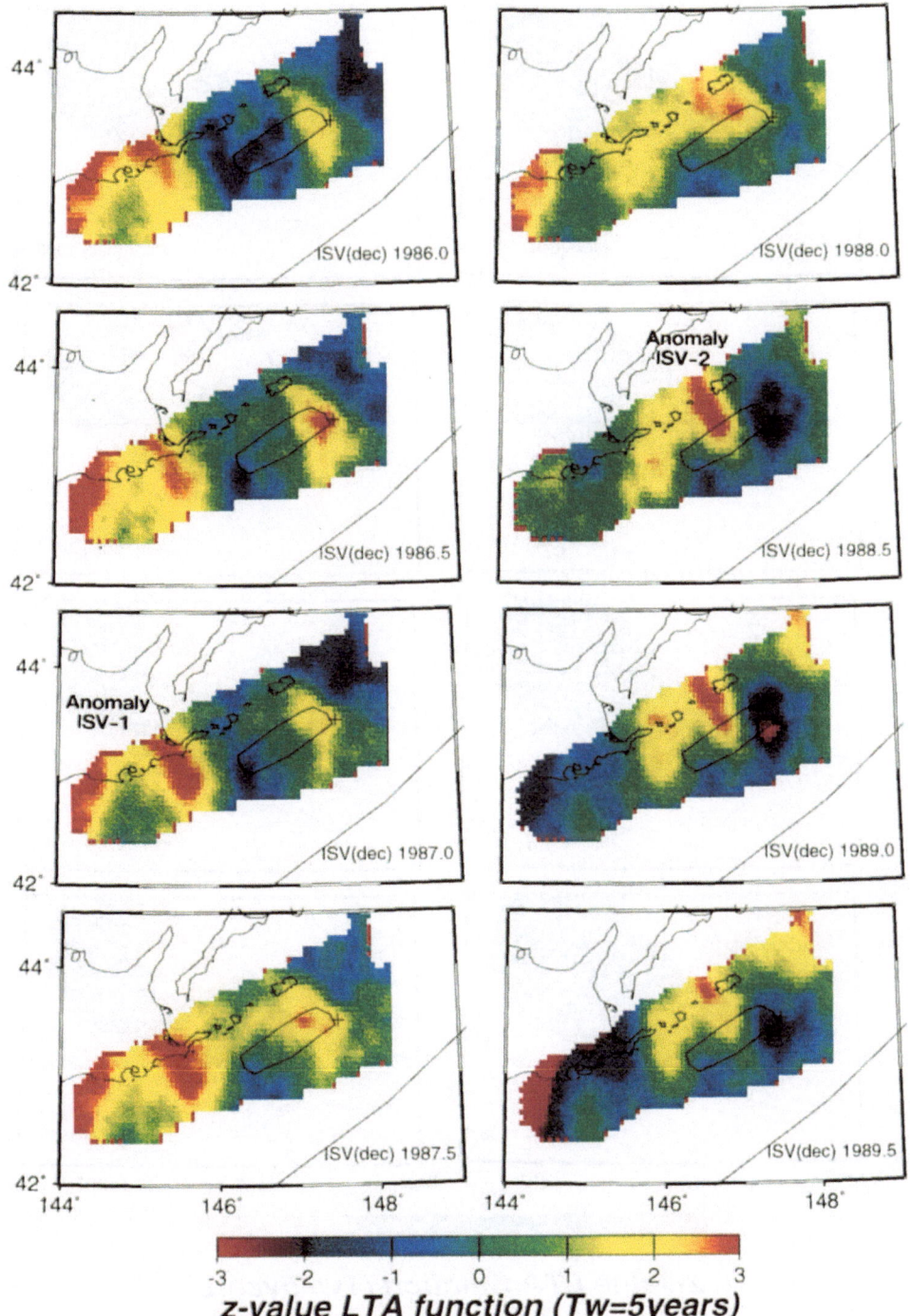

z-value LTA function (Tw=5years)

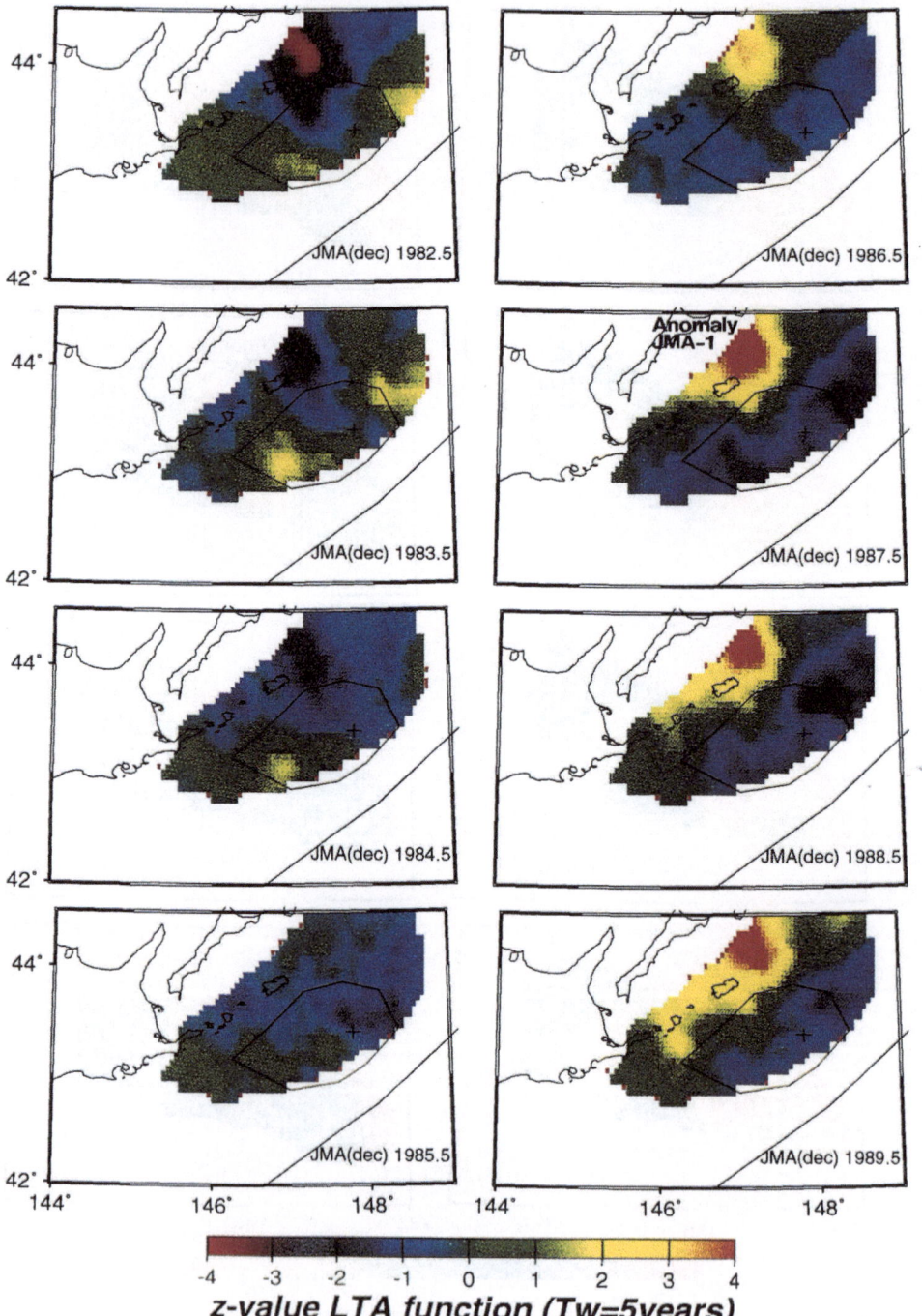

z-value LTA function (Tw=5years)

seismicity rate was detected in a circle centered on the epicenter. This pattern is clearly delineated on the time slice at 1989.5: A red area indicating a decrease of seismicity extends from the aftershock area of the mainshock except for a circular area centered on the epicenter.

The cumulative curves in Figures 9a and 9b of the number of earthquakes as a function of time manifest seismicity rate changes at specific nodes in the cases of Anomaly ISV-2 and Anomaly JMA-1, respectively. These nodes were positioned randomly by the gridding process, however they were selected for presentation here because they were located in the anomalous areas mapped in Figures 6 and 7. The statistical functions, LTA, displayed in Figure 9 are the Z-values obtained by a comparison of the mean rate within a sliding time window and the long-term average, defined by overall seismicity, including the sliding time window, in the same volume. For ISV and JMA the Z-value peaked with $Z_{max} = 4.0$ at 1988.8 and with $Z_{max} = 5.2$ at 1989.3, respectively. Since a peak of Z-value reveals the time when the decrease of seismic rate starts, both clear anomalies were found to start at around 1989 and persist to the mainshock in 1994. For ISC we found a high Z anomaly with the shape of ring on the map at 1989.5 referring as Anomaly ISC-1. Therefore we selected earthquakes in the ring and made a cumulative number plot (Fig. 9c). A decrease of seismicity was clearly found in the ring and the Z-value took the maximum of $Z_{max} = 6.8$ at 1989.5, which means that the decrease of seismicity started at 1989.5 and continued to the mainshock. On the other hand the rate was found to be rather constant in the circular area centered on the epicenter. Though the artificial shape of the ring made the statistical significance of this anomaly minor, the decrease in the ring should remain to be an outstanding candidate for a precursory quiescence.

We should be particularly attentive to the synchronization in time and space. All anomalies such as Anomaly ISV-2, Anomaly JMA-1 and Anomaly ISC-1 started around the year 1989 and lasted up until the mainshock in 1994, and they also existed very close to the focal area. However it is fact, from a statistical point of view, that the significance of quiescence was characterized by fairly low Z-values. Here Z_{max} values are typically of the order of three (one standard deviation). The largest Z_{max} value (6.8) was found for the anomaly of a particular shape brought to light with the ISC catalog.

Figure 7

Time slices of Z-values distribution every year between 1982.5 and 1989.5 using the JMA declustered catalog. The length of time window T_w is 5 years. Only grid points with a radius of resolution circle smaller than 80 km were selected. Their number is 414. The epicenter of the mainshock (+) and the aftershock area (a polygon bounded by a thin line) is indicated. Red color (positive Z-value) represents a decrease in seismicity rate.

4.2 Alarm Cube

The temporal and spatial correlation of the quiescence, mapped in Figures 6–8 and characterized in Figure 9, with the Kurile mainshock may have no significance, if similar quiescence anomalies occur frequently at locations in space and time where no mainshocks exist. Therefore we search the matrix of Z-values (generated by the LTA functions at all nodes) for high Z-values, which may approach, or exceed, the Z-values recorded by the anomaly before the Kurile earthquake. The display of the high Z-values that could be false alarms is made in the alarm-cubes (WIEMER, 1996; WYSS et al., 1996; WYSS and MAR-TIROSYAN, 1999). In these 3-D figures (Fig. 10) horizontal axes are the spatial coordinates of the target area and the vertical axis is time. Alarms are defined as instances of Z-values larger than the selected alarm-level at any node and any time. Figure 10 illustrates that the results are stable, regardless of the seismic catalogs. The ISV's alarm cube includes two outstanding groups of anomalies: a group around the western end of the target area (Alarm ISV-1) and a group close to the epicenter of the mainshock (Alarm ISV-2). The JMA's alarm cube includes only one group of anomalies before and near the Kurile mainshock (Alarm JMA-1). Alarm ISV-1 obviously corresponds to Anomaly ISV-1 judged as a man-made change above. Alarm ISV-2 corresponds to Anomaly ISV-2 consistent with Anomaly JMA-1 and Alarm JMA-1. In the ISC's alarm cube, there were three groups of anomalies starting in 1984 (Alarm ISC-1), 1986 (Alarm ISC-2) and 1989 (Alarm ISC-3). Alarm ISC-1 and Alarm ISC-2 continued through 1989 and 1991, respectively. However no mainshock occurred. Alarm ISC-3 was ongoing through 1994 and the $M = 8.3$ Kurile mainshock occurred. Although all three alarm groups were located near the mainshock, the extent of the quiet area was very different. Alarm ISC-1 and Alarm ISC-2 were represented by only one or two bars in the alarm cube. Alarm ISC-3 appears as a cluster of bars close to the mainshock area. Since the area of Alarm ISC-3 was substantially larger than that of Alarm ISC-1 and ISC-2, Alarm ISC-3 can be considered more significant than Alarm ISC-1 and ISC-2. Moreover, only Alarm ISC-3 was consistent with Alarm ISV-2 and Alarm JMA-1 in time. Therefore we concluded that the alarm, which started around 1989 and existed close to the focal area, showed the only quiescence not depending on the seismic catalogs.

Figure 8

Time slices of Z-values distribution every year between 1982.5 and 1989.5 using the ISC declustered catalog. The length of time window T_w is 5 years. Only grid points with a radius of resolution circle smaller than 80 km were selected. Their number is 573. The epicenter of the mainshock (+) and the aftershock area (a polygon bounded by a thin line) is indicated. Red color (positive Z-value) represents a decrease in seismicity rate.

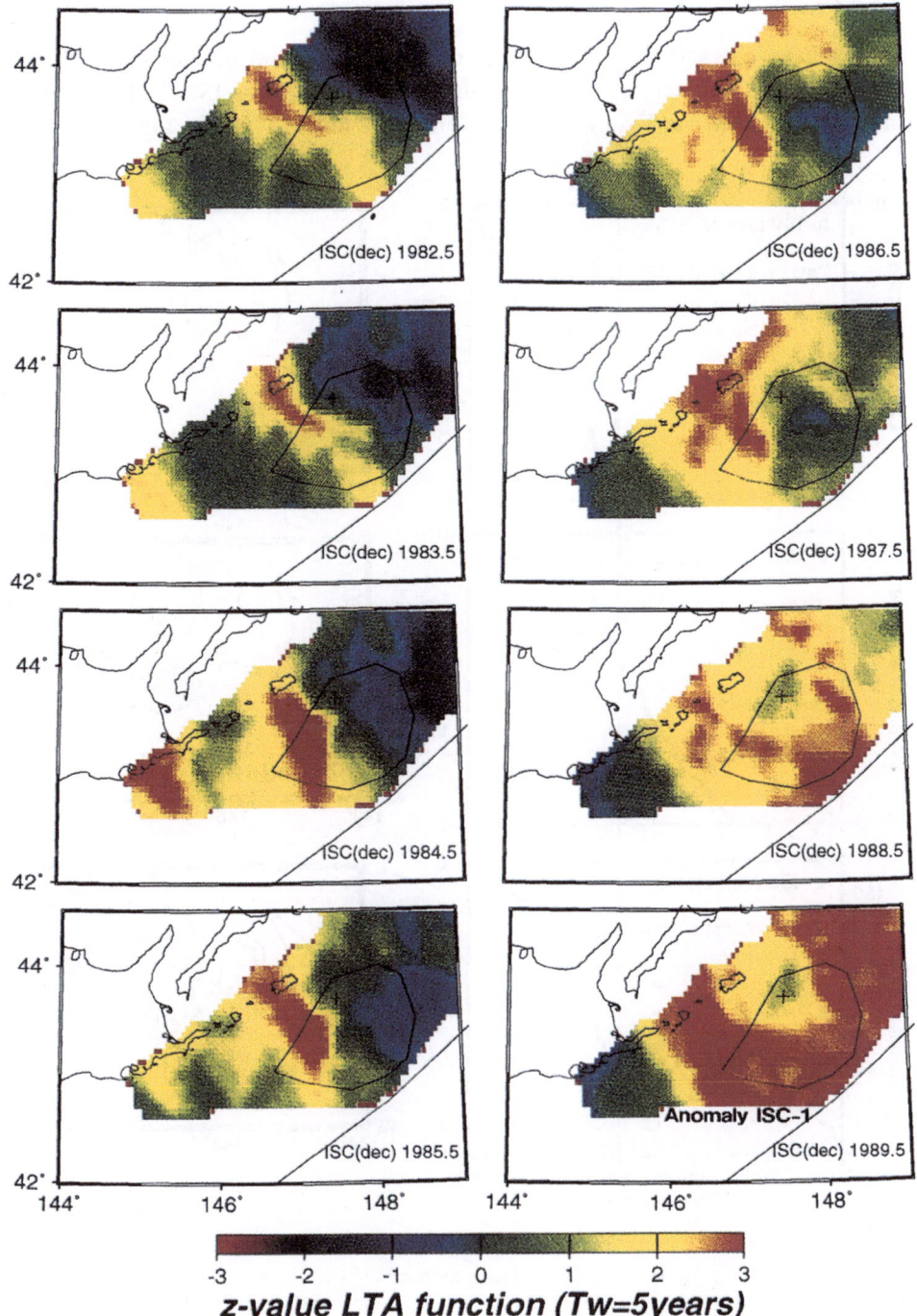

z-value LTA function (Tw=5years)

4.3 Effect of Declustering Process

We have checked that the process of declustering did not affect our results. Figure 11 is a counterpart of Figure 9. Although Z_{max} for ISC in Figure 11c is smaller than that in Figure 9c, we were also able to detect that the seismic

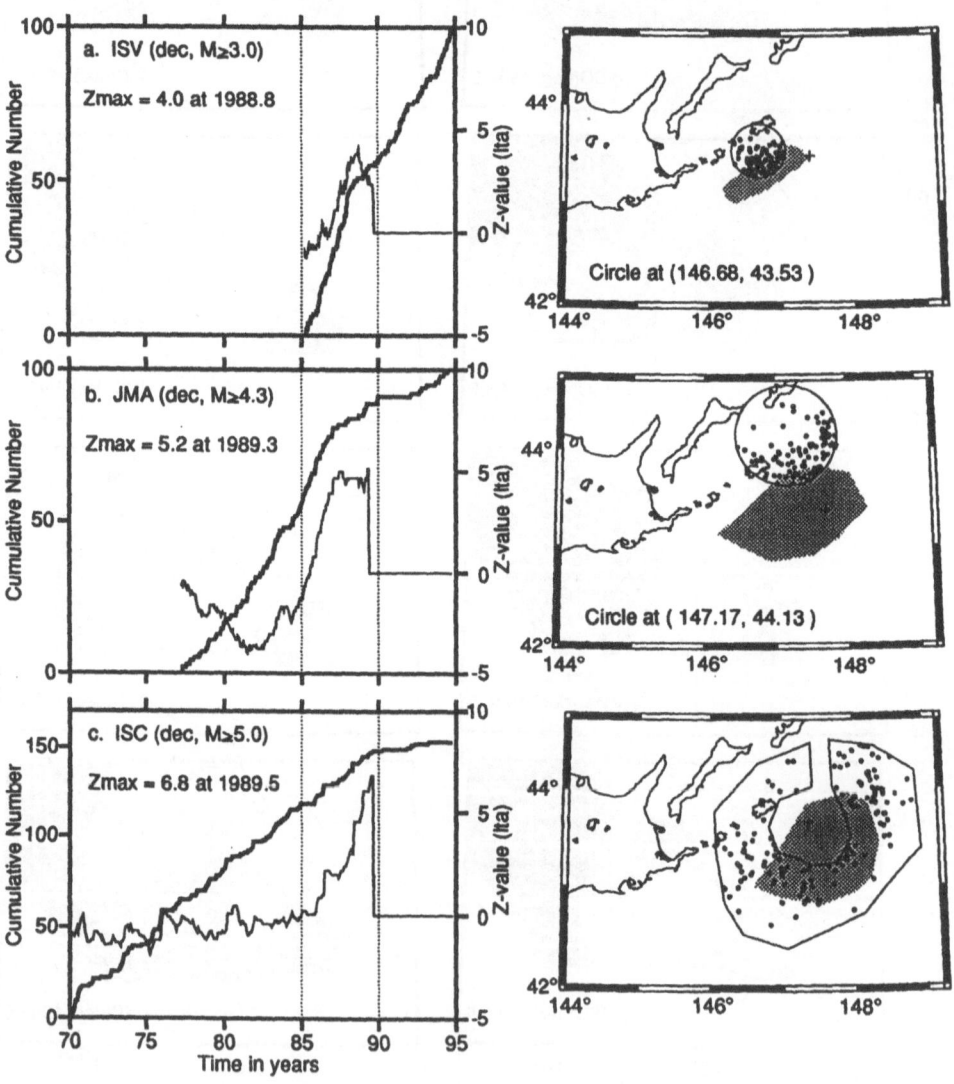

Figure 9

Cumulative number plots for anomalous areas detected in Figures 6–8. (a) Anomaly ISV-2, (b) Anomaly JMA-1 and (c) Anomaly ISC-1 (see in the text). Bold and thin lines in the cumulative number plots show cumulative numbers and Z-values as a function of time, respectively. Maps at the right of each cumulative number plots show the anomalous areas and aftershock areas (hatched portions) and the epicenters of the mainshock (+).

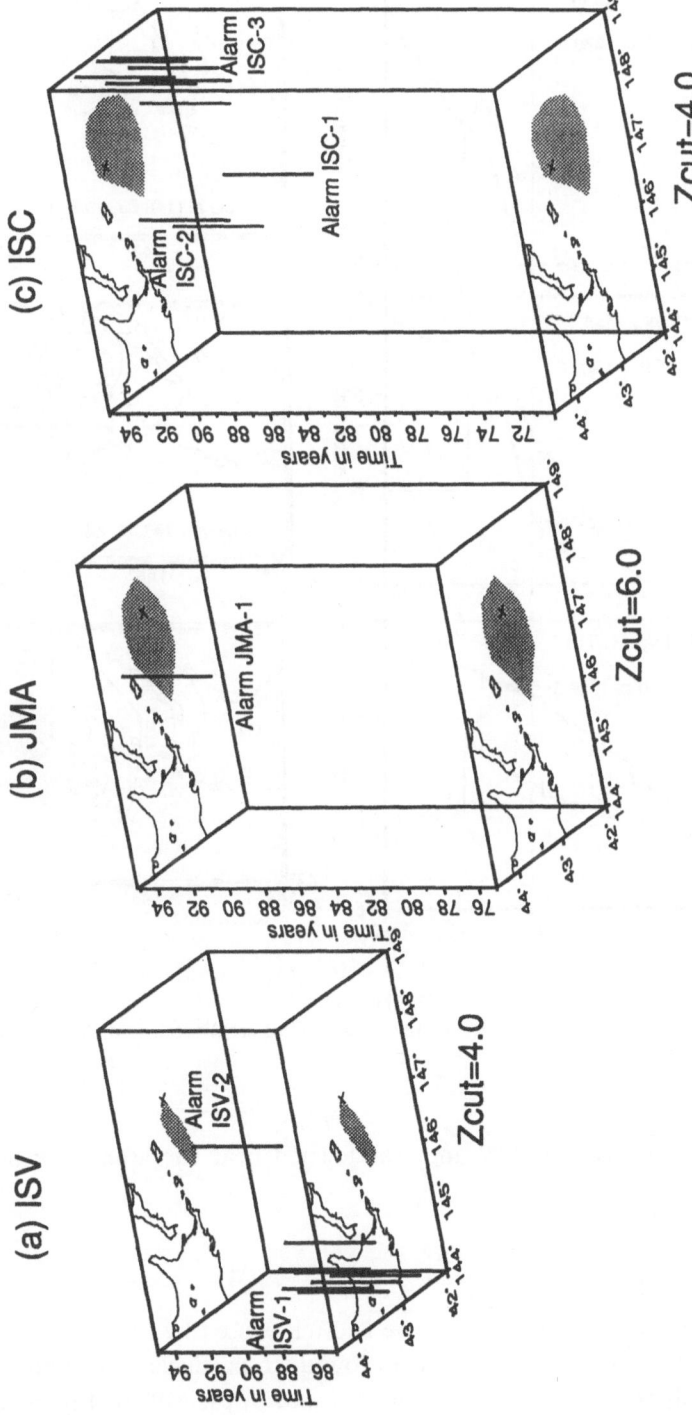

Figure 10

Alarm cubes for (a) ISV, (b) JMA and (c) ISC. Vertical bold lines with a time length of 5 years indicate alarms with Z-values larger than Z_{cut}. Hatched areas and (+) s show the aftershock area and the epicenter of the mainshock, respectively.

460 Kei Katsumata and Minoru Kasahara Pure appl. geophys.,

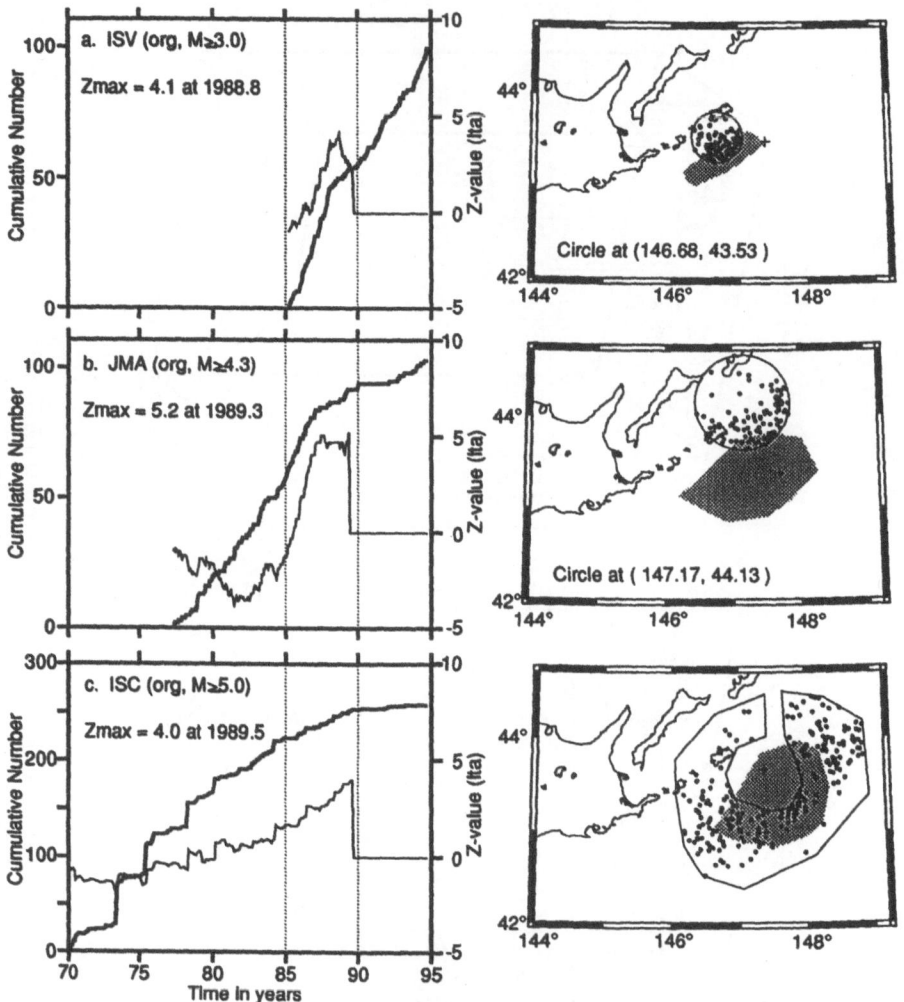

Figure 11
Same plots as Figure 9 except for the use of the original (clustered) catalogs of (a) ISV, (b) JMA and (c) ISC.

quiescence started around 1989 and was located near the aftershock area of the mainshock.

4.4 Magnitude Signatures

Since an apparent seismic quiescence is often caused by man-made changes, we should carefully examine the quiescences found. In particular, magnitude shift and stretch critically affect the results (WIEMER and WYSS, 1994). Figure 12 shows magnitude signatures for ISV comparing two time windows: before and after the start of quiescence. In these plots we used all earthquakes regardless of magnitude. In the quiescence area with high Z-value, b-values were similar in the two time windows (Fig. 12d) and the number of earthquakes decreased in all magnitude bands between $M = 1$ and 4. Therefore, we concluded that this quiescence was not

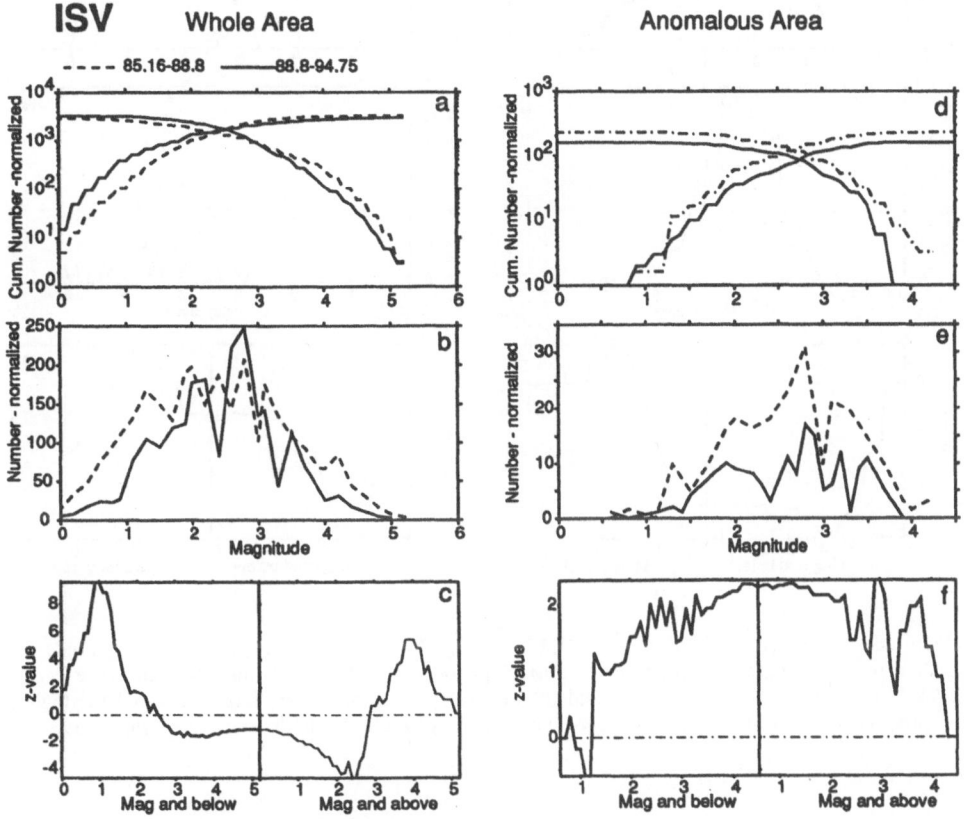

Figure 12

Magnitude signatures for the declustered ISV catalog. All earthquakes regardless of magnitude are used. Broken and bold lines indicate the normal period (1985.16–1988.8) and the quiescence period (1988.8–1994.75), respectivley. (a) and (d) are plots of magnitude vs. the cumulative number for the entire target area and the high-Z (Anomaly ISV-2) area, respectively. (b) and (e) are plots of magnitude vs. frequency for the entire target area and the high-Z (Anomaly ISV-2) area, respectively. (c) and (f) are plots of magnitude vs. Z-value for the entire target area and the high-Z (Anomaly ISV-2) area, respectively. The label "Mag and below" on horizontal axes in (c) and (f) means that Z-values are calculated for earthquakes with a magnitude equal to or less than the values indicated on the axis scale. The label "normalized" in (a), (b), (d) and (e) means that the number of earthquakes is normalized in time because two time periods comparing each other have a different time length. For instance, assume that the length of Period 1 is one year and that of Period 2 is two years. If the seismicity rate is constant, we will observe two times earthquakes in Period 2 more than in Period 1. Thus the number of earthquakes in Period 2 is divided by two in the normalization process.

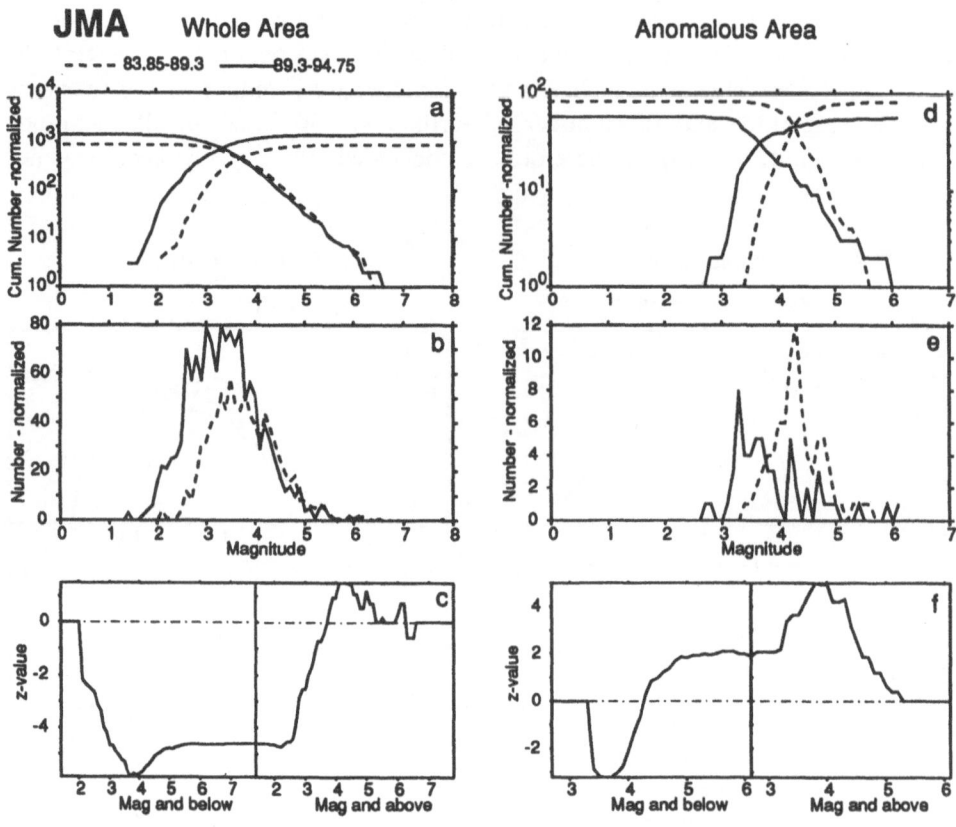

Figure 13

Magnitude signatures for the declustered JMA catalog. All earthquakes regardless of magnitude are used. Broken and bold lines indicate the normal period (1983.85–1989.3) and the quiescence period (1989.3–1994.75), respectively. (a)–(c) and (d)–(f) are plots for the entire target area and the high-Z (Anomaly JMA-1) area, respectively. The meaning of each plot is the same as in Figure 12.

an artifact caused by man-made changes. Figure 13 shows magnitude signatures for JMA comparing two time windows: before and after the start of quiescence. In the anomalous area with high Z-value, b-values in the magnitude-frequency plot (Fig. 13d) decreased from 0.91 to 0.77. The decrease of earthquakes larger than $M = 3.5$–4.0 and the increase of earthquakes smaller than $M = 3.5$–4.0 induced this change in b-value. It seems that a systematic shift in magnitude-frequency distribution occurred in 1989. In the entire target area, the detection was drastically advanced in earthquakes smaller than $M = 4$ (Fig. 13b). The b-values and the number of located earthquakes larger than $M = 4$, however, did not change (Figs. 13a and 13b), that means that there was no significant magnitude shift and stretch in the entire target area in 1989. Thus the change in b-value occurred only in the anomalous area. Therefore, we concluded that the decrease of earthquakes larger than $M = 4$ was probably not caused by any man-made change. Figure 14 displays magnitude signatures for ISC in the ring area which has the unusual high Z-value. Three time windows were selected to compare the seismicity: before the quiescence (1984.25–1989.5, Period 1), the time between the starting point of the quiescence and the mainshock (1989.5–1994.75, Period 2), and after the mainshock (1994.75–1995.1, Period 3). Note that earthquakes in Period 3 were not included in the data

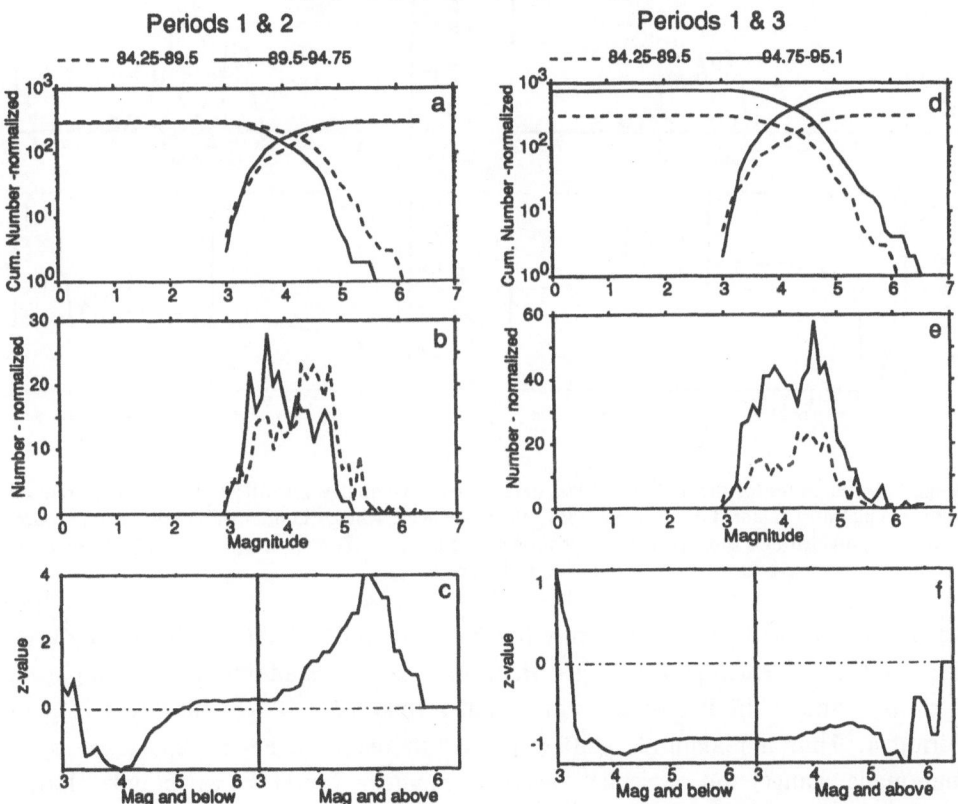

Figure 14

Magnitude signatures for the declustered ISC catalog. All earthquakes regardless of magnitude are used in the Anomaly area ISC-1. Three time windows were selected to compare the seismicity: before the quiescence (1984.25–1989.5, Period 1), the time between the starting point of the quiescence and the mainshock (1989.5–1994.75, Period 2), and after the mainshock (1994.75–1995.1, Period 3). In (a)–(c) Period 1 (broken lines) is compared with Period 2 (solid lines). In (d)–(f) Period 1 (broken lines) is compared with Period 3 (solid lines). The meaning of each plot is the same as in Figure 12.

ISC Normal Area

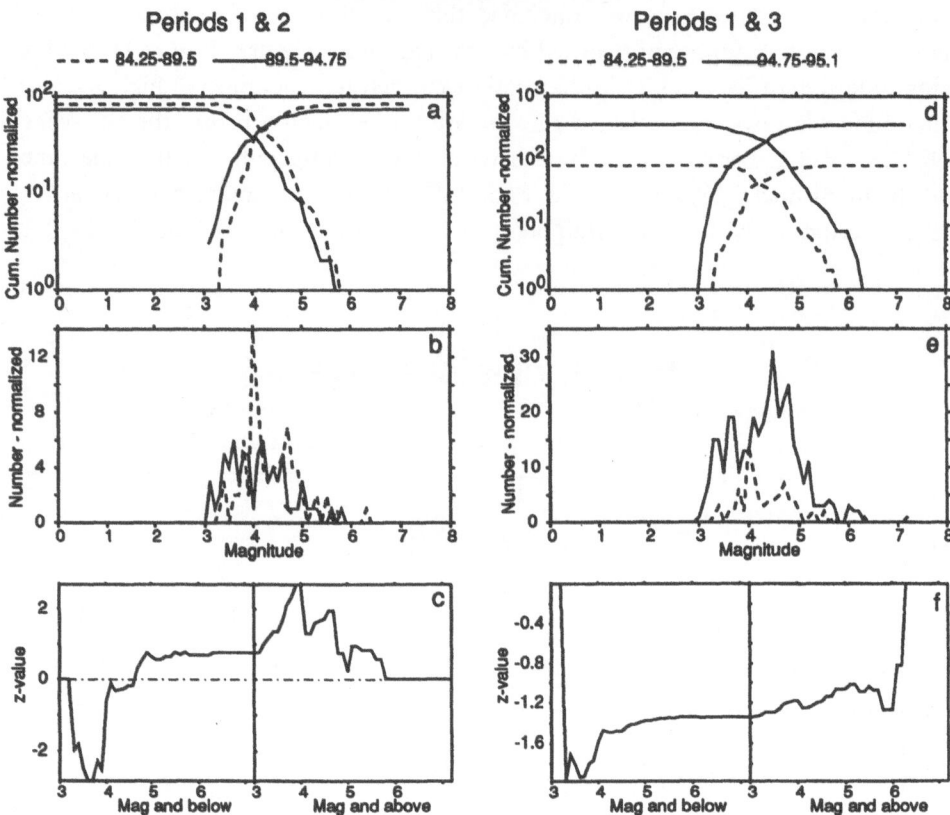

Figure 15
Magnitude signatures for the declustered ISC catalog. All earthquakes regardless of magnitude are used in the circular area centered on the epicenter of the mainshock, which did not exhibit seismic quiescence. The time is divided into three periods as mentioned in Figure 14. The meaning of each plot is the same as in Figure 12.

set to make Z-maps, but added only for the magnitude signature. In the ring area the decrease of earthquakes larger than $M = 4$ is outstanding when comparing Periods 1 and 2. In Period 2 events smaller than $M = 4$ were more detected in Period 1. Thus a magnitude shift appeared to occur in Period 2, caused by a man-made change, and a seismic quiescence might be found for earthquakes larger than $M = 4$. However, this is not plausible. Comparing Periods 1 and 3, such type of magnitude shift was not found. The b-values were the same (Fig. 14d) and earthquakes larger than $M = 4$ were similarly detected in Period 1 and Period 3. The shape of magnitude-number plots in Figure 14e in Period 1 is similar to that in Period 3. Figure 15 presents magnitude signatures for the circular area centered on the epicenter of the mainshock, which did not exhibit the seismic quiescence.

There was no significant decrease of seismicity in Period 2 compared with Period 1, and no significant change in b-values (Fig. 15a) and a pattern of frequency-magnitude plots (Fig. 15b). Comparing Period 1 with Period 3 we recognized neither a clear magnitude shift nor a magnitude stretch (Figs. 15d–f). Therefore we have concluded that the quiescence in the ring area in Period 2 was not caused by any man-made change but rather change in geophysical conditions in and around the focal area of the 1994 Kurile mainshock.

4.5 Comparing JMA and ISC Catalogs

The seismic quiescence associated with Anomaly ISC-1 and Alarm ISC-3 also has been detected using the JMA catalog. In Figure 16 we plotted the cumulative number vs. time using the JMA catalog for earthquakes larger than $M = 4.8$ in the ring area shown in Figure 9c. The Z-value clearly peaked at 1989.3, which is the same pattern as that pictured in Figure 9c. We should next discuss whether the JMA catalog is independent of the ISC catalog. JMA certainly provides arrival times of seismic waves with ISC and the data are used to calculate hypocenters. However, in the case of earthquakes with $M = 5.0$ and larger, the data from JMA are only a part of reports from seismic stations worldwide. For instance, ISC used 141 seismic stations including 35 stations of JMA to locate the $M = 5.0$ earthquake in the study area on 19 November, 1991. Therefore, we assumed that the ISC catalog is independent of the JMA catalog for earthquakes with $M = 5.0$ and larger. Thus the seismic quiescence in the ring area has been confirmed by two independent seismic catalogs.

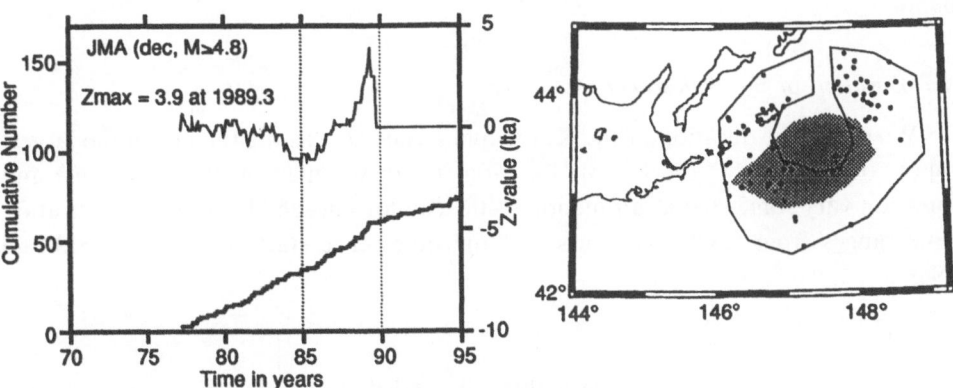

Figure 16
Cumulative number plot for earthquakes larger than $M = 4.8$ in the area of Anomaly ISC-1 using the JMA catalog.

Table 3

Parameters of precursory quiescence to the $M_w = 8.3$ Kurile 1994 earthquake

Catalog	Sample size	Window length	Relative significance	Start	Duration	Radius 75%
	N	T_w (years)	Z_{max}	t_Q (date)	T_Q (years)	r (km)
ISV	100	5	4.0	88.8	6.0	24
JMA	100	5	5.2	89.3	5.5	53
ISC	153	5	6.8	89.5	5.3	–

4.6 Spatial Extent of the Quiescence Anomaly

We define the spatial extent of the quiescence anomaly by the area in which a 75% rate decrease occurred for the ISV and JMA catalogs. This criterion (WIEMER and WYSS, 1994) is arbitrary. In this paper, we define the area of 75% rate decrease employing the same method as WYSS and MARTIROSYAN (1999): we increase the radius of a circle, centered at the coordinates of the largest Z-value, until the rate decrease is 75%. If more than one node has the same Z_{max}, we select the one that leads to the larger anomaly volume. As mentioned by WYSS and MARTIROSYAN (1999), this method is advantageous in that it is clearly defined and can be repeated. However, it is disadvantageous in that it does not map the quiescence volume optimally because we should accept a circle as the anomaly shape, though the shape is not usually defined by a circle. Depending on the catalogs chosen, the radii of the 75% circular quiescence anomaly before the Kurile earthquake are 24 km for ISV and 53 km for JMA (Table 3). Since the ISC catalog includes a number of earthquakes too small to apply this method, we did not estimate a radius for ISC. However, the quiescence area of ISC is evidently larger than that of ISV and JMA. The area of a seismic quiescence possibly changes as a function of the magnitude band.

4.7 Duration of the Quiescence Anomaly

We define the duration of the Kurile quiescence, T_Q, as the time from the largest report of Z-value for the LTA(t) function, t_Q, to the mainshock. Onset time and duration vary somewhat as a function of the catalog chosen (Table 3). The duration time ranges from 5.3 to 6.0 years and the quiescence starts between 1988.8 and 1989.5.

5. Discussion and Conclusions

The three catalogs used in this study have differing quality from the standpoint of homogeneous reporting in the area of Figure 2. Though the ISC catalog provides long-term data, earthquakes smaller than $M = 5.0$ are not reported homogeneously. Though the ISV catalog includes smaller events down to $M = 3.0$, the period of observation is shorter. Regardless of these differences we have found that results obtained from all three catalogs are mostly consistent with each other: the seismic activity started to decrease 5.3–6.0 years before the 1994 Kurile mainshock near the ruptured area. The seismic quiescence detected in this study is unusually significant and not caused by man-made change.

Nonetheless there are problems and limitations for the analysis in this study: (1) The geographical extension of the area considered in this paper (5° longitude by 2.5° latitude) appears insufficiently large in comparison with the presumable extension of the source area of an earthquake of magnitude $M_w = 8.3$. Moreover, the actual size of the area where the density of data was high enough for the analysis to be carried out, is less than 50% of the total. Therefore, strictly speaking, it is not possible to compare the seismicity pattern in the area affected by the source preparation process with other areas distant from the source. This circumstance is particularly evident with the ISC data, where the quiescence occupies most of the studied area. In this case, the significance of the anomaly as a precursor, estimated by the ration of the alarm volume and the total time-space volume, is not very good. (2) The temporal extension of the ISV catalog (9.6 years) appears rather short in comparison with the time length of he observed anomaly (6.0 years). In this circumstance it is hard to recognize the quiescence as an anomalous behavior if neither the JMA nor ISC catalog manifested the quiescence. (3) There is the different size and shape of the aftershock area among the three catalogs. Though the aftershock areas in JMA and ISC are approximate, the area of ISV appears to be smaller than JMA and ISC. KATSUMATA et al. (1995) found that some aftershocks were induced after the mainshock (Type 2 activity) and the largest aftershock and its aftershocks were located at the eastern end of the seismic fault ruptured by the mainshock (Type 3 activity). In this paper we plotted an area that excluded Types 2 and 3 activities on ISV maps. The aftershock areas on JMA and ISC maps include all aftershocks within one month after the mainshock because we hardly distinguished the detailed distributions such as Types 2 and 3. (4) There exists the different size and shape of the quiescence area among the three catalogs. Our hypothesis holds that as a magnitude band shifts larger, a quiescence area becomes wider. Another hypothesis maintains that a quiescence area should have constant size, shape and location regardless of magnitude band. If this is correct, the differences among the three catalogs reduce the reliability of the comparison. However both hypotheses should be tested and proved by future studies.

TAKANAMI et al. (1996) have found that a seismic quiescence started three years before the Kurile mainshock in the vicinity of the ruptured area. They used only the ISV catalog and earthquakes larger than $M = 3.5$ due to the incompleteness of the catalog. Our results are mostly consistent with TAKANAMI et al. (1996). However,

TAKANAMI *et al.* (1996) did not mention whether or not the quiescence they found was a man-made change.

Previous works on a seismic quiescence generally used only one seismic catalog and did not compare two or three independent catalogs. In these cases we can hardly estimate whether the quiescence is a fact or a man-made change. For instance TAYLOR *et al.* (1991) used only the ISV catalog to find a seismic quiescence before the $M_{JMA} = 7.1$ Urakawa-Oki earthquake, WIEMER and WYSS (1994) used only a seismic catalog of the California Institute of Technology for the $M = 7.5$ Landers earthquake, WYSS and BURFORD (1985) used only the central California seismograph network for the $M_L = 4.6$ San Andreas earthquake, KISSLINGER (1988) utilized only the central Aleutians (Adak) seismic network for the $M_w = 8.0$ Andreanof Islands earthquake and WYSS *et al.* (1981) used only the Hawaiian Volcano Observatory for the $M = 7.2$ Kalapana earthquake. In this paper we have analyzed three independent seismic catalogs, including ISV, JMA and ISC, and found that all three catalogs clearly manifested the seismic quiescences prior to the $M_w = 8.3$ Kurile earthquake. This fact strongly suggests that the seismic quiescences detected in this paper are more reliable than those in previous studies.

KATSUMATA and KASAHARA (1999) have reported that the eastern Hokkaido and the southern Kurile Islands around the focal area exhibited steep subsidences synchronized with the seismic quiescence detected in this study. Earthquake instability models predict that a preseismic stable sliding the time scale occurs several years before an unstable (earthquake) faulting, and may affect the background seismicity (SHIBAZAKI and MATSU'URA, 1992; DIETERICH, 1992; KATO *et al.*, 1997). KATSUMATA and KASAHARA (1999) proposed the preseismic stable sliding on the same fault as the mainshock to model the subsidences and the seismic quiescence.

Moreover, ISV reported to the Coordinating Committee for Earthquake Prediction in August 1994 that six strain meters in the eastern Hokkaido had displayed clear changes since 1991 (HOKKAIDO UNIVERSITY, 1994). At that time they were not able to sound the alarm because of lack of data: they had definitely recognized neither the seismic quiescence detected in this study nor the crustal deformation anomaly pointed out by KATSUMATA and KASAHARA (1999). If they had known all these anomalies and had suitably accomplished a computer simulation using an earthquake instability model, they might have forecasted that a sizeable earthquake would have occurred in the southern Kurile Islands within several years.

In conclusion, the Kurile earthquake in 1994 provided us with an important lesson: simultaneous detection of seismic quiescence and crustal deformation anomaly are keys to the success of intermediate-term earthquake prediction. Moreover, if we use two or three independent catalogs, we provably detect a seismic quiescence before a mainshock, even though the catalogs are not perfectly complete. Therefore we should more carefully monitor seismicity and crustal deformation to detect intermediate-term anomalies prior to a great earthquake in the near future.

Acknowledgements

We thank V.G. Kossobokov at Intl. Inst. Earthquake PT&MG, T. Takanami, Y. Motoya and N. Wada at ISV, M. Wyss at University of Alaska, Fairbanks and two anonymous referees for helpful comments. S. Wiemer provided us with the program ZMAP.

REFERENCES

DIETERICH, J. H. (1992), *Earthquake Nucleation on Faults with Rate- and State-dependent Strength*, Tectonophysics *211*, 115–134.

FURUKAWA, N. (1995), *Quick Aftershock Relocation of the 1994 Shikotan Earthquake and its Fault Planes*, Geophys. Res. Lett. *22*, 3159–3162.

HABERMANN, R. E., *The Quantitative Recognition and Evaluation of Seismic Quiescence: Applications to Earthquake Prediction and Subduction Zone Tectonics*, Ph.D. Thesis (University of Colorado, Boulder 1981a).

HABERMANN, R. E., *Precursory seismicity patterns: Stalking the mature seismic gap*. In *Earthquake Prediction* (eds. D. W. Simpson and P. G. Richards), Maurice Ewing Series (Amer. Geophys. Union 4 1981b) pp. 29–42.

HABERMANN, R. E. (1987), *Man-made Change of Seismicity Rates*, Bull. Seismol. Soc. Am. *77*, 141–159.

HABERMANN, R. E. (1991), *Seismicity Rate Variations and Systematic Changes in Magnitudes in Teleseismic Catalogs*, Tectonophysics *193*, 277–289.

HABERMANN R. E., and WYSS, M. (1984), *Seismic Quiescence and Earthquake Prediction on the Calaveras Fault, California*, Abstract, EOS *65*, 988.

HIRATA, N., and MATSU'URA, M. (1987), *Maximum-likelihood Estimation of Hypocenter with Origin Time Eliminated Using Nonlinear Inversion Technique*, Phys. Earth. Planet. Int. *47*, 50–61.

HOKKAIDO UNIVERSITY (1994), *Continuous Observation of Crustal Deformation in Hokkaido Region— Strain Accumulation of 9 Stations for the Period from May, 1987 to November, 1992*, Report of the Coordinating Committee for Earthquake Prediction *52*, 45–55 (in Japanese).

ICHIKAWA, M. (1978), *A New Subroutine Program for Determination of Earthquake Parameters and Local Travel-time Tables for Events near the Southern Kurile Trench* (in Japanese with English abstract), Quarterly J. Seismology *43*, 11–19.

ICHIKAWA, Y. (1987), *Change of JMA Hypocenter Data and Some Problems*, Quarterly J. Seismology *51*, 47–56.

KANBAYASHI, Y., and ICHIKAWA, M. (1977), *A Method for Determining Magnitude of Shallow Earthquakes Occurring in and near Japan*, Quarterly J. Seismology *41*, 57–61.

KATO, N., OHTAKE, M., and HIRASAWA, T. (1997), *Possible Mechanism of Precursory Seismic Quiescence: Regional Relaxation due to Preseismic Sliding*, Pure appl. geophys. *150*, 249–267.

KATSUMATA, K., ICHIYANAGI, M., MIWA, M., KASAHARA, M., and MIYAMACHI, H. (1995), *Aftershock Distribution of the October 4, 1994 M_w 8.3 Kurile Islands Earthquake Determined by a Local Seismic Network in Hokkaido, Japan*, Geophys. Res. Lett. *22*, 1321–1324.

KATSUMATA, K., and KASAHARA, M. (1999), *Precursors to the 1994 Kurile Earthquake ($M_w = 8.3$)*, in preparation.

KIKUCHI, M., and KANAMORI, H. (1995), *The Shikotan Earthquake of October 4, 1994: Lithospheric Earthquake*, Geophys. Res. Lett. *22*, 1025–1028.

KISSLINGER, C. (1988), *Prediction of the May 7, 1986 Andreanof Islands Earthquake*, Bull. Seismol. Soc. Am. *78*, 218–229.

McNALLY, K., *Plate subduction and prediction of earthquakes along the middle America trench*. In *Earthquake Prediction* (eds. D. W. Simpson and P. G. Richards), Maurice Ewing Series (Am. Geophys. Union 4 1981) pp. 63–72.

MOGI, K. (1969), *Some Features of Recent Seismic Activity in and near Japan (2), Activity before and after Great Earthquakes*, Bull. Earthq. Res. Inst., Univ. Tokyo *47*, 395–417.

OHTAKE, M., MATSUMOTO, T., and LATHAM, G. V. (1977), *Seismic Gap near Oaxaca, Southern Mexico as a Probable Precursor to a Large Earthquake*, Pure appl. geophys. *115*, 375–385.

OZAWA, S. (1996), *Geodetic Inversion for the Fault Model of the 1994 Shikotan Earthquake*, Geophys. Res. Lett. *23*, 2009–2012.

SHIBAZAKI, B., and MATSU'URA, M. (1992), *Spontaneous Process for Nucleation, Dynamic Propagation, and Stop of Earthquake Rupture*, Geophys. Res. Lett. *19*, 1189–1192.

TAKANAMI, T., SACKS, I. S., SNOKE, A., MOTOYA, Y., and ICHIYANAGI, M. (1996), *Seismic Quiescence before the Hokkaido-Toho-Oki Earthquake of October 4, 1994*, J. Phys. Earth *44*, 193–203.

TAKEUCHI, H. (1983), *Magnitude Determination of Small Shallow Earthquakes with JMA Electromagnetic Seismograph Model 76*, Quarterly J. Seismology *47*, 112–116.

TANIOKA, Y., RUFF, L., and SATAKE, K. (1995), *The Great Kurile Earthquake of October 4, 1994 Tore the Slab*, Geophys. Res. Lett. *22*, 1661–1664.

TAYLOR, D. W. A., SNOKE, J. A., SACKS, I. S., and TAKANAMI, T. (1991), *Seismic Quiescence before the Urakawa-Oki Earthquake*, Bull. Seismol. Soc. Am. *81*, 1255–1271.

TSUJI, H., HATANAKA, Y., SAGIYA, T., and HASHIMOTO, M. (1995), *Coseismic Crustal Deformation from the 1994 Hokkaido-Toho-Oki Earthquake Monitored by a Nationwide Continuous GPS Array in Japan*, Geophys. Res. Lett. *22*, 1669–1672.

URABE, T., and TSUKADA, S. (1992), *WIN—A Workstation Program for Processing Waveform Data from Microearthquake Networks* (abstract in Japanese), Programme and Abstracts, Seism. Soc. Japan *2*, 331.

WATANABE, H. (1971), *Determination of Earthquake Magnitude at Regional Distance in and near Japan*, Zisin (Bull. Seism. Soc. Japan) *32*, 281–296. (in Japanese with an English abstract).

WIEMER, S., *Analysis of Seismicity: New Techniques and Case Studies*, Dissertation Thesis (University of Alaska, Fairbanks, Alaska 1996) 151 pp.

WIEMER, S., and WYSS, M. (1994), *Seismic Quiescence before the Landers (M = 7.5) and Big Bear (M = 6.5) 1992 Earthquakes*, Bull. Seismol. Soc. Am. *84*, 900–916.

WYSS, M. (1986), *Seismic Quiescence Precursor to the 1983 Kaoiki (M_s = 6.6), Hawaii Earthquake*, Bull. Seismol. Soc. Am. *76*, 785–800.

WYSS, M., KLEIN, F. W., and JOHNSTON, A. C. (1981), *Precursors to the Kalapana M = 7.2 Earthquake*, J. Geophys. Res. *86*, 3881–3900.

WYSS, M., HABERMANN, R. E., and GRIESSER, J. C. (1984), *Seismic Quiescence and Asperities in the Tonga-Kermadec Arc*, J. Geophys. Res. *89*, 9293–9304.

WYSS, M., and BURFORD, R. O. (1985), *Current Episodes of Seismic Quiescence along the San Andreas Fault between San Juan Baustista and Stone Canyon, California: Possible Precursors to Local Moderate Mainshocks*, U.S. Geol. Survey Open-file Report *85–754*, 367–426.

WYSS, M., and BURFORD, R. O. (1987), *A Predicted Earthquake on the San Andreas Fault, California*, Nature *329*, 323–325.

WYSS, M., and HABERMANN, R. E. (1988), *Precursory Quiescence before the August 1982 Stone Canyon, San Andreas Fault, Earthquake*, Pure appl. geophys. *126*, 333–356.

WYSS, M., SHIMAZAKI, K., and URABE, T. (1996), *Quantitative Mapping of a Precursory Quiescence to the Izu-Oshima 1990 (M = 6.5) Earthquake, Japan*, Geophys. J. Int. *127*, 735–743.

WYSS, M., and MARTIROSYAN, A. H. (1999), *Seismic Quiescence before the M = 7, 1988, Spitak Earthquake, Armenia*, submitted to GJI.

(Revised October 14, 1998, revised January 4, 1999, accepted January 22, 1999)

Pure appl. geophys. 155 (1999) 471–507
0033–4553/99/040471–37 $ 1.50 + 0.20/0

❙Pure and Applied Geophysics

Seismicity Analysis through Point-process Modeling: A Review

Yosihiko Ogata[1]

Abstract—The occurrence times of earthquakes can be considered to be a point process, and suitable modeling of the conditional intensity function of a point process is useful for the investigation of various statistical features of seismic activity. This manuscript summarizes likelihood based methods of analysis of point processes, and reviews useful models for particular analyses of seismicity. Most of the analyses can be implemented by the computer programs published by the author and collaborators.

Key words: Causal relationship, ETAS model, modified Omori formula, relative quiescence, seasonality of seismicity, space-time models.

1. Introduction

Lists of earthquakes are published regularly by the seismological services of most countries in which earthquakes occur with frequency. These lists supply at least the epicenter of each shock, focal depth, origin time and instrumental magnitude.

Such records from a self-contained seismic region reveal time series of extremely complex structure. Large fluctuations in the numbers of shocks per time unit, complicated sequences of shocks related to each other, dependence on activity in other seismic regions, fluctuations of seismicity on a larger time scale, and changes in the detection level of shocks, all appear to be characteristic features of such records. In this manuscript the origin times are mainly considered to be modeled by point processes, with other elements being largely ignored, except that the ETAS model and its extensions use data of magnitudes and epicenters.

Modeling of point processes for such data was pioneered by D. Vere-Jones (e.g., Vere-Jones and Davies, 1966; Vere-Jones, 1970, 1978). However, the maximum likelihood estimation method was not fully feasible by that time, despite the theoretical recognition that the method is sensible for parameter estimation and is sensitive for the testing of models.

[1] The Institute of Statistical Mathematics, 4-6-7 Minami-Azabu Minato-ku, Tokyo, Japan.

A key to the likelihood theory of point processes is the conditional intensity function which is, roughly speaking, the derivative of the probability forecast of an event occurring at a time t

$$\lambda(t \mid F_t) = \lim_{\Delta \to 0} \text{Prob}\{\text{An event occurs in } (t, t+\Delta) \mid F_t\}/\Delta, \qquad (1)$$

where F_t is information over the time interval $(0, t)$ of available observations, including the history of the point process itself at time t. A conditional intensity function characterizes a point process completely (cf. Appendix A.1 in this paper). Once the conditional intensity function is given, simulation of the corresponding point process is easily performed by the so-called "thinning technique" (cf. Appendix A.2). Also, once the conditional intensity function is given, the joint density distribution for the realization of occurrence data in $(0, T)$ can be recorded, which is used to obtain the maximum likelihood estimates (cf. Appendix A.3). Consequently, it is important to obtain good parametric models of conditional intensity function. Further, the conditional intensity function is used for the residual analysis of the point process to ascertain the goodness-of-fit (cf. Appendix A.5).

The principal aim of this manuscript is to review a class of parametric models for statistical analysis of earthquake catalogues and for assessment of earthquake risk in a geophysical area. Firstly, in Section 2, I will display a class of models which decompose the seismic activity into components of evolutionary trend, clustering and periodicity, and also another class of models which analyze the causal relationship between earthquake sequences from two seismic regions. Thereafter, Section 3 illustrates that the traditional aftershock analysis can be formulated by a nonstationary Poisson process and implemented by the maximum likelihood method. This together with the residual analysis makes it possible to analyze an aftershock sequence more efficiently and thoroughly. An extensive review of the development of aftershock analyses since the pioneering working by Omori is given in UTSU et al. (1995). In Section 4, I will review the ETAS model which was originally proposed to describe the standard seismic activity in a geophysical region, although this was also confirmed to better fit the majority of aftershock sequences from the simplest to the complex cases, including swarms. In Section 5, space-time extensions of the ETAS model are compared, based on the data sets in the areas of both inter- and intra-seismicity. The last section concludes the paper and discusses the author's outlook regarding future study.

Most of the models and methods described in this paper are implemented by the computer programs in the package TIMSAC-84 (AKAIKE et al., 1984) and also by SASeis in Chapter 2 of the IASPEI Software Library for Personal Computers (UTSU and OGATA, 1997).

2. Detection of Seasonality for Shallow Earthquakes and Causal Relationship between Seismic Activities in Distinctive Areas

2.1. Trend/Cyclic/Clustering Decomposition of Earthquake Risk

In statistical seismology the analysis of cyclicity in seismicity such as seasonal, monthly, diurnal and semidiurnal, has been attracting the interest of many investigators. The most common method is the testing of uniformity of superposed point process on a circle of the cyclicity. One may use statistics on the circle such as the periodogram by SCHUSTER (1897). AKI (1956) critically summarizes research of various seismic cyclicities until that time. KAWASUMI (1970), SHIMAZAKI (1971), VERE-JONES and OZAKI (1982), and VERE-JONES (1985) discuss the significance level of periodicity in the periodogram for the same historical data in the southern Kanto area, Japan, compiled by Kawasumi. Another basic method is testing the uniformity of stacked data on the circle (the so-called directional data, FISHER, 1993). For example, OIKE (1977) uses histogram to suggest the seasonality of shallow seismicity in southwest Japan, compared with the change of a precipitation pattern.

Nonetheless, these testing procedures must be carried out assuming no contamination of the data either by the presence of clustered events or by any trend of either real or artificial seismicity owing to a change of detection rate. Works of careful analyses by such conventional tests usually have to use substantially reduced data, owing to the declustering and cut-off of smaller events necessary to guarantee the homogeneous detection of the data. In order to overcome these limitations, OGATA (1983b) sugested the following model in the term of the conditional intensity in (1),

$$\lambda_\theta(t \mid F_t) = a_0 + P_J(t) + C_K(t) + \sum_{t_i < t} g_M(t - t_i). \tag{2}$$

The second term on the right-hand side of (2) represents the evolutionary trend where

$$P_J(t) = \sum_{j=1}^{J} a_j \phi_j(t/T), \qquad 0 < t < T, \tag{3}$$

T is the total length of the observed interval and $\phi_j(\cdot)$ is a polynomial of order j. We introduce these components either for the genuine seismic trend or the evolutionary change of the detection rate of shocks, or both. The third term of (2) is the Fourier expansion

$$C_K(t) = \sum_{k=1}^{K} \{b_{2k-1} \cos(2k\pi t/T_0) + b_{2k} \sin(2k\pi t/T_0)\}, \tag{4}$$

for cyclic effects with a given fixed cycle length T_0. The last term in (2) stands for the clustering effects such as aftershocks and earthquake swarms. The function

$g_M(x)$ measures the increase in clustering due to a shock. We may call this function a *response function* of a shock, and parameterize it as

$$g_M(x) = \sum_{m=1}^{M} c_m x^{m-1} e^{-\alpha x}. \tag{5}$$

If the scaling parameter α is fixed in (5), then the model in (2) is linearly parameterized (cf. Appendix A.4). This model is used to examine and establish the existence of each component by the comparison of AIC values (cf. Appendix A.3) among a possible set of (J, K, M) configurations. For example, if the configuration with $K = 0$ is selected, that is to say, $C_0(t) = 0$, then this suggests that no periodicity of T_0 exists in the seismic activity. Otherwise, its shape is estimated by the maximum likelihood method (cf. Appendix A.3) for the selected configuration of $(\hat{J}, \hat{K}, \hat{M})$.

Applying this model to shallow seismicity in the inner zone of southwest Japan, OGATA (1983b) showed the existence and the shape of a seasonal component which is quite similar to the derivative of the averaged and smoothed precipitation curve in that area, where peaks are found at the typhoon season around early September and at the beginning of spring. As suggested by OIKE (1977), it is understood that the drastic change of rainfall relates to the drastic change of pore pressure owing to the increase of underground water level which can trigger shallow earthquakes with some probability. A similar feature is seen in the microseismicity around the Australian Capital Territory (ACT) including Canberra, Australia (see Fig. 1, after OGATA and KATSURA, 1986). The seasonality looks similar to the Sydney rainfall pattern rather than Canberra's. This might suggest a correlation between earthquake occurrence around ACT and the migration of underground water beneath the southeastern highlands from eastern Australia.

The computer program (LINLIN) for the present model is included in TIM-SAC-84 (AKAIKE *et al.*, 1984) and also in the IASPEI Software Library (UTSU and OGATA, 1997). MATSUMURA (1986) extensively applied the program to the ISC data for some decades in the presence of increased detection of events, and found the seasonality in a number of areas of mid-latitude where clear seasonality in precipitation patterns are associated. MA LI and VERE-JONES (1997) also show clear seasonality in shallow seismic activity with $M_L \geq 5$ in a region bounded by the parallels 36°S and 45°S and in the longitude range 172°E–179°E in New Zealand.

Figure 1

The trend/cyclic/clustering decomposition of the micro-earthquake activity with $M \geq 2.5$ around Canberra, Australia. Top diagram shows the trend and magnitude versus time plot, the middle left shows the response function of the clustering effect, and the middle right shows the seasonal effect. The real lines are by the maximum likelihood estimates (MLE) with the overall minimum AIC assuming the clustering effect (middle left), while the dotted lines are the MLE without the clustering effect. The seasonal pattern of the seismicity is similar to the seasonal precipitation pattern in Sydney (representing the East Coast area) in the bottom diagram.

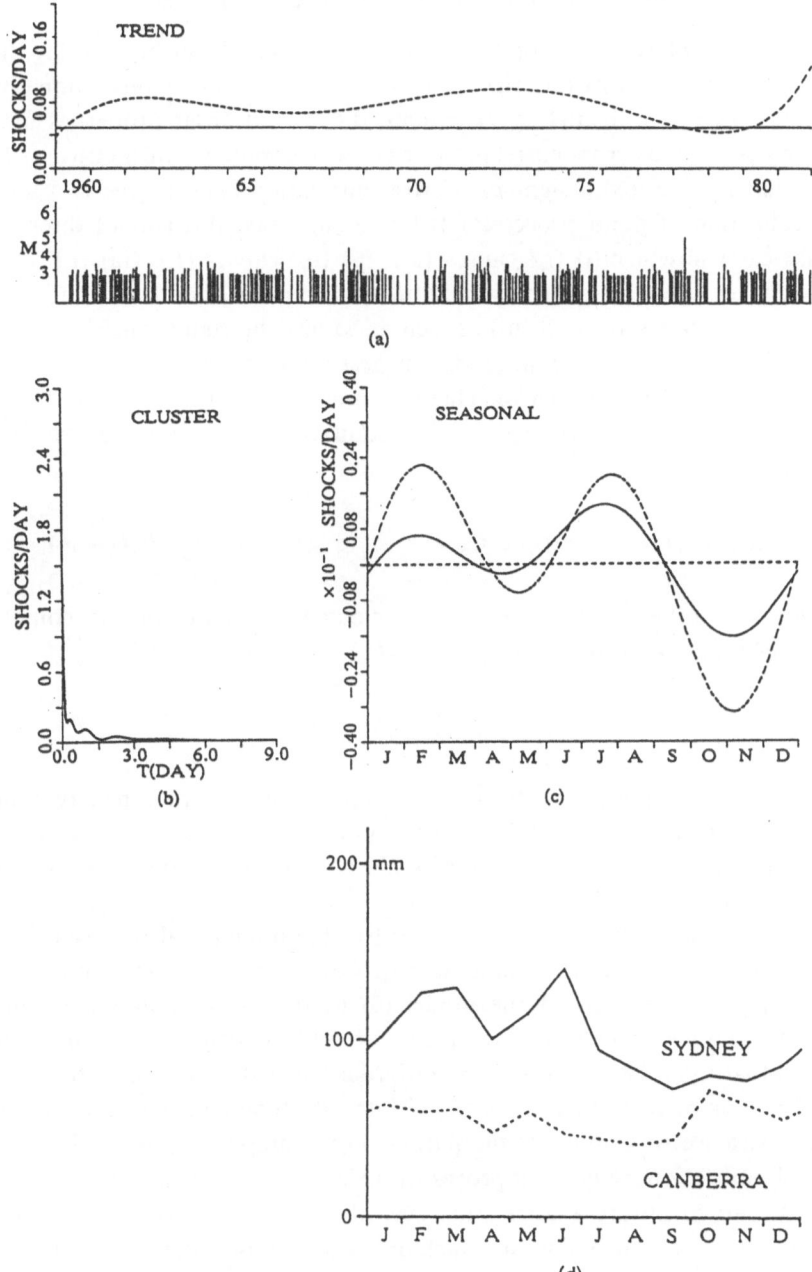

(a)

(b)

(c)

(d)

2.2. *Detection of Causal Relations between Two Series of Events*

It is also an interesting subject for the prediction of earthquakes to find a statistical causality of a seismic risk with other observations. Conventionally this problem has been investigated by examining the second-order properties between point processes such as cross correlation or cross spectra (e.g., BRILLINGER, 1988). However, having detected a significant cross correlation between the two series of events (realization of point processes) $\{t_i\}$ and $\{u_j\}$, say, this cannot discriminate among the cases in which (i) $\{t_i\}$ causes $\{u_j\}$, (ii) $\{u_j\}$ causes $\{t_i\}$, (iii) $\{t_i\}$ and $\{u_j\}$ cause each other, and possibly (iv) some other process causes both $\{t_i\}$ and $\{u_j\}$. To discriminate among (i), (ii) and (iii) as well as to test the significance, OGATA and AKAIKE (1982) and OGATA et al. (1982) suggest a parametric model, applying the minimum AIC procedure based on Hawkes' mutually-exciting process (HAWKES, 1971). Further, for examining case (iv) in addition, OGATA (1983b) extends the model in (2) to

$$\lambda_\theta(t \mid F_t) = a_0 + P_J(t) + C_K(t) + \sum_{t_i < t} g_M(t - t_i) + \sum_{u_j < t} h_N(t - u_j), \qquad (6)$$

where $\{u_j\}$ is another series of events considered as inputs of the conditional intensity function. The response function $h_N(x)$ is parameterized by

$$h_N(x) = \sum_{n=1}^{N} d_n x^{n-1} e^{-\beta x}. \qquad (7)$$

If there is no causal relation from $\{u_j\}$ to the conditional intensity function $\lambda_\theta(t \mid F_t)$, or to the occurrence of $\{t_j\}$, then $h_N(x) = 0$ is expected. Otherwise, we are interested in ascertaining the influential scale and the approximate shape of the response function.

A parallel model to (6) for $\{u_j\}$ versus $\{t_i\}$ as the inputs is also considered. The detection of mutual or one-way stimulation between point processes can be reduced to the independent analysis of the model (6) to the extent that the parameters between two marginal conditional intensities are independent to each other. It has been noted that the existence of the second-order correlation between the processes $\{t_i\}$ and $\{u_i\}$, or between their conditional intensity function, does not necessarily imply the existence of a mutual stimulation. For example, a similar shape in the intensity of nonstationary Poisson processes (i.e., $P_J(t) + C_K(t)$ in (6)) possibly with some delay, can constitute a correlation even if $h_N(x) = 0$ holds, which stands for the case (iv). There are also cases in which unknown factors stimulate both $\{t_i\}$ and $\{u_j\}$.

UTSU (1975) discussed the correlation between the deep earthquakes beneath Hida (the region neighboring Takayama City, central Japan) and the shallow earthquakes in central Kanto (the region around Tokyo). He tested the independence of the Kanto earthquakes from those of the Hida area, assuming the former shocks to form a stationary Poisson process, and concluded that there was a significant dependence which could be attributed to the mechanical connection between the two seismic regions which belong to the same segment of the Pacific plate underthrusting at the northeastern Japan arc. The data compiled by Utsu are composed of 61 earthquakes with the magnitude $M \geq 5.5$ in central Kanto and 16 earthquakes of $M \geq 5.0$ in Hida during the 51 years from 1924 through 1974.

Since the earthquakes in the Kanto District seem to have slight clustering, OGATA et al. (1982) fitted the model (6) with $P_J(t) = C_K(t) = 0$ against the Hida shocks as inputs to estimate the scale and shape of the response functions $g_M(\cdot)$ and $h_N(\cdot)$. A similar analysis was made for Hida seismicity against Kanto shocks as inputs. The results of these analyses demonstrated that the earthquakes in the central Kanto region are not only self-exciting but also receive significant one-way stimulation from earthquake occurrences in the Hida region, and that Hida shocks are identified as a stationary Poisson process. In personal communication Utsu imparted to me that such a correlation cannot be seen in the data after the period studied (i.e., 1975 ~). However, YOSHIDA (1994) indicates that, in order to see the correlation, the intermediate-depth interplate events between the Pacific and the North American plates and also between the Pacific and the Philippine Sea plates in the Kanto area should be discriminatingly selected from the intraplate events within the North American and Philippine Sea plates and interplate events between North American and Philippine Sea plates.

Furthermore, OGATA and KATSURA (1986) considered a similar problem for the two wider regions (see Fig. 2) with lower magnitude thresholds for the period from 1926 through September 1983, responding to the criticism that the data size of the Hida shocks may not be enough to conclude the independency from Kanto shocks. Consequently, the causality between the two areas remains identical to the case of the above Utsu's data except that proper seismicity trends (see Fig. 2 again) are revealed in both areas. The AIC selected far more complicated models than the previous models owing to data size larger than the Utsu's data.

Model (6) is also applied to the shallow and deep earthquakes in a region covering the North Island area, New Zealand, from 1946 through 1980. It is clearly shown that earthquake occurrences in the deep region significantly receive one-way stimulation from earthquake occurrences in the shallow region. This is a different outcome from the above cases in Japan where the existence of the opposite one-way stimulation is concluded from Utsu's data (UTSU, 1975) and the extended data. After this analysis I learned that these two types of relationship between the shallow and deep seismicity in the eastern Pacific region were already discussed by MOGI (1973). Particularly he found that the seismic activity in the Mariana and Tonga areas gradually migrated from shallow to deep regions within the descending slab, while the opposite migration is found in the Kurile-Kamchatka and northern Japan island-arc regions. The migration rate or speed along the deep seismic zone

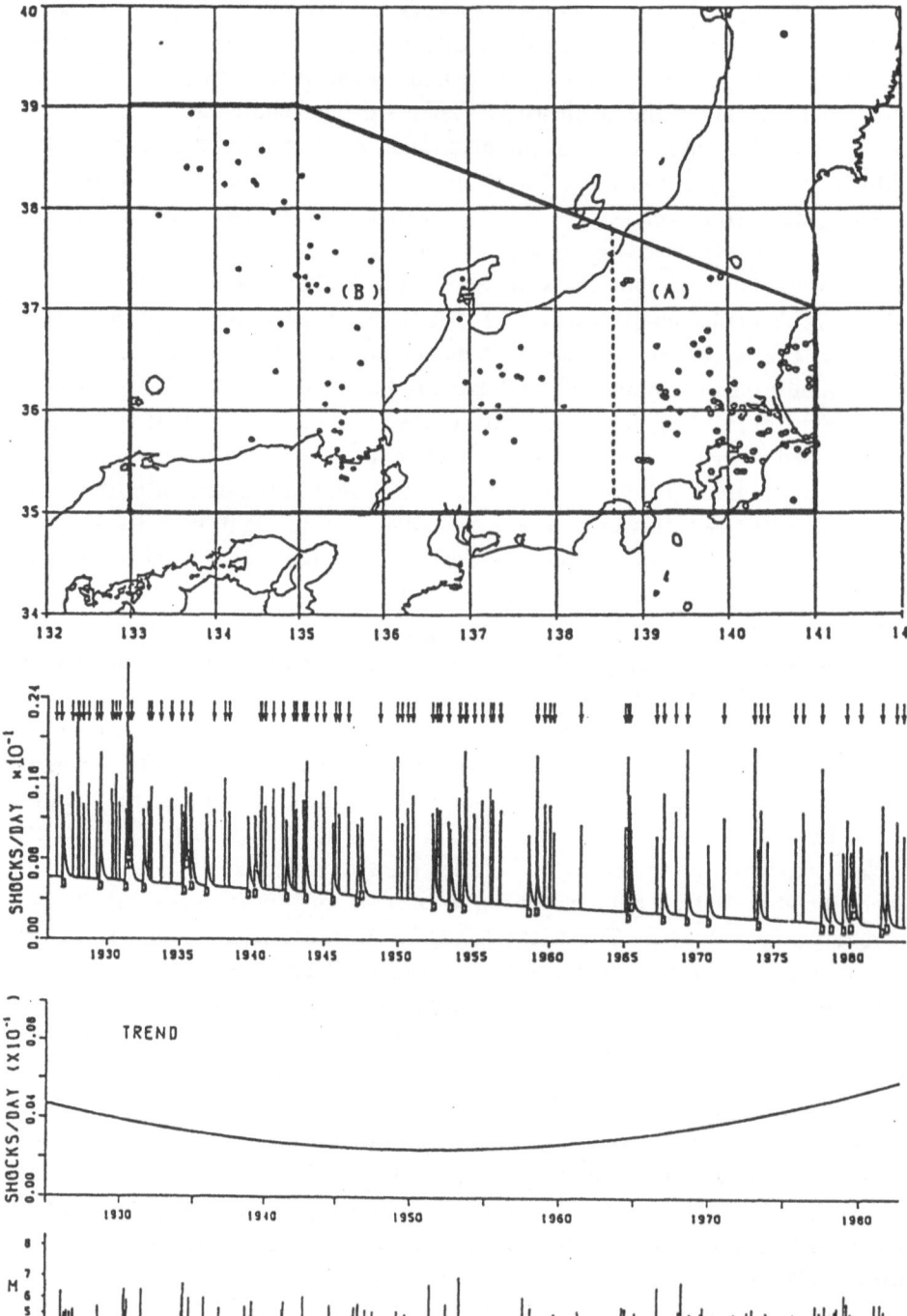

of the Tonga arc suggested by Mogi is about 45 km per year with which our estimated response impulse function $h_N(\cdot)$ in (6) is consistent. These may indicate that similar tendency holds within the Tonga-Kermadec-New Zealand tectonic zone.

The present model was also used by DE NATALE et al. (1988) for the investigation of the causal relation between the Calabrian and North-Aegean earthquake sequences. Furthermore, this model can also be used to investigate a causality in the case where the input process can be occurrence times of anomaly observations. In other words, the model is also applicable to test the statistical relation between suspected precursor signals and large earthquakes. For example, NISHIZAWA et al. (1994) examined the causality between seismo-electric precursor signals (SES) at a station in Greece, based on the VAN method, and earthquake events in a region around the station listed in the USGS catalog for the period March 1988 to December 1989. They concluded stimulation of the SES events to the earthquake sequences with $M_b \geq 4.0$ and $M_b \geq 4.5$, but no causal effect from the sequence $M_b \geq 4.5$ to the SES sequence.

In the case where suspected precursor signals are given by analogue data $\{\xi(u)\}$ such as a record of stress changes, we can apply the modified version of the model

$$\lambda_\theta(t \mid F_t) = a_0 + \sum_{t_i < t} g_M(t - t_i) + \int_0^t h_N(t - u)\xi(u)\,du.$$

See OGATA and AKAIKE (1982) for the computing algorithm and simulated experiment.

3. Analysis of Aftershock Occurrence

3.1. Modified Omori Formula

The frequency of aftershocks per unit time interval (one hour, one day, one month, etc.) is well represented by the modified Omori formula (UTSU, 1961)

$$n(t) = K(t + c)^{-p} \qquad (K, c, p: \text{ parameters}) \tag{8}$$

Figure 2
Examination of a causal relation between two seismic regions in central Japan. Top diagram shows earthquake epicenters of the considered regions on the subducting Pacific plate: open circles and solid circles stand for the shallower (≤ 200 km) and deeper (≥ 200 km) shocks in the regions (A) and (B), respectively. The diagram in the second row delineates the estimated change of the intensity rate of the activity in the region (A). The arrows and symbol D indicate the occurrence times of (A) shocks ($M \geq 5.5$) and (B) shocks ($M \geq 5.0$), respectively. This shows stimulations by the deeper events associated with the decreasing trend, which is probably due to the aftershock activity of the 1923 Kanto great event of M_J 7.9. The diagram in the third row shows the estimated change of intensity rate of (B) shocks in the bottom diagram. This indicates the trend approximated by the quadratic curve but with no stimulation from the shallower (A) events. After all, the one way stimulation from (B) events to (A) events are indicated.

where t is the lapse time from the occurrence of the mainshock, and K depends on the magnitude of the mainshock and the lower bound of the magnitude of aftershocks counted, while p is known to be independent of these. According to UTSU (1969), for the case of Japanese aftershock sequences, the constant c is small, 1 day at most, and p varies in the range 0.9–1.9 with values between 1.0 and 1.4 most frequently. In the case of $p = 1$, formula (8) is called the original Omori formula (OMORI, 1894), while for general p, it is called the modified Omori formula (UTSU, 1957, 1961). The value p is thought to reflect mechanical conditions of the earth's crust. For example, MOGI (1962) demonstrated a certain systematic regional variation of p value in Japan which could be attributed to regional variation of surface heat-flow values. He thinks that aftershock activity decays faster, namely the stress relaxes faster, in regions of higher crustal temperature. Later, a similar correlation is discussed by KISSLINGER and JONES (1991) and CREAMER and KISSLINGER (1993).

Traditionally, estimates of the parameter p had been obtained since UTSU (1961) in the following way. Plot $n(t)$ versus the lapse time t on a log-log scaled plane and then fit an asymptotic straight line; the slope of the line is an estimate for p. The values of c can be determined by a naïve empirical technique. Such analysis was based on the time series of counted numbers of aftershocks, while directly based on the occurrence times of aftershocks, OGATA (1983a,b), proposed the following method to estimate the three parameters of the modified Omori formula.

Consider occurrence times of aftershock sequence $\{t_1, t_2, \ldots, t_N\}$ in a time interval $[S, T]$, in which the origin of the time axis, $t = 0$, corresponds to the occurrence time of the mainshock. Assume that the aftershock sequence is distributed according to a nonstationary Poisson process with the intensity function

$$\lambda(t; \theta) = K(t + c)^{-p}, \qquad \theta = (K, c, p), \tag{9}$$

which represents the modified Omori formula. Then from (32) in Appendix A.3 the log-likelihood function of the aftershock sequence is written by

$$\log L(K, c, p; S, T) = N \log K - p \sum_{i=1}^{N} \log(t_i + c) - K\Lambda(c, p; S, T), \tag{10}$$

where

$$\Lambda(c, p; S, T) = \begin{cases} [(T + c)^{1-p} - (S + c)^{1-p}]/(1 - p) & \text{for } p \neq 1 \\ \log(T + c) - \log(S + c) & \text{for } p = 1. \end{cases}$$

Maximizing the log-likelihood function (10) with respect to the parameters $\theta = (K, c, p)$ we obtain the maximum likelihood estimates (MLE) $\hat{\theta} = (\hat{K}, \hat{c}, \hat{p})$. The maximum likelihood procedure also has the advantage of providing us estimates for the standard errors of the MLE as a byproduct. Specifically, the inverse of the Fisher information matrix $J(\hat{\theta}; S, T)^{-1}$ provides the variance-covariance matrix of the errors of the MLE, in which

$$J(\theta; S, T) = \int_S^T \frac{1}{\lambda(t; \theta)} \frac{\partial \lambda(t; \theta)}{\partial \theta'} \frac{\partial \lambda(t; \theta)}{\partial \theta} dt$$

$$= \int_S^T \begin{bmatrix} K^{-1}(t+c)^{-p} & -p(t+c)^{-p-1} & -(t+c)^{-p} \ln(t+c) \\ * & Kp^2(t+c)^{-p-2} & Kp(t+c)^{-p-1} \ln(t+c) \\ * & * & K(t+c)^{-p} \ln^2(t+c) \end{bmatrix} dt. \quad (11)$$

This is carried out by the program AFT in the package SASeis (UTSU and OGATA, 1997) and also ASPAR by REASENBERG (1994).

The maximum likelihood estimation of the modified Omori formula makes possible the practical forecasting of the probability of large aftershocks. REASENBERG and JONES (1989, 1990, 1994) provided the routine procedure by the joint use of this formula and the Gutenberg-Richter law of magnitude frequency.

3.2. Complex Aftershock Sequences and Model Selection

Initially, we always have the problem of determining the starting time S in the calculation of the log-likelihood (10) for aftershock occurrence data of a large earthquake. If the aftershocks were detected homogeneously throughout the entire observation period $[0, T]$, then we can formulate $S = 0$. However, the rate of missing events immediately after the mainshock can be large compared to the rest of the period. Another possibility may be that the beginning stages of some aftershock sequences are too complicated to be described by (9). In such cases, if the MLE is computed with $S = 0$, the estimates are very likely to be biased. Therefore we must find a suitable S both to avoid significant bias and to gain accurate estimation of p and c. One of the feasible methods for such a task is the comparison between

$$\text{AIC} = (-2) \max_\theta \ln L(\theta; 0, T) + 2 \times 3$$

and

$$\text{AIC} = (-2) \max_{\theta_1} \ln L(\theta_1; 0, S) + (-2) \max_{\theta_2} \ln L(\theta_2; S, T) + 2 \times 6 \quad (12)$$

for some candidates of S. Here we automatically take prescribed values of S, namely $S = i/8$ ($i = 1, 2, \ldots, 8$) days. Among the choices S we will adopt the one which has the smallest AIC value. Alternatively S is determined subjectively by inspecting the magnitude versus time plots by eyes.

Next, if an earthquake sequence on an observed time interval does not consist of a single pure aftershock sequence, the model (9) is not accurate. The sequence of shocks may contain secondary aftershocks (UTSU, 1970). Another possibility may be that the modified Omori formula would not hold in some later stage [see UTSU (1957) and MOGI (1962) for example]. Thus, testing these problems is reduced to the matter of model selection against alternative models such as

$$\lambda(t) = \begin{cases} K_1(t + c_1)^{-p_1} & \text{for } 0 < t \le t_0, \\ K_1(t + c_1)^{-p_1} + K_2(t - t_0 + c_2)^{-p_2} & \text{for } t_0 < t, \end{cases} \tag{13}$$

and

$$\lambda(t) = \begin{cases} K_1(t + c)^{-p} & \text{for } 0 < t \le t_0, \\ K_2 e^{-\alpha t} & \text{for } t_0 < t, \end{cases} \tag{14}$$

where $\theta = (K_1, K_2, c_1, c_2, p_1, p_2)$ for the case in (13) and $\theta = (K_1, K_2, c, p, \alpha)$ for the case in (14).

Furthermore, when we are interested in examining whether a pair of aftershock sequences from distinctive areas or time periods have the same value p or not, we will calculate both

$$\text{AIC}_0 = (-2) \max_{K_1, p, c_1, K_2, c_2} \{\ln L_1(K_1, p, c_1) + \ln L_2(K_2, p, c_2)\} + 2 \times 5$$

and

$$\text{AIC}_1 = (-2)\left\{ \max_{K_1, p_1, c_1} \ln L_1(K_1, p_1, c_1) + \max_{K_2, p_2, c_2} \ln L_2(K_2, p_2, c_2) \right\} + 2 \times 6, \tag{15}$$

where L_1 and L_2 stand for the likelihood of the first and second series of aftershocks, respectively. Then we choose the model which provides the smaller AIC values. See OGATA (1983a) for further details.

3.3. Residual Analysis of Aftershock Activity

The integration of the conditional intensity with respect to time in (35) of Appendix A.5 generates a transformation to a frequency-linearized time τ where the occurrence of earthquakes becomes the standard stationary Poisson process (i.e., with a constant intensity), if the intensity function $\lambda(t)$ substituted for the conditional intensity in (35) is the true one for the data. For example, the frequency-linearized time for the model of a single aftershock sequence in (9) can be given by

$$\tau = \Lambda(t) = \int_0^t K(s + c)^{-p} \, ds$$

$$= \begin{cases} K \ln(t + c)/c & \text{for } p = 1.0 \\ K[(t + c)^{1-p} - c^{1-p}]/(1 - p) & \text{for } p \ne 1.0. \end{cases}$$

Using this transformation, it can be seen that the modified Omori formula holds as long as the occurrence rate of aftershocks surpasses that of background activity around the aftershock region; for example, see Figure 3. That is to say, cumulative numbers of aftershocks against the transformed time aligns on a straight line until the rate of occurrence becomes lower than the level of the background activity of

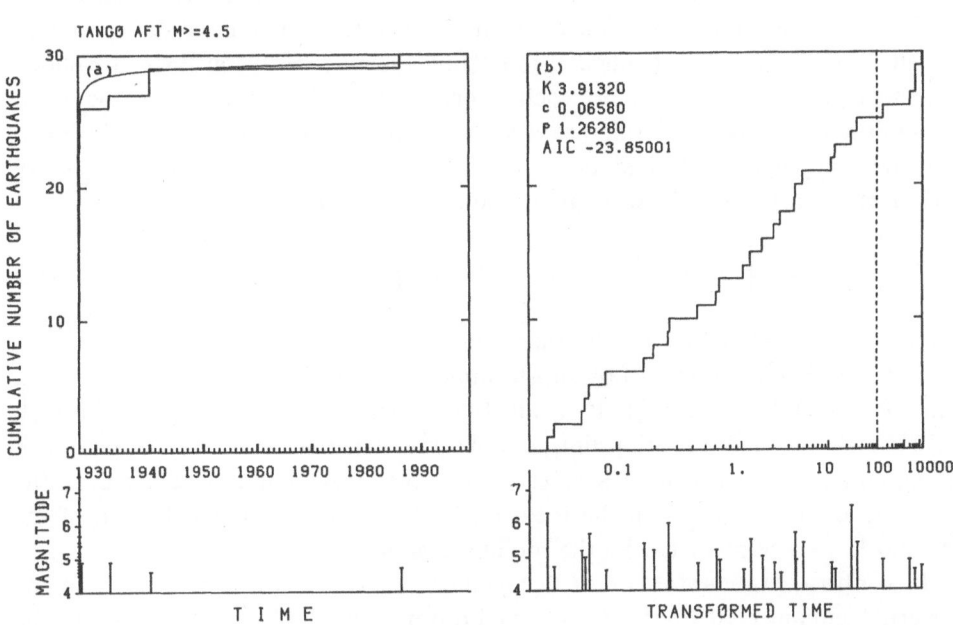

N=30   T=26200.0 TO=0.000

TANG0 AFT M>=4.5

Figure 3

Diagrams of cumulative numbers and magnitudes ($M_J \geq 4.5$) of the aftershocks of the 1927 Tango earthquakes of M_J 7.2 versus time to the present (Nov. 1998): (a) ordinary and (b) the transformed times. Here, the modified Omori intensity function is fitted for the aftershocks until 100 days after the mainshock, and the expected cumulative (smooth curve in (a)) and cumulative of the residual process is extrapolated to the remaining time interval of 100–26,200 days. These evidence that the aftershock activity of the Tango earthquake still persists according to the modified Omori law.

the area, the time of which is independent of threshold magnitude of the data to the extent that the data are complete or homogeneous.

Therefore, if the level of the background seismicity around an aftershock region is very low, then the intensity trend of aftershocks of a major earthquake decays according to the formula in (9) for a long time span, even longer than one century (OGATA, 1989). UTSU (1969) demonstrated this fact by the log–log plot of frequency rate versus the elapsed years for the felt aftershock of the Nobi earthquake of 1891; also see UTSU *et al.* (1995).

It is often observed that a sequence of aftershocks contains secondary after-shocks (UTSU, 1970). If a secondary aftershock sequence starts at time t_0 after the mainshock occurrence, then the intensity function $\lambda(t)$ can be given in (13), and the MLE of θ is also obtained by maximizing the log-likelihood in (32). For example, OGATA and SHIMAZAKI (1984) investigated the aftershock sequence of the 1965 Rat Islands earthquake of M_w 8.7, over a time span of some fifteen years. The selection of a statistical model based on AIC showed that the entire aftershock sequence is better modeled by the superposition of the primary and secondary aftershocks associated with the largest aftershock, the 30 March 1965 event of M_w 7.6. The difference in both parameters p and c between the main and the secondary aftershock sequences is insignificant, indicating that the decays in their aftershock sequences were very similar to each other despite the contrast of the focal mechanism and tectonic location between the mainshock and the largest aftershock. The frequency-linearized time in (35) for a sequence including the secondary aftershock sequence which starts at the time t_0, is given as

$$\tau = \int_0^t K_1(s + c_1)^{-p_1} \, ds + H(t - t_0) \int_{t_0}^t K_2(s - t_0 + c_2)^{-p_2} \, ds, \qquad (16)$$

where $H(t - t_0) = 1$ for $t > t_0 = 0$ otherwise.

The transformation in (16) by substituting the MLE illustrates that a transition from aftershock to normal activity is identified 2200 days after the 1965 mainshock occurrence. After the transition time, we observe a significant upward deviation of the cumulative number curve against τ from the straight line, which means the departure of the seismicity in the focal region from the extrapolated trend of the aftershock activity predicted by the modified Omori formula.

In contrast, HABERMANN (1983) concluded that a decrease of activity in the western Aleutians (170°E–180°E) occurred during early 1967 and that it marks the end of the Rat Islands aftershock sequence. This conclusion differs from the result of OGATA and SHIMAZAKI (1984) which indicates the aftershock sequence persisted at least six years. In examining at time-dependent seismicity patterns, Habermann used a declustering algorithm which should have removed all the dependent events, i.e., aftershocks. However, since the algorithm did not perform very well in removing all the aftershocks for the 1965 Rat Islands sequence, he interpreted the decrease in seismicity around 1967 as the end of the aftershock sequence. In OGATA and SHIMAZAKI (1984), an algorithm which can predict the aftershock sequence is used to show that the sequence lasted considerably longer. The techniques allow straightforward and objective detection of a transition from aftershock to normal activity. Defining the normal activity is essential in the detection of a seismic quiescence which may take place prior to the next major event.

Furthermore, the cumulative number of such residual processes sometimes deviates downwards significantly in some stage, which means that the rate of aftershocks decays more rapidly in the corresponding stage than the expected rate according to the modified Omori formula. Using the methods and models described in the last two subsections, MATSU'URA (1986) inclusively investigated this phenomenon in aftershock sequence data of Japanese major earthquakes and found that, in many cases, the largest aftershocks were likely to take place in the recovery stage from the quiescence relative to the ordinary decreasing rate expected by the modified Omori formula. The reported phenomenon was nominated as one of the "Preliminary Candidate for Outstanding Geophysical Precursors to Earthquakes" by the IASPEI Sub-Commission on Earthquake Prediction (M. Wyss, Chairman) at the IASPEI Istanbul assembly in August 1989. Similar results have been reported in aftershock sequences in China (ZHAO *et al.*, 1989).

4. The ETAS Model and Relative Quiescence

4.1. The Epidemic-type Aftershock-sequences Model

In many cases, aftershock activity appears complex in such a way that

$$\lambda(t \mid H_t) = \mu + \sum_{\tau_m < t} \frac{K_m}{(t - \tau_m + c_m)^{p_m}}, \tag{17}$$

where μ stands for the rate of background activity, and $\{\tau_m\}$ are origin times of triggering events of secondary aftershock, and the sum $\Sigma_{\tau_m < t}$ is taken over the trigger events $\{m\}$ which occurred before time t. Conspicuous triggering events $\{\tau_m\}$ can be determined by the residual analysis as described in the previous subsection.

This model also can be applied to characterize general seismic activity in which triggering events include the mainshocks. This is called the restricted trigger model in OGATA (1988), in which the triggering events are identified before estimation. The original trigger model was suggested by VERE-JONES and DAVIES (1966), in which a triggering event is a hidden variable realized by a probability. Therefore, in order to see a goodness-of-fit, the theoretical second-order moment of a point process such as the auto-correlations function and the spectrum are compared with those from data. The parameter estimation is carried out by maximizing the log spectral-likelihood (e.g., HAWKES and ADAMOPOULOS, 1973; OGATA and ABE, 1991; OGATA and KATSURA, 1991), while identification of trigger events is inevitable for the ordinary maximum likelihood estimation procedure of the trigger models.

Extending these models further, OGATA (1985, 1988, 1989) demonstrated that the ordinary seismic activity of a wide region is remarkably well described, in terms of the conditional intensity, by the superposition of a constant rate for background seismicity and the modified Omori functions of *any* shocks i which occurred at time t_i, in such a way that

$$\lambda(t \mid H_t) = \mu + \sum_{t_i < t} \frac{K_i}{(t - t_i + c)^p}, \tag{18}$$

where μ is an occurrence rate for the background seismic activity. The sum $\Sigma_{t_i < t}$ is taken for *all* shocks i which occurred before time t, and the parameter K_i for each shock i contributes to the size of the corresponding offspring, or aftershocks in a wide sense. The crucial point of the model here is that the parameter K_i is dependent on its magnitude M_i as well as the cut-off magnitude M_0 of the data set according to the following exponential function form

$$K_i = K_0 \, e^{\alpha(M_i - M_0)}. \tag{19}$$

This form is based on the empirical formula obtained by UTSU and SEKI (1955) regarding the linear relation between the logarithms of aftershock areas and the magnitudes M of the mainshock (also see UTSU, 1971, for data supporting the relation). This relation suggests that the number N of aftershocks with magnitudes over a threshold M_0 for a fixed time span is roughly estimated as

$$N \propto \exp\{\alpha(M - M_0)\}, \tag{20}$$

for a constant α, where M is the magnitude of the mainshock. The relation in (20) also appears to be true for a set of the secondary aftershocks (UTSU, 1970, Section 8.1, and also OGATA, 1983a, 1989). Furthermore, YAMANAKA and SHIMAZAKI (1990) discuss the distinction of the proportional constant relative to the difference between the mainshock's magnitude and threshold magnitude M_0 in (20), or K_0 in (19), between intra- and interplate earthquakes. Namely, it is larger in the case of intraplate events than in the case of interplate events.

Model (18) with (19) for the ordinary seismicity in terms of the occurrence rate of shocks is hereafter called *Epidemic Type Aftershock-Sequences (ETAS) model*. Note here that $H_t = \{(t_i, M_i); \, t_i < t\}$ is the history of occurrence times $\{t_i\}$ up to time t and their corresponding magnitudes $\{M_i\}$. The original epidemic model, called birth and death process, was considered by KENDALL (1949) in application to population genetics in epidemiology. HAWKES (1971) modeled the birth process, allowing immigration at a constant rate μ per unit time, in terms of the conditional intensity rate depending on the history of occurrence times.

Compared to the restricted trigger model, the ETAS model is extremely useful, especially when the discrimination between mainshocks (triggering events) and aftershocks (triggered events) is not easy, as in earthquake swarms. Even with ordinary seismicity, the rigorous discrimination of these is impossible. OGATA (1985, 1988) compares the goodness-of-fit of the ETAS model with possible alternative models including the restricted trigger models to demonstrate the best fit. Further, while the ordinary seismicity is long-range correlated and approximately self-similar in time and space (e.g., OGATA, 1987; OGATA and KATSURA, 1991; OGATA and ABE, 1991), RAMSELAAR (1990) shows that ETAS model is self-similar in an approximate sense. Incidentally, the inverse power-law response function in (18) causes heavy computation of the log-likelihood, unlike the Laguerre type polynomials (5) and (7) in (2) and (6), respectively (OGATA and AKAIKE, 1982). To overcome the intensive computation OGATA et al. (1993) uses the accurate approximation by superposition of exponential functions with different scales using the double exponential integration formula. It is worthwhile to note that seismic activity is also spatially long-range correlated (KAGAN and KNOPOFF,

1980), and a certain explicity form for an extension of the ETAS to the spatial seismicity is confirmed, which will be described in Section 5.

In summary, among the parameters $\theta = (\mu, K_0, c, \alpha, p)$ of ETAS model in (18) with (19), the last two parameters α and p are extremely useful for characterizing the temporal pattern of seismicity. The p value in (18) indicates the decay rate of aftershocks, and the α value measures magnitude sensitivity of an earthquake in generating its offspring, or aftershocks in a wide sense. For example, OGATA (1987) evidences that swarm-type activity has a smaller α value than that of ordinary mainshock and aftershock activities and that α is likely to be large if there are few conspicuously large aftershocks in an aftershock sequence.

4.2. Change-point Problem and Model Selection

The question of whether or not the temporal patterns of seismicity changed before and after a time T_0 in a given data set on a time interval $[S, T]$ can be reduced to a problem of model selection (cf. Appendix A.3), namely, whether or not a model throughout the time interval $[S, T]$ fits better than two different models on the respective intervals $[S, T_0]$ and $[T_0, T]$ for some T_0. Thus, we calculate the following AIC values,

$$\text{AIC}_0 = (-2) \max_{\theta_0} \ln L(\theta_0; S, T) + 2k_0,$$

$$\text{AIC}_1 = (-2) \max_{\theta_1} \ln L(\theta_1; S, T_0) + 2k_1,$$

$$\text{AIC}_2 = (-2) \max_{\theta_2} \ln L(\theta_2; T_0, T) + 2k_2;$$

where $k_0 = \dim(\theta_0)$, $k_1 = \dim(\theta_1)$ and $k_2 = \dim(\theta_2)$ are the number of adjusted parameters in the models fitted to the data from the corresponding time span. For example, $k = 5$ if all the parameters of the ETAS are used, however sometimes $k = 4$ when for example either $\mu = 0$ or $p = 1$ is assumed. In particular, when the number of events in the latter time span $[T_0, T]$ is extremely small, we may apply the stationary Poisson process, $\lambda(t \mid H_t) = \mu$, so that $k_2 = 1$ in such case. We should note here that the log-likelihood function $\ln L(\theta; T_0, T)$ includes the data H_{T_0} in the previous time span $[S, T_0]$, as well as these in $[T_0, T]$, since $\lambda_\theta(t \mid H_t)$ is the function of the history from the time origin. Also, in order to minimize the bias owing to the effect of the previous large events before the considered time span $[S, T]$, it is strongly recommended either to set the starting time S to be the occurrence time of the last large earthquake or to take account of the history of large events in a previous time interval $[R, S]$ for a suitable time R.

If a potential change-point T_0 is set based on information other than the data, such as a scientific reason or some other independent setting from the data, AIC_0 is compared with the following AIC_{12} of the alternative model for the changed seismicity at the time T_0 where

$$\text{AIC}_{12} = \text{AIC}_1 + \text{AIC}_2$$

(see OZAKI and TONG, 1975; KITAGAWA and AKAIKE, 1978). Otherwise, i.e., if the T_0 is searched and set based on the data, AIC_0 is compared with

$$AIC_{12} = AIC_1 + AIC_2 + 2q(N),$$

where the quantity $q(N)$ corresponds to the contribution of T_0 as an adjusted parameter and is dependent on the number of events N in the interval $[S, T]$. The $q(N)$ is obtained by a Monte Carlo experiment (OGATA, 1992), and accurately given by the Padé approximant

$$q(N) = 1 + \frac{7.6623\left(\dfrac{N}{10}\right) + 1.9688\left(\dfrac{N}{10}\right)^2 + 0.022822\left(\dfrac{N}{10}\right)^3}{1 + 5.0900\left(\dfrac{N}{10}\right) + 0.95595\left(\dfrac{N}{10}\right)^2 + 0.0090963\left(\dfrac{N}{10}\right)^3}$$

for $10 \le N \le 2000$). Adjusting the parameter T_0 belongs to the so-called change-point problem and the contribution $q(N)$ is extraordinary because the ordinary contribution to the AIC is unity for a parameter which satisfies a standard large sample theory.

The inequality $AIC_{12} < AIC_0$ indicates a significant distinction between the seismicity patterns in the two divided time spans (see Fig. 4 for an example). Otherwise, we conclude that no change-point exists there. This procedure can be continued for detection of further possible seismicity changes within the interval $[S, T_0]$.

Figure 4

Comparison of two cases which examine the existence of change-point. Aftershock sequence ($M \ge 2.7$) of the 1997 Satuma (north-west Kagoshima Prefecture) earthquake (M_J 6.5) for a time span up to 47.88 days when the large event of M_J 6.2 made a parallel rupture zone to that of the former event in the very neighborhood. Upper diagrams show cumulative numbers and magnitudes versus ordinary (left) and transformed time (left) where a single ETAS model (four parameters) is applied to the entire time span except within 0.03 days after the mainshock when the detection rate of events with $M \ge 2.7$ is inferior. The thin smoother curve (left) and straight line (right) show the expected cumulative numbers obtained by the integral of the estimated intensity function. From these diagrams it is easily seen that the goodness-of-fit is poor, however the fit of the modified Omori curve is even worse due to the clusters within the aftershock sequence. Lower diagrams show the case in which two different models are applied to the divided time spans at the suspected change-point, namely 0.03–5.37 days and 5.37–47.88 days; the two ETAS models are applied to the data in the former and latter time intervals, independently. Since the total number of events in the entire time span is $N = 493$, we have $q(N) = 3.15$ for the penalty of AIC of the change-point in this case. It is shown that the lower case is chosen by the difference $AIC_0 - AIC_{12} = 5.35$. The thin smoother curve (left) and straight line (right) show the expected cumulative numbers obtained by the integral of the intensity function estimated for the data in the time span until the change-point. From these we see that the aftershocks after the change-point are substantially fewer than expected by the extrapolation.

4.3. Residual Point Process and Relative Quiescence

Assume we have the maximum likelihood estimate (MLE) $\hat{\theta} = (\hat{\mu}, \hat{K}_0, \hat{c}, \hat{\alpha}, \hat{p})$ of the ETAS model for a data set. Next consider the time-change, using the increasing function (35) in Appendix A.5, such that $\tau_i = \Lambda_{\hat{\theta}}(t_i)$. Thus, origin times $\{t_i\}$ are transformed 1-to-1 into the residual point process (RPP) $\{\tau_i\}$. Since we do not

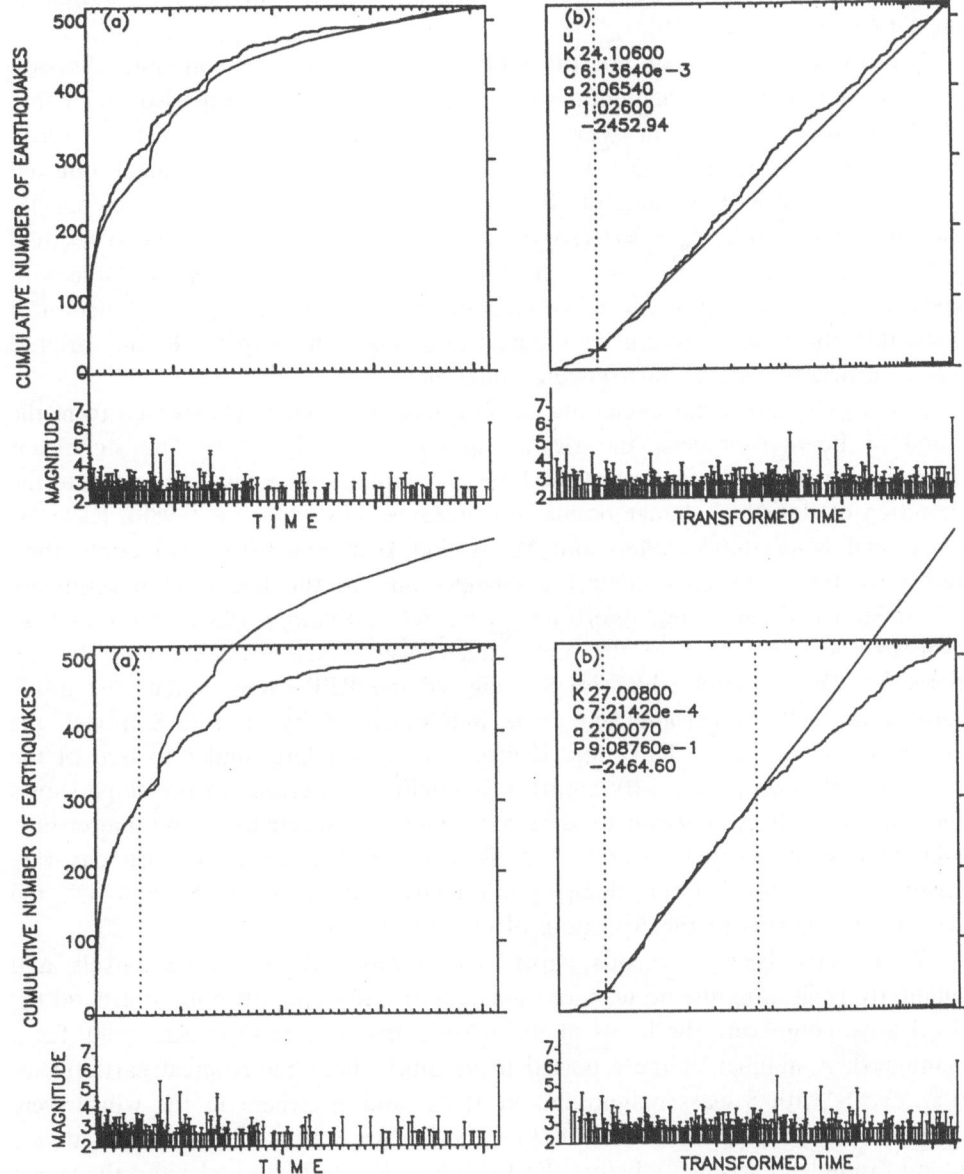

know the true model nor its parameter value, we use the MLE $\hat{\theta}$ for the parameter value of the assumed conditional intensity model in (35). If the model provides a good fit to the seismicity, then the RPP is approximately the standard stationary Poisson process. If, on the contrary, we find a significant deviation of any characteristic property of the RPP from that expected from the stationary Poisson, this suggests some discrepancy between the model and the data, such as nonhomogeneity of the data or the existence of seismic quiescence, which is not modeled by the ETAS in (18) with (19).

To examine whether or not the RPP obeys the standard stationary Poisson process, we initially see the cumulative curve of the RPP in comparison with that of the uniform distribution. Various other tests against the stationary Poisson process are described in OGATA (1988). In particular, to find local unusual characteristics in RPP invalidating the stationary Poisson assumption, we consider the number of points $\xi_h(\tau) = N(\tau - h, \tau)$ in the interval $(\tau - h, \tau)$. If the residual process $\{\tau_i\}$ is stationary Poisson, then $\xi_h(\tau)$ is a Poisson random variable with mean h for each τ. Setting $h = 8$, for example, the time series of ξ as a function of τ should behave like a stationary Gaussian process with mean $h = 8$ and variance $h = 8$, obviously with a short-range correlation.

Among the significant deviations of RPP from that which is expected from the standard Poisson process, the *relative quiescence* is shown by the significant decrease of the occurrence rate of the RPP. Such quantitative analysis relates to the detection of statistical change-points. For instance, HABERMANN (1988), REASENBERG and MATTHEWS (1988) and WYSS and HABERMANN (1988) apply their respective test statistics to detect a change-point in the background seismicity (obtained by declustering algorithms; e.g., REASENBERG, 1985, 1994) and to evaluate the significance of the naïve quiescence against the stationary Poisson processes. OGATA (1992) likewise investigated the RPP events versus the transformed time. If the seismic activity is well described by an ETAS model, the cumulative curve of the RPP data is nearly a straight line similar to that of the stationary Poisson process. By contrast, a significant decrease of the slope shows the integrated effect of seismicity decreases from the expected rate, no matter how high or low the original seismicity is. To test the significance of seismicity rate change, in addition to the change-point analysis discussed in Section 4.2, the readers are referred to the Appendix of OGATA (1992).

There have been numerous papers which reported quiescence naively and qualitatively showing the occurrence time data or a few epicentral maps around the focal area, comparing the levels of the seismic activity (see OHTAKE, 1980, for a summary). A number of the reported naïve quiescences before great earthquakes (i.e., events with 8 class magnitude) in Japan and elsewhere in the world were examined by applying the ETAS (OGATA, 1992): Diagrams in Figure 5 illustrate the oldest example of seismicity before the 1923 great Kanto earthquake and the recent example of seismicity before the 1995 off the east coast of Hokkaido (or off the coast of Kurile Island) great event of M_J 8.3. All naively recognized quiescences are clearly confirmed to be relative quiescences, too, which are revealed in seismicity for far wider areas than the reported focal areas. In many cases, these quiescences last several years prior to the occurrence of great earthquakes, which occasionally take place in the recovering stage of the seismicity. Some relative quiescences, which are not clear in ordinary occurrence data, can also be found by the present method. It especially is shown that even in an active stage of seismicity, a decrease from the expected level can take place. Indeed, it is desirable to predict whether an earthquake

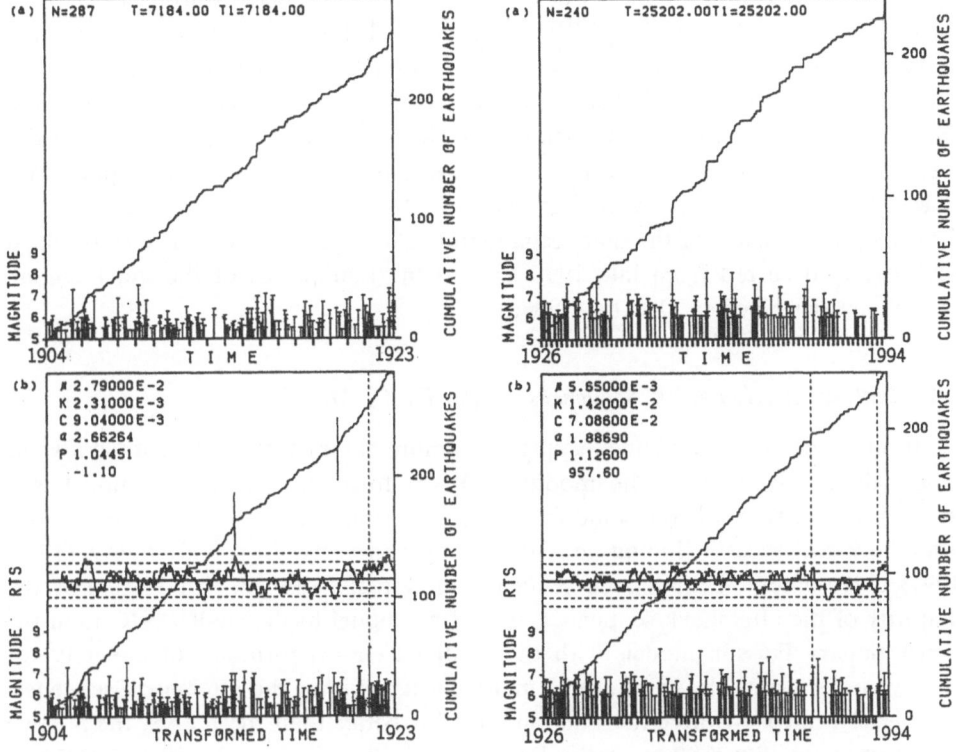

Figure 5

Relative quiescences before great earthquakes. Left diagrams display cumulative numbers and magnitudes of earthquakes with $M \geq 5.4$ and depth ≤ 100 km in the region 32–38°N and 136–144°E versus ordinary time (top figure) and the transformed time (bottom figure) from 1904 until September 1, 1923 when the Kanto earthquake (M_J 7.9) took place (the vertical dotted line). The residual time series (RTS) $\zeta_h(\tau)$ show the substantially long range of bias below the average ($h = 8$), which indicates the relative quiescence. Correspondingly, the vertical segments on the cumulative lines indicate the estimated onset and termination times of relative quiescence. Right diagrams exhibit cumulative numbers and magnitudes of earthquakes with $M_J \geq 6.0$ and depth ≤ 100 km in the region in 41–46°N and 141–151°E versus ordinary time (top figure) and the transformed time (bottom figure) from 1926 through October 1994. The relative quiescence is seen in the RTS and cumulative of the residual process in the period between the two vertical dotted lines, where the right one coincides with the occurrence time of the earthquakes of M_J 8.3 off the east coast of Hokkaido.

swarms or aftershock activity is nearing to an end or is a precursor of a forthcoming larger earthquake (OGATA, 1989, 1993). Existence or nonexistence of the relative quiescence will be particularly useful for the predictive discrimination.

In fact, we saw no relative quiescence for about 20 years up to 1990 in the wide areas including the Tokai and Boso gaps (OGATA, 1992), which suggests a low probability that the gaps will rupture within several years. On the contrary, a few authors cited in OGATA (1998a) manifested a very significant quiescence for the previous twenty years since the early or mid-1970s in the seismicity of certain areas in or around the Tokai region. However, it ensues that these quiescences are seemingly ones owing to magnitude shifts which took place during 1975–1976 throughout Japan and vicinity. The magnitudes below M_J 5.0 are substantially underestimated after the period. This shift is found and estimated by a statistical comparison of magnitudes between the JMA and USGS catalogs (OGATA, 1998a). On the other hand, the recent seismicity with a level of $M_J \geq 5.5$ in a far wider region including central and western Japan exhibits a significant relative quiescence since 1992, which may be related to the 1995 Kobe earthquake of M_J 7.2 which occurred in the center of the quiet region and also to many M 6 class events which successively occurred from late 1996 around the boundaries of the quiet region. Details are described in OGATA (1998a).

4.4. *Analysis of Aftershock Sequences by the ETAS Model*

In many cases an aftershock sequence is more complex than the simple inverse power decay expected by the modified Omori formula. The ETAS model is a generalized version of the modified Omori formula and fits well to various aftershock sequences including non-volcanic type swarms. GUO and OGATA (1995, 1997) investigated 34 aftershock sequences in Japan from 1971 to 1995. For the majority of the aftershock sequences, the ETAS model has a smaller AIC than the nonstationary Poisson models with the modified Omori formulae of either (9) or (13), which suggests the existence of many clusters within the aftershock sequence. Conversely, in cases where the Poisson model is better than the ETAS, the p value approaches equality to that of the ETAS model. This means that the ETAS is a higher-order approximation of the aftershock occurrences.

Using the ETAS model, GUO and OGATA (1997) analyzed the correlation between characteristic parameters of statistical models, such as the b value of the Gutenberg-Richter relation, the p and α values, and the fractal dimension D of the hypocenter distribution. All the parameters are estimated using maximum likelihood methods with their error assessments. Most of the scatter plots between the maximum likelihood estimates of the characteristic parameters are seen to be either positively or negatively correlated. The contrasting correlation patterns are revealed between the parameters for the intraplate and interplate earthquakes except the two pairs $(\hat{b}, \hat{D})$ and $(\hat{\alpha}, \hat{p})$ where similar patterns are found. We then focus our attention on these patterns as a source of interesting contrasts between the two earthquake groups, namely aftershocks of interplate and intraplate earthquakes. In particular, the significant dependence of these parameters on depth appears to be a key to understand the correlation pattern for interplate aftershocks, while a different interpretation is made for intraplate aftershocks because no dependence on depth is seen.

5. Space-time Extensions of the ETAS Model

5.1. Space-time Conditional Intensity

Denote by $P_{\Delta t, \Delta x, \Delta y}(t, x, y \mid H_t)$ a history-dependent probability that an earthquake occurs in a small time interval between t and $t + \Delta t$, in a small region $[x, x + \Delta x) \times [y, y + \Delta y)$, where $H_t = \{(t_i, x_i, y_i, M_i); t_i < t\}$ is the history of occurrence times $\{t_i\}$ up to time t, their corresponding epicenters $\{(x_i, y_i)\}$ and magnitudes $\{M_i\}$. Thereafter the conditional intensity function $\lambda(t, x, y \mid H_t)$ of the space-time point process can be defined as

$$\lambda(t, x, y \mid H_t) = \lim_{\Delta t, \Delta x, \Delta y \to 0} \frac{P_{\Delta t, \Delta x, \Delta y}(t, x, y \mid H_t)}{\Delta t \Delta x \Delta y}.$$

As far as the stationarity is assumed, Hawkes' type self-exciting point-process model (HAWKES, 1971) is naturally extended to the following form

$$\lambda(t, x, y \mid H_t) = \mu(x, y) + \sum_{\{i: t_i < t\}} g(t - t_i, x - x_i, y - y_i; M_i) \qquad (21)$$

for $(t, x, y) \in [0, T] \times A$.

An important aspect of the space-time statistical modeling is the parametric forms of the response function $g(\cdot, \cdot, \cdot; \cdot)$. MUSMECI and VERE-JONES (1992) suggest a diffusion-type function for $g(\cdot, \cdot, \cdot; \cdot)$, KAGAN (1991) suggested other parametric forms based on investigations of the second-order moment features in time and space of various hypocenter catalogs (e.g., KAGAN and KNOPOFF, 1978, 1980), and RATHBUN (1993) applied the standard bivariate normal density function. For the case of nonhomogeneous seismicity in space, MUSMECI and VERE-JONES (1992) introduce a kernel-type function fitting to $\mu(x, y)$.

OGATA (1998b) extended the ETAS model to three cases in which the response functions are provided by

$$g_\phi(t, x, y; M) = \frac{K_0}{(t + c)^p} \exp\left\{ -\frac{1}{2} \frac{x^2 + y^2}{d \, e^{\alpha(M - M_0)}} \right\}, \qquad (22)$$

$$g_\phi(t, x, y; M) = \frac{K_0}{(t + c)^p} \cdot \frac{e^{\alpha(M - M_0)}}{(x^2 + y^2 + d)^q}, \qquad (23)$$

and

$$g_\phi(t, x, y; M) = \frac{K_0}{(t+c)^p} \left(\frac{x^2 + y^2}{e^{\alpha(M-M_0)}} + d \right)^{-q}, \qquad (24)$$

with $\phi = (K_0, c, \alpha, p, d)$ in (22), and $\phi = (K_0, c, \alpha, p, d, q)$ in (23) and (24). The models (22) and (23) are the same ones as presented by OGATA (1993).

All the above response functions can be rewritten in the common standard form

$$g_\phi(t, x, y; M) = \kappa(M) \times \frac{(p-1)c^{p-1}}{(t+c)^p} \times \left[\frac{1}{\pi\sigma(M)} \cdot f\left\{ \frac{x^2 + y^2}{\sigma(M)} \right\} \right], \qquad (25)$$

where $\kappa(M) \propto e^{\alpha M}$ is a cluster size factor (expected number of aftershocks for the event of magnitude M), $(p-1)c^{p-1}/(t+c)^p$ is the time probability density distribution, and $\pi^{-1}\sigma(M)^{-1}f\{(x^2+y^2)/\sigma(M)\}$ is the space probability density distribution (equivalently $\int_0^\infty f(r)2\pi r \, dr = 1$) in which the scale factor $\sigma(M)$ is allowed to depend on magnitude M. Thereafter the main contrasts in modeling the response function taking the points of our questions into consideration are:

1. between the functional forms allowed for $f(\cdot)$; short-range decay (i.e., normal etc.) versus long-range decay (i.e., inverse power law), and
2. between either $\sigma(M) = $ const. (independent of M) or $\sigma(M) = e^{\alpha M}$.

It should be noted that the exponential form for $\kappa(M)$ implies that the superposition of the present space-time models with respect to space is the ETAS model in time, and that the exponential form for $\sigma(M)$ is consistent with the Utsu-Seki law (UTSU and SEKI, 1955) of the aftershock areas in relation to the magnitude of the mainshock.

Among the possible space-time extensions of the ETAS model in (22)–(24), the model with the response function (24) and its extensions to heterogeneous background intensity $\mu(x, y)$ or the anisotropic functions $f(\cdots)$ is best fitted to both the data sets from the plate boundary and intraplate region. The result remains the same for the data with different threshold magnitudes of earthquakes. The difference of AIC increases as the size of the data increases, which suggests that the results clarify for the data with the lower threshold magnitude levels. After all the result indicates that:

1. the functional form for $f(\cdot)$ in (25) extends long range (i.e., inverse power law) rather than short range (i.e., normal etc.), and
2. the scale factor depends on the magnitude in the form $\sigma(M) = e^{\alpha M}$.

Thus, conclusions and consequences of this study are as follows:

1. The clusters in space extend beyond the traditional aftershock regions, a considerably more diffuse boundary with power-law decay rather than a somewhat well-defined region with a fairly sharp boundary.
2. There may be two components (near-field and far-field) with different characteristics; the near-field component corresponds to the traditional aftershock area, and the far-field component may relate to the so-called immigrations of earthquake activity or a causal relation between the distant regions.

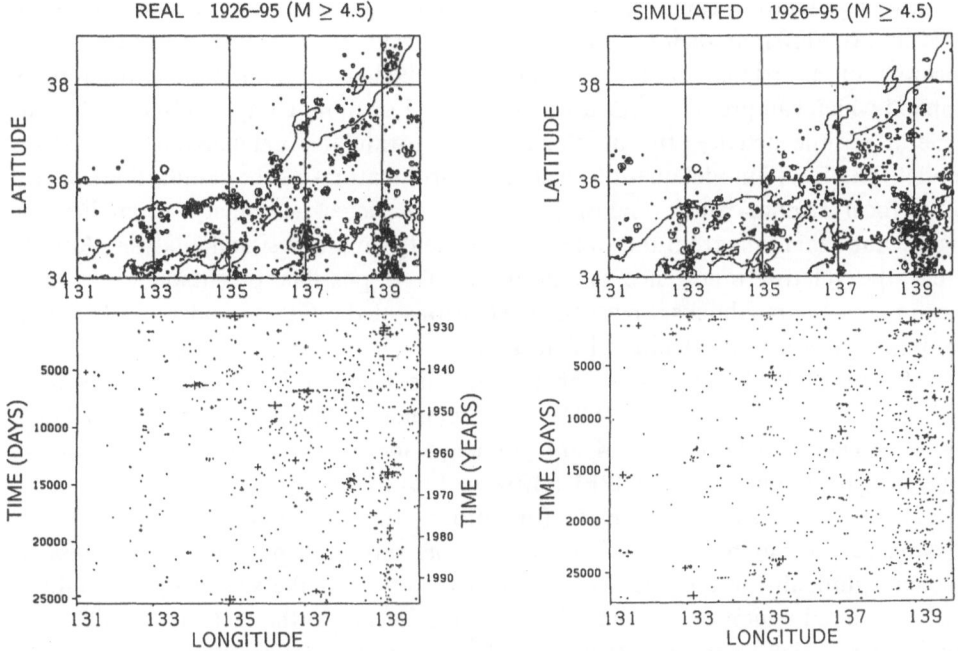

Figure 6

Left diagrams show shallow earthquakes with $M \geq 4.5$ in the central and western Honshu areas, central Japan, spanning 1926–1995; (a) epicenter distribution, and (b) time versus space (longitude) plot, while the right diagrams show simulated events by the selected model (21) with (24). Size of circle and plus sign correspond to the magnitude of the earthquake.

3. The cluster regions scale with magnitudes firmly according to the Utsu-Seki formula.

Further practical extensions of the models are suggested to apply the aniotropic features of earthquake clustering, and the above conclusions are unchanged for the extended models.

The simulation of the space-time point process is considered by applying an extension of the thinning algorithm (Appendix A.2 and OGATA, 1998). Figure 6 compares space-time patterns of the real seismicity and the simulated point process of the MLE in (21) with the response function in (24).

6. Concluding Remarks

Point-process models characterized by parameterized conditional intensity functions can be useful tools to investigate seismic activity through a hypocenter catalog. The goodness-of-fit of the model is measured by AIC, which can lead to a conclusion among competing hypotheses. As examples, we have seen the models for the investigation of seismic seasonality and the causal relationship between seismicities in two different regions. For an efficient estimation of the modified Omori formula of aftershock decay, the maximum likelihood estimation procedure is applied which can provide error estimates, also. For more complex aftershocks and general seismic activity, the modified Omori formula is extended to the ETAS model. The ETAS model quantitatively measures features of the standard seismicity of a focal region, which makes it possible to detect seismicity anomalies, or significant deviations from the standard seismicity, by the residual analysis. Among them the relative quiescence is important as a possible precursor to a large earthquake or aftershocks. Space-time extensions of the ETAS model are discussed to derive several spatio-temporal features.

The FORTRAN programs for the models used in Section 2 are given in TIMSAC-84 (AKAIKE et al., 1984), and Personal Computer programs package, SASeis, related to Sections 2–4 are published in Chapter 2 (UTSU and OGATA, 1997) of the Volume 6 of IASPEI Software Library.

There are many requirements for future study of point-process modeling. Firstly, as the number of data increases or as threshold magnitudes decrease, it becomes difficult for a single ETAS model to represent the seismicity throughout the considered region or volume. This is mainly owing to the fact that significantly different seismicity patterns often take place, even in neighboring regions. Consequently, the ETAS model frequently fits poorly to data with a large number of small events of microearthquakes. One way to avoid such difficulty could be a careful seismic zoning, which however causes a different predicament. The other approach, which could be more promising, is to consider Bayesian modeling in which the parameters are functions of the location. For example, OGATA and KATSURA (1988, 1993) and OGATA et al. (1991) analyze the b value and related parameters of magnitude frequency as the function of location where the smoothness constraint of the function is assumed and determined objectively. This procedure can be applied to the ETAS model, as well.

Secondly, it will be required to model the intensity function in relation to the stress-field change in the crust. Primarily, such studies include the stress-release model by VERE-JONES (1978) and its versions which are being proposed (LIU et al., 1999; IMOTO et al., 1999). However, the stress in these models has been hidden variables. Since we have good data sets of focal mechanisms and the global positioning system (GPS), the models using such information as an exogeneous variable will be useful; an example of such may be $\xi(t)$ in the model provided in the last paragraph of Section 2. On the other hand, spatial statistical features of seismicity using data sets which include focal mechanisms have been investigated by KAGAN (1992a,b), KAGAN and KNOPOFF (1985), and KAGAN and JACKSON (1998).

Finally, our ultimate objective of the point-process modeling will be the practical application of probability forecasts of large earthquakes using information of anomalous phenomena such as the relative quiescence. Also, an evaluation method for the probability forecasting should be established so as to compare various prediction algorithms. An attempt has been made in OGATA *et al.* (1995, 1996) to make real-time probability discrimination of foreshocks, making use of space, time and magnitude pattern. Additionally the use of AIC difference is proposed to evaluate the probability forecasting.

Acknowledgements

The referee's comments were useful in composing the revised version.

Appendix A. Statistical Methods for Point Processes

This Appendix section describes a set of statistical methods for analyzing various point-process data. These are (1) efficient simulation method by the *thinning algorithm*, (2) model comparison and parameter estimation by the AIC and the *maximum likelihood method*, and (3) graphical diagnostic checks of models and data by the *residual point process*. These methods are closely related to each other and based on the idea of the *conditional intensity* of a point process. Parametric models regarding the conditional intensity function is the keystone of the present approach.

A.1. Specifications of Point Processes and the Conditional Intensity Function

There were two traditional approaches to describe a point process, or time series of events. Consider a series of events $\{t_i; 0 < t_1 < t_2 < \cdots \}$ occurring at random on the half-line $(0, \infty)$. To describe this sequence we can take the time differences, $X_i = t_i - t_{i-1}$, between the consecutive points, and we then consider $\{X_i\}$ as a positive valued stochastic process. When this is distributed independently and identically, we call this a *renewal process*. Further, if its marginal is exponential distribution, then this is the *stationary Poisson process*. The *Wold process* is an extension of the renewal process to the case where $\{X_i\}$ is a Markov process of any order.

Another traditional specification of point processes is based on the counting of points on intervals. Let $N(a, b)$ be the number of points in an interval (a, b) on the real line such that this is a nonnegative integer valued random variable. To specify a point process is to give consistently defined finite-dimensional joint distribution of such random variables for any corresponding mutually disjoint intervals (DALEY and VERE-JONES, 1972). For example, a stationary Poisson process is defined by the following two conditions: 1) for any interval (a, b), $N(a, b)$ is the Poisson random variable with mean $\lambda(b - a)$ for a positive constant λ, and 2) random variables $N(a_k, b_k)$ are mutually independent if their corresponding intervals (a_k, b_k), $k = 1, 2, \ldots, K$ are disjoint to each other. The probability generating functional is a useful tool to analyze point processes in this specification. This is particularly useful for clustering point processes such as the Neyman-Scott cluster processes whose parents are distributed according to a Poisson process: see VERE-JONES (1970), and DALEY and VERE-JONES (1972) for details.

The intensity specification of point processes is developed closely relating to the information theory. Consider a prediction of an event occurring on a small time interval. That is to say, assume a point process on the half-line $(0, \infty)$, and divide it into small intervals of length δ. Thereafter, we get a stochastic process $\{\xi_k\}$, where $\xi_k = N[(k-1)\delta, k\delta)$ is k-th random variable on the subinterval $[(k-1)\delta, k\delta)$. For a small enough δ we may assume that $\{\xi_k\}$ is a binary process (i.e., 0 or 1 valued). If the considered point process is a stationary Poisson, then $\{\xi_k\}$ is identically and independently distributed, that is, a Bernoulli series. Otherwise, it is easily seen that, in general, the joint probability of the sequence is essentially determined by a sequence of conditional probabilities $P\{\xi_k = 1 \mid \xi_1, \ldots, \xi_{k-1}\}$, $k = 1, 2, \ldots$, on the history of events. Roughly speaking, as a derivative of the conditional probability with respect to the time, the *conditional intensity function* $\lambda(t \mid F_t)$ is defined by

$$P\{N(t, t + \delta) = 1 \mid F_t\} = \lambda(t \mid F_t)\delta + o(\delta);$$

that is

$$\lambda(t \mid F_t) = \lim_{\Delta \to 0} P\{\text{an event in } [t, t + \Delta] \mid F_t\}/\Delta, \qquad (26)$$

where F_t is a set of observations extending to time t, including the history of the occurrence times of the events $H_t = \{t_i; t_i < t\}$ (see RUBIN, 1972, for example). The conditional intensity completely characterizes the corresponding point process (see LIPTZER and SHIRYAEV, 1978). For example, if the conditional intensity depends only on the elapsed time since the last occurrence t_L, then the corresponding point process is a renewal process, and the conditional intensity coincides with so-called *hazard rate*, which is frequently used in the field of reliability theory, in such a way that

$$\lambda(t \mid H_t) = v(t - t_L) = \frac{f(t - t_L)}{1 - F(t - t_L)}, \qquad (27)$$

where $F(x)$ is the probability distribution function of the interval length of consecutive points and $f(x)$ is its density function. A conditional intensity function which is only dependent on the lengths of the last n intervals determines the Wold process of $(n-1)$-th order. Also, if the conditional intensity is independent of the history but dependent only on the current time t, like $\lambda(t \mid H_t) = v(t)$ for any nonnegative function $v(t)$ of t, then this supplies a *non-stationary Poisson process*.

Of course, a constant conditional intensity provides a stationary Poisson process. There are many other interesting classes of point processes which are defined by certain conditional intensity functions such as the self-correcting (or Stress-Release) processes, $\lambda(t \mid H_t) = v(\alpha t - N(0, t))$, (see ISHAM and WESTCOTT, 1979; OGATA and VERE-JONES, 1984; VERE-JONES, 1978; and VERE-JONES and OGATA, 1984), and Hawkes' self-exciting process

$$\lambda(t \mid H_t) = \mu + \int_0^t g(t - s) \, dN_s = \mu + \sum_{t_i < t} g(t - t_i),$$

(see HAWKES, 1971; HAWKES and OAKES, 1974). The last process is very interesting and useful from a statistical prediction viewpoint, because, like the auto-regressive time series, the expectation of an event occurring is given by a linear combination of past occurrences, where the so-called *impulse response function* $g(\cdot)$ measures the weights of such combinations (see OGATA and AKAIKE, 1982; and OGATA and KATSURA, 1986, for example).

A.2. Thinning Simulation Method

The conditional intensity function is mathematically proved to characterize the corresponding point process completely. Practical applications of this fact include simulation of point-process data using a conditional intensity function. In principle we can obtain a sample point process, or a series of events, by making use of the interval distributions. That is to say, consider a conditional distribution $F(t \mid t_1, \ldots, t_n)$ of an event occurring at time t, given a history of events which occurred at times $t_1, t_2, \ldots, t_n$. Then the relation between his distribution and the corresponding conditional intensity is given by the equation,

$$\lambda(t \mid H_t) = \frac{f(t \mid t_1, \ldots, t_n)}{1 - F(t \mid t_1, \ldots, t_n)}, \tag{28}$$

where f is the density function of F (see LIPTZER and SHIRYAEV, 1978, for example). This is an extension of equation (27) related to the hazard rate of renewal process. Solving this equation we have

$$F(t \mid t_1, \ldots, t_n) = 1 - \exp\left\{ -\int_{t_n}^t \lambda(s \mid t_1, \ldots, t_n) \, ds \right\}. \tag{29}$$

For example, consider the simulation of a stationary Poisson process. Since the conditional intensity rate is constant, say λ, we derive from (29) that the distribution function of the intervals is exponential, and the sample events are then simulated by the inversion of uniform random variable U_i in such a way that $t_i = t_{i-1} - \log(U_i)/\lambda$. For general purposes, including renewal processes or nonstationary Poisson processes, we must solve the equation

$$F(t \mid t_1, \ldots, t_n) = U_{n+1} \tag{30}$$

of t, given a uniform random variable U_{n+1} (see OZAKI, 1978). However, this simulation procedure is neither efficient nor very practical because of two technical reasons: firstly we must use numerical integration in (29), and secondly we must use a numerical iteration algorithm to solve the equation (30).

LEWIS and SHEDLER (1979) introduced the *thinning method* for the simulation of a nonstationary Poisson process. The basic algorithm is very simple:

1. given an intensity function $\lambda(t)$, take a finite upper bound, say Λ, such that $\lambda(t) \leq \Lambda$, $0 \leq t \leq T$;
2. simulate trial events $\{t_i\}$ of the stationary Poisson with intensity Λ; and
3. for each event t_i we keep it with probability $\lambda(t_i)/\Lambda$, otherwise we eliminate it. Then the remaining events form samples of the nonstationary Poisson process with intensity $\lambda(t)$. Actually this method is very similar to the so-called rejection method for simulating samples from a probability density function, $f(t) = \lambda(t)/\int_0^T \lambda(t)\,dt$, when the total number of accepted points is fixed. Nevertheless, the idea of thinning can be extended to general point processes (OGATA, 1981). For example, a useful algorithm for simulation of events is as follows:

1. set $n = 0$, $t_n = 0$;
2. set $i = 1$, $r_1 = 0$;
3. take any r such that $r_i < r \leq T$ and find a constant Λ_i such that $\lambda(t \mid H_t) \leq \Lambda_i$ holds for all $t \in [r_i, r]$;
4. set $j = 0$, $\tau_0 = 0$;
5. set j equal to $j+1$ and generate a uniform random variable U_j, and calculate the trial event τ_j such that $\tau_j = \tau_{j-1} - \log(U_j)/\Lambda_i$; if $\tau_j > T$ then the simulation terminates; otherwise go to the next.
6. if $r_i + \tau_j > r$, set i equal to $i+1$, set $r_i = r$, and go to step 3; otherwise go to the next step;
7. set j equal to $j+1$ and generate a uniform random variable U_j; if $\lambda(t_n + r_i + \tau_j)/\Lambda_i > U_j$ then go to step 5, otherwise go to the next step;
8. set n equal to $n+1$, then set $t_n = t_{n-1} + r_i + \tau_j$ and go to step 2.
9. set i equal to $i+1$, $r_i = t_n$ and go to step 3.

The thinning method can be further extended to the case of *multivariate*, or *multichannel*, point processes as well as marked point processes with general state including space-time point processes (MUSMECI and VERE-JONES, 1992; OGATA, 1998b). For simplicity consider the bivariate point process $\{N_1(\cdot), N_2(\cdot)\}$ whose joint intensity rate is given by $\{\lambda_1(t \mid F_t), \lambda_2(t \mid F_t)\}$, where F_t is the history of both components of the process. See HAWKES (1971) as an example of mutually interactive multivariate point processes. Note that the *superposition* of the components, $N_1(\cdot) + N_2(\cdot)$, is specified by the direct sum

$$\lambda(t \mid F_t) = \lambda_1(t \mid F_t) + \lambda_2(t \mid F_t) \tag{31}$$

of the components of the corresponding conditional intensity. Subsequently the simulation of the bivariate point process is carried out in a similar manner to the above algorithm except for the following remarks:

- A trial point $\tau = t_n + r_i + \tau_j$ is obtained as a univariate Poisson sample with intensity Λ_i, which is an upper bound of the superposed intensity (31) on the determined interval.
- The trial point τ has three choices; that is, accepted as the first component $N_1(\cdot)$ with probability $\lambda_1(\tau \mid F_\tau)/\Lambda$, or accepted as the second component $N_2(\cdot)$ with probability $\lambda_2(\tau \mid F_\tau)/\Lambda$, otherwise deleted.

A mathematical justification of the thinning simulation is presented in OGATA (1981) together with details of the algorithms and numerical performances.

A.3. The Maximum Likelihood and Minimum AIC Procedures

Despite theoretical results advocating the maximum likelihood estimation procedure for point processes (OGATA, 1978; OGATA and VERE-JONES, 1984; and KUTOYANTS, 1979, 1982), the modeling and its numerical implementation were not quite developed by the time. This was mainly due to the fact that neither numerical algorithms of the nonlinear optimization method were familiar nor was any suitable criterion available for the comparison of models. Currently, owing to the progress in computer technology, the major difficulties in the numerical aspect are disappearing.

Given a set of occurrence data $t_1, t_2, \ldots, t_n$ in an observed time interval $[0, T]$ and a parameterized conditional intensity $\lambda_\theta(t \mid F_t)$, the likelihood is written in the form

$$L_T(\theta \mid t_1, t_2, \ldots, t_n; 0, T) = \left\{ \prod_{i=1}^{n} \lambda_\theta(t_i \mid F_{t_i}) \right\} \exp\left\{ -\int_0^T \lambda_\theta(t \mid F_t)\, dt \right\}.$$

The maximum likelihood estimate of θ is the value of the parameter vector which maximizes the likelihood or its logarithm

$$\log L_T(\theta \mid t_1, \ldots, t_n; 0, T) = \sum_{i=1}^{n} \log \lambda_\theta(t_i \mid F_{t_i}) - \int_0^T \lambda_\theta(t \mid F_t)\, dt. \qquad (32)$$

If the integral in (32) can be expressed analytically in θ, then the gradients of the log-likelihood function can be easily obtained. In such case the maximization of the function can be carried out by using a standard nonlinear optimization technique such as in FLETCHER and POWELL (1963). See AKAIKE et al. (1984) for the FORTRAN program.

Assume that we have to choose the best model among proposed competing models. The *Akaike Information Criterion* (AKAIKE, 1974, 1977),

$$\text{AIC} = (-2)(\text{maximum log-likelihood}) + 2(\text{number of parameters})$$

is very suitable for such comparisons. Here "log" denotes natural logarithm, and a model with a smaller AIC is considered to be a better fit.

A.4. Linearly Parameterized Intensity Models

OGATA (1978, 1983b) recommended a systematic use of the following parameterizations for the conditional intensity function

$$\lambda_\theta(t \mid F_t) = \sum_{k=1}^{K} \theta_k \cdot Q_k(t \mid F_t), \tag{33}$$

$$\log \lambda_\theta(t \mid F_t) = \sum_{k=1}^{K} \theta_k \cdot Q_k(t \mid F_t), \tag{34}$$

where $\theta = (\theta_1, \ldots, \theta_K)$ and each $Q_k(t \mid F_t)$ is independent of the parameter θ. A main advantage of such parameterization is that the log-likelihood function (32) has at most one maximum, regardless of how the dimension of the parameter increases. This is because the second derivative of log-likelihood (the Hessian matrix) is everywhere negative-definite in θ. In the case of (33), if the analytic calculation of the integral $\int Q_k(t \mid F_t) \, dt$ is feasible, then the maximization algorithm for the log-likelihood function (32) is implemented efficiently. However, the major disadvantage of (33) is that the conditional intensity function may be negative for some values of t. The larger the number of parameters, the more readily this occurs. This causes difficulty in the numerical process of maximizing the log-likelihood, because the negative values in the conditional intensity function contributes to the seeming increase of log-likelihood due to the second term of (32).

In the log-linear case in (34), the analytic calculation of the integral is not generally feasible, except in particular cases (e.g., LEWIS, 1970). However, for the slowly varying intensity model such as the exponential polynomial rate for a trend (e.g., MACLEAN, 1974), numerical integration well approximates the integral so that the maximum likelihood procedure becomes feasible. This parameterization has the advantage of nonnegative value intensity without any constraint within the entire parameter space. A FORTRAN program EPTREN in the packages TIM-SAC-84 (AKAIKE et al., 1984) and SASeis (UTSU and OGATA, 1997) include automatic and objective analyses for trend and cyclicity of point process data by the use of the minimum AIC procedure. BERMAN and TURNER (1992) recommended the use of GLIM for an alternative calculation procedure of the maximum likelihoods for the models with the form of (33) and (34).

A.5. Residual Analysis of Point Processes

The AIC is useful for the comparison of competing models. However, having obtained the best model among those proposed, there is still the possibility of the existence of a better fitted model. The *residual analysis* will be useful for checking such a case (OGATA, 1988). Presume that point-process data $t_1, \ldots, t_N$ are simulated from the conditional intensity $\lambda(t \mid H_t)$ (see Appendix A.2 for the simulation algorithm by thinning). Then consider the integral

$$\Lambda(t) = \int_0^t \lambda(s \mid H_s) \, ds \tag{35}$$

of the conditional intensity function, which is an increasing function since $\lambda(s \mid H_s)$ is nonnegative. Consider the transformation of time $\tau = \Lambda(t)$ from t to τ so that the original occurrence data $t_1, \ldots, t_N$ are transformed 1-to-1 into $\tau_1, \ldots, \tau_N$. Then, it is known that $\tau_1, \ldots, \tau_N$ are distributed as the standard stationary Poisson process (i.e., with the constant intensity 1; see PAPANGELOU, 1972).

A similar transformation is considered using the estimated intensity $\lambda_{\hat{\theta}}(t \mid H_t)$ for the integrand, and we have the corresponding transformed data $\hat{\tau}_1, \ldots, \hat{\tau}_N$ which we call *residual process*. If the estimated conditional intensity is a good approximation of the true intensity, then the residual process is expected to behave like the standard stationary Poisson process. In explanation, a deviation of any statistical property of the residual process $\hat{\tau}_1, \ldots, \hat{\tau}_N$ from that expected from the stationary Poisson process implies the existence of features in the data $t_1, \ldots, t_N$ which were not considered by the model $\lambda_{\hat{\theta}}(t \mid H_t)$. Any conventional graphic tests for complete randomness, or the stationary Poisson, such as those in COX and LEWIS (1966), can be useful for such *residual analysis* (OGATA, 1988). When a model is reasonably good, the residual data can often provide new findings which were hard to discern from the original data.

REFERENCES

AKAIKE, H. (1974), *A New Look at the Statistical Model Identification*, IEEE Trans. Autom. Control *19*, 716–723.

AKAIKE, H., *On entropy maximization principle*. In *Applications of Statistics* (ed. Krishnaiah, P. R.) (North-Holland, Amsterdam 1977) pp. 27–41.

AKAIKE, H., OZAKI, T., ISHIGURO, M., OGATA, Y., KITAGAWA, G., TAMURA, Y. H., ARAHATA, E., KATSURA, K., and TAMURA, Y. (1984), *Time Series Analysis and Control Program Package*, *TIMSAC-84*, The Institute of Statistical Mathematics, Tokyo.

AKI, K. (1956), *A Review of Statistical Seismology*, Zisin (J. Seismol. Soc. Japan), Ser. 2, *8*, 205–228, in Japanese.

BERMAN, M., and TURNER, T. R. (1992), *Approximating Point-process Likelihoods*, Applied Statist. (J. Roy. Statist. Soc. C) *41*, 31–38.

BRILLINGER, D. (1988), *Some Statistical Methods for Random Process Data from Seismology and Neurophysiology*, The 1983 Wold Memorial Lectures, Ann. Statist. *16*, 1–54.

COX, D., and LEWIS, P. A. W., *The Statistical Analysis of Series Events* (Methuen, London 1966).

CREAMER, F. H., and KISSLINGER, C. (1993), *The Relation between Temperature and the Omori Decay Parameter for Aftershock Sequences near Japan*, EOS, 74, 43, Supplement, p. 417.

DALEY, D. J., and VERE-JONES, D., *A summary of the theory of point processes*. In *Stochastic Point Processes: Statistical Analysis, Theory and Appplications* (ed. Lewis, P. A. W.) (Wiley, New York 1972).

DE NATALE, G., MUSMECI, F., and ZOLLO, A. (1988), *A Linear Intensity Model to Investigate the Causal Relation between Calabrian and North-Aegean Earthquake Sequences*, Geophys. J. *95*, 285–293.

FISHER, N., *Statistical Analysis of Circular Data* (Cambridge University Press, Cambridge 1993).

FLETCHER, R., and POWELL, M. J. D. (1963), *A Rapidly Convergent Method for Minimization*, Comput. J. *6*, 163–168.

GUO, Z., and OGATA, Y. (1995), *Correlation between Characteristic Parameters of Aftershock Distributions in Time, Space and Magnitude*, Geophys. Res. Lett. *22*, 993–996.

GUO, Z., and OGATA, Y. (1997), *Statistical Relations between the Parameters of Aftershocks in Time, Space and Magnitude*, J. Geophys. Res. *102* (B2), 2857–2873.

HABERMANN, R. E. (1983), *Teleseismic Detection in the Aleutian Island Arc*, J. Geophys. Res. *88*, 5056–5064.

HABERMANN, R. E. (1988), *Precursory Seismic Quiescence: Past, Present, and Future*, Pure appl. geophys. *126*, 279–318.

HAWKES, A. G. (1971), *Point Spectra of Some Mutually Exciting Point Processes*, J. Roy. Statist. Soc. *B33*, 438–443.

HAWKES, A. G., and ADAMOPOULOS, L. (1973), *Cluster Models for Earthquakes–Regional Comparisons*, Bulletin of the International Statistical Institute *45*, Book 3, 454–461.

HAWKES, A. G., and OAKES, D. A. (1974), *A Cluster Process Representation of Self-exciting Process*, J. Appl. Probab. *11*, 493–503.

IMOTO, M., MAEDA, K., and YOSHIDA, A. (1999). *Use of statistical Models to Analyze Periodic Seismicity Observed for Clusters in the Kanto Region*, Central Japan, this volume.

ISHAM, V., and WESTCOTT, M. (1979), *A self-correcting Point Process*, Stoc. Proc. Appl. *8*, 335–348.

KAGAN, Y. Y. (1991), *Likelihood Analysis of Earthquake Catalogues*, J. Geophys. Res. *106*, 135–148.

KAGAN, Y. Y. (1992a), *Correlations of Earthquake Focal Mechanisms*, Geophys. J. Int. *110*, 305–320.

KAGAN, Y. Y. (1992b), *On the Geometry of an Earthquake Fault System*, PEPI *71*, 15–35.

KAGAN, Y. Y., and KNOPOFF, L. (1978), *Statistical Study of the Occurrence of Shallow Earthquake*, Geophys. J. R. Astron. Soc. *55*, 67–86.

KAGAN, Y. Y., and KNOPOFF, L. (1980), *Spatial Distribution of Earthquakes: The Two-point Correlation Function*, Geophys. J. R. Astron. Soc. *62*, 303–320.

KAGAN, Y. Y., and KNOPOFF, L. (1985), *The first-order Statistical Moment of the Seismic Moment Tensor*, Geophys. J. R. Astron. Soc. *81*, 429–444.

KAGAN, Y. Y., and JACKSON, D. D. (1998), *Spatial Aftershock Distribution: Effect of Normal Stress*, J. Geophys. Res. *103*, 24,453–24,465.

KAWASUMI, H. (1970), *Proofs of 69 Year Periodicity and Imminence of Destructive Earthquakes in Southern Kwanto District and Problems in the Countermeasures Thereof*, Chigaku Zasshi (J. Geography) *79(3)*, 115–138, in Japanese.

KENDALL, D. G. (1949), *Stochastic Processes and Population Growth*, J. Roy. Statist. Soc. *11*, 230–264.

KISSLINGER, C., and JONES, L. M. (1991), *Properties of Aftershocks in Southern California*, J. Geophys. Res. *96*, 11,947–11,958.

KITAGAWA, G., and AKAIKE, H. (1978), *A Procedure for the Modeling of Non-stationary Time Series*, Ann. Inst. Statist. Math. *30*, 351–363.

KUTOYANTS, Yu. A. (1979), *Local Asymptotic Normality for Processes of Poisson Type*, Izvest. Akad. Arm. Nauk. Ser. Matematika *14*, 3–20.

KUTOYANTS, Yu. A. (1982), *Multidimensional Parameter Estimation of the Intensity Function of Inhomogeneous Poisson Processes*, Probl. Control Inf. Theory *11* (4), 325–334.

LEWIS, P. A. W. (1970), *Remarks on the Theory, Computation and Application of the Spectral Analysis of Series of Events*, J. Sound Vib. *12*, 353–375.

LEWIS, P. A. W., and SHEDLER, G. S. (1979), *Simulation of Non-homogeneous Poisson Processes by Thinning*, Naval Res. Logistics Quart. *26*, 403–413.

LIPTZER and SHIRYAEV, *Statistics of Random Processes* (Springer-Verlag, Berlin 1978).

LIU, J., CHEN, Y., SHI, Y., and VERE-JONES, D. (1999), *Coupled Stress Release Model for Time-dependent Seismicity*, this volume.

MA, LI, and VERE-JONES, D. (1997), *Application of M 8 and Lin-Lin Algorithms to New Zealand Earthquake Data*, N.Z. J. Geol. Geophys. *40*, 77–89.

MATSUMURA, K. (1986), *On Regional Characteristics of Seasonal Variation of Shallow Earthquake Activities in the World*, Bull. Disas. Prev. Inst., Kyoto Univ. *36*, 43–98.

MATSU'URA, R. S. (1986), *Precursory Quiescence and Recovery of Aftershock Activities before Some Large Aftershocks*, Bull. Earthq. Res. Inst. Univ. Tokyo *61*, 1–65.

MACLEAN, C. J. (1974), *Estimation and Testing of an Exponential Polynomial Rate Function within the Non-stationary Poisson Process*, Biometrika *61*, 81–86.

MOGI, K. (1962), *On the Time Distribution of Aftershocks Accompanying the Recent Major Earthquakes in and near Japan*, Bull. Earthq. Res. Inst. Univ. of Tokyo *40*, 107–124.

MOGI, K. (1973), *Relationship between Deep and Shallow Seismicity in the Western Pacific Region*, Tectonophysics *17*, 1–22.

MUSMECI, F., and VERE-JONES, D. (1992), *A space-time Clustering Model for Historical Earthquakes*, Ann. Inst. Statist. Math. *44*, 1–11.

NISHIZAWA, O., LEI, X., and NAGATO, T., *Hazard function analysis of seismo-electric signals in Greece*. In *Electromagnetic Phenomena Related to Earthquake Prediction* (eds. Hayakawa, M., and Fujinawa, Y.) (Terra Publishing Company, Tokyo 1994) pp. 459–474.

OHTAKE, M. (1980), *Earthquake Prediction Based on the Seismic Gap with Special Item to the Oaxaca, Mexico Earthquake* (in Japanese with English summary), Report of the National Research Center for Disaster Prevention *23*, 65–110.

OGATA, Y. (1978), *Asymptotic Behaviour of the Maximum Likelihood Estimators for the Stationary Point Processes*, Ann. Inst. Statist. Math. A *30*, 243–261.

OGATA, Y. (1981), *On Lewis' Simulation Method for Point Processes*, IEEE Trans. Inform. Theory *IT-30*, 23–31.

OGATA, Y. (1983a), *Estimation of the Parameters in the Modified Omori Formula for Aftershock Frequencies by the Maximum Likelihood Procedure*, J. Phys. Earth *31*, 115–124.

OGATA, Y. (1983b), *Likelihood Analysis of Point Processes and its Applications to Seismological Data*, Bull. Int. Statist. Inst. *50*, Book 2, 943–961.

OGATA, Y. (1985), *Statistical Models for Earthquake Occurrences and Residual Analysis for Point Processes*, Research Memo. (Technical report), No. 288, Inst. Statist. Math., Tokyo.

OGATA, Y. (1987), *Long term dependence of earthquake occurrences and statistical models for standard seismic activity*. In *Suri Zisin Gaku (Mathematical Seismology) II* (ed. Saito, M.) (Cooperative Research Report 3, Inst. Statist. Math., Tokyo 1987) pp. 115–124 (in Japanese).

OGATA, Y. (1988), *Statistical Models for Earthquake Occurrences and Residual Analysis for Point Processes*, J. Amer. Statist. Assoc. *83*, 9–27.

OGATA, Y. (1989), *Statistical Model for Standard Seismicity and Detection of Anomalies by Residual Analysis for Point Process*, Tectonophysics *169*, 1–16.

OGATA, Y. (1992), *Detection of Precursory Relative Quiescence before Great Earthquakes through a Statistical Model*, J. Geophys. Res. *97*, 19,845–19,871.

OGATA, Y. (1993), *Space-time Modeling of Earthquake Occurrences*, Bull. Int. Statist. Inst. *55*, Contributed papers, Book 2, 249–250.

OGATA, Y. (1998a), *Quiescence Relative to the ETAS Model*, Zisin (J. Seismol. Soc. Japan), Ser. 2, *10*, 35–45 (in Japanese with English summary).

OGATA, Y. (1998b), *Space-time Point-process Models for Earthquake Occurrences*, Ann. Inst. Math. Statist. *50*, 379–402.

OGATA, Y., and AKAIKE, H. (1982), *On Linear Intensity Models for Mixed Doubly Stochastic Poisson and Self-exciting Point Processes*, J. Royal Statist. Soc. B *44*, 102–107.

OGATA, Y., AKAIKE, H., and KATSURA, K. (1982), *The Application of Linear Intensity Models to the Investigation of Causal Relations between a Point Process and Another Stochastic Process*, Ann. Inst. Statist. Math. *34B*, 373–387.

OGATA, Y., and SHIMAZAKI, K. (1984), *Transition from Aftershock to Normal Activity*, Bull. Seismol. Soc. Am. *74*, 1757–1765.

OGATA, Y., and VERE-JONES, D. (1984), *Inference for Earthquake Models: A Self-correcting Model*, Stoch. Processes Appl. *17*, 337–347.

OGATA, Y., and KATSURA, K. (1986), *Point-process Model with Linearly Parameterized Intensity for the Application to Earthquake Data*, J. Appl. Probab. *23A*, 291–310.

OGATA, Y., and KATSURA, K. (1988), *Likelihood Analysis of Spatial Inhomogeneity for Marked Point Patterns*, Ann. Inst. Statist. Math. *40*, 29–40.

OGATA, Y., and ABE, K. (1991), *Some Statistical Features of Long-term Variation of the Global and Regional Seismic Activity*, Int. Statist. Rev. *59*, 139–161.

OGATA, Y., IMOTO, M., and KATSURA, K. (1991), *Three-dimensional Spatial Variation of b Values of Magnitude Frequency Distribution beneath the Kanto District, Japan*, Geophys. J. Int. *10*, 135–146.

OGATA, Y., and KATSURA, K. (1991), *Maximum Likelihood Estimates of the Fractal Dimension for Spatial Patterns*, Biometrika *78*, 463–467.

OGATA, Y., and KATSURA, K. (1993), *Analysis of Temporal and Spatial Heterogeneity of Magnitude Frequency Distribution Inferred from Earthquake Catalogues,* Geophys. J. Int. *113*, 727–738.

OGATA, Y., MATSU'URA, R. S., and KATSURA, K. (1993), *Fast Likelihood Computation of Epidemic Type Aftershock-sequence Model*, Geophys. Res. Lett. *20* (19), 2143–2146.

OGATA, Y., UTSU, T., and KATSURA, K. (1995), *Statistical Features of Foreshocks in Comparison with other Earthquake Clusters*, Geophys. J. Int. *121*, 233–254.

OGATA, Y., UTSU, T., and KATSURA, K. (1996), *Statistical Discrimination of Foreshocks from other Earthquake Clusters*, Geophys. J. Int. *127*, 17–30.

OIKE, K. (1977), *On the Relation between Rainfall and the Occurrence of Earthquakes*, Bull Disas. Prev. Res. Inst. *20* (B1), 35–45 (in Japanese).

OMORI, F. (1894), *On the Aftershocks of Earthquake*, J. Coll. Sci. Imp. Univ. Tokyo *7*, 111–200.

OZAKI, T. (1978), *Maximum Likelihood Estimation of Hawkes' Self-exciting Point Processes*, Ann. Inst. Stat. Math. *30*, 145–155.

OZAKI, T., and TONG, H. (1975), *On the Fitting of Non-stationary Auto-regressive Models in Time Series Analysis*, Proceeding of the 8-th Hawaii Intern. Conf. on System Science, Western Periodical Company.

PAPANGELOU, F. (1972), *Integrability of Expected Increments of Point Processes and Related Random Change of Scale*, Trans. Amer. Math. Soc. *165*, 483–506.

RAMSELAAR, P. A. (1990), *The Mean Behaviour of the Ogata Earthquake Process*, Master's Thesis, Dept. Math., Univ. Utrecht.

REASENBERG, P. A. (1985), *Second-order Moment of Central California Seismicity*, J. Geophys. Res. *90*, 5479–5493.

REASENBERG, P. A. (1994), *Computer Programs ASPAR, GSAS and APROB for the Statistical Modeling of Aftershock Sequences and Estimation of Aftershock Hazard*, U.S.G.S. Open File Report 94–221.

REASENBERG, P. A., and MATTHEWS, M. V. (1988), *Precursory Seismic Quiescence: A Preliminary Assessment of the Hypothesis*, Pure appl. geophys. *126*, 373–406.

REASENBERG, P. A., and JONES, L. M. (1989), *Earthquake Hazard after a Mainshock in California*, Science *243*, 1173–1176.

REASENBERG, P. A., and JONES, L. M. (1990), *California Aftershock Hazard Forecast*, Science *247*, 345–346.

REASENBERG, P. A., and JONES, L. M. (1994), *Earthquake Aftershocks: Update*, Science *265*, 1251–1252.

RUBIN, I. (1972), *Regular Point Processes and their Detection*, IEEE Trans. Inform. Theory *IT-18*, 547–557.

RATHBUN, S. L. (1993), *Modeling Marked spatio-temporal Point Patterns*, Bull. Int. Statist. Inst. *55*, Book 2, 379–396.

SHIMAZAKI, K. (1971), *On Periodicity of Earthquake Occurrences*, Kagaku (Natural Sciences) *41*, Iwanami Publ. Co., Tokyo, 688–689 (in Japanese).

SCHUSTER, A. (1897), *On Lunar and Solar Periodicities of Earthquakes*, Proc. Roy. Soc. *61*, 455–465.

UTSU, T. (1957), *Magnitude of Earthquakes and Occurrence of their Aftershocks*, Zisin (J. Seismol. Soc. Japan), Ser. 2 *10*, 35–45 (in Japanese with English summary).

UTSU, T. (1961), *A Statistical Study on the Occurrence of Aftershocks*, Geophys. Mag. *30*, 521–605.

UTSU, T. (1969), *Aftershocks and Earthquake Statistics (I): Some Parameters which Characterize an Aftershock Sequence and their Interaction*, J. Faculty Sci., Hokkaido Univ., Ser. VIII *3*, 129–195.

UTSU, T. (1970), *Aftershocks and Earthquake Statistics (II): Further Investigation of Aftershocks and other Earthquakes Sequence Based on a New Classification of Earthquake Sequences*, J. Faculty Sci., Hokkaido Univ., Ser. VII *3*, 379–441.

U<small>STU</small>, T. (1971), *Aftershocks and Earthquake Statistics (III): Analyses of the Distribution of Earthquakes in Magnitude, Time, and Space with Special Consideration to Clustering Characteristics to Earthquake Occurrence (1)*, J. Faculty Sci., Hokkaido Univ., Ser. VIII *3*, 379–441.

U<small>TSU</small>, T. (1975), *Correlation between Shallow Earthquakes in Kwanto Region and Intermediate Earthquakes in Hida Region, Central Japan*, Zisin (J. Seismol. Soc., Japan) Ser. 2 *(28)*, 303–311 (in Japanese).

U<small>TSU</small>, T., and S<small>EKI</small>, A. (1955), *Relation between the Area of Aftershock Region and the Energy of the Mainshock*, Zisin (J. Seismol. Soc. Japan), Ser. 2, ii *7*, 233–240 (in Japanese).

U<small>TSU</small>, T., O<small>GATA</small>, Y., and M<small>ATSU'URA</small>, R. S. (1995), *The Centenary of the Omori Formula for a Decay Law of the Aftershock Activity*, J. Phys. Earth *43*, 1–33.

U<small>TSU</small>, T., and O<small>GATA</small>, Y. (1997), *Statistical analysis of seismicity*. In *Algorithms for Earthquake Statistics and Prediction, IASPEI Software Library 6*, 13-94, International Association of Seismology and Physics of the Earth's Interior in collaboration with the Seismological Society of America.

V<small>ERE</small>-J<small>ONES</small>, D. (1970), *Stochastic Models for Earthquake Occurrence (with discussion)*, J. Roy. Stat. Soc. B *32*, 1–62.

V<small>ERE</small>-J<small>ONES</small>, D. (1978), *Earthquake Prediction—A Statistician's View*, J. Phys. Earth *26*, 129–146.

V<small>ERE</small>-J<small>ONES</small>, D. (1985), *The Detection and Estimation of Periodicities in Point Process Data*, Technical Report, ISOR, Victoria Univ. of Wellington.

V<small>ERE</small>-J<small>ONES</small>, D., and D<small>AVIES</small>, R. B. (1966), *A Statistical Survey of Earthquakes in the Main Seismic Region of New Zealand, Part 2, Time Series Analyses*, N.Z. J. Geol. Geophys. *9*, 251–284.

V<small>ERE</small>-J<small>ONES</small>, D., and O<small>GATA</small>, Y. (1984), *On the Moments of a Self-correcting Process*, J. Appl. Probab. *21*, 335–352.

V<small>ERE</small>-J<small>ONES</small>, D., and O<small>ZAKI</small>, T. (1982), *Some Examples of Statistical Estimation Applied to Earthquake Data, 1. Cyclic Poisson and Self-exciting Models*, Ann. Inst. Statist. Math. *34B*, 189–207.

W<small>YSS</small>, M., and H<small>ABERMANN</small>, R. E. (1988), *Precursory Seismic Quiescence*, Pure appl. geophys. *126*, 319–332.

Y<small>AMANAKA</small>, Y., and S<small>HIMAZAKI</small>, K. (1990), *Scaling Relationship between the Number of Aftershocks and the Size of the Mainshock*, J. Phys. Earth *38*, 305–324.

Y<small>OSHIDA</small>, A. (1994), *Re-examination of the Correlation between Earthquakes in Kanto Region and Intermediate-depth Earthquakes in Hida Region, Central Japan*, Chigaku Zasshi (J. Geography) *103* (3), 201–206 (in Japanese).

Z<small>HAO</small>, Z., M<small>ATSUMURA</small>, K., and O<small>IKE</small>, K. (1989), *Precursory Change of Aftershock Activity before Large Aftershock: A Case Study for Recent Earthquakes in China*, J. Phys. Earth *37*, 155–177.

(Received July 31, 1998, revised December 27, 1998, accepted February 3, 1999)

To access this journal online:
http://www.birkhauser.ch

Pure appl. geophys. 155 (1999) 509–535
0033–4553/99/040509–27 $ 1.50 + 0.20/0

❘ Pure and Applied Geophysics

Representation and Analysis of the Earthquake Size Distribution: A Historical Review and Some New Approaches

Tokuji Utsu[1]

Abstract—The size distribution of earthquakes has been investigated since the early 20th century. In 1932 Wadati assumed a power-law distribution $n(E) = kE^{-w}$ for earthquake energy E and estimated the w value to be $1.7 \sim 2.1$. Since the introduction of the magnitude-frequency relation by Gutenberg and Richter in 1944 in the form of $\log n(M) = a - bM$, the spatial or temporal variation (or stability) of b value has been a frequently discussed subject in seismicity studies. The $\log n(M)$ versus M plots for some data sets exhibit considerable deviation from a straight line. Many modifications of the G-R relation have been proposed to represent such character. The modified equations include the truncated G-R equation, two-range G-R equation, equations with various additional terms to the original G-R equation. The gamma distribution of seismic moments is equivalent to one of these equations.

In this paper we examine which equation is the most suitable to magnitude data from Japan and the world using AIC. In some cases, the original G-R equation is the most suitable, however in some cases other equations fit far better. The AIC is also a powerful tool to test the significance of the difference in parameter values between two sets of magnitude data under the assumption that the magnitudes are distributed according to a specified equation. Even if there is no significant difference in b value between two data sets (the G-R relation is assumed), we may find a significant difference between the same data sets under the assumption of another relation. To represent a character of the size distribution, there are indexes other than parameters in the magnitude-frequency distribution. The η value is one of such numbers. Although it is certain that these indexes vary among different data sets and are usable to represent a certain feature of seismicity, the usefulness of these indexes in some practical problems such as foreshock discrimination has not yet been established.

Key words: Earthquake statistics, size distribution, b value, η value, AIC.

1. Historical Review

1.1. Early Studies in Japan

It has been recognized since the early years of seismology that smaller earthquakes are considerably more frequent than larger ones. Omori (1902) illustrated a table of the frequency distribution of maximum amplitudes recorded by a

---

[1] University of Tokyo and Institute of Statistical Mathematics, Tokyo, Japan.
Present address: 9-5-18 Kitami, Setagaya-ku, Tokyo 157-0067, Japan.

seismometer in Tokyo. If his data were plotted in a double-logarithmic diagram, they fit a straight line indicating a power-law distribution. However, the power-law relation for the distribution of amplitudes

$$n(A) = kA^{-m} \tag{1}$$

was first reported in a paper by ISHIMOTO and IIDA (1939) as a result of the observation with a newly designed seismograph.

ENYA (1908a) applied a lognormal distribution to the frequency of maximum velocities recorded at Tokyo station. ENYA (1908b) also tried to represent the distribution of earthquakes with respect to the radius R of the felt region. Before the introduction of the earthquake magnitude, R (or area S) of the felt region was only a measure of earthquake size routinely reported in the Bulletins of the Central Meteorological Observatory, Tokyo since 1885. However, the formula proposed by ENYA (1908b) for the distribution of R was too complicated to be of practical use. OMORI (1908) also discussed the frequency distribution of earthquakes with respect to the felt area S, although no specific distribution functions were suggested.

1.2. Power-law Distribution of Earthquake Energies

WADATI (1932) published a paper titled *"On the Frequency Distribution of Earthquakes."* This paper received slight attention because of its vague title and the language used. In his paper he assumed that the earthquake energy E has a distribution in the form

$$n(E) = kE^{-w} \tag{2}$$

(k and w are constants) and tried to estimate the w value from the observed frequency distribution of $S - P$ times recorded at the Tokyo station. The distribution of $S - P$ is controlled by the spatial distribution of earthquakes around the station, the attenuation of seismic waves, and the frequency distribution of earthquakes with respect to energy. He obtained $w = 1.7$ and $w = 2.1$ under the assumption that the hypocenters were distributed uniformly on a horizontal line and a horizontal surface, respectively. Only geometrical spreading was assumed for attenuation, however he noted that the w value might become smaller if the effect of absorption was included. It should be noted that Wadati's estimates of w are close to 5/3, now generally accepted for the index of the power-law distribution of seismic energies or moments. $w = 5/3$ corresponds to $b = 1$ in Equation (4), since $w = b/1.5 + 1$, where 1.5 is the coefficient of a well-known formula connecting the magnitude M and the seismic energy (or moment) E

$$\log E = 1.5M + \text{constant}. \tag{3}$$

In the present paper, $\log X$ denotes $\log_{10} X$, while $\ln X$ denotes $\log_e X$. We use the same notation for both moment and energy, because they are proportional.

1.3. Exponential Distribution of Earthquake Magnitudes

In the first paper on the instrumental magnitude scale, RICHTER (1935) noted that the number of shocks falls off very rapidly for the higher magnitudes. GUTENBERG and RICHTER (1941) suggested an exponential distribution for earthquake magnitude. The famous equation

$$\log n(M) = a - bM \tag{4}$$

was used by GUTENBERG and RICHTER (1944, 1949). The coefficient b usually takes a value around 1.0. Since then, this formula has been used by many investigators. The b value has been considered as an important parameter which characterizes the seismicity of a region.

Throughout this paper, $n(M)\,dM$ represents the frequency of earthquakes having magnitude between $M - dM/2$ and $M + dM/2$, and $N(M)$ represents the number of earthquakes with magnitude M and larger. If we use the earthquakes with magnitude M_z and larger, the density function for the G-R formula (4) is written in the form

$$f(X) = B \exp(-BX) \quad (X \geq 0) \tag{5}$$

where $X = M - M_z$ and $B = b \ln 10$.

1.4. Temporal and Spatial Variability and Stability of the b Value

Under the assumption that the magnitudes are distributed in accordance with the G-R formula, the b value is only the parameter which characterizes the distribution. If we have a complete magnitude data for earthquakes with magnitude M_z and larger, $M_1, M_2, \ldots, M_N$, the b value is usually calculated from the equation (UTSU, 1965)

$$B = E[X]^{-1} \quad \text{i.e.,} \quad b = (\log e)N \Big/ \sum_{i=1}^{N} (M_i - M_z) \tag{6}$$

where $E[\cdot]$ denotes the expectancy and N is the total number of earthquakes. This is the maximum likelihood estimate (MLE) of b (AKI, 1965).

Even if the above assumption is not valid, we can determine the b value from Equation (6). Under the above assumption the significance of the difference in b values between two earthquake groups can be tested by using the F distribution or more easily by using AIC (see Section 3).

The spatial or temporal variation of b value has been one of the frequently discussed topics in seismicity studies, since GUTENBERG and RICHTER (1949) estimated the b values for earthquakes occurring in various regions of the world. Numerous papers were published dealing with the b values or m values in Equation (1) ($m = b + 1$, ASADA et al., 1951). Some tried to relate the spatial variation to tectonics, degree of fracturing, material properties, degree of stress concentration, etc. Some tried to relate the temporal variation to changes in stress level, pore-fluid pressure, fracture growth condition, etc. which might be connected with the occurrence of large earthquakes. Recent papers include OGATA et al. (1991), FROHLICH and DAVIS (1993), OGATA and KATSURA (1993), KÁRNÍK and KLÍMA (1993), OKAL and KIRBY (1995), ÖNCEL et al. (1996), WIEMER and BENOIT (1996), WIEMER and McNUTT (1997), MOLCHAN et al. (1997), MORI and ABERCROMBIE (1997), WIEMER and WYSS (1997), WYSS et al. (1997) for spatial variations, and SMITH (1986), IMOTO (1987), JIN and AKI (1989), OGATA and ABE (1991), IMOTO (1991), HENDERSON et al. (1992, 1994), TRIFU and SHUMILA (1996) for temporal variations. It is also known that some volcanic earthquakes have quite unique magnitude distribution (e.g., OKADA et al., 1981; MAIN, 1987).

Uncertainties of the published b values are often quite large. The b values are affected by various factors; properties of the magnitude scale used, magnitude range of adopted data, method of determination, data completeness, etc. Care must be taken to accept the geographic variations (in this connection, see UTSU, 1971; FROHLICH and DAVIS, 1993; KAGAN, 1997).

Some authors are of the opinion that the b value for tectonic earthquakes in general does not differ significantly from a universal value. Some of the reported variations in b value must be real (e.g., earthquakes with normal faulting have larger b values, FROHLICH and DAVIS, 1993), however there may be many cases in which occurrence or non-occurrence of relatively few numbers of large events by chance causes an apparent variation in b value. The spatial stability of the magnitude distribution has been suggested or emphasized by SUZUKI (1959), RIZNICHENKO (1959), ALLEN et al. (1965), BLOOM and ERDMAN (1980), KAGAN (1991, 1997), among others. The temporal stability is also mentioned in some papers.

If there is no significant difference in b value between different earthquake groups, this does not always mean that the earthquakes have the same size distribution. We can calculate the b value for any earthquake group by the use of Equation (6), whether the magnitude distribution fits the G-R relation or not. We often find two earthquake groups for which the b values are nearly equal but the patterns of the magnitude distribution are quite different. If we assume a distribution function other than the G-R relation, we may find the significant difference between the two groups. We will discuss this problem in Section 3.

1.5. Modifications of the G-R Relation

The $\log n(M)$ versus M plots for some magnitude data exhibit considerable deviation from a straight line expected from the G-R relation. The deviation is either the convex type (Fig. 1, curve $\eta < 2$) or the concave type (curve $\eta > 2$). Many modified equations have been proposed to represent such data. The problem of selecting the equation best representing a given set of data will be discussed in Section 2.

1.5.1. Convex Type Equations. Most of the modified equations have been designed for the convex type distribution. They have the form

$$\log n(M) = a - bM - \phi(M) \quad (b > 0) \tag{7}$$

where $\phi(M)$ is an increasing function of M. Most of the functions proposed hitherto belong to one of the following four groups.

The first group uses a polynomial $\phi(M) = k(M - c)^n$ $(n = 1, 2, \ldots)$.

For $n = 1$, only the case in which $\phi(M)$ is truncated at $M = c$ ($\phi(M) = 0$ for $M < c$) is meaningful. This case corresponds to a two-range G-R relation

$$\log n(M) = a_1 - b_1 M \quad (M \leq c), \tag{8a}$$

$$\log n(M) = a_2 - b_2 M \quad (M \geq c). \tag{8b}$$

A condition $a_1 - b_1 c = a_2 - b_2 c$ is required for the continuity of $n(M)$ at $M = c$. Such two-range expression was used by GUTENBERG (1956), PACHEKO and SYKES (1992), OKAL and ROMANOWICZ (1994), SORNETTE *et al.* (1996), among others. TRIEP and SYKES (1997) used the two-range $\log N(M)$ versus M relation, which is different from Equations (8a,b) as $N(M)$ represents the cumulative number. This relation looks somewhat strange because it has a discontinuity of the gradient of the cumulative curve.

The truncated G-R equation

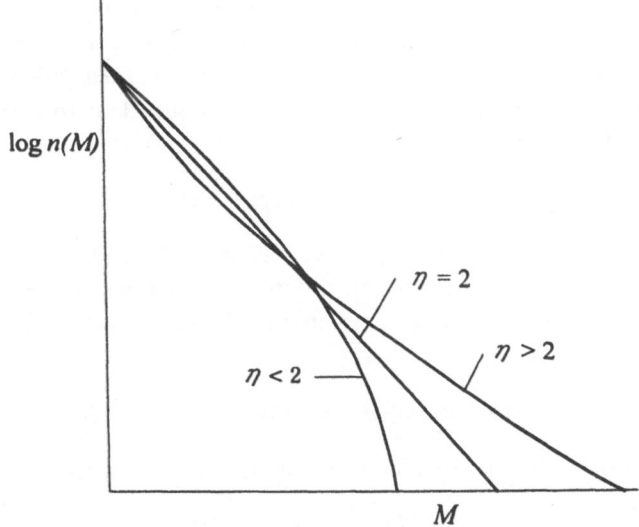

Figure 1
Schematic diagram of the magnitude distribution of earthquakes. Two types of the deviation from the G-R relation ($\eta = 2$), convex type ($\eta < 2$) and concave type ($\eta > 2$) are shown. For the definition of η, see Section 4.2.

$$\log n(M) = a - bM \quad (M \le c) \tag{9a}$$

$$n(M) = 0 \quad (M > c) \tag{9b}$$

may be a variation of Equations (8a,b), i.e., $b_2 \to \infty$. The truncated G-R equation was used by RIZNICHENKO (1964, 1966), PAGE (1968), CORNELL and VANMARCHE (1969), KAGAN (1969), OKADA (1970), COSENTINO and LUZIO (1976), COSENTINO et al. (1977), UTSU (1978), BERRILL and DAVIS (1980), among others. An equation used by LATOUSSAKIS and DRAKOPOULOS (1987), $N(M) = -A_1 + A_2 \exp(-A_3 M)$ is the same as the truncated G-R equation, because Equation (9a) can be transformed to $N(M) = 10^4 \{\exp(-BM) - \exp(-Bc)\}$ where $A = a - \log B$ and $B = b \ln 10$.

If $n = 2$, $n(M)$ becomes a normal distribution. This distribution was used by NIAZI (1964), NEUNHÖFER (1969), OLSSON (1986), and SPEIDEL and MATTSON (1993). This corresponds to a lognormal distribution of energy E, which was used by LOMNITZ (1964) and KAGAN (1969). Since no complete data are available below a certain magnitude level M_z (or energy level E_z), the distribution must be truncated at this level. This left-hand truncation makes the density function for $n(M)$ fairly complex and the maximum likelihood estimation of the parameters is not easy.

PURCARU (1975) considered the equation for the cumulative frequency

$$\log N(M) = a - bM - k(c - M)^3. \tag{10}$$

This does not belong to any of the four groups treated here.

The second group uses an exponential function $\phi(M) = k \exp(hM)$ $(k > 0, h > 0)$. SAITO et al. (1973) obtained Equation (11) for the frequency distribution of the size E of events generated by a branching model (equivalent to a site percolation model) proposed by OTSUKA (1972).

$$n(E) = kE^{-3/2} \exp(-aE) \tag{11}$$

where k and a are constants. This equation was derived also by VERE-JONES (1976, 1977) and MARUYAMA (1978) through different procedures. If the size is proportional to the energy and the energy is related to the magnitude M by Equation (3), this yields the equation

$$\log n(M) = a - 0.75M - k10^{1.5M}. \tag{12}$$

Since Equation (12) is constrained too tightly, a relaxed form

$$\log n(M) = a - bM - k10^{2bM} \tag{13}$$

has been considered. We shall call (13) the SAITO et al. equation.

KAGAN and KNOPOFF (1984) and KAGAN (1991, 1993, 1997) used an equation for seismic moment distribution, which has the form of a gamma distribution

$$n(E) = \kappa E^{-1-\beta} \exp(-E/E_m) \tag{14}$$

where κ, β, and E_m are constants. This is equivalent to a generalized form of Equation (12)

$$\log n(M) = a - bM - k10^{1.5M} \tag{15}$$

where $b = 1.5\beta$. We shall call Equation (15) the generalized SAITO et al. equation ($h = 1.5$), since this is a special case of a more general equation

$$\log n(M) = a - bM - k10^{hM}. \tag{16}$$

An equation was obtained by MAIN and BURTON (1984a, 1986) based on the entropy maximization principle. This equation which can be written as

$$n(M) = a \exp(-\lambda_1 M - \lambda_2 E) \tag{17}$$

is the same as the generalized SAITO et al. equation ($h = 1.5$), if moment E is converted to magnitude M using Equation (3).

An equation in the form

$$N(M) = \exp\{A - c \exp(BM)\} \tag{18}$$

was proposed by LOMNITZ-ADLER and LOMNITZ (1978, 1979). This can be transformed to

$$\log n(M) = a + bM - c10^{bM} \quad (b > 0). \tag{19}$$

Since the sign of the second term bM is plus, $n(M)$ decreases as M decreases for small M. Due to this unique character, the Lomnitz-Adler and Lomnitz equation fits better than most other equations to incomplete data sets in which small earthquakes are missing. Incomplete data sets should not be used unless special consideration is given such as described by OGATA and KATSURA (1993).

The third group uses a logarithmic function $\phi(M) = -k \log(c - M)$ for $M < c$. $n(M) = 0$ for $M > c$. UTSU (1971, 1978) proposed the equation

$$\log n(M) = a - bM + \log(c - M) \quad (M < c). \tag{20}$$

This is equivalent to the power-law distribution of energy with a logarithmic taper

$$n(E) = \kappa E^{-1-\beta} \log(E_m/E) \quad (E \le E_m) \tag{21}$$

where κ, β, and E_m are constants. The equation of MAKJANIĆ (1972, 1980) can be written as $N(M) = N\{(c - M)/(c - M_z)\}^{k+1}$. This is equivalent to

$$\log n(M) = a + k \log(c - M) \quad (M < c). \tag{22}$$

The equation proposed by PURCARU (1975)

$$\log N(M) = a - bM + k \log(c - M) \quad (M < c) \tag{23}$$

is not the generalization of Equation (20) or (22), since this represents the cumulative frequency.

The fourth group uses a function $\phi(M) = \log[1 - \exp\{-h(c - M)\}]$. $n(M) = 0$ for $M > c$. The equation used by ANDERSON and LUCO (1983) and MAIN and BURTON (1984b), $n(M) = A\{\exp(-BM) - \exp(-Bc)\}$ for $M < c$, can be written as

$$\log n(M) = a - bM + \log\{1 - 10^{-b(c - M)}\} \quad (M < c), \tag{24}$$

i.e., $h = b \ln 10$ in this case. This is a special case ($M_z = m$) of a more general distribution used by CAPUTO (1976)

$$n(M) = \lambda_1 \exp(-BM) \quad (M_z < M \leq m) \tag{25a}$$

$$n(M) = \lambda_2 \{\exp(-BM) - \exp(-Bc)\} \quad (m \leq M < c). \tag{25b}$$

Expressions in Caputo's paper are complicated but they are equivalent to (25a,b). Here, we call (24) the Caputo equation.

SEINO et al. (1989) used the general form of this group

$$\log n(M) = a - bM - \log[1 - \exp\{-h(c - M)\}] \quad (M < c). \tag{26}$$

It is interesting that this equation degenerates into the G-R relation (4), when $c \to \infty$, into the Utsu Equation (20) when $h \to 0$, and into the truncated G-R Equation (7) when $h \to \infty$. When $h = b \ln 10$, Equation (26) coincides with the Caputo Equation (24).

1.5.2. Concave Type Equations. The equations for the concave type distributions are few. Here we consider an equation

$$n(M) = n_1(M) + n_2(M) \tag{27a}$$

where

$$\log n_1(M) = a_1 - b_1 M \tag{27b}$$

$$\log n_2(M) = a_2 - b_2 M \quad (b_1 > b_2 > 0). \tag{27c}$$

It is readily seen that $n_1(M) = n_2(M)$ at $M = m$, where $m = (a_1 - a_2)/(b_1 - b_2)$. $n_1(M)$ and $n_2(M)$ predominates in (27a) for $M \ll m$ and $M \gg m$, respectively. We call (27a,b,c) the combined G-R equation. If this equation is applied to a data set of convex type, the MLEs of b_1 and b_2 become equal, indicating that the equation degenerates into a single G-R relation.

The two-range G-R relation (8a,b) is concave if $b_1 > b_2$. If the second range is truncated at $M = d(> c)$, i.e., $n(M) = 0$ for $M > d$ in Equation (8b), b_2 may take a negative value. It is possible that Equations (27b) and (27c) are both truncated at different magnitude levels (WARD, 1996).

1.5.3. Other Equations. Many other equations have been proposed in seismological papers, but the MLEs of the parameter values for most of these equations are not easy (though not impossible) to compute, because of the complicated form of the likelihood function. These include the truncated normal distribution (mentioned already), the truncated lognormal distribution (SACUIU and ZORILESCU, 1970; RANALLI, 1975), and various functions (SHLIEN and TOKSÖZ, 1970; MERZ and CORNELL, 1973, 1981; GUARNIERI-BOTTI et al., 1981; CORNELL and WINTERSTEIN, 1998; RUNDLE, 1993, etc.).

2. Parameter Estimation for the Frequency-magnitude Relations: Selection of the Most Suitable One

2.1. General

We assume that we have complete magnitude data for N earthquakes with magnitude M_z and larger, $M_1, M_2, \ldots, M_N$. These are considered as random samples from a population whose magnitude distribution is represented by a density function $f(X)$ $(X = M - M_z)$. The MLEs of the parameters θ_i $(i = 1, 2, \ldots, v)$ in the density function are the values for θ_i which maximize the log-likelihood function

$$\ln L = \sum_{i=1}^{N} \ln f(X_i) \quad (X_i = M_i - M_z). \tag{28}$$

The density function for Equations (4), (8), (9), (12), (13), (15), (16), (19), (20), (22), (24), (26), and (27) are shown below. The MLEs can be computed either by solving simultaneous equations $\partial \ln L / \partial \theta_i = 0$ $(i = 1, 2, \ldots, v)$, or by maximizing $\ln L$ by using some nonlinear optimization procedure.

Once the MLEs are obtained we can compute AIC (Akaike information criterion, AKAIKE, 1974)

$$\text{AIC} = -2 \ln L_m + 2v \tag{29}$$

where L_m is the maximum of L and v is the number of parameters in the density function $f(X)$.

When AIC values are calculated for each of these distributions for a given data set, we can judge which is the most suitable distribution for the data set. The one which provides the smallest AIC is the best, though the difference less than about 2 in AIC is considered insignificant.

2.2. Density Functions

In the following density functions (a) to (n), $B = b \ln 10$ and $C = c - M_z$ unless otherwise noted, where b and c are the parameters in the respective distribution and M_z is the threshold magnitude.

(a) Gutenberg–Richter Equation (4)

$$f(X) = B \exp(-BX) \quad (X \geq 0).$$

(b) Truncated G-R Equation (9)

$$f(X) = B \exp(-BX)/\{1 - \exp(-BC)\} \quad (0 \leq X \leq C).$$

(c) Utsu Equation (20)

$$f(X) = \exp(-BX)(C - X)B^2/\{\exp(-BC) + BC - 1\} \quad (0 \leq X \leq C).$$

(d) Makjanić Equation (22)

$$f(X) = (1 - X/C)^{1/k - 1}/(kC) \quad (0 \leq X \leq C).$$

(e) Saito $et\ al.$ Equation ($b = 0.75$) (12): $B = 0.75 \ln 10$ in (f).
(f) Saito $et\ al.$ Equation (13)

$$f(X) = 2BC^{-0.5} \exp\{-BX - C \exp(2BX)\}/G(0.5, C) \quad (X \geq 0)$$

where $C = c\ (\ln 10) \exp(2BM_z)$ and $G(\cdot, \cdot)$ denotes the incomplete gamma function.
(g) Generalized Saito $et\ al.$ Equation ($h = 1.5$) (15): $H = 1.5 \ln 10$ in (h).
(h) Generalized Saito $et\ al.$ Equation (16)

$$f(X) = HC^{-B/H} \exp\{-BX - C \exp(HX)\}/G(-B/H, C) \quad (X \geq 0)$$

where $C = c\ (\ln 10) \exp(HM_z)$, $H = h \ln 10$.
(i) Caputo Equation (24)

$$f(X) = B\{\exp(-BX) - \exp(-BC)\}/\{1 - (1 + BC) \exp(-BC)\} \quad (0 \leq X \leq C).$$

(j) Seino $et\ al.$ Equation (26)

$$f(X) = \exp(-BX)[1 - \exp\{-h(C - X)\}]/F \quad (0 \leq X \leq C)$$

where $F = \{1 - \exp(-BC)\}/B + \{\exp(-BC) - \exp(-hC)\}/(B - h)$.
(k) Lomnitz-Adler and Lomnitz Equation (19)

$$f(X) = BC \exp(BX) \exp[-C\{\exp(BX) - 1\}] \quad (X \geq 0)$$

where $C = c(\ln 10) \exp(BM_z)$.

(l) Combined G-R Equation (27)

$$f(X) = \lambda B_1 \exp(-B_1 X) + (1 - \lambda)B_2 \exp(-B_2 X) \quad (1 > \lambda > 0, X \geq 0).$$

(m) Two-range G-R Equation (8)

$$f(X) = \lambda B_1 \exp(-B_1 X) \quad (0 \leq X \leq X_c)$$

$$f(X) = \mu B_2 \exp(-B_2 X) \quad (X \geq X_c)$$

where $\lambda = \{1 - (1 - B_1/B_2)\}^{-1} \exp(B_1 X_c)$ and $\mu = \lambda(B_1/B_2) \exp(B_2 X_c)/\exp(B_1 X_c)$.

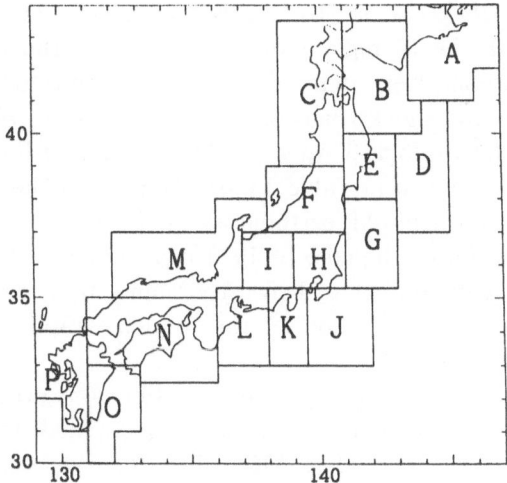

Figure 2
Index map of 16 regions of Japan.

To obtain MLEs of X_c, B_1, and B_2, we assume a certain value for X_c and calculate the B_1 and B_2 values which maximize

$$\ln L = N_1 \ln(\lambda B_1) - B_1 \sum_{X_i \le X_c} X_i + N_2 \ln(\mu B_2) - B_2 \sum_{X_i > X_c} X_i \qquad (30)$$

where N_1 is the number of events with $X \le X_c$ and N_2 is the number of events with $X > X_c$. We search a value for X_c which maximizes $\ln L$ by changing X_c in a systematic manner. Although the number of the parameters v is 3, the penalty term of AIC must be larger than $2v$ ($=6$) by α, because the parameter X_c represents the changing point of the slope (for this problem, see OGATA, 1992). The increment of the penalty α is about 6 at most.

(n) Truncated two-range G-R Equation

$$f(X) = \lambda B_1 \exp(-B_1 X) \quad (0 \le X \le X_c)$$

$$f(X) = \mu B_2 \exp(-B_2 X) \quad (X_c \le X \le C)$$

where $\lambda = [1 - \exp(-B_1 X_c) + [1 - \exp\{(B_2(C - X_c)\}](B_1/B_2) \exp(-B_1 X_c)]^{-1}$ and $\mu = \lambda(B_1/B_2) \exp(B_2 X_c)/\exp(B_1 X_c)$. The same comment on the penalty for AIC as (m) is needed in this case.

2.3. Results from Selected Data Sets

To demonstrate the applicability of our approach, the following data sets (2.3.1–2.3.4) are analyzed.

2.3.1. Shallow Earthquakes in 16 Regions of Japan (1926–1997, $0 \le h \le$ 100 km). We deal with the 16 regions of Japan (A to P shown in Fig. 2). The boundaries of the regions were determined with seismotectonic considerations. They do not cross the aftershock zones of major earthquakes. Magnitudes are M_J given by the Japan Meteorological Agency (JMA). The threshold magnitudes M_z differ among regions as shown in Table 3 (Section 3.1). The largest earthquake in this period was the Tokachi-oki earthquake of 1952 in region A ($M_J = 8.2$, $M_w = 8.0$), or the Sanriku earthquake of 1933 in region D ($M_J = 8.1$, $M_w = 8.4$). There is a small systematic difference between M_J and M_w (or M_s) (UTSU, 1982), so care is needed in the comparison of the results, with the results from the worldwide data using M_w or M_s.

The MLEs of the parameters in the 14 density functions (a) to (n) presented in Section 2.2 and the corresponding AIC values are computed. Here the results for region B are shown in Figure 3. The AIC values for each distribution for each region are shown in Figure 4, where δAIC represents the difference from the smallest AIC for each region. From this figure we notice the following points.

(1) The G-R relation (a) is most suitable for 8 regions.

(2) For the remaining 8 regions except region C, one of the convex equations provides the smallest AIC. For 6 regions, δAIC for the G-R relation is larger than about 2 or more.

(3) For 11 regions δAIC for the truncated G-R relation (b) is less than about 1. No other distributions perform so well. The mean of δAIC for (b) is 1.94, which is the smallest among the 14 equations tested. The second smallest mean δAIC, 2.57, is attained by the Utsu equation (c). Since mean values of δAIC for (d), (f), (g), (i), (k) fall in the range 2.61–2.72, it can be said that no appreciable difference in the performance seems to exist among these 6 two-parameter equations.

(4) One-parameter equations (a) and (e) and three-or-more-parameter equations (h), (j), (l), (m), and (n) have larger mean δAIC.

(5) In computing the parameters, we often encounter the case in which some parameter value increases infinitely or converges to zero during the iteration. This means that the equation degenerates into the G-R relation or another equation with fewer parameters.

2.3.2. Shallow Earthquakes in the World (1904–1980, $M_s \ge 7.0$). Shallow earthquakes in the world (depth less than about 65 km). The magnitudes are M_s given by ABE (1981) with corrections by ABE (1984), ABE and NOGUCHI (1983a,b). M_s for great earthquakes ($M_w \ge 8.5$) is replaced by M_w taken mostly from KANAMORI (1977). The largest one is the M_w 9.5 Chilean earthquake of 1960. Other corrections (e.g., PACHECO and SYKES, 1992) are not considered.

The whole data (W) are divided into two groups (H and L). Group H includes the quakes in the high-latitude zone (south of 38°S and north of 38°N) and group L those in the low-latitude zone (38°S to 38°N). This is to reconfirm the result by MOGI (1979) who demonstrated that the magnitude distribution is quite different between high- and low-latitude zones. Mogi selected the zone boundary at 40°S and 40°N, but here we use 38°S and 38°N to include the 1960 Chilean sequence in the high latitude group. This sequence is located south of 40°S in the catalog used by Mogi, however they are located north of 40°S in the catalog used here.

The AIC values for the 14 equations (a) to (n) for groups W, H, and L are shown in Table 1. The smallest AIC values and the AIC values not different from the smallest value by 0.5 are shown in bold italic letters.

In Figure 5 (left) the magnitude distributions are shown for group W, H, and L. The curves for the G-R equation and equation of the smallest AIC are drawn in the figure.

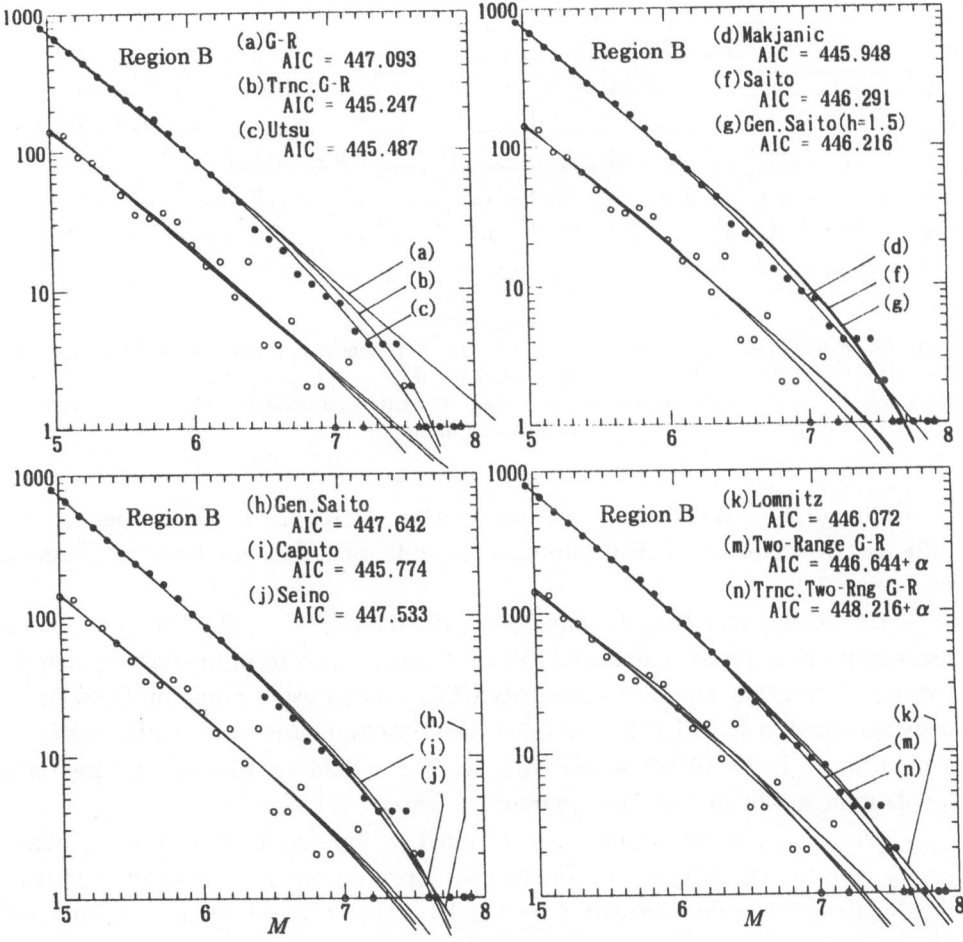

Figure 3

Magnitude-frequency diagrams for shallow earthquakes in region B, 1926–1997. Solid and open circles represent $N(M)$ and $n(M)\,dM$ ($dM = 0.1$), respectively. Three different curves fitted to the data are drawn in each diagram.

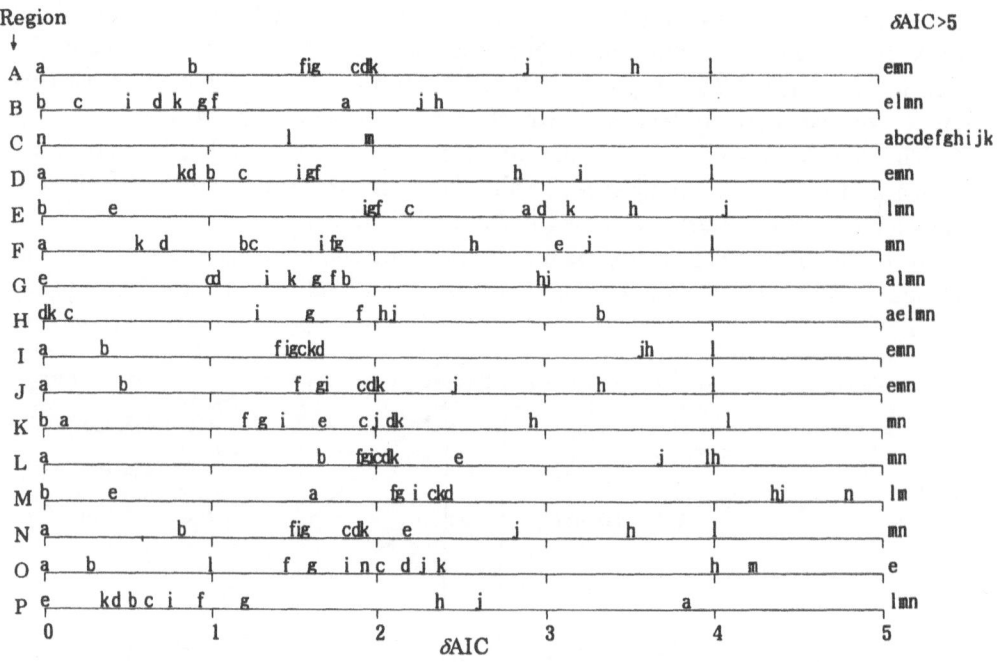

Figure 4

δ AIC for 14 distributions (a) to (n) (see Section 2.2) for 16 regions A to P (see Fig. 2). δ AIC represents
the difference in AIC from the smallest AIC for each region. If two or more AIC values are nearly equal,
the letters are slightly displaced so as not to overlap. The letters on the right (δ AIC > 5) are arranged
in alphabetical order.

Although the two b values, $b = 1.040$ for H, $b = 1.102$ for L, are not significantly
different, the shape of the distribution is very different. This point will be discussed
in Section 3.

2.3.3. Shallow and Deep Earthquakes in the World (1977–1997, $M_w \geq 5.50$). All
magnitudes used are M_w calculated to two decimal places from the seismic moment
(M_0 (dyn-cm) in the Harvard University CMT catalog using Equation (3) with the
constant equal to 16.1. The largest one is the 1977 Sumba Is. earthquake which has
a moment of 3.59×10^{28} dyn-cm ($M_w = 8.31$). The data are divided into high and
low-latitude groups in the same manner as before.

Table 2 lists the AIC values. For W and L groups the G-R equation provides
a very poor fit. The AIC values for the G-R equation differ by more than 24 from
the smallest AIC values provided by the Utsu and Caputo equations. The AIC
values for the other equations are larger by more than about 2. For H group the
truncated G-R and Utsu equations give the smallest AIC.

Figure 5 (right) shows the magnitude distributions for groups W, H, and L. The
curves for the G-R and Utsu equations are drawn.

Table 1

AIC values for shallow earthquakes of $M_s \geq 7.0$ occurring during 1904–1980 in the whole world (W) and high- and low-latitude zones (H and L) for 14 equations (a) to (n) given in Section 2.2. For values with asterisk, increments of penalty of about 6 (at most) should be added to compare with other AIC values. N is the number of earthquakes

Zone (N)	(a) G-R	(b) T. G-R	(c) Utsu	(d) Mak	(e) Sai-1	(f) Sai-2	(g) Sai-3
W (764)	134.40	133.65	132.01	*130.36*	179.63	134.52	134.06
H (232)	*60.90*	61.82	62.91	62.90	77.42	62.62	62.73
L (532)	74.95	*57.72*	*57.35*	58.71	58.74	58.05	57.85

Zone	(h) Sai-4	(i) Capu	(j) Seino	(k) Lom	(l) C. G-R	(m) R. G-R	(n) TR. G-R
W	132.15	133.66	134.48	*130.13*	138.40	129.59*	130.61*
H	64.90	62.80	63.82	63.06	64.64	64.18*	63.98*
L	59.83	*57.45*	59.35	60.45	78.95	60.51*	60.81*

2.3.4. Aftershocks of the 1995 Hyogoken–Nanbu (Kobe) Earthquake ($M_J \geq$ 2.5). This devastating earthquake (January 16, 1995, $M_J = 7.2$) produced relatively weak but remarkably regular aftershock activity. The data are taken from the preliminary catalog of JMA for the first 1,000 days. 4157 shocks of focal depths less than 40 km (most shocks are less than 20 km) and of $M_J \geq 2.0$ occurred in the quadrangular region defined by the four points (34.55°N, 134.65°E), (35.0°N, 135.3°E), (34.75°N, 135.55°E), and (34.3°N, 134.9°E). These shocks are regarded as

Table 2

AIC values for earthquakes of $M_w \geq 5.50$ occurring during 1977–1997 in the whole world (W) and high- and low-latitude zones (H and L) for 14 equations (a) to (n). For values with asterisk, see Table 1

Zone (N)	(a) G-R	(b) T. G-R	(c) Utsu	(d) Mak	(e) Sai-1	(f) Sai-2	(g) Sai-3
W (7340)	2746.05	2720.71	*2714.38*	2719.66	2754.86	2717.19	2716.16
H (1580)	659.80	*654.24*	*654.58*	655.16	700.08	655.45	655.57
L (5760)	2087.33	2068.69	*2062.65*	2067.46	2053.65	2064.46	2063.15

Zone	(h) Sai-4	(i) Capu	(j) Seino	(k) Lom	(l) C. G-R	(m) R. G-R	(n) TR. G-R
W	2717.13	*2714.48*	2715.08	2720.77	2750.05	2724.60*	2714.66*
H	657.18	655.98	656.72	655.17	663.80	657.19*	656.94*
L	2064.12	*2062.39*	2063.19	2068.57	2091.33	2068.30*	2063.62*

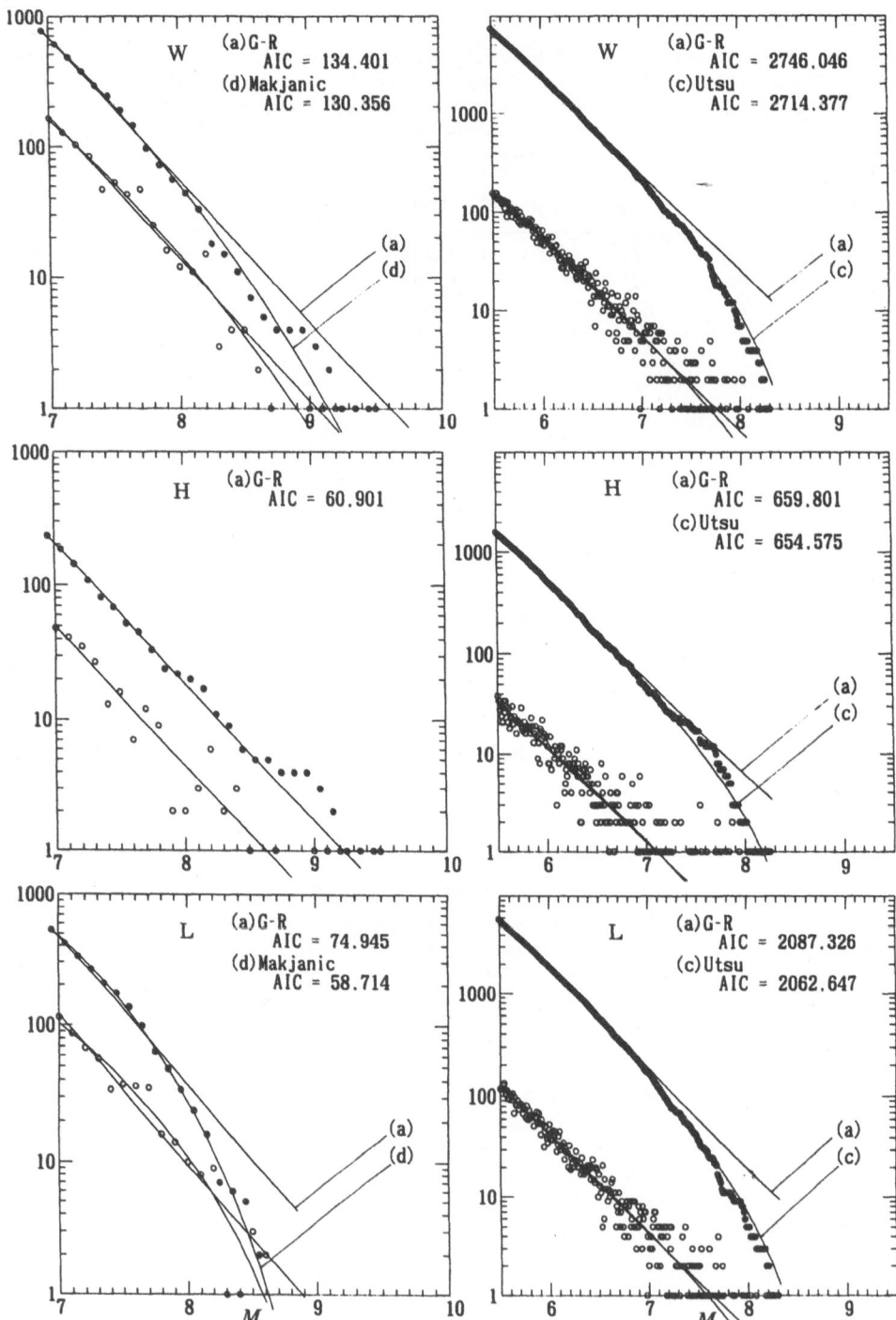

aftershocks in this study. The magnitude distribution of the aftershocks of $M_J \geq 2.5$ is studied, dividing the sequence into four periods I (0.1–1 day), II (1–10 days), III (10–100 days), and IV (100–1000 days).

The G-R relation provides the smallest AIC for periods II and III. For period I, the Lomnitz-Adler and Lomnitz equation has the smallest AIC. Perhaps some shocks of M near 2.5 are missing. For $M_J \geq 3.0$ the truncated G-R equation has the smallest AIC. For period IV, the Makjanić equation provides the smallest AIC, but the AIC values for the Utsu, Caputo, Lomnitz-Adler and Lomnitz, and generalized Saito ($h = 1.5$) equations do not differ by more than 0.5.

The b value of the G-R relation for periods I, II, III, and IV are 1.096 ($N = 154$, $M_J \geq 3.0$), 1.104 ($N = 455$, $M_J \geq 2.5$), 1.230 ($N = 356$), and 1.061 ($N = 258$), respectively. A significant difference in b values is not found among the four periods (using the method described in Section 3). The temporal stability of the magnitude-frequency relation in this aftershock sequence is also supported by the nearly constant p value for different threshold magnitudes. The p value is an index in the modified Omori formula introduced by Utsu (1961). The MLEs of p for aftershocks with $M_J \geq 2.0$, $M_J \geq 2.5$, $M_J \geq 3.0$, $M_J \geq 3.5$, $M_J \geq 4.0$ and $M_J \geq 4.5$ are 1.134, 1.160, 1.116, 1.169, 1.301, and 1.197, respectively (the data between 0.1 day and 1000 days from the mainshock are used). Such magnitude stability in aftershock sequences was first pointed out by Utsu (1962).

3. Difference in the Size Distribution Between Two Groups of Earthquakes

3.1. Significance Test of the Difference in b Value Between Two Groups .

When we obtain fairly different MLEs of b, b_1 and b_2 ($b_1 > b_2$), for two groups of earthquakes, we have the problem of deciding whether this difference is statistically significant or not. If the earthquakes in both groups are random samples from the same population obeying the G-R relation, b_1/b_2 has the F distribution with $2N_1$ and $2N_2$ degrees of freedom (N_1 and N_2 are the number of earthquakes in each group). A significance test using this property was introduced by Utsu (1966).

A similar test can be performed by using AIC (Utsu, 1992). We use two hypotheses, a null hypothesis that the two groups have the same b value (MLE of b for the combined group is b_0) and an alternative one that the b values are

---

Figure 5

Magnitude-frequency distribution for world earthquakes. Solid and open circles represent $N(M)$ and $n(M) \, dM$, respectively. Top: the whole world. Center: the high-latitude zone, Bottom: the low-latitude zone. Left: Shallow earthquakes of $M_s \geq 7.0$ ($dM = 0.1$), 1904–1980. M_s for earthquakes of $M_w \geq 8.5$ has been replaced by M_w. For the high-latitude zone, the Makjanić equation degenerates into the G-R equation as $c \to \infty$. Right: Earthquakes of all depths, $M_w \geq 5.50$ ($dM = 0.01$), 1977–1997.

Table 3

b values for 16 regions A to P in Japan during 1926–1997. The significance of the difference between every combination of regions is shown by ◎ (highly significant difference), ○ (significant difference), and × (no significant difference)

N	M_z	b	Region (A to P)
224	5.5	0.957	A
806	5.0	0.896	× B ◎: ΔAIC ≥ 5
127	5.0	1.072	× × C ○: 5 > ΔAIC ≥ 2
195	5.7	1.067	× ○ × D ×: ΔAIC < 2
185	5.0	0.761	○ ○ ◎ ◎ E
183	4.5	0.879	× × × × × F
290	5.5	0.932	× × × × ○ × G
820	4.5	0.934	× × × × ○ × × H
185	4.5	1.014	× × × × ◎ × × × I
158	5.2	1.043	× × × × ◎ × × × × J
139	5.0	0.919	× × × × × × × × × × K
78	5.0	0.913	× × × × × × × × × × × L
311	4.5	0.797	○ × ◎ ◎ × × × ○ ○ ◎ × × M
461	4.5	0.819	× × ○ ◎ × × × ○ ○ ○ × × × N
217	4.5	0.952	× × × × ○ × × × × × × × ○ × O
96	4.5	0.806	× × ○ ○ × × × × × ○ × × × × × P

different (MLEs of b are b_1 and b_2). AIC for the former and latter hypothesis is denoted by AIC_0 and AIC_{12}, respectively. It is easy to show that

$$AIC_0 = -2(N_1 + N_2)\ln(b_0 \ln 10) + 2(N_1 + N_2) + 2 \tag{31}$$

$$AIC_{12} = AIC_1 + AIC_2 = -2N_1 \ln(b_1 \ln 10) + 2N_1 - 2N_2 \ln(b_2 \ln 10) + 2N_2 + 4. \tag{32}$$

If AIC_{12} is significantly smaller than AIC_0, we can reject the null hypothesis and believe that the two groups have different b values. Usually the difference in AIC is considered significant if ΔAIC ($= AIC_0 - AIC_{12}$) exceeds about 2. If $\Delta AIC > 5$, the difference is highly significant. If the present case, by using Equation (6), ΔAIC can be written as

$$\Delta AIC = -2(N_1 + N_2)\ln(N_1 + N_2) + 2N_1 \ln(N_1 + N_2 b_1/b_2)$$
$$+ 2N_2 \ln(N_1 b_2/b_1 + N_2) - 2. \tag{33}$$

It is noted here that we can test the two data sets with different magnitude threshold M_z, because the G-R relation we assume is perfectly self-similar.

As an example, Table 3 shows the result of the test of the difference in b value among the regions A to P in Japan (Section 2.3.1). It is seen that most of the double and single circles are related to the small b values in regions E and M.

3.2. Difference in Size Distribution by Using a Formula Other than the G-R Relation

It is possible that the MLEs of b for two groups are nearly equal (no significant difference based on the test under the assumption of the G-R relation), but the patterns of the distribution are quite different. The distributions for high- and low-latitude zones (Section 2.3.2, Fig. 5 left) provide a good example. In this case, putting $b_1 = 1.1023$, $N_1 = 532$, $b_2 = 1.0398$, $N_2 = 232$ in Equation (33), we obtain $\Delta AIC = -1.45$ (this is confirmed from $\Delta AIC = AIC_0 - AIC_1 - AIC_2$ using the value shown in Table 1). This ΔAIC is smaller than 1 (even smaller than 0). Therefore we conclude that the b values for the two zones do not differ significantly.

Since the whole world's data fit the Makjanić equation best, we use this equation as representative of the population. From Table 1 $AIC_0 = 130.36$, $AIC_1 = 62.90$, $AIC_2 = 58.71$, then $\Delta AIC = 8.75$. Since this is larger than 5, the difference between H and L groups is highly significant. This indicates that the comparison of the b values of the G-R relation only is not always enough to find a variation in the size distribution. ΔAIC values for other distributions except (1) are also larger than 5.

To test the significance of the difference between H and L groups for the data set of 1977–1997 (Section 2.3.3, Fig. 5 right), ΔAIC is calculated for the 14 formulas (a) to (n). We find that ΔAIC is smaller than 2 for all formulas (smaller than 0 in most cases). This indicates that there is no significant difference under the assumption of any of the 14 formulas. The striking difference between H and L groups found for the data set of 1904–1980 is mainly due to the occurrence of several great earthquakes during 1952–1965.

4. Other Indexes for the Earthquake Size Distribution

4.1. Examples of the Indexes

The character of the size distribution of earthquakes is very often indicated by the b value of the G-R relation, or in some cases by the parameter values in other magnitude-frequency relations. Some investigators suggested the use of indexes other than these parameters. These are the η value (UTSU, 1978), the H value (OUCHI and YOKOTA, 1979), the R index (KAYANO, 1982), the C value (OKUDA et al., 1992) among others.

The η value indicates the degree of departure of the $\log n(M)$ versus M plots from a straight line (the G-R relation) to either concave side or convex side (see Fig. 1). The H, R, and C values provide some measure to indicate the diversity of the earthquake sizes or the deviation from the G-R relation. Here the η value and the R index will be discussed in detail.

4.2. η Value

The η value was introduced by UTSU (1978). It is defined by

$$\eta = E[X^2]/E[X]^2 \tag{34}$$

where $E[\cdot]$ denotes the expectancy and $X = M - M_z$ (M_z is the threshold magnitude). The theoretical value of η for the G-R relation (a) is 2. $\eta < 2$ for concave distributions and $\eta > 2$ for convex distributions. For example, theoretical η for the truncated G-R relation (b) is given by

$$\eta = [2 - BC(BC+2)/\{\exp(BC)-1\}]/[1 - BC/\{\exp(BC)-1\}] \tag{35}$$

and for the Utsu equation (c)

$$\eta = \frac{\{\exp(-BC)(B^2C^2+4BC+6)+2BC-6\}\{\exp(-BC)+BC-1)\}}{\{\exp(-BC)(BC+2)+BC-2\}^2}. \tag{36}$$

These are smaller than 2 for any positive values of B and C. As an example of concave distributions, η for the combined G-R equation (l) is given by

$$\eta = 2\{\lambda B_1^{-2}+(1-\lambda)B_2^{-2}\}/\{\lambda B_1^{-1}+(1-\lambda)B_2^{-1}\}^2 \tag{37}$$

which is larger than 2 if $B_1 \neq B_2$.

It is apparent that earthquake swarms usually have smaller η values compared with mainshock–aftershock sequences. UTSU (1988) suggested that foreshock sequences tend to have smaller η values as compared with the aftershock sequences of the same mainshocks. Examination of more data indicates that this property of foreshock sequences is not so clear, as will be shown in the next section. η values were also used by ZHANG and HUANG (1990) and OKUDA et al. (1992).

The formula (6) for computing the b value by UTSU (1965) was obtained by the method of moment, which equates the theoretical and empirical first moments. If we use the nth moment, we obtain another equation for estimating b

$$b_{(n)} = (\log e)\{N(n!)/\Sigma(M_i - M_z)^n\}^{1/n} \tag{38}$$

$b_{(1)}$ is identical to the MLE of b. Since the weight given to the high end range of magnitude increases with increasing n, $b_{(n)} - b_{(1)}$ is positive and increases with n for the convex distribution. We obtain $\eta = 2(b_{(1)}/b_{(2)})^2$ from (38).

4.3. R Index

The R index (relative entropy) used by KAYANO (1982) is defined by

$$R = \left(\sum_{i=1}^{K} p_i \log_2 p_i\right)\Big/\log_2 K, \quad p_i = E_i\Big/\sum_{i=1}^{K} E_i \tag{39}$$

where E_i is the energy of the ith largest earthquake in a sequence. K is an integer larger than 1. If $E_1 = E_2 = \cdots = E_K$, $R = 1$. If $E_1 \gg \Sigma_{i=2}^{K} E_i$, $R \approx 0$. Usually R is large for swarms and small for mainshock–aftershock sequences.

4.4. b, η, and R Values for Earthquake Sequences

Here a result of the study of b, η, and R values for earthquake sequences in Japan is shown. We call the largest shock in a sequence "the mainshock," and the shocks occurring before and after the mainshock, "foreshocks" and "aftershocks," respectively. The method for identifying earthquake sequences is the same as MBC described in OGATA et al. (1995).

The earthquakes of $M \geq 3.0$ listed in the JMA catalog for the period from January 1926 through September 1997 are used as the database. There are 432 sequences consisting of 10 or more shocks of $M \geq 3.0$, whose mainshock magnitude is 5.0 or larger. Among these 432 sequences, 48 sequences include 10 or more foreshocks of $M \geq 3.0$, and 326 sequences include 10 or more aftershocks of $M \geq 3.0$. The b, η, and R values have been computed using the largest 10 shocks in each whole sequence (W), in each foreshock sequence (F), and in each aftershock sequence (A). These foreshock sequences are divided into two groups, (F1 and F2) based on the magnitude difference between the mainshock and the largest fore-shock. The difference is larger than 0.45 for F1 and smaller than 0.45 for F2.

It is natural that the b and R values are small and the η value is large for W group, because the mainshock usually has a magnitude exceedingly larger than those of the other shocks in the same sequence. There seems to be some difference between F1 and F2 groups, although more data should be collected to draw a reliable conclusion. Many tables similar to Table 4 have been prepared for various combinations of M_m (lowest limit for mainshock magnitude), M_z (lowest limit of foreshock and aftershock magnitudes), δM (magnitude difference for dividing into F1 and F2), and K (the number of the largest earthquakes used in computing b, η, and R values), but no particularly interesting results have been found.

Table 4

Mean b, η, and R values (± standard deviation) for foreshock sequences (F1 and F2), aftershock sequences (A), and the whole sequences (W) which include foreshocks, the mainshock, and aftershocks. $M_m = 5.0$, $M_z = 3.0$, $dM = 0.45$, and $K = 10$

Group	N_G	$\bar{b}$	$\bar{\eta}$	$\bar{R}$
F1	35	0.991 ± 0.354	1.660 ± 0.356	0.646 ± 0.211
F2	13	0.746 ± 0.173	1.833 ± 0.432	0.469 ± 0.234
W	432	0.733 ± 0.413	2.055 ± 0.357	0.255 ± 0.241
A	326	1.024 ± 0.217	1.706 ± 0.498	0.592 ± 0.215

For the purpose of probabilistic earthquake prediction, the earliest K shocks of $M \geq M_z$ forming a sequence must be used. The algorithm MBC is not adequate for this purpose because it uses the information pertaining to the mainshock. This approach requires a rather complex procedure and the final results are not yet achieved.

5. Conclusions

In the history of the study of the size distribution of earthquakes, introduction of the power-law distribution of earthquake energy by WADATI (1932) and the exponential distribution of earthquake magnitude by GUTENBERG and RICHTER (1941, 1944) is of prime importance. These two distributions are equivalent if a linear relationship between the logarithm of energy and the magnitude is accepted. The power-law distribution of amplitude by ISHIMOTO and IIDA (1939) is also equivalent to the above distributions under some natural assumptions.

The $\log n(M)$ versus M plots or the $\log n(E)$ versus $\log E$ plots (E denotes earthquake energy or moment) for some data sets display considerable curvature, especially near the high end of magnitude range. To represent such data, natural modification of the power-law distribution is the truncation at some level E_m (i.e., $n(E) = 0$ for $E > E_m$), or the multiplication of an exponential taper $\exp(-E/E_m)$ or a logarithmic taper $\log(E_m/E)$ (for $E \leq E_m$). In the magnitude domain, these tapers yield the generalized Saito $et\ al.$ equation ($h = 1.5$) and the Utsu equation, respectively.

In addition to the above, various modifications of the G-R relation have been proposed. The question of which is the most suitable relation among these candidates for a given set of data can be answered by using AIC. Application of this method to many data sets indicates that the original G-R relation is the most suitable (has the smallest AIC) for some data sets, although for other data sets one of the modified formulas is found to be the most suitable. In some cases several modified formulas have AIC within 1 unit of the smallest AIC. There is no appreciable difference in performance between these formulas and the most suitable one.

To establish the spatial or temporal variation of b value, the statistical significance of the difference between different groups of earthquakes must be tested. The test can be done easily by the use of AIC. Of course the test is based on the assumption that the magnitude distribution obeys the G-R relation. Under the assumption of distribution functions other than the G-R relation, we can perform a similar significance test. It is possible that two earthquake groups have nearly equal b values, however the magnitude distribution is significantly different if another distribution function is adopted.

Some indexes other than the parameters in the formulas for magnitude-frequency distribution have been proposed. It is certain that these indexes indicate some characteristic feature of the size distribution of earthquakes concerned. However, it is not yet established that these indexes are useful in some practical problems such as detection of precursory change in seismicity, foreshock discrimination, etc.

In this paper, we are concerned only with the statistical problems of representing and analyzing the size distribution of earthquakes. It is beyond the scope of this paper to discuss the mechanism responsible for the size distribution and share the knowledge of the distribution for the studies of the complexity of seismic activity, though numerous papers taking this approach have been published. For reviews of some of these studies, see MAIN (1996), TURCOTTE (1997, Chapters 4 and 16) and KOYAMA (1997, Chapter 8).

Acknowledgments

The author thanks two anonymous reviewers for careful reviews and helpful suggestions.

REFERENCES

ABE, K. (1981), *Magnitudes of Large Shallow Earthquakes from 1904 to 1980*, Phys. Earth Planet. Inter. 27, 72–92.

ABE, K. (1984), *Compliments to "Magnitudes of Large Shallow Earthquakes from 1904 to 1980,"* Phys. Earth Planet. Inter. 34, 17–23.

ABE, K., and NOGUCHI, S. (1983a), *Determination of Magnitude for Large Shallow Earthquakes 1898–1917*, Phys. Earth Planet. Inter. 32, 45–59.

ABE, K., and NOGUCHI, S. (1983b), *Revision of Magnitudes of Large Earthquakes, 1897–1912*, Phys. Earth Planet. Inter. 33, 1–11.

AKAIKE, H. (1974), *A New Look at the Statistical Model Identification*, IEEE Trans. Automatic Control 19, 716–723.

AKI, K. (1965), *Maximum Likelihood Estimates of b in the Formula* $\log N = a - bM$ *and its Confidence Limits*, Bull. Earthq. Res. Inst. Univ. Tokyo 43, 237–239.

ALLEN, C. R., ST. AMAND, P., RICHTER, C. F., and NORDQUIST, J. M. (1965), *Relationship Between Seismicity and Geologic Structure in the Southern California Region*, Bull. Seismol. Soc. Am. 55, 753–797.

ANDERSON, J. G., and LUCO, J. E. (1983), *Consequences of Slip Constraints on Earthquake Occurrence Relations*, Bull. Seismol. Soc. Am. 76, 273–290.

ASADA, T., SUZUKI, Z., and TOMODA, Y. (1951), *Notes on the Energy and Frequency of Earthquakes*, Bull. Earthq. Res. Inst. Univ. Tokyo 29, 289–293.

BERRILL, J. B., and DAVIS, R. D. (1980), *Maximum Entropy and the Magnitude Distribution*, Bull. Seismol. Soc. Am. 70, 1823–1831.

BLOOM, E. D., and ERDMAN, R. C. (1980), *The Observation of a Universal Shape Regularity in Earthquake Frequency Magnitude Distributions*, Bull. Seismol. Soc. Am. 70, 349–362.

CAPUTO, M. (1976), *Model and Observed Seismicity Represented in a Two-dimensional Space*, Ann. Geofis. 29, 277–288.

CORNELL, C. A., and VANMARCHE, E. H. (1969), *The Major Influence on Seismic Risk*, Proc. 4th World Conf. Earthq. Engin. A1, 69–83.

CORNELL, C. A., and WINTERSTEIN, S. R. (1988), *Temporal and Magnitude Dependence in Earthquake Recurrence Models*, Bull. Seismol. Soc. Am. 78, 1522–1537.

COSENTINO, P., and LUZIO, D. (1976), *A Generalization of the Frequency-magnitude Relation in the Hypothesis of a Maximum Regional Magnitude*, Ann. Geofis. *29*, 3–8.

COSENTINO, P., FICARRA, V., and LUZIO, D. (1977), *Truncated Exponential Frequency-magnitude Relationship in Earthquake Statistics*, Bull. Seismol. Soc. Am. *67*, 1615–1623.

ENYA, M. (1908a), *Relation Between the Maximum Velocity of Motion and the Frequency of the Earthquakes Observed in Tokyo*, Rep. Earthq. Invest. Com. *61*, 77–78 (in Japanese).

ENYA, M. (1908b), *Relation Between the Frequency of Large Earthquakes and that of Small Ones*, Rep. Earthq. Invest. Com. *61*, 71–75 (in Japanese).

FROHLICH, C., and DAVIS, S. S. (1993), *Teleseismic b Values; or, Much Ado About 1.0*, J. Geophys. Res. *98*, 631–644.

GUARNIERI-BOTTI, L., PASQUALE, V., and ANGHINOLFI, M. (1980), *A New General Frequency-magnitude Relationship*, Pure appl. geophys. *119*, 196–206.

GUTENBERG, B. (1956), *The Energy of Earthquakes*, Quart. J. Geol. Soc. London *112*, 1–14.

GUTENBERG, B., and RICHTER, C. F. (1941), *Seismicity of the Earth*, Geol. Soc. Am. Spec. Pap. *No. 34*, 1–131.

GUTENBERG, B., and RICHTER, C. F. (1944), *Frequency of Earthquakes in California*, Bull. Seismol. Soc. Am. *34*, 185–188.

GUTENBERG, B., and RICHTER, C. F., *Seismicity of the Earth and Associated Phenomena* (Princeton Univ. Press, 1949) 273 pp.

HENDERSON, J., MAIN, I., MEREDITH, P., and SAMMONDS, P. (1992), *The Evolution of Seismicity of Parkfield: Observation, Experiment and a Fracture-mechanical Interpretation*, J. Struct. Geol. *14*, 905–913.

HENDERSON, J., MAIN, I. G., PEARCE, R. G., and TAKEYA, M. (1994), *Seismicity in North-eastern Brazil: Fractal Clustering and the Evolution of the b Value*, Geophys. J. Int. *116*, 217–226.

IMOTO, M. (1987), *A Bayesian Method for Estimating Earthquake Magnitude Distribution and Changes in the Distribution with Time and Space in New Zealand*, New Zealand J. Geol. Geophys. *30*, 1063–1116.

IMOTO, M. (1991), *Changes in the Magnitude-frequency b Value Prior to Large (M ≥ 6) Earthquakes in Japan*, Tectonophys. *193*, 311–325.

ISHIMOTO, M., and IIDA, K. (1939), *Observations of Earthquakes Registered with the Microseismograph Constructed Recently (I)*, Bull. Earthq. Res. Inst. Univ. Tokyo *17*, 443–478 (in Japanese).

JIN, A., and AKI, K. (1989), *Spatial and Temporal Correlation Between Coda Q^{-1} and Seismicity and its Physical Mechanism*, J. Geophys. Res. *94*, 14,041–14,059.

KAGAN, Y. Y. (1969), *A Study of the Energy of the Seismoacoustic Pulses Arising during Bursts in a Coal Bed*, Izv. Phys. Solid Earth, 85–91 (English translation).

KAGAN, Y. Y. (1991), *Seismic Moment Distribution*, Geophys. J. Int. *106*, 123–134.

KAGAN, Y. Y. (1993), *Statistics of Characteristic Earthquakes*, Bull. Seismol. Soc. Am. *83*, 4–24.

KAGAN, Y. Y. (1997), *Seismic Moment-frequency Relation for Shallow Earthquakes: Regional Comparison*, J. Geophys. Res. *102*, 2835–2852.

KAGAN, Y. Y., and KNOPOFF, L. (1984), *A Stochastic Model of Earthquake Occurrence*, Proc. 8th Int. Conf. Earthq. Eng. *1*, 295–302.

KANAMORI, H. (1977), *The Energy Release in Great Earthquakes*, J. Geophys. Res. *82*, 2981–2987.

KÁRNÍK, V., and KLÍMA, K. (1993), *Magnitude-frequency Distribution in the European–Mediterranean Earthquake Regions*, Tectonophys. *220*, 309–323.

KAYANO, I. (1982), *A Characteristic of Earthquake Sequences—An Index for Energy Distribution of Major Earthquakes Belonging to Earthquake Sequences*, Bull. Earthq. Res. Inst. Univ. Tokyo *57*, 317–336 (in Japanese).

KOYAMA, J., *The Complex Faulting Process of Earthquakes* (Kluwer Academic Press, 1997) 208 pp.

LATOUSSAKIS, J., and DRAKOPOULOS, J. (1987), *A Modified Formula for Frequency-magnitude Distribution*, Pure appl. geophys. *25*, 753–764.

LOMNITZ, C. (1964), *Estimation Problems in Earthquake Series*, Tectonophys. *2*, 193–203.

LOMNITZ-ADLER, J., and LOMNITZ, C. (1978), *A New Magnitude-frequency Relation*, Tectonophys. *49*, 237–245.

LOMNITZ-ADLER, J., and LOMNITZ, C. (1979), *A Modified Form of the Gutenberg-Richter Magnitude-frequency Relation*, Bull. Seismol. Soc. Am. *69*, 1209–1214.

MAIN, I. G. (1987), *A Characteristic Earthquake Model of the Seismicity Preceding the Eruption of Mount St. Helens on 18 May 1980*, Phys. Earth Planet. Inter. *49*, 283–293.

MAIN, I. G. (1996), *Statistical Physics, Seismogenesis, and Seismic Hazard*, Rev. Geophys. *34*, 433–462.

MAIN, I. G., and BURTON, P. W. (1984a), *Information Theory and the Earthquake Frequency-magnitude Distribution*, Bull. Seismol. Soc. Am. *74*, 140–1426.

MAIN, I. G., and BURTON, P. W. (1984b), *Physical Links Between Crustal Deformation, Seismic Moment and Seismic Hazard for Regions of Varying Seismicity*, Geophys. J. R. Astr. Soc. *79*, 469–488.

MAIN, I. G., and BURTON, P. W. (1986), *Long-term Earthquake Recurrence Constrained by Tectonic Seismic Moment Release Rates*, Bull. Seismol. Soc. Am. *76*, 297–304.

MAKJANIĆ, B. (1972), *A Contribution to the Statistical Analysis of Zagreb Earthquakes in the Period 1869–1968*, Pure appl. geophys. *95*, 80–88.

MAKJANIĆ, B. (1980), *On the Frequency Distribution of Earthquake Magnitude and Intensity*, Bull. Seismol. Soc. Am. *70*, 2253–2260.

MARUYAMA, T. (1978), *Frequency Distribution of the Sizes of Fractures Generated in the Branching Process—Elementary Analysis*, Bull. Earthq. Res. Inst. Univ. Tokyo *53*, 407–421 (in Japanese).

MERZ, H. A., and CORNELL, C. A. (1973), *Seismic Risk Analysis Based on a Quadratic Magnitude-frequency Law*, Bull. Seismol. Soc. Am. *63*, 1999–2006.

MOGI, K. (1979), *Global Variation of Seismic Activity*, Tectonophys. *57*, T43–T50.

MOLCHAN, G., KRONROD, T., and PANZA, G. P. (1997), *Multiscale Seismicity Model for Seismic Risk*, Bull. Seismol. Soc. Am. *87*, 1220–1229.

MORI, J., and ABERCROMBIE, R. E. (1997), *Depth Dependence of Earthquake Frequency-magnitude Distributions in California: Implications for the Rupture Initiation*, J. Geophys. Res. *102*, 15,081–15,090.

NEUNHÖFER, H. (1969), *Non-linear Energy-frequency Curves in Statistics of Earthquakes*, Pure appl. geophys. *72*, 76–83.

NIAZI, M. (1964), *Seismicity of Northern California and Western Nevada*, Bull. Seismol. Soc. Am. *54*, 845–850.

OGATA, Y. (1992), *Detection of Precursory Relative Quiescence Before Great Earthquakes Through a Statistical Model*, J. Geophys. Res. *97*, 19,845–19,871.

OGATA, Y., and ABE, K. (1991), *Some Statistical Features of the Long-term Variation of the Global and Regional Seismic Activity*, Int. Statist. Rev. *59*, 139–161.

OGATA, Y., IMOTO, M., and KATSURA, K. (1991), *3-D Spatial Variation of b Values of Magnitude-frequency Distribution Beneath the Kanto District, Japan*, Geophys. J. Int. *104*, 135–146.

OGATA, Y., and KATSURA, K. (1993), *Analysis of Temporal and Spatial Heterogeneity of Magnitude Frequency Distribution Inferred from Earthquake Catalogues*, Geophys. J. Int. *113*, 727–738.

OGATA, Y., UTSU, T., and KATSURA, K. (1995), *Statistical Features of Foreshocks in Comparison with Other Earthquake Clusters*, Geophys. J. Int. *121*, 233–254.

OKADA, H., WATANABE, H., YAMASHITA, H., and YOKOYAMA, I. (1981), *Seismological Significance of the 1977–1978 Eruptions and Magma Intrusion Process of Usu Volcano, Hokkaido*, J. Volc. Geotherm. Res. *9*, 311–334.

OKADA, M. (1970), *Magnitude-frequency Relationship of Earthquakes and Estimation of Supremum—A Statistical Study*, Kenkyujiho (J. Meteorol. Res. JMA) *22*, 8–19 (in Japanese).

OKAL, E. A., and ROMANOWICZ, B. A. (1994), *On the Variation of b Value with Earthquake Size*, Phys. Earth Planet. Inter. *87*, 55–76.

OKAL, E. A., and KIRBY, S. H. (1995), *Frequency-moment Distribution of Deep Earthquakes; Implications for the Seismogenic Zone at the Bottom of Slabs*, Phys. Earth Planet. Inter. *92*, 169–187.

OKUDA, S., OUCHI, T., and TERASHIMA, T. (1992), *Deviation of Magnitude Frequency Distribution of Earthquakes from the Gutenberg–Richter Law: Detection of Precursory Anomalies Prior to Large Earthquakes*, Phys. Earth Planet. Inter. *73*, 229–238.

OLSSON, R. (1986), *Are Earthquake Magnitudes Exponentially or Normally Distributed?*, Tectonophys. *82*, 379–382.

OMORI, F. (1902), *Macroseismic Measurements in Tokyo, II and III*, Pub. Earthq. Invest. Com. *11*, 1–95.

OMORI, F. (1908), *List of Stronger Japan Earthquakes, 1902–1907*, Bull. Earthq. Inv. Com. *2*, 58–88.

ÖNCEL, A. O., MAIN, I., ALPTEKIN, Ö., and COWIE, P. (1996), *Spatial Variations of the Fractal Properties of Seismicity in the Anatolian Fault Zones*, Tectonophys. *275*, 189–202.

OTSUKA, M. (1972), *A Chain-Reaction-Type Source Model as a Tool to Interpret the Magnitude-frequency Relation of Earthquakes*, J. Phys. Earth *20*, 35–45.

OUCHI, T., and YOKOTA, T. (1979), *A New Scale of Magnitude-frequency Distribution of Earthquakes*, Zisin (J. Seism. Soc. Japan) Ser. 2, *32*, 415–421 (in Japanese).

PACHECO, J. F., and SYKES, L. R. (1992), *Seismic Moment Catalog of Large Shallow Earthquakes, 1900–1989*, Bull. Seismol. Soc. Am. *82*, 1306–1349.

PAGE, R. (1968), *Aftershocks and Microaftershocks of the Great Alaska Earthquake of 1964*, Bull. Seismol. Soc. Am. *58*, 1131–1168.

PURCARU, G. (1975), *A New Magnitude-frequency Relation for Earthquakes and a Classification of Relation Types*, Geophys. J. R. Astr. Soc. *42*, 61–79.

RANALLI, G. (1975), *A Test of the Lognormal Distribution of Earthquake Magnitude*, Veroff. Zentralinst. Phys. Erde, DAW *31*, 163–180.

RICHTER, C. F. (1935), *An Instrumental Magnitude Scale*, Bull. Seismol. Soc. Am. *25*, 1–32.

RIZNICHENKO, YU. V. (1959), *On Quantitative Determination and Mapping of Seismicity*, Ann. Geofis. *12*, 227–237.

RIZNICHENKO, YU. V. (1964), *The Investigation of Seismic Activity by the Method of Earthquake Summation*, Izv. Phys. Solid Earth 589–593 (English translation).

RIZNICHENKO, YU. V. (1966), *Problems of Earthquake Physics*, Izv. Phys. Solid Earth 73–86 (English translation).

RUNDLE, J. B. (1993), *Magnitude-frequency Relations for Earthquakes Using a Statistical Mechanical Approach*, J. Geophys. Res. *98*, 21,943–21,949.

SACUIU, I., and ZORILESCU, D. (1970), *Statistical Analysis of Seismic Data on Earthquakes in the Area of Vrancea Focus*, Bull. Seismol. Soc. Am. *60*, 1089–1099.

SAITO, M., KIKUCHI, M., and KUDO, K. (1973), *Analytical Solution of Go-game Model of Earthquakes*, Zisin (J. Seism. Soc. Japan), Ser. 2, *26*, 19–25 (in Japanese).

SEINO, M., FUKUI, K., and CHUREI, M. (1989), *Magnitude vs. Frequency Distributions of Earthquakes with Upper Bound Magnitude*, Zisin (J. Seism. Soc. Japan), Ser. 2, *42*, 73–80 (in Japanese).

SHLIEN, S., and TOKSÖZ, M. N. (1970), *Frequency-magnitude Statistics of Earthquake Occurrences*, Earthq. Notes *41* (1), 5–18.

SMITH, W. D. (1986), *Evidence for Precursory Changes in the Frequency-magnitude b Value*, Geophys. J. Roy. Astr. Soc. *86*, 815–838.

SORNETTE, D., KNOPOFF, L., KAGAN, Y. Y., and VANNESTE, C. (1996), *Rank-ordering Statistics of Extreme Events: Application to the Distribution of Large Earthquakes*, J. Geophys. Res. *101*, 13,883–13,893.

SPEIDEL, D. H., and MATTSON, P. H. (1993), *The Polymodal Frequency-magnitude Relationship of Earthquakes*, Bull. Seismol. Soc. Am. *83*, 1893–1901.

SUZUKI, Z. (1959), *A Statistical Study on the Occurrence of Small Earthquakes, IV*, Sci. Rep. Tohoku Univ., Ser. 5, *11*, 10–54.

TRIEP, E. G., and SYKES, L. R. (1997), *Frequency of Occurrence of Moderate to Great Earthquakes in Intracontinental Regions: Implication for Changes in Stress, Earthquake Prediction, and Hazard Assessments*, J. Geophys. Res. *102*, 9923–9948.

TRIFU, C.-I., and SHUMILA, V. I. (1996), *A Method for Multidimensional Analysis of Earthquake Frequency-magnitude Distribution with an Application to the Vrancea Region of Romania*, Tectonophys. *261*, 9–22.

TURCOTTE, D. L., *Fractals and Chaos in Geology and Geophysics*, 2nd Ed. (Cambridge University Press 1997) 398 pp.

UTSU, T. (1961), *A Statistical Study on the Occurrence of Aftershocks*, Geophys. Mag. *30*, 521–605.

UTSU, T. (1962), *On the Nature of Three Alaskan Aftershock Sequences of 1957 and 1958*, Bull. Seismol. Soc. Am. *52*, 279–297.

UTSU, T. (1965), *A Method for Determining the Value of b in a Formula $\log n = a - bM$ Showing the Magnitude-frequency Relation for Earthquakes*, Geophys. Bull. Hokkaido Univ. *13*, 99–103 (in Japanese).

UTSU, T. (1966), *A Statistical Significance Test of the Difference in b Value Between Two Earthquake Groups*, J. Phys. Earth *14*, 37–40.

UTSU, T. (1971), *Aftershocks and Earthquake Statistics (III)*, J. Fac. Sci. Hokkaido Univ. Ser. VII *3*, 379–441.

UTSU, T. (1978), *Estimation of Parameter Values in Formulas for Magnitude-frequency Relation of Earthquake Occurrence*, Zisin (J. Seism. Soc. Japan), Ser. 2, *31*, 367–382 (in Japanese).

UTSU, T. (1982), *Relationships Between Earthquake Magnitude Scales*, Bull. Earthq. Res. Inst. Univ. Tokyo *57*, 465–497 (in Japanese).

UTSU, T. (1988), *η Value for Magnitude Distribution and Earthquake Prediction*, Rep. Coord. Comm. Earthq. Pred. *39*, 280–386 (in Japanese).

UTSU, T. (1992), *Introduction to Seismicity*, Surijishingaku (Mathematical Seismology) (VII), Inst. Statis. Math. 139–157 (in Japanese).

VERE-JONES, D. (1976), *A Branching Model for Crack Propagation*, Pure appl. geophys. *114*, 711–725.

VERE-JONES, D. (1977), *Statistical Theories of Crack Propagation*, J. Math. Geol. *9*, 455–481.

WADATI, K. (1932), *On the Frequency Distribution of Earthquakes*, Kishoshushi (J. Meteorol. Soc. Japan), Ser. 2, *10*, 559–568 (in Japanese).

WARD, S. N. (1996), *A Synthetic Seismicity Model for Southern California: Cycles, Probabilities, and Hazard*, J. Geophys. Res. *101*, 22,393–22,418.

WIEMER, S., and BENOIT, J. (1996), *Mapping the b Value Anomaly at 100 km Depth in the Alaska and New Zealand Subduction Zones*, Geophys. Res. Lett. *23*, 1557–1560.

WIEMER, S., and McNUTT, S. (1997), *Variations in the Frequency-magnitude Distribution with Depth in Two Volcanic Areas: Mount St. Helens, Washington, and Mt. Spurr, Alaska*, Geophys. Res. Lett. *24*, 189–192.

WIEMER, S., and WYSS, M. (1997), *Mapping the Frequency-magnitude Distribution in Asperities: An Improved Technique to Calculate Recurrence Times?*, J. Geophys. Res. *102*, 15,115–15,128.

WYSS, M., SHIMAZAKI, K., and WIEMER, S. (1997), *Mapping Active Magma Chamber by b Values Beneath the Off-Ito Volcano, Japan*, J. Geophys. Res. *102*, 20,413–20,422.

ZHANG, Y.-X., and HUANG, D.-Y. (1990), *On the η Value as a Parameter for Earthquake Prediction*, Earthq. Res. China *4*, 63–70.

(Received June 27, 1998, revised/accepted November 11, 1998)

 To access this journal online:
http://www.birkhauser.ch

Pure appl. geophys. 155 (1999) 537–573
0033–4553/99/040537–37 $ 1.50 + 0.20/0

┃Pure and Applied Geophysics

Universality of the Seismic Moment-frequency Relation

Y. Y. KAGAN[1]

Abstract—We analyze the seismic moment-frequency relation in various depth ranges and for different seismic regions, using Flinn-Engdahl's regionalization of global seismicity. Three earthquake lists of centroid-moment tensor data have been used: the Harvard catalog, the USGS catalog, and the HUANG *et al.* (1997) catalog of deep earthquakes. The results confirm the universality of the β-values and the maximum moment for shallow earthquakes in continental regions, as well as at and near continental boundaries. Moreover, we show that although fluctuations in earthquake size distribution increase with depth, the β-values for earthquakes in the depth range of 0–500 km exhibit no statistically significant regional variations. The regional variations are significant only for deep events near the 660 km boundary. For declustered shallow earthquake catalogs and deeper events, we show that the worldwide β-values have the same value of 0.60 ± 0.02. This finding suggests that the β-value is a universal constant. We investigate the statistical correlations between the numbers of seismic events in different depth ranges and the correlation of the tectonic deformation rate and seismic activity (the number of earthquakes above a certain threshold level per year). The high level of these correlations suggests that seismic activity indicates tectonic deformation rate in subduction zones. Combined with the universality of the β-value, this finding implies little if any variation in maximum earthquake seismic moment among various subduction zones. If we assume that earthquakes of maximum size are similar in different depth ranges and the seismic efficiency coefficient, χ, is close to 100% for shallow seismicity, then we can estimate χ for deeper earthquakes: for intermediate earthquakes $\chi \approx 5\%$, and $\chi \approx 1\%$ for deep events. These results may lead to new theoretical understanding of the earthquake process and better estimates of seismic hazard.

Key words: Gutenberg-Richter relation, maximum seismic moment, shallow, intermediate, and deep earthquakes, fractals.

1. Introduction

A previous paper (KAGAN, 1997) investigated worldwide seismic moment-frequency relation for shallow earthquakes. We showed that the size distribution of moderate and large earthquakes can be approximated by a power-law: a transformation of the Gutenberg-Richter (G-R) relation. For the largest earthquakes, we use an exponential taper (the gamma distribution) to describe the size distribution appropriate for finite seismic energy release. The exponent of the power-law (β) does not show any discernible variation among continental regions and their

[1] Institute of Geophysics and Planetary Physics, University of California, Los Angeles, CA 90095-1576, U.S.A. E-mail: ykagan@ucla.edu

boundaries (subduction zones). For brevity we call all these non-spreading zones *Compressive*. Due to insufficient data, the exact form of the moment-frequency relation for mid-ocean ridges has not been determined. This result suggests that the β-value is a universal constant characterizing occurrence of shallow earthquakes. Moreover, comparing the tectonic deformation rate with the seismic activity level implies that for *Compressive* regions the maximum earthquake size is similar. The hypothesis of the maximum size universality cannot be rejected using available data (KAGAN, 1997).

In this paper the scalar seismic moment M is used to describe earthquake size; for comparison with previous investigations and for brevity, the moment magnitude m is used:

$$m = \frac{2}{3} \lg M - 6, \tag{1}$$

where $\lg = \log_{10}$. The moment is measured in Newton meters (Nm), and, unless, specifically mentioned, the magnitude is the moment magnitude as defined in (1).

MOLCHAN *et al.* (1996, 1997) claim to have found statistically significant variations in the b-values for moment magnitudes of shallow earthquakes in the Harvard catalog. [Although MOLCHAN *et al.* (1996) is a more detailed exposition of their tests, here we only refer to the points which coincide in both English and Russian versions.] According to (1) $\beta = 2/3b$. Their five statistical tests of the b-value variability (Tables 1–5 in MOLCHAN *et al.*, 1996; and Table 1 in MOLCHAN *et al.*, 1997) involve a comparison of the b-values:

(a) for subduction zones versus mid-ocean ridges; (b) for two magnitude intervals in subduction zones ($m = 5.55–7.56$ versus $m = 7.57–8.90$); (c) two subsets of mid-ocean ridges; (d) two subsets of subduction zones; and (e) three depth intervals, i.e., 0–15, 16–33, and 34–70 km.

Let us consider these tests.

● (a) The difference in the b-values for subduction zones and mid-ocean earthquakes has been pointed out by OKAL and ROMANOWICZ (1994) and by KAGAN (1997). The exact cause of this difference is not yet clear. KAGAN (1997) argues that the disparity may be due to a mixture of several earthquake mid-ocean ridge populations with a significantly different maximum magnitude value. The thickness of the brittle crust increases as the newly formed crust moves from a mid-ocean ridge; thus one expects a significant variation in the maximum magnitude over relatively short distances. Because of the relatively small number of mid-ocean earthquakes in the Harvard catalog, it is difficult to determine their size distribution. Thus, the universality of the β-value can now be proposed only for the *Compressive* regions. In this paper we analyze the mid-ocean earthquakes only briefly.

● (b) The b-value changes for very large earthquakes ($m > 8$), because of the finite size of tectonic plates. For these earthquakes the energy conservation principle requires the G-R relation to be modified (KAGAN, 1997). For very large earthquakes, the original G-R law cannot be used as the null hypothesis to be tested against the earthquake size distribution. Thus, one should not expect the size distributions in these magnitude intervals (moderate-large versus very large events) to be similar.

● (c, d) KRONROD (1984, 1985) proposed a more detailed regionalization of worldwide seismicity than did Flinn-Engdahl (FLINN et al., 1974; YOUNG et al., 1996). Since these regions have been selected independently from the Harvard catalog data, it would be of interest to see whether we detect any variation of the moment-frequency parameters in these zones. Unfortunately, these regions are not specified by a computer algorithm, making independent testing of seismicity variations difficult. Although KRONROD (1984, 1985) discusses various selections of zone groups, the regions analyzed by MOLCHAN et al. (1996, 1997) seem to be combined in groups unlike KRONROD's (1985). In such cases, it is important to present the analysis results for *all* zone groups. Otherwise one cannot be sure whether the new groups have been selected after a preliminary inspection of the results.

● (e) Determining the depth dependence of the b-values presents a serious challenge. Depth values in the Harvard catalog are determined to the first decimal place; MOLCHAN et al. (1996, 1997) apparently use discretized (integer) depth values. Although, we reproduced most of MOLCHAN's et al. (1997, Table 1) b-values, thus ensuring similarity of our techniques, the depth discretization prevents a more complete comparison of the results.

There are two reasons that the earthquake depth distribution is significantly biased near the Earth's surface. 1) Earthquake coordinates available in the Harvard catalog are for the seismic moment centroid (DZIEWONSKI et al., 1998) in the center of the moment release volume. Thus, the centroid depth should be at least 15–30 km, especially for great earthquakes, since the largest earthquakes rupture the whole brittle crust. 2) The second reason is that, if the centroid solution does not converge in the depth coordinate, the depth is assigned by a seismologist, usually at 10 or 15 km. The probability of a poor depth convergence is higher for weaker events, because their signal-to-noise ratio is smaller. Therefore, the earthquake population in standard catalogs in the upper 15 km has an artificially high proportion of small events. Thus, the universality of the β-value distribution should not be claimed or tested for earthquakes in the crust's upper 20–30 km.

The results of MOLCHAN et al. (1996, 1997) which imply the strongest rejection of the β-value universality are for cases (a) and (e). The significance level for those tests is less than 0.05%. However, as discussed in the previous paragraphs, these tests are the most questionable in design. Thus, the β-value universality for shallow earthquakes in *Compressive* regions cannot be considered disproved.

In this work we explore whether the parameter β differs from region to region for earthquakes deeper than 70 km, and whether deep earthquakes have a different β-value than shallow events. (For brevity we sometimes refer to events below 70 km as *deeper* earthquakes.) Several previous studies (FROHLICH, 1989, 1998; FROHLICH and DAVIS, 1993; OKAL and KIRBY, 1995; PACHECO *et al.*, 1992; WIENS and GILBERT, 1996) have suggested significant differences in these values.

2. Data

We study the earthquake size distribution for the global catalog of moment tensor inversions compiled by the Harvard group (DZIEWONSKI *et al.*, 1998, and references therein). The catalog contains 14,325 solutions spanning the period from 1977/1/1 to 1997/6/30. We analyze the completeness of the catalog by a procedure

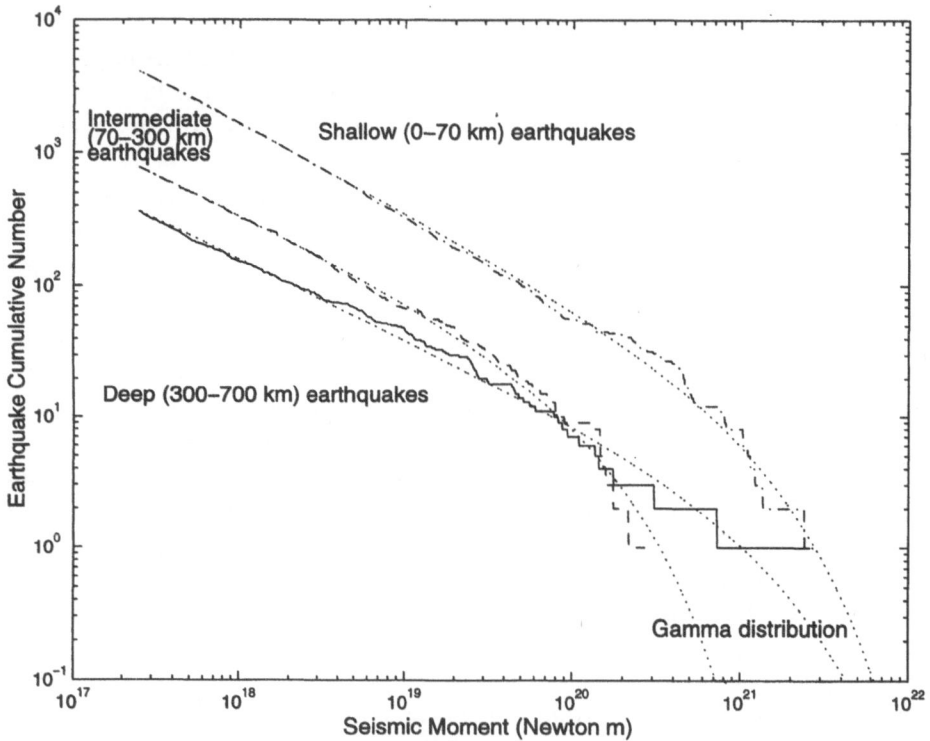

Figure 1

Earthquake cumulative number versus log seismic moment for the global earthquake distribution in the 1982/1/1–1997/7/1 Harvard catalog. The curves show the numbers of events with the moment larger than or equal to M. We also show the approximation of curves by the gamma distribution (Eq. (3)) which is the G-R law restricted at large magnitudes by an exponential taper. The slope of the linear part of the curves corresponds to β-values (Eq. (3)) 0.657 ± 0.017, 0.573 ± 0.046, 0.580 ± 0.035, and the maximum moment $M_{xg} = 3.5 \times 10^{21}$, 4.0×10^{20}, 2.2×10^{21} Nm, for shallow, intermediate, and deep earthquakes, respectively. The 95% confidence limits for the maximum magnitude are similarly 8.1–8.7 (8.37), 7.5–8.3 (7.73), and 7.7–∞ (8.23), where the values in parentheses are used in the graph.

similar to that described by KAGAN (1997, Figure 1). The results suggest that the detection threshold for seismic moment, M_c, decreased with time (see also MOLCHAN et al., 1996, 1997): $M_c = 10^{17.7}$ for 1977–1981, and $M_c = 10^{17.4}$ for 1982–1997. For moment magnitude, these cutoffs correspond to $m_c = 5.8$ and $m_c = 5.6$, respectively (see Eq. (1)). There were 4063 $m \geq 5.6$ earthquakes in 1982–1997, and 3394 $m \geq 5.8$ events in 1977–1997, of which 785 earthquakes occurred in 1977–1981. Although it is possible to determine the parameters of the moment-frequency relation using several different moment thresholds, the complexity of the calculations and the loss of interpretative clarity do not compensate for the small additional information (about 19%) than one obtains by combining two cutoffs. Thus, in this work we used the Harvard data in the time period 1982/1/1– 1997/6/30, and magnitude cutoff $m_c = 5.6$.

The global catalog issued by the USGS (SIPKIN and ZIRBES, 1997, and references therein) spans 1980/1/1–1996/3/1, and contains 1,836 solutions. Unfortunately, the moment threshold changes significantly during the catalog time span; hence we selected only 954 earthquakes with $m_c = 6.2$ (starting on 1983/1/1) here.

HELFFRICH (1997) compares moment tensor solutions in the Harvard and USGS catalogs and finds them strongly correlated, the coefficient of correlation increasing for deeper earthquakes. He determines a standard deviation for a logarithm (lg) of the ratio of two scalar seismic moments for paired events in both catalogs. These uncertainties are ± 0.15, ± 0.06, and ± 0.07, for shallow, intermediate, and deep earthquakes, respectively. Such results underscore again the superiority of the scalar seismic moment as the measure of earthquake size.

We have also used a catalog supplied by HUANG et al. (1997), who determined centroid-moment tensor solutions for about 100 deep earthquakes in 1962–1976, applying the inversion technology of the Harvard group to analog records. The magnitude cutoff is $m_c = 6.3$.

In all moment tensor catalogs the depth value is listed to the first decimal place. For brevity we display depth intervals as rounded-off values, e.g., 0–35, 35–70, to mean 0–35.0, 35.1–70.0, etc.

For comparison we analyzed the magnitude-frequency relations for the PDE worldwide catalog of earthquakes (U.S. GEOLOGICAL SURVEY, 1997). The available catalog ends on May 1, 1997. The catalog measures earthquake size, using several magnitude scales, of which m_b and M_S are provided for most of the moderate and large events. Since regular reporting of the M_S magnitude only started in mid-1968, we use the catalog starting with 1969/1/1. The catalog contains 30,102 shallow earthquakes with $m_b \geq 5$ from 1969 to 1997.

We use the Flinn-Engdahl algorithm to subdivide seismicity into groups of earthquakes, because it is apparently the only appropriate regionalization available. The requirements for such regionalization include its availability in a computer code. Moreover, region boundaries should be formulated before the starting date of the Harvard catalog (see the Introduction section): to avoid a bias, the boundaries should have been chosen before earthquake locations were known.

3. Frequency-moment Relation

The earthquake size distribution is usually described by the G-R magnitude-frequency law

$$\lg N(m) = a - bm, \tag{2}$$

where $N(m)$ is the number of earthquakes with magnitude $\geq m$, and $b \approx 1$.

As in earlier work (KAGAN, 1997), we use the gamma law: a version of the G-R distribution with an exponential roll-off at large moments to describe earthquake size distribution (see also MAIN, 1996; UTSU and OGATA, 1997, p. 81). The gamma distribution has a probability density

$$\phi(M) = C^{-1} M_c^\beta M^{-1-\beta} \exp(-M/M_{xg}), \tag{3}$$

where C is a normalizing coefficient and M_{xg} is a maximum moment parameter (the effective maximum seismic moment).

Since the notion of the maximum moment is a common source of misunderstanding, a few comments on its representation are in order. Contrary to the usual determination of the maximum moment M_{max}, M_{xg} has a different meaning here. Where M_{max} represents a 'hard' limit (that is, no earthquake can be larger than M_{max}), the latter is a 'soft' limit. As we see in (3), in the gamma distribution, some earthquakes may have the moment $M > M_{xg}$, but with a probability density that decays for $M > M_{xg}$ considerably faster than the standard G-R law. The difference between these two notions may not be as large as it seems: M_{max} in reality is an experimentally determined parameter. Thus, it should have statistical uncertainties and be treated as a continuously distributed random variable. To avoid awkward wording, we continue to use the term 'maximum magnitude' (or moment) for m_{xg} or M_{xg}, respectively.

Figure 1 displays cumulative histograms for the scalar seismic moment of earthquakes in the Harvard catalog for three depth ranges. The curves display a scale-invariant segment (linear in the log-log plot) for small and moderate values of the seismic moment. At large M, the curve for shallow events is bent downward. It is more difficult to see the bending of curves for deeper earthquakes, since the number of earthquakes is relatively small. One can see the difference in the distributions for deep earthquakes in Figure 4a in KAGAN (1994) and in Figure 1. Two very large earthquakes (the 1994 Bolivia and Tonga events, see, for instance, WIENS and McGUIRE, 1995) fundamentally modified the curve at large moment values.

We determined the maximum likelihood values for M_{xg} and β as well as their standard deviations σ_β and σ_m, for the worldwide earthquake distribution in several depth ranges (KAGAN, 1991b). These values are listed in the caption for Figure 1. Comparison with the other evaluation of these parameters (KAGAN, 1994, 1997, p. 168) illustrates that the new M_{xg}-value is outside the 95% confidence interval only for the maximum moment of deep earthquakes. This interval is estimated by KAGAN (1994) to be $M_{xg} = 5.2 \times 10^{19} - 2.0 \times 10^{21}$ Nm.

As in KAGAN (1991b, 1997), to determine M_{xg} and β we obtain maps of the log likelihood function for two parameters of the gamma distribution (3). Three examples of these diagrams are shown in Figures 2a,b,c; the 95% confidence level corresponds to the contour value 0.0. Two diagrams (Figs. 2a,b) display the dependence of parameter estimates on catalog data. Prior to the occurrence of the 1994 Bolivia and Tonga shocks, the empirical moment-frequency relation and the likelihood function (Fig. 2a) suggested that $m_{xg} \leq 8.3$; after these earthquakes only the lower confidence limit for m_{xg} can be determined (Fig. 2b). Although both 95% confidence areas intersect near the point $\beta = 0.6$, $m_{xg} = 8.0$, the occurrence of these two large earthquakes significantly changed the shape of the likelihood function. The confidence interval for β is smaller and the m_{xg} estimate extends to infinity in Figure 2b. One reason for the tightening of the β-value error bounds in Figure 2b is the extension of the magnitude threshold to 5.6. This provided an additional

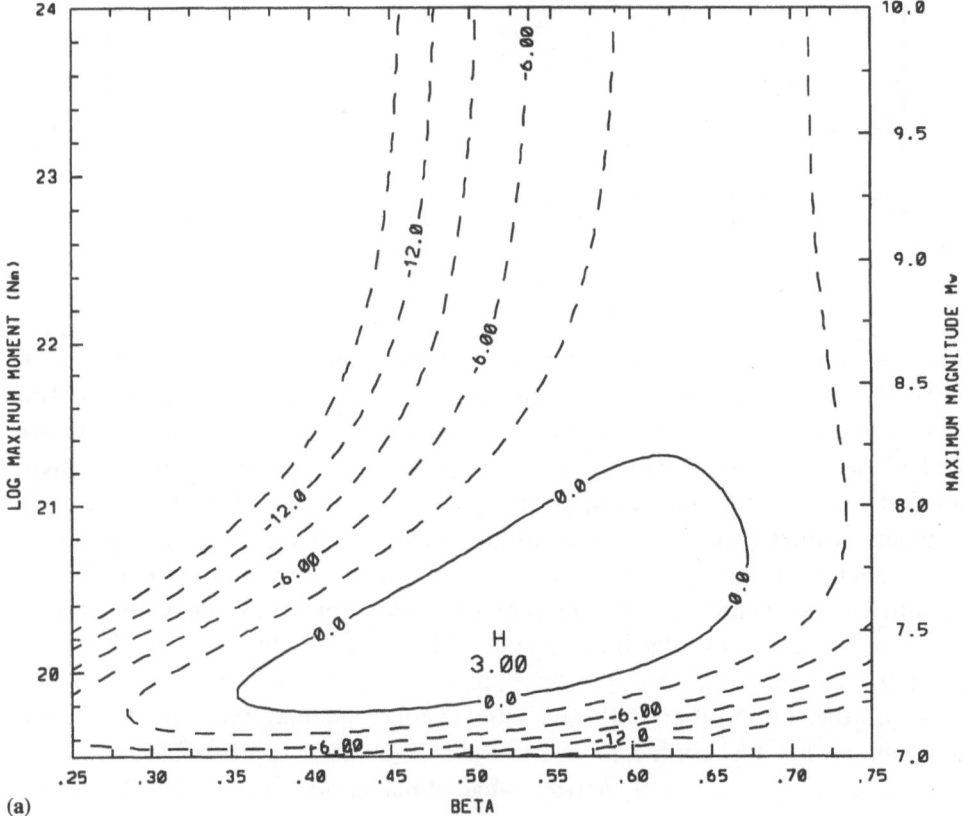

(a)

Fig. 2.

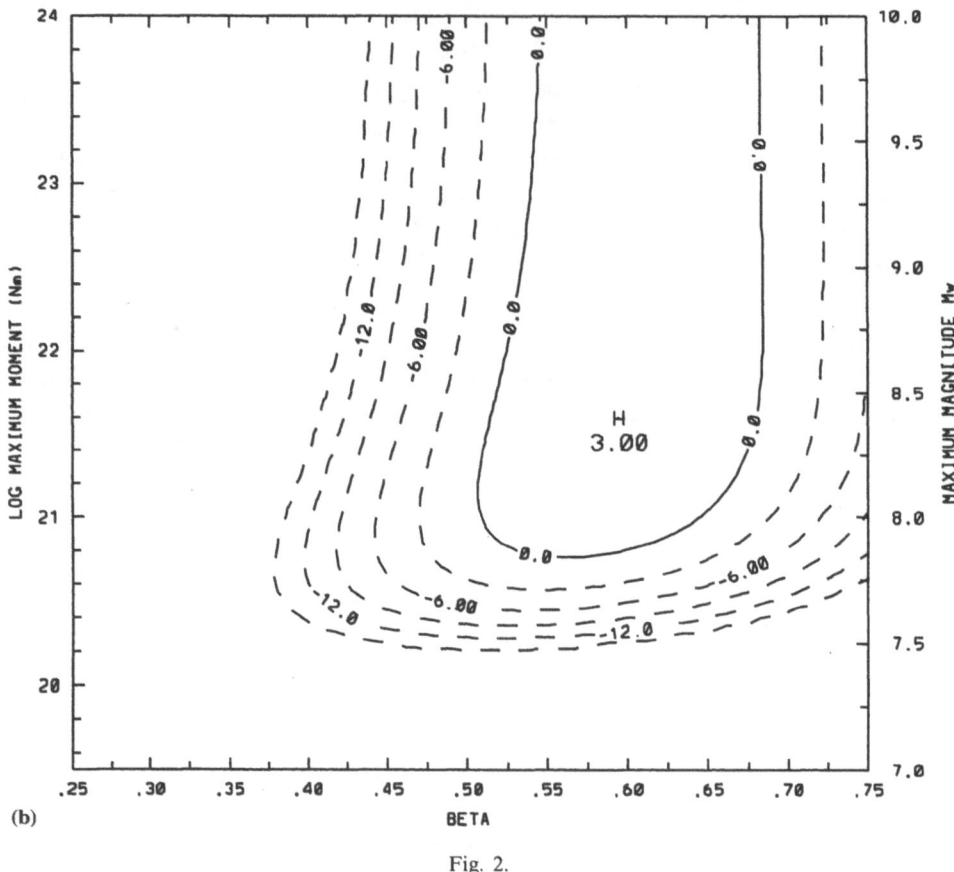

Fig. 2.

power-law interval to the moment-frequency curve, resulting in a better estimate of β. The reason for the discrepancy in the m_{xg} estimate is the occurrence of the two very large earthquakes, mentioned above. Almost all of the earthquake data are used to determine the β-value and its confidence limits, whereas the maximum moment estimate depends strongly on a few large earthquakes. Therefore, the statistical properties of the estimate are not well defined. A similar change in the M_{xg} estimates can be seen in KAGAN (1997, Figs. 3d,e), where an occurrence of one earthquake significantly modified the confidence areas for the maximum magnitude.

These features can be illustrated again by the likelihood function for HUANG et al. (1997) catalog (Fig. 2c). The 95% contour covers the value of $\beta \to 0$ and $m_{xg} \to \infty$; this means that both the pure exponential and the pure power-law distributions for the seismic moment could acceptably approximate data. The reason is obvious—a relatively narrow range of magnitudes (6.3–8.1) in the catalog does not permit reliable determination of both parameters in (3). In general, Figures 2a,b,c demonstrate the difficulties of using a straightforward statistical

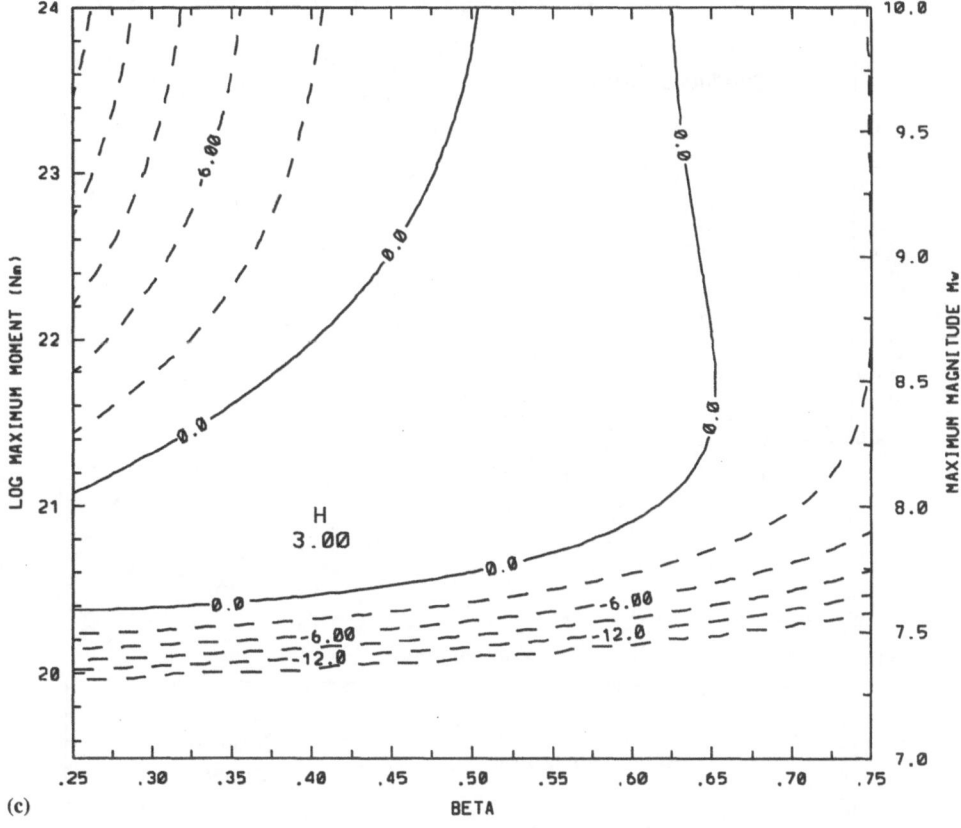

Figure 2
Log-likelihood function for distribution of scalar seismic moment of earthquakes: (a) Deep earthquakes
in the Harvard catalog, 1977/1/1–1994/1/1, $m_c = 5.8$; (b) deep earthquakes in the Harvard catalog,
1982/1/1–1997/7/1, $m_c = 5.6$; (c) deep earthquakes in the HUANG et al. (1997) catalog, 1962/1/1–
1977/1/1, $m_c = 6.3$.

approach (the maximum likelihood method) to determine the maximum magnitude,
m_{xg}. In the rest of the paper, we use the seismic moment conservation principle (see
Section 5 below) to determine m_{xg}.

Comparison of β-values for three depth intervals in Figure 1 suggests that the
slope of the curves in the log-log plot varies in a statistically significant manner. As
a statistical test, we compare the difference of two β-values with their standard
deviations σ_β. The ratio

$$z = \frac{\beta_1 - \beta_2}{\sqrt{\sigma_1^2 + \sigma_2^2}} \tag{4}$$

is distributed for a large number of events ($n > 30$) according to a normal
(Gaussian) distribution with a standard deviation of 1.0. We obtain $z = 1.71$ for the

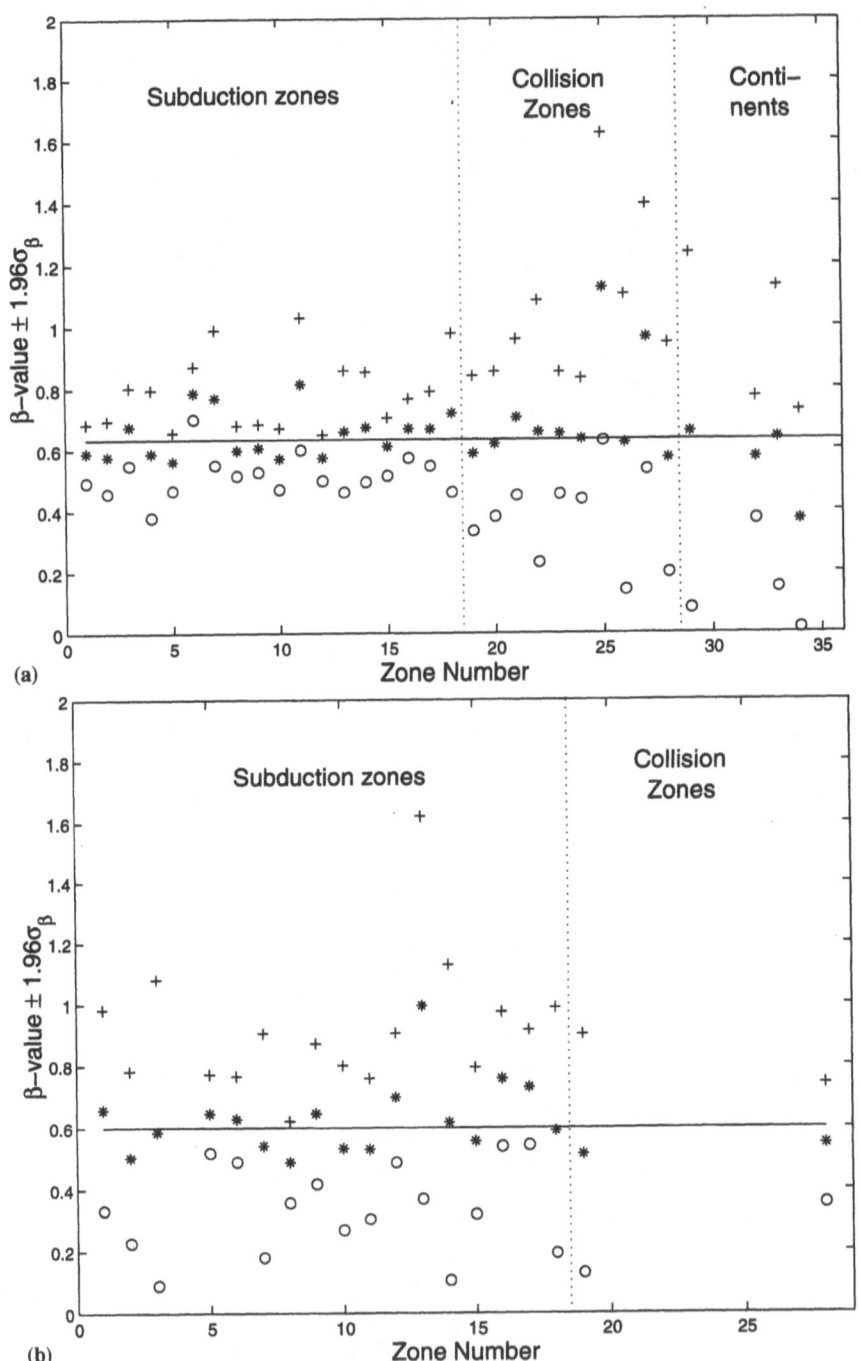

(a)

(b)

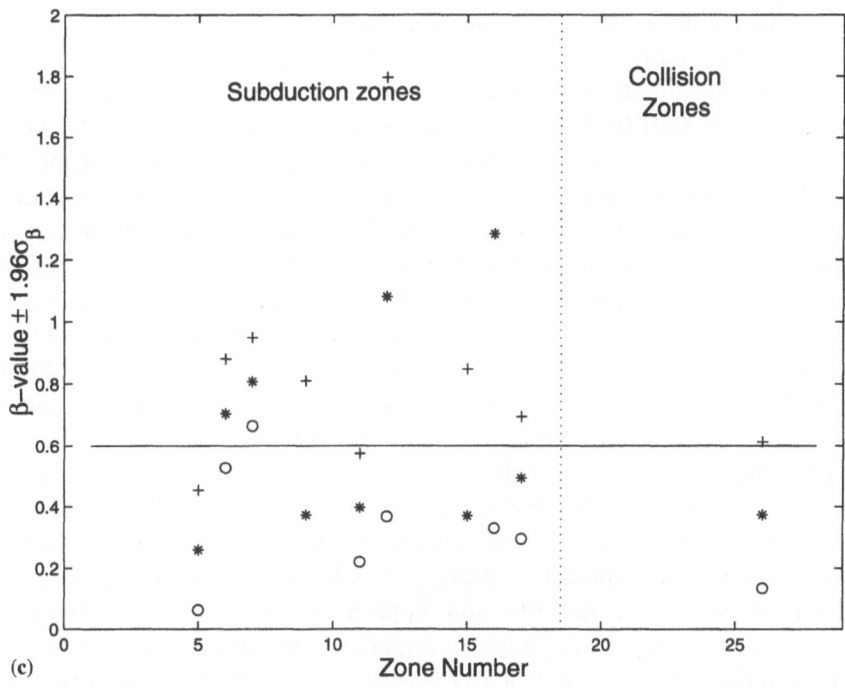

(c)

Figure 3

Regional distribution of β-values in seismic zones for the Harvard catalog. Solid line is the worldwide β average. (a) Shallow (0–70 km) earthquakes; (b) intermediate (70–300 km) earthquakes; (c) deep (300–700 km) earthquakes; Legend: star, β estimate; circle, $\beta - 1.96\sigma_\beta$; cross, $\beta + 1.96\sigma_\beta$.

difference between shallow and intermediate earthquakes, and $z = 1.98$ for the difference between shallow and deep events. One explanation for this disparity is that the maximum likelihood estimate of β may be biased when evaluated in plots like Figure 2, where it is seen as correlated with the estimates of m_{xg}-values. Thus, the change in the m_{xg}-value caused by the occurrence of a few large earthquakes may induce the change in the β-estimate. On a more heuristic level, fitting a straight line to a quasi-linear part of a curve such as in Figure 1 may be influenced by the presence of a few large earthquakes.

We carry out another test by approximating the earthquake size distribution in the moment range $10^{17.4} - 10^{20.7}$ (magnitude range is 5.6–7.8), i.e., before the curves are influenced by the maximum magnitude, by the power-law (Pareto) distribution truncated at both ends. In Section 5.2 we present our arguments for the equality of the maximum magnitude in different depth ranges. A view of Figure 1 seems to suggest that m_{xg} is smaller for deeper earthquakes compared to shallow events, however the likelihood functions shown in Figure 2 and similar diagrams for other subcatalogs indicate that the upper limit on the maximum magnitude is undefined for deeper earthquakes; thus we make the simplest assumption of m_{xg} equality.

Even if the maximum magnitude varies for shallow and deeper earthquakes, this would not strongly influence the estimates of β.

We use the maximum likelihood procedure (DEEMER and VOTAW, 1955) to evaluate β and its statistical errors. We obtained $\beta = 0.661 \pm 0.011$, $\beta = 0.619 \pm 0.025$, and $\beta = 0.597 \pm 0.036$, for shallow, intermediate, and deep earthquakes, respectively. [We use here three significant digits for β to easily verify calculations.] The new z-values are 1.53 and 1.71 for shallow-intermediate and shallow-deep comparison. The z-values are smaller than those obtained for curves in Figure 1, nonetheless they still significantly differ at the 90% confidence level.

There is one more reason for the difference in the β-values between shallow and deeper earthquakes. Worldwide values for shallow events are influenced by mid-ocean ridge earthquakes, which have relatively high β-values (OKAL and RO-MANOWICZ, 1994; KAGAN, 1997, see more in the Introduction section). We determine β-values separately for the *Compressive* earthquakes and the mid-ocean ridge events: $\beta = 0.638 \pm 0.015$ and $\beta = 0.833 \pm 0.036$, respectively. Similar values for $m_c = 5.8$ evaluated in KAGAN (1997) are $\beta = 0.635 \pm 0.015$ and $\beta = 0.920 \pm 0.048$. Therefore, with a smaller magnitude threshold, the β-value for mid-ocean earthquakes decreases significantly and approaches the value characteristic of *Compressive* earthquakes. Thus, ridge earthquakes may obey a universal moment-frequency relation similar to that of *Compressive* events. Lack of sufficient data in available catalogs makes it impossible to confirm this conjecture.

Accordingly, in the remainder of this paper we explore only the earthquakes in the *Compressive* regions, i.e., those regions that are of most interest for seismic hazard analysis. Another subset of earthquakes, that associated with volcanic activity must be deleted from consideration: the seismogenic zones near volcanoes are of limited size, hence the moment-frequency relation should be significantly different from those corresponding to earthquakes due to global tectonic activity (MAIN, 1996). However, removal of volcanic earthquakes from a catalog presents considerable difficulty, and since the number of these earthquakes is relatively small, we did not attempt to delete them.

4. Regional and Depth Variation of β

4.1. Distribution of β

In Tables 1–3 we display the β-values and their error estimates for three depth ranges. The values of $\beta \pm \sigma_\beta$ are listed for all *Compressive* regions and region categories with the number of events greater or equal to 5. Although one may doubt the usefulness of the β-values determined for regions of low seismicity, we may still test whether a statistically significant difference exists in the β's even with their large error bars. We evaluate the β-values in the magnitude range 5.6–7.8 (see

Table 1

Shallow earthquakes in the Harvard catalog (82/1/1–97/7/1): List of seismic zones

FE Flinn-Engdahl No. Seismic Region Name	n $M \geq 10^{17.4}$	$\beta \pm \sigma_\beta$	Depth (km) $H \pm \sigma_H$	$\dot{M}$ $\times 10^{20}$	$m_{xg} \pm \sigma_m$
1 1 Alaska-Aleutian Arc	192	0.59 ± 0.05	25.6 ± 11.6	2.00	8.55 ± 0.28
2 5 Mexico-Guatemala	121	0.58 ± 0.06	25.5 ± 15.4	0.84	8.23 ± 0.29
3 6 Central America	128	0.68 ± 0.07	24.0 ± 14.1	0.88	8.22 ± 0.29
4 7 Caribbean Loop	40	0.59 ± 0.11	21.6 ± 11.7	0.37	8.45 ± 0.32
5 8 Andean S. America	185	0.56 ± 0.05	29.0 ± 15.7	3.00	8.90 ± 0.28
6 12 Kermadec-Tonga-Samoa	358	0.79 ± 0.04	23.2 ± 14.3	2.10	8.09 ± 0.28
7 13 Fiji Is	52	0.77 ± 0.11	14.9 ± 3.9	0.81	8.86 ± 0.31
8 14 New Hebrides Is	258	0.60 ± 0.04	26.3 ± 14.2	1.70	8.18 ± 0.28
9 15 Bismarck-Solomon Is	285	0.61 ± 0.04	31.0 ± 16.5	1.74	8.12 ± 0.28
10 16 New Guinea	166	0.57 ± 0.05	24.2 ± 13.5	3.00	8.98 ± 0.29
11 18 Guam-Japan	61	0.81 ± 0.11	21.5 ± 10.8	0.90	8.82 ± 0.31
12 19 Japan-Kamchatka	298	0.57 ± 0.04	30.2 ± 13.6	3.00	8.52 ± 0.28
13 20 S.E. Japan-Ryukyu Is	51	0.66 ± 0.10	28.1 ± 12.4	0.64	8.69 ± 0.31
14 21 Taiwan	64	0.67 ± 0.09	24.1 ± 11.8	0.54	8.38 ± 0.31
15 22 Philippines	199	0.61 ± 0.05	26.8 ± 13.7	1.25	8.15 ± 0.28
16 23 Borneo-Celebes	219	0.67 ± 0.05	30.9 ± 13.6	1.47	8.20 ± 0.28
17 24 Sunda Arc	140	0.67 ± 0.06	34.4 ± 17.0	2.30	8.91 ± 0.29
18 46 Andaman Is-Sumatra	34	0.72 ± 0.13	34.2 ± 16.5	0.94	9.32 ± 0.33
SUBDUCTION ZONES	2851	0.634 ± 0.013	27.1 ± 14.7	27.48	8.49 ± 0.27
19 25 Burma-S.E. Asia	27	0.59 ± 0.13	26.7 ± 19.0	0.52	9.03 ± 0.35
20 26 India-Tibet-Szechwan-Yunan	33	0.62 ± 0.12	18.5 ± 8.7	0.45	8.76 ± 0.33
21 27 S. Sinkiang-Kansu	34	0.70 ± 0.13	17.4 ± 5.7	0.21	8.13 ± 0.33
22 28 Alma-Ata-Baikal	11	0.66 ± 0.22	19.5 ± 7.1		
23 29 W. Asia	50	0.65 ± 0.10	21.2 ± 11.1	0.38	8.30 ± 0.32
24 30 Middle East-Crimea-Balkans	49	0.63 ± 0.10	18.8 ± 10.2	0.29	8.10 ± 0.32
25 31 W. Mediterranean	20	1.13 ± 0.25	17.2 ± 6.5	0.15	8.29 ± 0.37
26 41 E. Asia	8	0.62 ± 0.25	24.1 ± 10.4		
27 47 Baluchistan	20	0.96 ± 0.22	17.3 ± 6.1	0.37	9.00 ± 0.37
28 48 Hindu Kush, Pamir	12	0.57 ± 0.19	18.0 ± 8.5		
COLLISION ZONES	264	0.681 ± 0.046	19.7 ± 10.8	2.37	8.43 ± 0.28
29 34 E. North America	6	0.66 ± 0.30	16.9 ± 4.8		
30 35 E. South America	0				
31 36 N.W. Europe	0				
32 37 Africa	42	0.57 ± 0.10	16.0 ± 4.9		
33 38 Australia	8	0.64 ± 0.25	16.4 ± 3.1		
34 42 N.E. Asia-Greenland	9	0.37 ± 0.18	14.4 ± 2.0		
35 49 N. Asia	0				
36 50 Antarctica	0				
INTRACONTINENTAL REGIONS	65	0.554 ± 0.080	15.9 ± 4.5		
GLOBAL COLLISION	3180	0.636 ± 0.013	26.3 ± 14.5	29.85	8.41 ± 0.28

FE, Flinn-Engdahl seismic region; n, the number of shallow (depth limit 0–70 km) events in the Harvard catalog. Seismic moment M and moment rate $\dot{M}$ are measured in Nm and Nm/yr, respectively.

above). The *Compressive* worldwide average of β is also shown in Tables 1–3. We also display the average depth of the moment centroids with the corresponding standard deviation.

As expected for shallow earthquakes, the average centroid depth is the largest for subduction zones and smallest for intracontinental regions; the zones of continental collision are in an intermediate range. The depth standard deviation increases for deeper earthquakes, both in absolute values and relative to the thickness of each layer. This feature of the hypocenter distribution corresponds to a higher spatial fluctuation of deeper seismicity (see the Discussion section).

Table 2

Intermediate earthquakes: list of seismic zones (82/1/1–97/7/1)

	FE No.	Flinn-Engdahl Seismic Region Name	n $M \geq 10^{17.4}$	$\beta \pm \sigma_\beta$	Depth (km) $H \pm \sigma_H$
1	1	Alaska-Aleutian Arc	19	0.66 ± 0.17	139.9 ± 54.2
2	5	Mexico-Guatemala	19	0.50 ± 0.14	111.4 ± 33.0
3	6	Central America	7	0.58 ± 0.25	131.6 ± 58.4
4	7	Caribbean Loop	4		123.5 ± 28.4
5	8	Andean S. America	123	0.65 ± 0.06	147.0 ± 53.6
6	12	Kermadec-Tonga-Samoa	100	0.63 ± 0.07	173.7 ± 57.6
7	13	Fiji Is	12	0.54 ± 0.19	210.6 ± 33.0
8	14	New Hebrides Is	82	0.49 ± 0.07	156.9 ± 44.2
9	15	Bismarck-Solomon Is	38	0.65 ± 0.12	121.9 ± 38.5
10	16	New Guinea	22	0.53 ± 0.14	135.5 ± 60.8
11	18	Guam-Japan	30	0.53 ± 0.12	158.5 ± 51.0
12	19	Japan-Kamchatka	50	0.70 ± 0.11	123.4 ± 41.8
13	20	S.E. Japan-Ryukyu Is	10	0.99 ± 0.32	133.4 ± 37.7
14	21	Taiwan	7	0.62 ± 0.26	129.0 ± 39.5
15	22	Philippines	29	0.56 ± 0.12	143.1 ± 53.1
16	23	Borneo-Celebes	52	0.76 ± 0.11	129.0 ± 39.3
17	24	Sunda Arc	67	0.73 ± 0.10	140.0 ± 55.5
18	46	Andaman Is-Sumatra	11	0.59 ± 0.20	112.3 ± 45.6
		SUBDUCTION ZONES	682	0.624 ± 0.027	145.6 ± 53.2
19	25	Burma-S.E. Asia	10	0.52 ± 0.20	113.8 ± 28.1
20	26	India-Tibet-Szechwan-Yunan	1		101.8 ± 0.00
21	27	S. Sinkiang-Kansu	0		
22	28	Alma-Ata-Baikal	0		
23	29	W. Asia	1		157.2 ± 0.00
24	30	Middle East-Crimea-Balkans	4		93.8 ± 22.9
25	31	W. Mediterranean	2		182.6 ± 112.2
26	41	E. Asia	0		
27	47	Baluchistan	1		105.0 ± 0.00
28	48	Hindu Kush, Pamir	43	0.55 ± 0.10	
		COLLISION ZONES	62	0.501 ± 0.078	154.5 ± 58.6
		GLOBAL ZONES	745	0.612 ± 0.025	146.2 ± 53.8

FE, Flinn-Engdahl seismic region; n, the number of intermediate (depth limit 70–300 km) events.

Table 3

Deep earthquakes: list of seismic zones (82/1/1–97/7/1)

	FE No.	Flinn-Engdahl Seismic Region Name	n $M \geq 10^{17.4}$	$\beta \pm \sigma_\beta$	Depth (km) $H \pm \sigma_H$
1	1	Alaska-Aleutian Arc	0		
2	5	Mexico-Guatemala	0		
3	6	Central America	0		
4	7	Caribbean Loop	0		
5	8	Andean S. America	26	0.26 ± 0.10	597.5 ± 23.7
6	12	Kermadec-Tonga-Samoa	71	0.70 ± 0.09	522.7 ± 93.2
7	13	Fiji Is	136	0.81 ± 0.07	553.6 ± 84.5
8	14	New Hebrides Is	3		621.1 ± 32.6
9	15	Bismarck-Solomon Is	6	0.37 ± 0.22	425.9 ± 57.6
10	16	New Guinea	0		
11	18	Guam-Japan	38	0.40 ± 0.09	481.3 ± 75.1
12	19	Japan-Kamchatka	9	1.08 ± 0.36	394.2 ± 108.3
13	20	S.E. Japan-Ryukyu Is	4		418.4 ± 77.3
14	21	Taiwan	0		
15	22	Philippines	5	0.37 ± 0.24	560.6 ± 56.5
16	23	Borneo-Celebes	7	1.28 ± 0.49	451.1 ± 120.2
17	24	Sunda Arc	37	0.50 ± 0.10	539.3 ± 101.6
18	46	Andaman Is-Sumatra	0		
		SUBDUCTION ZONES	342	0.614 ± 0.037	531.5 ± 95.4
19	25	Burma-S.E. Asia	0		
20	26	India-Tibet-Szechwan-Yunan	0		
21	27	S. Sinkiang-Kansu	0		
22	28	Alma-Ata-Baikal	0		
23	29	W. Asia	0		
24	30	Middle East-Crimea-Balkans	0		
25	31	W. Mediterranean	0		
26	41	E. Asia	20	0.37 ± 0.12	508.7 ± 97.9
27	47	Baluchistan	0		
28	48	Hindu Kush, Pamir	0		
		COLLISION ZONES	20	0.373 ± 0.122	508.7 ± 97.9
		GLOBAL ZONES	362	0.597 ± 0.036	530.3 ± 95.7

FE, Flinn-Engdahl seismic region; n, the number of deep (depth limit 300–700 km) events.

Figure 3 displays the β value estimates for all *Compressive* Flinn-Engdahl seismic zones; 1.96 standard deviations have been added and subtracted from the estimate to indicate the 95% confidence intervals. The worldwide averages are within the confidence interval for almost all individual regions, implying that the region variations are not statistically significant at the 95% confidence level. The 95% confidence level actually means that in one case out of 20, the confidence interval could be outside the variable's real range because of random fluctuations. For shallow earthquakes the confidence bounds are outside the average value for only one Flinn-Engdahl's region (Kermadec-Tonga-Samoa). All intervals intersect the worldwide average for intermediate shocks (Fig. 3b). Only deep earthquakes display large fluctuations which cannot be explained as the random variations due to a small event number. For three regions the β-intervals lie outside the confidence intervals: Andean South America, Fiji Islands, and Guam-Japan (Fig. 3c).

In Table 4 we display the β-values for several selections of catalogs and depth intervals. The last column shows the number of 'Outliers,' i.e., the number of 95% confidence intervals outside the β average values (see Fig. 3).

Another stricter test of the β-value universality (the Bartlett test) has been applied to regional β distribution (KAGAN, 1997; MOLCHAN et al., 1997). A small significance level (ε) suggests that the null hypothesis (all the β-values belong to the same population and their estimate difference is caused by random fluctuations) should be rejected. The results of these tests show that (a) the significance level decreases as the number of earthquakes available for analysis increases; (b) for declustered catalogs ε increases compared to the original catalog (see the next subsection for more detail); (c) deep earthquakes consistently disagree with the null

Table 4

Depth dependence of β-values and Bartlett's test of β-value homogeneity

Catalog	Depth range	n	$\beta \pm \sigma_\beta$	Zone #	Sum S	Signif. ε %	5% Outliers
O	0–70	3180	0.636 ± 0.013	32	48.2	2.5	1/0
L	0–70	2530	0.606 ± 0.014	32	43.1	7.6	2/0
O	70–300	745	0.613 ± 0.025	19	15.2	65	0/0
O	300–700	362	0.597 ± 0.036	10	58.4	<0.05	1/2
O_1	0–70	1497	0.638 ± 0.018	29	38.7	8.8	0/2
O_2	0–70	1683	0.634 ± 0.017	29	36.5	13.6	1/0
O_1	300–700	169	0.613 ± 0.053	6	26.7	<0.05	1/1
O_2	300–700	191	0.583 ± 0.048	7	32.1	<0.05	0/1
O	0–35	2357	0.638 ± 0.015	32	65.3	<0.05	0/1
L	0–35	1885	0.608 ± 0.016	32	59.6	0.3	1/2
O	35–70	823	0.628 ± 0.025	18	38.0	0.3	1/2
O	70–150	449	0.680 ± 0.035	17	16.8	40	0/0
O	150–300	296	0.525 ± 0.037	13	5.0	96	0/0
O	300–500	117	0.554 ± 0.060	8	11.3	13	0/1
O	500–700	245	0.619 ± 0.045	6	46.4	<0.05	1/1
U	0–70	576	0.617 ± 0.034	22	25.6	23	0/0
U	70–300	177	0.701 ± 0.064	13	14.6	27	0/0
U	300–700	76	0.512 ± 0.087	6	5.8	33	0/0
H	300–700	80	0.427 ± 0.084	6	13.9	1.8	0/1

O—original Harvard catalog; O_1—original Harvard catalog, first half in time; O_2—original Harvard catalog, second half in time; L—declustered Harvard catalog optimized by maximum likelihood procedure; U—USGS (SIPKIN and ZIRBES, 1997) catalog; H—HUANG et al. (1997) catalog; n, the number of earthquakes in a catalog; S—see KAGAN (1997), eq. (6); 5% outliers—the number of zones with 95% confidence intervals above (o/) or below /+) the average β-value (see Fig. 3).

hypothesis, whereas rejection of universality for shallow events is not as strong. As we pointed out earlier (KAGAN, 1997), this test assumes that the G-R relation can be extended to infinite magnitudes. If we introduce the maximum magnitude, the ε-values should increase, making the passage of the null hypothesis more likely.

We will first explore whether differences in the worldwide β-values at various depth intervals described above and shown in Tables 1–3 can be explained by earthquake clustering. A major difference between shallow and deeper earthquakes is the presence of extensive aftershock sequences for strong shallow events (FROHLICH, 1989; KAGAN, 1991c; WIENS et al., 1997). Thus, we would like to ascertain what change in the worldwide β-values ensues if aftershocks are removed.

4.2. Catalog Declustering

Since there is no standard procedure for aftershock removal, we try three declustering methods to find a possible bias in each. For one scheme we apply a variant of REASENBERG's (1985) method for the aftershock identification. We calculate a time-distance window around each earthquake in a catalog. The windows are scaled in time and distance domain according to the earthquake magnitude.

In particular, the epicentral distance limits (in km) are calculated according to the formula (cf., REASENBERG, 1985)

$$R(m) = 2.5 \times 10^{(1.2m - 4)/3}. \tag{5}$$

The aftershock zone sizes are 4.6 km for $m = 4$ and 46 km for $m = 6.5$. For the time limits we use the following formula, a variant of that used by REASENBERG (1985):

$$T(m) = \frac{10}{3} \times 10^{2(m - 4)/3}. \tag{6}$$

Equation (6) yields a time limit of 3.33 days for $m = 4$ and 33.3 days for $m = 5.5$. If an earthquake is in the time-distance window of another larger preceding event, it is deleted from the declustered catalog.

MOLCHAN et al. (1996, 1997, p. 1223) propose the following windows for removing aftershocks: for $m \leq 6.5$ a time limit of 1 year is proposed, for $m > 6.5$ the time range is 2 years. The spatial window radius is 50, 60, 70, 100, and 200 km for the earthquake magnitude ranges 5.5–6.5, 6.5–7.0, 7.0–7.5, 7.5–8.0, and ≥ 8.0, respectively.

We use the results of the likelihood analysis of earthquake catalogs (KAGAN, 1991c) as a third declustering method. We approximate an earthquake occurrence by a multidimensional Poisson cluster process, i.e., origin time, earthquake size, and centroid location are modelled. In this model mainshocks are distributed according to the Poisson distribution, whereas dependent events in earthquake sequences are controlled by a distribution which is characterized by a few adjustable parameters.

Table 5

Effect of declustering on β-values, depth range 0–70 km, Harvard catalog

Declust. method	n $m \geq 5.6$	% deleted	$\beta \pm \sigma_\beta$
O	3168	0	0.635 ± 0.013
L	2530	20.6	0.606 ± 0.014
R	2263	28.6	0.622 ± 0.015
M	2204	30.4	0.573 ± 0.014

O—original Harvard catalog; declustered catalogs: L—optimized by maximum likelihood procedure, R—REASENBERG (1985) method; M—MOLCHAN et al. (1997) method.

These parameters are estimated by a maximum likelihood search, and the probability of each earthquake being an independent event is evaluated. Thereafter, we remove the event from the catalog, depending on its probability of being an aftershock (dependent earthquake). This declustering method differs from the two windowing techniques described above, since any arbitrariness in the window selection is eliminated.

A model of earthquake occurrence similar to KAGAN (1991c), but excluding the earthquake location, has been proposed by OGATA (1988) and called ETAS (Epidemic-Type Aftershock Sequence). Computer programs and the description of the algorithm are presented in the IASPEI software library (UTSU and OGATA, 1997).

In general, only shallow earthquakes and intermediate events with a hypocentral depth of 70–100 km have a significant number of aftershocks (FROHLICH and DAVIS, 1993; KAGAN, 1991c). Therefore, we should expect a significant decrease of the b- or β-values due to clustering for such earthquakes.

The results, shown in Table 5, suggest that for shallow earthquakes we obtain new values of β 5–7% smaller than those in Tables 1–3 for individual seismic regions. Therefore, we conjecture that the difference in worldwide β-values for shallow and deeper earthquakes can be explained by aftershock sequences. This would mean that for all depth ranges the worldwide β-values for *mainshocks* are similar.

4.3. Variability and Universality of the β-value

The observed regional variations of the β-values may have several explanations:

● (a) The differences can be attributed to random fluctuations due to an insufficient number of cataloged earthquakes in seismic zones. Only these fluctuations are analyzed by the statistical tests shown in Tables 1–5.

● (b) It is possible that systematic errors bias the β-value estimate. Such errors strongly influence the b-value estimate, which may differ significantly and without an obvious correlation for various tectonic regions, depending on the magnitude scale used (KAGAN, 1991b; FROHLICH and DAVIS, 1993; see more in the next subsection). Several factors may introduce a systematic bias in the b- or β-values. Two factors (the saturation of magnitude scales and magnitude discretization and round-off, especially in old data, where the values ending in 0.5 and 0.0 are favored over others) are specific to determining the magnitude. Other causes, such as insufficient knowledge of earth structure, non-uniform distribution of seismic stations, and other biases specific to a particular method of evaluating earthquake size, influence both the magnitude and the scalar seismic moment estimates. One possible systematic effect is a variable magnitude cutoff in different seismic zones. With limited amounts of data available, especially for the seismic moment, it is very difficult to evaluate the moment threshold for smaller regions, if the cutoff is higher than that assumed for global seismicity. 'Missing' earthquakes would lead to underestimation of the β-values. Some earthquakes, like close-in-time aftershocks, may be treated differently in various seismic zones and depth ranges (HELFFRICH, 1997, p. 742). Such differences can also cause systematic effects in earthquake statistics.

● (c) Finally, seismic regions may have different physical properties which may influence the β-values: the age of the subducting slab, the plate convergence rate, slab composition, the dip angle, etc. (MCCAFFREY, 1997b).

If these physical properties were largely responsible for the b- and, β-value variations, we would observe a high coefficient correlation for these parameters values in various seismic zones. Thus, we can test the cause of the β-value variations by looking at the correlation of β-values for different depth intervals or different catalogs. Low correlation values would suggest that the variations are due to either random or systematic errors, although some systematic fluctuations may be common to different techniques employed in the moment inversion.

Figure 4 shows the correlation of earthquake numbers in the Harvard catalog in Flinn-Engdahl zones for two shallow depth ranges 0–35 and 35–70 km. For these two sets we have the maximum numbers of comparable earthquakes. The correlation coefficient between these numbers is relatively large (0.78). Thus the level of seismicity in the upper layer predicts well the seismic activity in deeper parts of a slab. However, for the β-values in these layers the correlation coefficient is -0.29 which would suggest that the β fluctuations are caused by random factors. The statistical fluctuations due to small earthquake numbers cannot account for the low correlation between the β-values. Thus, we assume that significant differences in the regional parameter estimates are caused by systematic effects, not by variation of physical properties in seismic zones.

Figure 5 displays a scatter plot of the β-values in two depth intervals. Table 6 displays some correlations of β-estimates for various regions. We only evaluate the correlation for regions which have at least 5 earthquakes in both catalogs or in both depth ranges of comparison. Most correlation values displayed in Table 6 are quite low and show no particular pattern; the differences in the magnitude cutoff for three catalogs may also introduce some bias in our estimate. The positive correlation for deep earthquakes in three catalogs indicates that the β-variations for this depth range may not be random.

4.4 b-values in the PDE Catalog

We investigate the b-values for two magnitude scales in the PDE catalog mostly to study the systematic effects in the magnitude-frequency relation and confirm our conclusions regarding the causes of β-value variations. As indicated earlier, some systematic errors may be common to the magnitude and the seismic moment evaluation. The b-values for both magnitudes (m_b and M_S) differ drastically and, since both magnitude-frequency curves are concave, the b-values are uniquely definable only in a limited magnitude range. We selected the magnitude range of one unit (e.g., $5.0 \leq m_b < 6.0$) to determine the b-values (cf., FROHLICH and DAVIS,

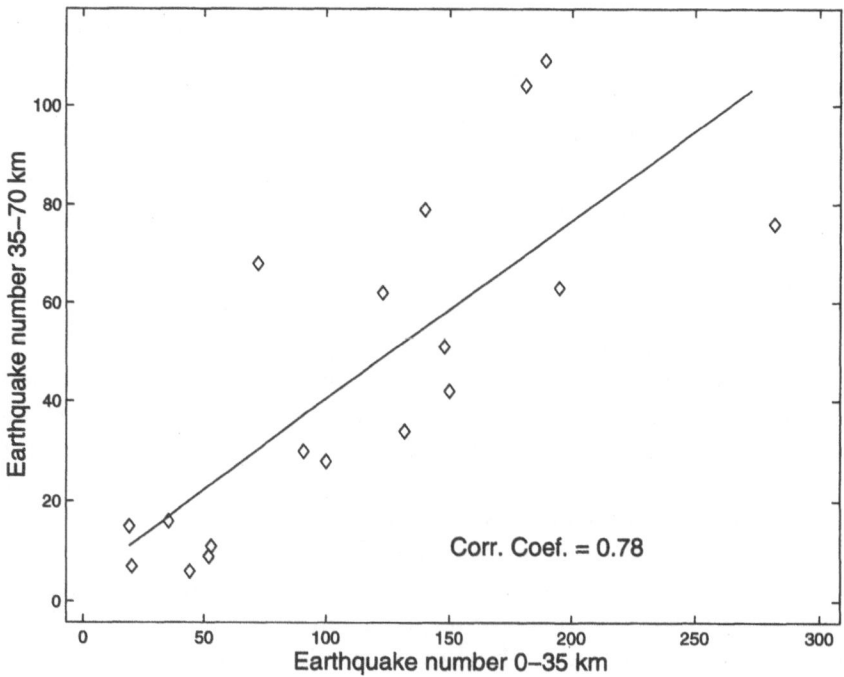

Figure 4
Correlation between the numbers of earthquakes in different depth intervals for the Harvard catalog.
Solid line is the linear regression approximation. Shallow earthquakes, depth 0–35 km versus 35–70 km,
Harvard catalog.

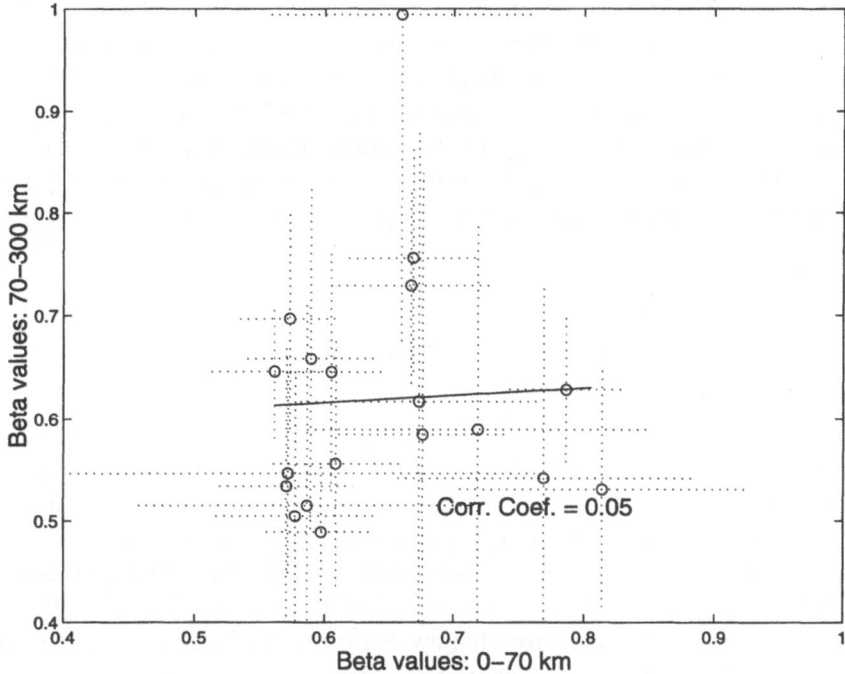

Figure 5

Correlation of β-values for earthquakes in different depth intervals for the Harvard catalog. Solid line is the linear regression approximation. Shallow (0–70 km) earthquakes versus intermediate (70–300 km), Harvard catalog.

1993). Depending on the magnitude range used, the b-values are 1.6–2.0 for $m_b \geq 5.0$, $m_b \geq 5.5$, or $m_b \geq 5.7$, and 0.9–1.05 for $M_S \geq 5.0$ or $M_S \geq 5.5$ (see also KAGAN, 1991b) for *Compressive* shallow earthquakes.

The numbers of 'Outliers' (defined in Table 4) are significantly higher compared to the seismic moment values, especially for m_b magnitude. The numbers are 8/4, 3/6, and 1/4, for $m_b \geq 5.0$, $m_b \geq 5.5$, and $m_b \geq 5.7$, respectively. This difference cannot be explained by a greater number of earthquakes in the PDE catalog and smaller confidence limits, since in the last case above, the total number of *Compressive* events is 3080, i.e., smaller than that for the Harvard catalog (see Table 1). For M_S magnitude, the number of outliers is 3/3 and 1/1 for $M_S \geq 5.0$ and $M_S \geq 5.5$, respectively. In the second case, the number of earthquakes is 3201, i.e., comparable to that in the Harvard catalog. The higher variability of the b-values for m_b is to be expected, since this body-wave magnitude is determined using teleseismic P-waves with the prevalent period of 1-s, whereas M_S magnitude is determined using 20-s surface waves (LAY and WALLACE, 1995, p. 16).

As in Figure 5 and Table 6, we determine the correlation between two sets of b-values in the PDE catalog. For the magnitude cutoff 5.0 the correlation coefficient is 0.09, for m_b and $M_S \geq 5.5$ the coefficient value is 0.20. These coefficient values indicate most probably that b-value variations are not caused by regional tectonic or physical factors. The relatively low correlation between b- and β-values displayed in Table 6 confirms this conclusion. Thus, it is quite possible that systematic biases play a major role in the b-value fluctuations, and the regional variations of the b-values are largely artifacts of systematic effects. In such a case, accumulation of additional data may not improve our estimate of the b-values.

5. Maximum Earthquake Size Variation

5.1. Maximum Earthquake Determination

Several methods can be used to estimate the maximum earthquake size, i.e., the maximum seismic moment, $M_{\max}$ (YEATS et al., 1997, p. 452)—see also our discussion of the difference between $M_{\max}$ and M_{xg} below Equation (3).

1. The simplest one is based on earthquake history (for example, RUFF and KANAMORI, 1980), i.e., the largest observed earthquake is considered as $M_{\max}$. Its drawbacks are clear, since even for shallow seismicity the available catalogs in all earthquake zones are considerably shorter than the recurrence time of the strongest earthquakes (cf., FROHLICH, 1998). Moreover, since the exact form of the statistical distribution for the largest earthquakes is unknown, a putative recurrence time can only be estimated with great uncertainty. Thus, we cannot be sure that such earthquakes are present in the record. For instance, since the publication of RUFF and KANAMORI (1980), two large shallow earthquakes occurred in the Marianas and Java subduction zones (1993/8/8 and 1994/6/2, respectively), their magnitude (7.8 in both cases, according to the Harvard catalog) significantly exceeded the maximum size of the earthquake quoted for these regions.

2. According to the characteristic earthquake hypothesis (see JACKSON et al., 1995; KAGAN, 1996), the maximum earthquake size can be determined by measuring the length of an earthquake fault or a fault segment which is assumed to fail in characteristic earthquakes. However, it has been observed that in many cases a

Table 6

Correlation of β-values in different depth intervals and catalogs

Harvard depth	Harvard		USGS			HUANG	PDE, 0–70 km			
	70–300	300–700	0–70	70–300	300–700	300–700	$m_b \geq 5.0$	≥ 5.5	$M_S \geq 5.0$	≥ 5.5
0–70	0.05	0.07	0.60	–	–	–	0.09	0.45	−0.26	0.41
70–300	–	0.50	–	−0.08	–	–	–	–	–	–
300–700	–	–	–	–	0.62	0.53	–	–	–	–

large earthquake would rupture through several faults or segments which have been defined as characteristic. Calculations based on the moment conservation principle often yield a recurrence time of characteristic earthquakes which is too small compared to historical record (KAGAN, 1996). Several tests of long-term predictions, based on the characteristic model, also indicate that the probabilities assigned to these earthquakes are too low (KAGAN, 1996). Moreover, for many tectonic regions the fault information is either unavailable or insufficient to determine the characteristic magnitude. This evidence suggests that the characteristic earthquake method applies only to specific fault systems, is based on subjective judgements, and, generally, significantly underestimates the value of $M_{\max}$.

3. Another method to determine the maximum size is statistical and illustrated in Figure 2 (see also KAGAN, 1991b, 1997). Although this technique provides confidence limits for the estimate, these limits are not useful, unless we apply the method to worldwide earthquake data. An occurrence of a very large earthquake may also change our estimate of $M_{\max}$ or M_{xg}. This revision is not as drastic as in estimates based on earthquake history (see the first paragraph); however, as Figure 2 demonstrates, the changes in the M_{xg}-values can be substantial.

4. The last method to evaluate maximum earthquake size uses the moment conservation principle (KAGAN, 1997). We surmise that earthquake size distribution follows the modified G-R relation (3). We integrate the moment-frequency relation (Fig. 1), assuming the universality of β to compute the seismic moment rate. Under such assumptions, the total seismic moment rate depends only on maximum magnitude. This estimate of the seismic moment rate is compared with the tectonic rate calculated from the plate tectonic considerations (DeMETS et al., 1990). By matching these two rates, the maximum earthquake size M_{xg} can be calculated together with its error bounds. We calculate the maximum magnitude for the gamma distribution as

$$m_{xg} = \frac{2}{3(1-\beta)}\left[\lg \dot{M} + \lg\left(\frac{1-\beta}{\beta}\right) + \lg \chi - \beta \lg M_c - \lg \alpha_c - \lg \Gamma(2-\beta)\right] - 6,$$

(7)

where Γ is a gamma function, α_c is the yearly number of events above the threshold level, and $\dot{M}$ is the geologic rate of seismic moment release from plate motion models (Nm/yr). Formula (9) in KAGAN (1997) assumes that the seismic moment is measured in dyne-cm. The crucial assumption in these computations is the value of the seismic efficiency coefficient, χ, i.e., the part of the tectonic deformation rate which is accounted for by earthquakes (see subsection 5.3).

5.2. Maximum Earthquake Distributions

In Table 1 (the last column) and Figure 6, we display the distribution of the maximum magnitude that has been estimated, assuming that all plate tectonic

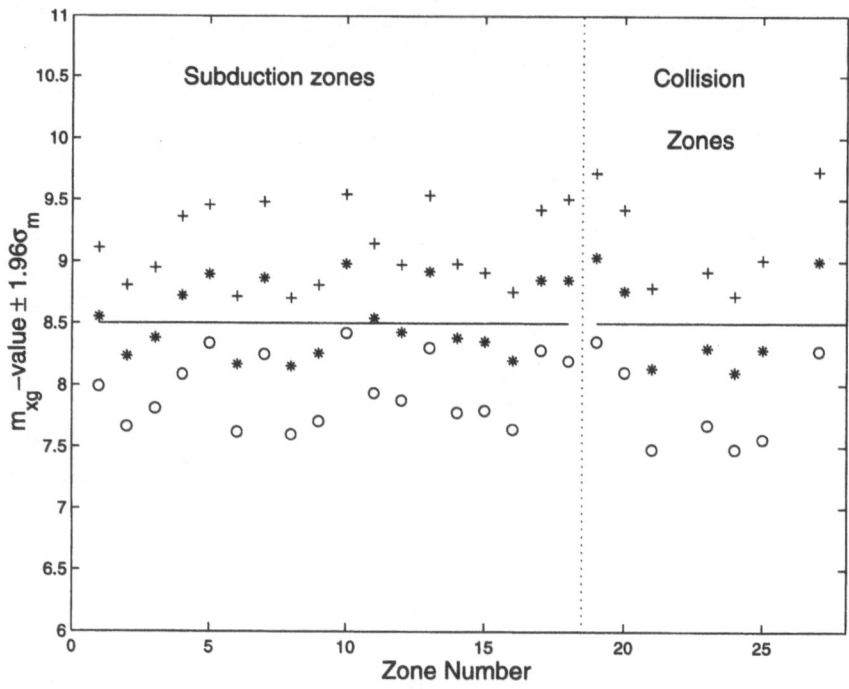

Figure 6

Regional distribution of m_{xg} values for shallow earthquakes for the Harvard catalog. Solid line is the worldwide m_{xg} average. Legend: star, m_{xg} estimate; circle, $m_{xg} - 1.96\sigma_m$; cross, $m_{xg} + 1.96\sigma_m$.

deformation is released by earthquakes ($\chi = 1$). In computations displayed in Figure 6 we use McCaffrey's (1997b) estimates of plate velocities which, generally provide results similar to those of Kagan (1997). Following Okal and Romanowicz (1994) and Kagan (1997), we take 30 km for subduction zones and 20 km for collision and intracontinental zones as the value of W: the seismic zone width. All confidence intervals for m_{xg} intersect the worldwide value of the maximum magnitude $m_{xg} = 8.5$. If we assume that all tectonic deformation is released in the upper 35 km of the crust, the worldwide estimate of m_{xg} is 8.6.

We do not correct for the bias in m_{xg} evaluation due to random errors in determining the scalar seismic moment (Kagan, 1997). The correction in the maximum magnitude is $+0.09$. In this paper, we are interested in the relative relations between the m_{xg}-values in various regions. The absolute values for the global m_{xg} are discussed by Kagan (1994, Fig. 4b); the major difficulty in its determination, if one uses the seismic moment conservation principle, is its dependence on values of W and the shear modulus μ not known with a high degree of accuracy (see more in subsection 5.3).

We cannot compute the maximum moment for deeper earthquakes using the moment conservation principle, since most tectonic deformation is released aseismically (KIRBY et al., 1996). However, we can construct diagrams similar to Figure 6 for these earthquakes, to investigate whether estimates of the maximum moment differ among various subduction zones. Such plots reveal that the consistency of seismic moment release, although it deteriorates compared to shallow seismicity, still conforms to the uniformity of the maximum moment for deeper earthquakes. For intermediate events, two regions (New Hebrides Islands and New Guinea) have M_{xg} confidence intervals outside the average maximum moment value (too low and too high, respectively). The maximum moment for deep earthquakes in two zones—Kermadec-Tonga-Samoa and Fiji Islands—is consistently too low for all tectonic rate estimates (KIRBY et al., 1996; KAGAN, 1997; McCAFFREY, 1997b).

In Figure 7, we present two cases of correlation between the tectonic deformation rate and the earthquake number in the *Compressive* Flinn-Engdahl zones. The tectonic deformation is estimated based on McCAFFREY (1997b) for Figure 7a and on KIRBY et al. (1996) for Figure 7b. Both plots manifest a significant correlation between these quantities; the correlation coefficient is larger than 0.8. The zones which exhibit the largest deviations in Figure 7a are (from left to right) Sunda Arc, Andean S. America, and New Guinea in the upper-left part of the plot, and Kermadec-Tonga-Samoa in the lower-right part. These regions display values larger and lower than the average values of the maximum magnitude, respectively (Table 1). However, it would be premature to infer any conclusions from m_{xg} fluctuations such as those in Figure 7. Since different estimates of maximum magnitude are not statistically significant, there is no point in discussing whether the m_{xg}-values for one region are higher than for another: these differences may be due to random fluctuations.

Table 7 summarizes the coefficients of correlation for various earthquake depths. The correlation decreases with depth for the following reasons: (a) our estimate of tectonic rate is based on the current plate velocity measurements, whereas deep seismicity results from the plate motion millions of years ago, and (b) tectonic movement at great depth is becoming more chaotic and difficult to estimate. However, the high level of correlation between earthquake numbers and tectonic deformation estimates confirms that maximum earthquake size is similar for various subduction regions. An alternative explanation for the high coefficient value, especially for shallow earthquakes, is to assume that the values for W, μ, and χ in (7) change simultaneously, so that they preserve the high correlation between the event number and the value of tectonic deformation rate. There is no available evidence to support the latter assumption.

However, the similar m_{xg}-values in the *Compressive* regions seem to contradict the available history of large earthquakes: earthquakes larger than m_{xg} occur in some regions; in other regions no very large earthquakes are recorded. As explained earlier (see below Eq. (3)), for the gamma distribution it is allowable for an earthquake to be larger than the m_{xg} estimate, a 'soft' limit. KAGAN (1997, p. 2850) discusses why the absence of observations of strong events in instrumental and

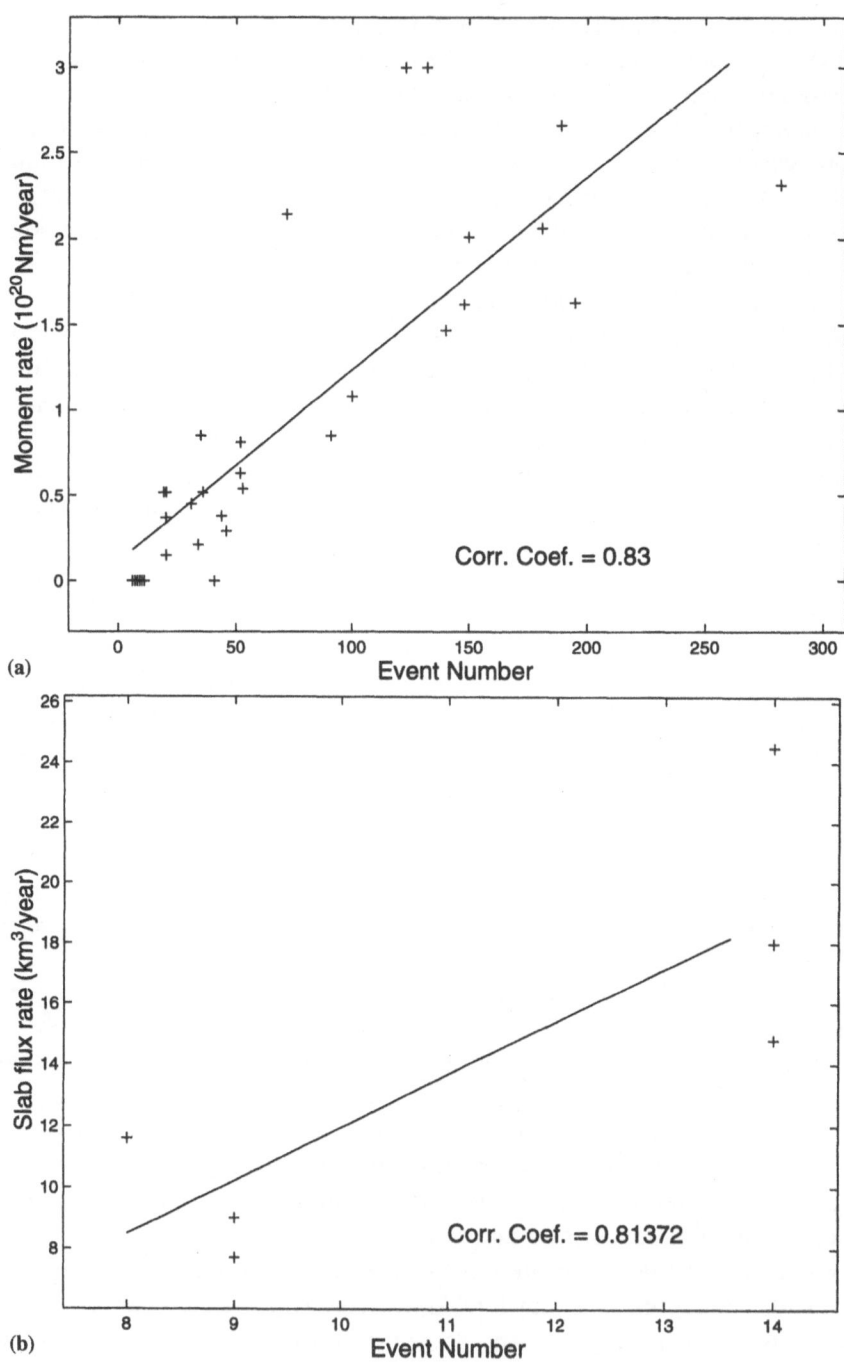

Figure 7
Correlation between the number of earthquakes and the moment rate. Solid line is the linear regression approximation. (a) Shallow earthquakes (depth 0–35 km), Harvard catalog; (b) deep earthquakes (depth 300–700 km), HUANG *et al.* (1997) catalog.

historic data should not be considered a refutation of the hypothesis of the m_{xg} similarity: the recurrence time of these very large earthquakes is on the order of hundreds and thousands of years. It significantly exceeds the available record.

The statistical analysis in this study does not prove that future earthquakes, for instance, in the Marianas (Japan-Guam in Table 1) arc, will be as large as the 1964 Alaskan earthquake or the 1960 Chilean quake. It only points out that the hypothesis that the Marianas will experience such large quakes in the future cannot be rejected with a high degree of confidence. Future data may show that the m_{xg} estimate for the Marianas is lower compared to that of Chile in a statistically significant manner, however our results suggest that the difference would be relatively small.

There is no major fundamental theory that requires maximum earthquakes in various seismic regions to significantly differ. Some models (e.g., that of UYEDA and KANAMORI, 1979) describe the Marianas as aseismic at shallow depths due to weak coupling between the slab and overriding arc. However, as the m_{xg}-value for Japan-Guam in Table 1 indicates, the number of earthquakes in that zone correlates well with the estimates of plate tectonic deformation, suggesting that at least in this respect the Marianas behave similarly to other subduction zones. Very large earthquakes occur unexpectedly in various regions and depth ranges (WIENS and McGUIRE, 1995; WIENS et al., 1998). To prove that the maximum magnitude varies significantly in different regions, an appropriate explanation should be found for the high correlation between earthquake numbers and the tectonic rate. This correlation is demonstrated in Figure 7 and Table 7, the stability of the m_{xg}-values (Table 1 and Fig. 6) is also a powerful argument for the universality of the maximum moment.

5.3. Seismic Efficiency Coefficient

Significant efforts have been made to compare the seismic moment rate with available estimates of tectonic deformation (PETERSON and SENO, 1984; PACHECO et al. 1993; SCHOLZ and CAMPOS, 1995; McCAFFREY, 1997a,b; PICHON et al. 1998). These investigators used a direct summation of the scalar seismic moment of cataloged earthquakes to infer the moment rate. Unfortunately, due to the power-law distribution of earthquake size, the sum is controlled by the largest earthquake (WYSS, 1973; KAGAN, 1996), therefore rate estimates based on the moment summation are highly unstable (McCAFFREY, 1997a). In this paper we evaluate the seismic moment rate, assuming that the moment is distributed according to the gamma law (3). Both estimates (summation and integration) yield similar results for very large regions (KAGAN, 1994). However, for smaller seismic zones integration is the only viable procedure.

Previously (KAGAN, 1997) we tried to evaluate the seismic efficiency coefficient, χ, for shallow earthquakes by comparing two estimates of M_{xg}: that obtained by purely statistical means and another by using the seismic moment conservation principle (items 3 and 4 in subsection 5.1). Both of the estimates have certain drawbacks. As explained earlier, the statistical estimate of M_{xg} depends on a few of the strongest earthquakes. As noted by KAGAN (1994), the Harvard catalog apparently spans a period of relatively low global seismic activity. Several exceptionally strong earthquakes occurred earlier in this century. Although their seismic moment cannot be estimated with an accuracy comparable to that of modern seismographic networks, the size of these earthquakes suggests that m_{xg} could be as high as 9.0–9.5 (KAGAN, 1994). In contrast, the tectonic rate estimate depends on the values of a few variables. One of these variables, the thickness of a seismogenic zone W, is subject to significantly different estimates. For example, Table 4 by McCAFFREY (1997b) provides the values of W estimated by three groups of investigators; these values sometimes vary by a factor of two. At the same time, TICHELAAR and RUFF (1993, Fig. 12) find that the W-values are similar (40 ± 5 km) for all subduction zones, except Mexico, for which they obtain $W = 20$ to 30 km. The reason for such disagreement can be traced to the highly inhomogeneous, fractal nature of the spatial distribution of seismicity. It is difficult to draw definitive, unambiguous boundaries for the depth distribution of earthquake hypocenters.

There are indications that tectonic deformation in the upper crust is released aseismically (see, for example, SHEN et al., 1994; HEKI et al., 1997; McCAFFREY, 1997b). However, comparing both estimates for the maximum earthquake size (KAGAN, 1997) as well as other evidence (McCAFFREY, 1997b; NORABUENA et al., 1998; PICHON et al. 1998) suggests that the seismic moment rate for shallow earthquakes should be a significant proportion of the tectonic rate: the χ-value should be of the order of at least 30–100%. Similar values of m_{xg} for different seismic regions in various depth ranges (subsection 5.2), suggest that the χ-value does not change significantly for *Compressive* and subduction zones. The high correlation coefficient between the earthquake numbers and plate tectonic rate (Fig. 7 and Table 7) calls for an explanation, and the only reasonable one is that β and m_{xg} are statistically the same for all regions considered.

We endeavor to quantify (at least at the order of the magnitude level) how the seismic efficiency coefficient χ depends on depth. Deeper earthquakes release only a small part of the tectonic deformation within their depth range (KIRBY et al., 1996). Thus we cannot estimate the maximum earthquake size by integrating the moment-frequency relation and comparing it with the plate motion. However, if we assume that the maximum earthquake size is similar at all depths, then we can estimate χ by calculating the seismic moment rate and comparing it to an approximate estimate of the subducting slab deformation in these depth intervals. There is no consensus (FROHLICH, 1989; STEIN, 1995; SILVER et al., 1995; GREEN and HOUSTON, 1995; LAY and WALLACE, 1995; KIRBY et al., 1996; NOTHARD et al., 1996; McGUIRE et al., 1997) on the rupture geometry and mechanism of the deeper earthquakes. This uncertainty makes estimation of the deformation rate ambiguous and highly nonunique. If we assume that deformation is proportional to layer thickness and the seismic efficiency coefficient, χ, is close to 100% for shallow seismicity, then for intermediate earthquakes $\chi \approx 5\%$ and for deep events $\chi \approx 1\%$ (Table 7).

If the χ-value is smaller for shallow earthquakes, the above estimates for deeper events need to be reduced accordingly. KIRBY et al. (1996, p. 296) suggest that mantle rigidity μ at the depth 300–700 km is significantly higher (~ 94 GPa) than that assumed in our calculations ($\mu = 30$ GPa, see KAGAN, 1997). If this value of μ is accepted, χ estimates should be reduced by a factor of about three for deeper earthquakes.

6. Discussion

Summarizing this investigation, we first address the robustness of results. Due to space limitations, only a small part of the obtained results can be reported in the paper. Several depth subdivisions have been investigated, to ascertain whether the β-value regularities, discussed above, are robust. We subdivided the Harvard catalog into two subcatalogs (cf., KAGAN, 1997, Fig. 4). We also compared (Table 7) several determinations of plate deformation; all these tests yield consistent results in the β- and the m_{xg}-values.

Table 7

Correlation of earthquake numbers with tectonic deformation

Catalog name	Depth range	K97	Correlation McC	Kirby	χ
Harvard	0–70	0.82	0.86		1.0
($m \geq 5.6$)	70–300	0.61	0.59		0.058
	300–700	−0.22	0.14	0.42	0.016
	0–35	0.79	0.83		1.0
	35–70	0.70	0.73		0.263
	70–150	0.66	0.64		0.065
	150–300	0.14	0.14		0.022
	300–500	−0.32	−0.09		0.006
	500–700	−0.07	0.20		0.013
HUANG	300–700	0.51	0.86	0.81	
($m \geq 6.3$)					

χ—seismic efficiency estimate; K97—KAGAN (1997); McC—MCCAFFREY (1997b); KIRBY—KIRBY et al. (1996); HUANG—HUANG et al. (1997) catalog.

6.1. b- and β-value Variations

There are two reasons why the universality of the *magnitude*-frequency relation, unlike the *moment*-frequency relation, cannot be proposed as a credible hypothesis. (a) The b-value depends on the magnitude scale used (see, for example, subsection 4.4, WYSS, 1973; FROHLICH and DAVIS, 1993). The magnitude is not a physical quantity; hence the interpretation of the b-value is generally speculative and depends on additional assumptions. (b) The decay of the high-magnitude distribution tail is produced mostly by a magnitude saturation effect (KANAMORI and ANDERSON, 1975; GELLER, 1976; KAGAN, 1991b).

Variations of the b-values can be as high as 50% or more, depending on the magnitude scale, depth range, and other factors (FROHLICH and DAVIS, 1993). In contrast, for shallow and intermediate earthquakes in sufficiently large continental regions, the fluctuations of the β-values are on the order of a few percent so that statistical estimates are relatively stable. The β-values for smaller zones exhibit broader variations, nonetheless they can be explained either by statistical fluctuations due to a few earthquakes, or by systematic effects in centroid-moment tensor inversions. However, the regional variations of the β-value appear real for deep earthquakes.

Several reasons can be proposed for higher regional fluctuations of the moment-frequency relation at depths greater than 70 km. The available earthquake catalog data span only a few decades. If we compare mantle convection to atmosphere turbulence which is about $10^8 - 10^9$ times faster than the mantle flow, a few decades for the former correspond to only a few seconds of air motion. Thus, in effect, we see a geologic 'snapshot' of mantle convection and resulting seismicity. If we were to study the statistical properties of the atmosphere for such small time spans, we would find great regional random variations. The only way to suppress the variations would be to use spatial averaging instead of temporal averaging. Global average values for the moment-frequency relation parameters are similar (Tables 1–5); therefore we conjecture that if we could observe regional seismicity for prolonged periods, we would see decreased regional random fluctuations. Might it similarly be possible that the depth distribution of deeper earthquakes which exhibits a bimodal pattern (FROHLICH, 1989; GREEN and HOUSTON, 1995; LAY and WALLACE, 1995), is also a transient feature?

Another reason for the higher variability of the β-value with depth is higher spatial fluctuations in deeper seismicity. Rock temperature and pressure are slowly changing continuous variables. Thus, if their influence on an earthquake occurrence is decisive, the hypocenter distribution would exhibit a slow, quasi-continuous decline with depth. Instead, deep seismicity is characterized by discontinuous, lacunary 'hot spots' of seismic activity (STEIN, 1995) and, as inspection of world-wide seismicity maps indicates, the degree of spatial intermittency increases with depth. This intermittency is best characterized quantitatively by the fractal dimension of earthquake hypocenters. GIARDINI (1988) and OKAL and KIRBY (1995, pp. 177–179) suggest that the β-value of deep earthquakes depends on the spatial fractal dimension of earthquakes, although they used a fractal dimension based on speculative assumptions, not on direct measurements (see also KAGAN, 1991b).

KAGAN (1991a) determined that the spatial correlation dimension of earthquake hypocenters is a function of depth and found that the dimension decreases from about 2.2 for shallow earthquakes to 1.6 for deep events. If seismicity is confined to relatively small volumes of subducting slab (hot-spots of activity), one should assume that the β-value is a highly fluctuating variable simply because earthquake size would be largely controlled by geometrical factors. However, these temporarily active volumes of seismic activity would not persist over geologically long periods of time. Consequently, we should expect higher spatial fluctuations in parameters for deeper seismicity to decrease if averaged over geologically long time periods. Shallow seismicity with a fractal dimension higher than the dimension of a plane (2.0) does not exhibit such large regional fluctuations.

Apparently the physical parameters of subducting slabs, such as temperature, slab age, and pressure can explain only the broad features of earthquake occurrence (FROHLICH, 1989; KIRBY et al., 1996; GREEN and HOUSTON, 1995). The details of earthquake distribution in space and time appear controlled by random factors which exhibit fractal behavior. Quasi-turbulent motion of fluids is known to produce patterns of fractally distributed regions of high strain (MANDELBROT, 1983). It is possible that mantle convection in the neighborhood of subducting slabs is very nonuniform, subjecting the slabs to strains which the slabs' colder material cannot accommodate plastically. The distribution of such regions might contain a fractal pattern as described above.

6.2. Universality of Moment-frequency Relation

Two kinds of seismic moment-frequency relation universality can be considered: weak universality for the long-term, worldwide seismicity distribution, and a more encompassing strong universality in which there is uniformity of earthquake size statistical distribution for arbitrary subdivision of seismicity into various seismic regions and time spans. Only the first type of universality can be proposed for deeper earthquakes. There are clear regional variations of β-values for very deep earthquakes (close to the 660 km seismic velocity discontinuity). Although the worldwide β-values for various depth ranges manifest uniformity, we can only conjecture that individual seismic regions would exhibit the universal β-value over geologically long time intervals.

The value of the maximum moment cannot be established for deeper earthquakes. Analyzing the data set of deep earthquakes for most of the 20th century, HUANG and OKAL (1998, p. 216) indicate that "large earthquakes ($M_0 \geq 10^{27}$ dyn-cm [$m > 7.3$]) can be expected at all depths, including 300 and 500 km, and in most subduction zones." They also demonstrate (Figs. 12 and 14, Table 5) that the maximum size of an observed earthquake is roughly proportional to the number of events in a corresponding depth range, suggesting that, given sufficient time, very large earthquakes may occur in all depth ranges.

Although the lack of statistically significant variation in M_{xg}-values for deeper earthquakes in different regions implies that maximum earthquake size is controlled by a general mechanism, the equality of M_{xg} in various depth ranges is still conjectural. A mechanism of sufficient generality must be proposed to explain the universality of the moment-frequency relation as well as the apparent regional variation of the β-values for deep earthquakes.

We are on firmer ground when we state that shallow earthquakes have a universal seismic moment-frequency relation. The β-values and the M_{xg}-values for the Flinn-Engdahl *Compressive* regions do not exhibit statistically significant variations, and the estimates of the maximum earthquake moment based on earthquake statistics and the moment conservation principle agree within reasonable bounds. Some β estimates diverge from a uniform universal value (see column ε in Table 4), however we found no regularities in the β-value deviations from the average of $\beta = 0.6$. Thus, we conclude that these β variations are possibly due to systematic effects in moment determination. Further investigations may reveal significant variations in the moment-frequency relation either for smaller regions, or for certain time intervals, although apparently such variations are not large for shallow events. If such regional or temporal variations of β were found, we conjecture that they would be on the order of a few percent for shallow earthquakes.

Many investigators studying the b-value behavior in local catalogs find that spatial fluctuations in b-value are substantial, sometimes of the order of 50% (see, for example, WIEMER and WYSS, 1997 and references therein). How can we reconcile these results with ours? There are several possible explanations for such a disagreement. (a) As we argue in subsection 4.4, the b-value determination apparently is strongly influenced by systematic errors. Thus for any b-value variations one needs to prove that this is not an artifact of some obvious or hidden magnitude bias. (b) It is possible that the universality paradigm is only valid on a regional scale (i.e., over a distance range of several hundreds of km); for smaller spatial scales β-values are heterogeneous. (c) Finally, although it is less likely, the universality of the moment-frequency relation may break down for smaller size earthquakes ($m \leq 6.0$). The questions above may soon be resolved due to the availability of seismic moment tensor solutions for small ($m \geq 3.5$) and moderate size earthquakes (ZHU and HELMBERGER, 1996; SONG and HELMBERGER, 1997, and references therein).

The results of these investigations may have significant implications for seismic hazard analysis and for theoretical modeling of seismicity. Many researchers (see, for example, KNOPOFF, 1996; LANGER *et al.*, 1996; RICE and BEN-ZION, 1996; RUNDLE *et al.*, 1997) try to reproduce earthquake behavior, including the size distribution of earthquakes. It is important to know whether the distribution is universal or not. Similarly, if the β and M_{xg} in Equation (3) have universal values, or, at least, do not vary significantly for shallow *Compressive* zones, we can adapt these results to evaluate seismic hazard. Applying the seismic moment conservation principle, and assuming that the seismic efficiency coefficient is close to 100%, we can use geodetic deformation data to calculate the number of earthquakes necessary to release the geological seismic moment rate (JACKSON *et al.*, 1995; KAGAN, 1997). This will allow calculations of future seismic hazard for areas with insufficient historical records of earthquakes, such as the eastern USA and other intracontinental areas. There is historic and paleoseismic evidence that these regions experienced earthquakes with magnitudes exceeding 8 (ARVIDSSON, 1996; JOHNSTON, 1996; KAGAN, 1997, and references therein). Therefore, it is important to evaluate hazard for these areas.

7. Conclusions

We studied parameters for earthquake size distribution and their standard errors for gobal catalogs of seismic moment solutions. The results can be summarized as follows:

1. Worldwide β-values for shallow, intermediate and deep mainshocks in nonspreading zones (*Compressive* tectonic environment) have a universal value of about 0.60.

2. Shallow and intermediate earthquakes do not exhibit statistically significant regional variation of the β-values; the regional variations appear statistically significant only for deep events near the 660 km mantle discontinuity.

3. Our analysis of the magnitude-frequency relations suggests that various biases dominate b-value fluctuations. Therefore, regional variations of the b-values are largely artifacts of systematic effects.

4. The earthquake maximum moment does not display statistically significant regional variation for any depth interval tested. For shallow earthquakes the universal value of the effective maximum moment magnitude is on the order of 8.5–9.0.

5. If we assume that the earthquake maximum moment is similar for all depth ranges, and for shallow events the seismic efficiency, χ, is close to 100%, then $\chi \approx 5\%$ for intermediate events and $\chi \approx 1\%$ for deep seismicity.

Acknowledgements

I am grateful to E. A. Okal of Northwestern University for providing data. I thank D. D. Jackson, H. Houston and D. Sornette of UCLA, I. G. Main of University of Edinburgh, R. Geller of Tokyo University, S. Uyeda of Tokai University, S. Duda of University of Hamburg, G. Molchan, A. Gusev, and G. S. Golitsyn of Russian Academy of Sciences for their useful comments. Remarks by reviewer C. Frohlich (University of Texas), by an anonymous reviewer, and by the Editor M. Wyss (University of Alaska) have been very helpful in revising the manuscript. This research was partially supported by the Southern California Earthquake Center (SCEC). SCEC is funded by NSF Cooperative Agreement EAR-8920136 and USGS Cooperative Agreement 14-08-0001-A0899 and 1434-HQ-97AG01718. The SCEC contribution number 455. Publication 5228, Institute of Geophysics and Planetary Physics, University of California, Los Angeles.

REFERENCES

ARVIDSSON, R. (1996), *Fennoscandian Earthquakes: Whole Crustal Rupturing Related to Postglacial Rebound*, Science 274, 744–746.

DEEMER, W. L., and VOTAW, D. F. (1955), *Estimation of Parameters of Truncated or Censored Exponential Distributions*, Ann. Math. Stat. 26, 498–504.

DEMETS, C., GORDON, R. G., ARGUS, D. F., and STEIN, S. (1990), *Current Plate Motions*, Geophys. J. Int. 101, 425–478.

DZIEWONSKI, A. M., EKSTRÖM, G., and MATERNOVSKAYA, N. N. (1998), *Centroid-moment Tensor Solutions for October–December, 1996*, Phys. Earth Planet. Inter. 105, 95–108.

FLINN, E. A., ENGDAHL, E. R., and HILL, A. R. (1974), *Seismic and Geographical Regionalization*, Bull. Seismol. Soc. Am. 64, 771–992.

FROHLICH, C. (1998), *Does Maximum Earthquake Size Depend on Focal Depth?*, Bull. Seismol. Soc. Am. 88, 329–336.

FROHLICH, C. (1989), *The Nature of Deep-focus Earthquakes*, Ann. Rev. Earth Planet. Sci. 17, 227–254.

FROHLICH, C., and DAVIS, S. D. (1993), *Teleseismic b Values; or, much Ado About 1.0*, J. Geophys. Res. 98, 631–644.

GELLER, R. J. (1976), *Scaling Relations for Earthquake Source Parameters and Magnitudes*, Bull. Seismol. Soc. Am. 66, 1501–1523.

GIARDINI, D. (1988), *Frequency Distribution and Quantification of Deep Earthquakes*, J. Geophys. Res. 93, 2095–2105.

GREEN, H. W., and HOUSTON, H. (1995), *The Mechanics of Deep Earthquakes*, Ann. Rev. Earth Planet. Sci. 23, 169–213.

HEKI, K., MIYAZAKI, S., and TSUJI, H. (1997), *Silent Fault Slip Following an Interplate Thrust Earthquake at the Japan Trench*, Nature 386, 595–598.

HELFFRICH, G. R. (1997), *How Good are Routinely Determined Focal Mechanisms? Empirical Statistics Based on a Comparison of Harvard, USGS and ERI Moment Tensors*, Geophys. J. Int. 131, 741–750.

HUANG, W.-C., and OKAL, E. A. (1998), *Centroid-moment Tensor Solutions for Deep Earthquakes Predating the Digital Era: Discussion and Inferences*, Phys. Earth Planet. Inter. 106, 191–218.

HUANG, W.-C., OKAL, E. A., EKSTRÖM, G., and SALGANIK, M. P. (1997), *Centroid-moment Tensor Solutions for Deep Earthquakes Predating the Digital Era: The Worldwide Standardized Seismograph Network Dataset (1962–1976)*, Phys. Earth Planet. Inter. 99, 121–129.

JACKSON, D. D., AKI, K., CORNELL, C. A., DIETERICH, J. H., HENYEY, T. L., MAHDYIAR, M., SCHWARTZ, D., and WARD, S. N. (1995), *Seismic Hazards in Southern California: Probable Earthquakes, 1994–2024*, Bull. Seismol. Soc. Am. 85, 379–439.

JOHNSTON, A. C. (1996), *A Wave in the Earth*, Science 274, 735.

KAGAN, Y. Y. (1991a), *Fractal Dimension of Brittle Fracture*, J. Nonlinear Sci. *1*, 1–16.

KAGAN, Y. Y. (1991b), *Seismic Moment Distribution*, Geophys. J. Int. *106*, 123–134.

KAGAN, Y. Y. (1991c), *Likelihood Analysis of Earthquake Catalogues*, Geophys. J. Int. *106*, 135–148.

KAGAN, Y. Y. (1994), *Observational Evidence for Earthquakes as a Nonlinear Dynamic Process*, Physica D *77*, 160–192.

KAGAN, Y. Y. (1996), *Comment on "The Gutenberg-Richter or Characteristic Earthquake Distribution, which is it?" by Steven G. Wesnousky*, Bull. Seismol. Soc. Am. *86*, 274–285.

KAGAN, Y. Y. (1997), *Seismic Moment-frequency Relation for Shallow Earthquakes: Regional Comparison*, J. Geophys. Res. *102*, 2835–2852.

KANAMORI, H., and ANDERSON, D. (1975), *Theoretical Basis of Some Empirical Relations in Seismology*, Bull. Seismol. Soc. Am. *65*, 1073–1095.

KIRBY, S. H., STEIN, S., OKAL, E. A., and RUBIE, D. C. (1996), *Metastable Mantle Phase Transformations and Deep Earthquakes in Subducting Oceanic Lithosphere*, Rev. Geophys. *34*, 261–306.

KNOPOFF, L. (1996), *The Organization of Seismicity on Fault Networks*, Proc. Nat. Acad. Sci. U.S.A. *93*, 3830–3837.

KRONROD, T. L. (1984), *Seismicity Parameters for the Main High-seismicity Regions of the World*, Vychislitel'naya Seismologiya *17*, 36–58 (Comput. Seismol., Engl. Transl. *17*, 35–54, 1984).

KRONROD, T. L. (1985), *Seismicity Parametrs in Tectonically Similar Regions*, Vychislitel'naya Seismologiya *18*, 154–164 (Comput. Seismol., Engl. Transl. *18*, 144–152, 1985).

LANGER, J. S., CARLSON, J. M., MYERS, C. R., and SHAW, B. E. (1996), *Slip Complexity in Dynamic Models of Earthquake Faults*, Proc. Nat. Acad. Sci. USA *93*, 3825–3829.

LAY, T., and WALLACE, T. C., *Modern Global Seismology* (Academic Press, San Diego 1995) 512 pp.

MAIN, I. G. (1996), *Statistical Physics, Seismogenesis, and Seismic Hazard*, Rev. Geophys. *34*, 433–462.

MANDELBROT, B. B., *The Fractal Geometry of Nature* (W. H. Freeman, San Francisco, Calif. 1983) 2nd edition, 468 pp.

McCAFFREY, R. (1997a), *Statistical Significance of the Seismic Coupling Coefficient*, Bull. Seismol. Soc. Am. *87*, 1069–1073.

McCAFFREY, R. (1997b), *Influence of Recurrence Times and Fault Zone Temperatures on the Age-rate Dependence of Subduction Zone Seismicity*, J. Geophys. Res. *102*, 22,839–22,854.

McGUIRE, J. J., WIENS, D. A., SHORE, P. J., and BEVIS, M. G. (1997), *The March 9, 1994 (M_w 7.6), Deep Tonga Earthquake: Rupture Outside the Seismically Active Slab*, J. Geophys. Res. *102*, 15,163–15,182.

MOLCHAN, G. M., KRONROD, T. L., DMITRIEVA, O. E., and NEKRASOVA, A. K. (1996), *Multiscale Model of Seismicity Applied to Problems of Seismic Risk: Italy*, Vychislitel'naya Seismologiya (Comput. Seismol.) *28*, 193–224, Nauka, Moscow (in Russian).

MOLCHAN, G., KRONROD, T., and PANZA, G. F. (1997), *Multiscale Seismicity Model for Seismic Risk*, Bull. Seismol. Soc. Am. *87*, 1220–1229.

NORABUENA, E., LEFFLER-GRIFFIN, L., MAO, A. L., DIXON, T., STEIN, S., SACKS, I. S., OCOLA, L., and ELLIS, M. (1998), *Space Geodetic Observations of Nazca-South America Convergence across the Central Andes*, Science *279*, 358–362.

NOTHARD, S., HAINES, J., JACKSON, J., and HOLT, B. (1996), *Distributed Deformation in the Subducting Lithosphere at Tonga*, Geophys. J. Int. *127*, 328–338.

OGATA, Y. (1988), *Statistical Models for Earthquake Occurrence and Residual Analysis for Point Processes*, J. Amer. Statist. Assoc. *83*, 9–27.

OKAL, E. A., and ROMANOWICZ, B. A. (1994), *On the Variation of b-values with Earthquake Size*, Phys. Earth Planet. Inter. *87*, 55–76.

OKAL, E. A., and KIRBY, S. H. (1995), *Frequency-moment Distribution of Deep Earthquakes; Implications for the Seismogenic Zone at the Bottom of Slabs*, Phys. Earth Planet. Inter. *92*, 169–187.

PACHECO, J. F., SCHOLZ, C. H., and SYKES, L. R. (1992), *Changes in Frequency-size Relationship from Small to Large Earthquakes*, Nature *355*, 71–73.

PACHECO, J. F., SYKES, L. R., and SCHOLZ, C. H. (1993), *Nature of Seismic Coupling along Simple Plate Boundaries of the Subduction type*, J. Geophys. Res. *98*, 14,133–14,159.

PETERSON, E. T., and SENO, T. (1984), *Factors Affecting Seismic Moment Release Rates in Subduction Zones*, J. Geophys. Res. *89*, 10,233–10,248.

PICHON, X. L., MAZZOTTI, S., HENRY, P., and HASHIMOTO, M. (1998), *Deformation of the Japanese Islands and Seismic Coupling: An Interpretation Based on GSI Permanent GPS Observations*, Geophys. J. Int. *134*, 501–514.

REASENBERG, P. (1985), *Second-order Moment of Central California Seismicity, 1969–1982*, J. Geophys. Res. *90*, 5479–5495.

RICE, J. R., and BEN-ZION, Y. (1996), *Slip Complexity in Earthquake Fault Models*, Proc. Nat. Acad. Sci. USA *93*, 3811–3818.

RUFF, L., and KANAMORI, H. (1980), *Seismicity and the Subduction Process*, Phys. Earth Planet. Inter. *23*, 240–252.

RUNDLE, J. B., KLEIN, W., GROSS, S., and FERGUSON, C. D. (1997), *Traveling Density Wave Models for Earthquakes and Driven Threshold Systems*, Phys. Rev. E *56*, 293–307.

SHEN, Z.-K., JACKSON, D. D., FENG, Y. J., CLINE, M., KIM, M., FANG, P., and BOCK, Y. (1994), *Postseismic Deformation Following the Landers Earthquake, California, 28 June 1992*, Bull. Seismol. Soc. Am. *84*, 780–791.

SCHOLZ, C. H., and CAMPOS, J. (1995), *On the Mechanism of Seismic Decoupling and Back Arc Spreading at Subduction Zones*, J. Geophys. Res. *100*, 22,103–22,115.

SILVER, P. G., BECK, S. L., WALLACE, C., MEADE, C., MYERS, S. C., JAMES, D. E., and KUEHNEL, R. (1995), *Rupture Characteristics of the Deep Bolivian Earthquake of 9 June 1994 and the Mechanism of Deep-focus Earthquakes*, Science *268*, 69–73.

SIPKIN, S. A., and ZIRBES, M. D. (1997), *Moment-tensor Solutions Estimated Using Optimal Filter Theory: Global Seismicity, 1995*, Phys. Earth Planet. Inter. *101*, 291–301.

SONG, X. J., and HELMBERGER, D. V. (1997), *Northridge Aftershocks, a Source Study with TERRAscope Data*, Bull. Seismol. Soc. Am. *87*, 1024–1034.

STEIN, S. (1995), *Deep Earthquakes—A Fault Too Big*, Science *268*, 49–50.

TICHELAAR, B. W., and RUFF, L. J. (1993), *Depth of Seismic Coupling along Subduction Zones*, J. Geophys. Res. *98*, 2017–2037.

U.S. GEOLOGICAL SURVEY (1997), *Preliminary Determination of Epicenters (PDE), Monthly Listings*, U.S. Dept. of Inter., Natl. Earthquake Inf. Cent., Denver.

UYEDA, S., and KANAMORI, H. (1979), *Back-arc Opening and the Mode of Subduction*, J. Geophys. Res. *84*, 1049–1061.

UTSU, T., and OGATA, Y., *Statistical analysis of seismicity*. In *IASPEI Software Library*, *6* (eds. Healy, J. H., Keilis-Borok, V. I., and Lee, W. H. K.) (Int. Assoc. of Seismol. and Phys. of the Earth's Inter. and Seismol. Soc. Am., El Cerrito, CA 1997) pp. 13–94 .

WIEMER, S., and WYSS, M. (1997), *Mapping the Frequency-magnitude Distribution in Asperities: An Improved Technique to Calculate Recurrence Times?*, J. Geophys. Res. *102*, 15,115–15,128.

WIENS, D. A., and McGUIRE, J. J. (1995), *The 1994 Bolivia and Tonga Events—Fundamentally Different Types of Deep Earthquakes*, Geophys. Res. Lett. *22*, 2245–2248.

WIENS, D. A., and GILBERT, H. J. (1996), *Effect of Slab Temperature on Deep-earthquake Aftershock Productivity and Magnitude-frequency Relations*, Nature *384*, 153–156.

WIENS, D. A., WYSESSION, M. E., and LAWVER, L. (1998), *Recent Oceanic Intraplate Earthquake in Balleny Sea was Largest Detected*, EOS Trans. AGU *79*(30), 353–354.

WIENS, D. A., GILBERT, H. J., HICKS, B., WYSESSION, M. E., and SHORE, P. J. (1997), *Aftershock Sequences of Moderate-sized Intermediate and Deep Earthquakes in the Tonga Subduction Zone*, Geophys. Res. Lett. *24*, 2059–2062.

WYSS, M. (1973), *Towards a Physical Understanding of the Earthquake Frequency Distribution*, Geophys. J. R. Astr. Soc. *31*, 341–359.

YEATS, R. S., SIEH, K., and ALLEN, C. R., *The Geology of Earthquakes* (Oxford University Press, New York 1997) 568 pp.

YOUNG, J. G., PRESGRAVE, B. W., AICHELE, H., WIENS, D. A., and FLINN, E. A. (1996), *The Flinn-Engdahl Regionalisation Scheme: The 1995 Revision*, Phys. Earth Planet. Inter. *96*, 223–297.

ZHU, L. P., and HELMBERGER, D. V. (1996), *Advancement in Source Estimation Techniques Using Broadband Regional Seismograms*, Bull. Seismol. Soc. Am. *86*, 1634–1641.

(Received August 19, 1998, revised December 23, 1998, accepted December 27, 1998)

 To access this journal online:
http://www.birkhauser.ch

Pure appl. geophys. 155 (1999) 575–607
0033–4553/99/040575–33 $ 1.50 + 0.20/0

⌐ Pure and Applied Geophysics

Physical Basis for Statistical Patterns in Complex Earthquake Populations: Models, Predictions and Tests

JOHN B. RUNDLE,[1] W. KLEIN[2] and SUSANNA GROSS[3]

Abstract—Understanding the physics of earthquakes and the space-time patterns they produce is illuminated by the use of coarse-grained models and simulations that capture the basic physical processes, and that are amenable to analysis. We present a summary of ideas that describe the nucleation, growth, and arrest of earthquakes on individual faults. Under shear loading, we find that faults reside in a metastable state near a classical spinodal that governs the nucleation and growth of slip events. The roughness of an associated stress distribution field $\Sigma(\mathbf{x}, t)$ determines whether slip events are confined within the initial high stress patch, or break away and grow to become very large. We find a critical value of roughness that is associated with a first-order, "order–disorder" transition. We also give a number of predictions, examples and applications of these ideas, and indicate how they might be tested through systematic observational programs.

Key words: Earthquakes, friction, threshold systems, stochastic resonance, nonequilibrium systems, driven dissipative systems.

1. Introduction

Earthquakes represent a class of nonlinear processes arising from the interaction between the underlying fluid driving processes of mantle convection, and the constraints imposed by deformation processes within the more brittle crustal regions of the earth. These processes, which are represented by a fracture of intact rock and frictional slip on pre-existing faults, result in space-time patterns of seismicity and earthquake rupture that are known to be recurrent over many thousands of years in the seismically active fault zones of the earth. In work carried out over the past several years, a physical picture is emerging of the earthquake

[1] Department of Physics and Colorado Center for Chaos and Complexity, Cooperative Institute for Research in Environmental Sciences, University of Colorado, Boulder, CO 80309, USA.
[2] Department of Physics and Center for Computational Science, Boston University, Boston, MA 02215, USA.
[3] Cooperative Institute for Research in Environmental Sciences, Colorado Center for Chaos and Complexity, University of Colorado, Boulder, CO 80309, USA.

source process that is based on applying ideas from statistical mechanics to these inherently nonequilibrium threshold systems. This picture, and the processes that are implied by it, have important implications for the physics of earthquake nucleation, growth and arrest, seismic clustering, the development of space-time patterns and correlations on systems of faults, and the possibility of earthquake forecasting (RUNDLE *et al.*, 1995, 1996a,b, 1997a–c, 1998a,b: KLEIN *et al.*, 1996, 1997; GOLDSTEIN *et al.*, 1998; FERGUSON *et al.*, 1998).

We have focussed upon understanding the emergent properties of systems of faults, and the resulting space-time correlations in the populations of earthquakes that result. In such efforts, the details of seismic wave propagation over a fault and into the surrounding medium are not important. Only the initial and the final slip and stress states of a fault that undergoes an earthquake event are important. How the fault makes the transition from the initial state to the final state is only important in a statistical sense, not in a deterministic sense, because the space-time correlations that average over many thousands of events are not sensitive to the time-dependent details of the slip process in individual earthquakes.

Because we are interested in space-time correlations, our research has been focussed on understanding recurrent fault models, in which a population of many earthquakes occurs over a long period of time. We have begun by considering the simplest possible planar fault models, with two basic friction laws, the rationale for which will be discussed below. Stress transfer interactions between different spatial sites on the faults have been characterized by simple nearest-neighbor interactions, up to complete elastic-viscoelastic stress Green's functions (RUNDLE, 1988). A new Traveling Density Wave Model (RUNDLE *et al.*, 1996, 1997) was developed based upon slip-weakening friction that includes the effects of disorder, wear and abrasion. We have recently been working with highly realistic, fully three-dimensional models for complex vertical strike-slip fault systems embedded in an elastic layer overlying a Maxwell viscoelastic half space, driven by persistent plate tectonic driving forces (RUNDLE *et al.*, 1998).

Using these models and simulations, we have identified a number of processes that arise directly from the space-time correlations generated by the dynamics. From the statistical-mechanical analysis of the simulations (RUNDLE *et al.*, 1996; KLEIN *et al.*, 1997), two nucleation processes were found to be important, "spinodal" and "arrested" nucleation. Furthermore, the growth and eventual arrest of earthquakes were found to depend on the statistical distribution of heterogeneity, leading to an order-disorder transition when the level of heterogeneity passes through a critical value. Finally, methods have been developed to analyze and understand the development of space-time correlations and the resulting space-time patterns in earthquake populations in arbitrarily complex fault systems. These "Pattern Dynamics" (RUNDLE *et al.*, 1998) methods use Principal Component Analysis as the starting point, and suggest new methods for the development of forecasting/prediction tools for earthquakes occurrence.

As this work has developed over the past several years, a number of specific predictions and consequences have emerged that demand systematic testing through observational and laboratory means. In this paper, we present a summary, together with a tabulation of the predictions that arise from it, in the hope that workers in the field will find the ideas useful as a starting point. We encourage them to test, and perhaps to falsify, these models and ideas. Note that we have collected all symbols and definitions into Table 1.

2. Simulations and Laboratory Experiments

The systematic investigation of earthquakes, seismicity, and the associated space-time clustering is enormously facilitated by the use of numerical simulations (see, e.g., RUNDLE and KLEIN, 1995a). To acquire an adequate data base over the hundreds of years and thousands of kilometers typifying earthquake recurrence (e.g., RICHTER, 1958; KANAMORI and KIKUCHI, 1993; PACHECO et al., 1992) requires the use of historical records, which are known to possess considerable uncertainty. Instrumental coverage of even recent events is often inconsistent, and network coverage and detection levels can change with time (HABERMAN, 1991). Understanding the details of the rupture process is further complicated by the spatial heterogeneity of elastic properties, the vagaries of near-field instrumental coverage, and other factors (see for example HEATON, 1990; KANAMORI and KIKUCHI, 1993). Numerical laboratories for generating simulated earthquakes are needed so that experiments can be carried out under controlled and exactly repeatable conditions to investigate the competing effects of the various physical processes. Beyond the simple Gutenberg–Richter relation, which integrates over space and time, one of the most fundamental observational properties of all earthquake fault system is the existence of space-time seismic clustering, and the implied space-time correlations. Because these observations will have important implications for the physical properties of the source over long time scales, it is mandatory to use models and simulations that produce such clustering (many do not: see, e.g., RUNDLE and KLEIN, 1995a).

A central ingredient of any examination of earthquakes is the physics of the frictional processes operating on the fault surface. While many laboratory experiments have been useful in characterizing aspects of sliding friction (see, e.g., RABINOWICZ, 1995; BOWDEN and TABOR, 1950; SCHOLZ, 1990 for summaries), the primary limitations in interpreting and applying results of these experiments to field data relate to questions of: (1) Spatial and temporal scales, in that experiments can only be conducted on samples far smaller than the source region of significant earthquakes (typically 10 s to 1000 s of km), and over times far shorter than a typical earthquake recurrence interval (typically 10 s to 1000 s of years; e.g., RICHTER, 1958; KANAMORI and KIKUCHI, 1993; PACHECO et al., 1992); (2) The typically uniform nature of the samples, which does not reflect the physical and chemical constitution of fault rocks, or the fractures and scale dependence of flaws and heterogeneities seen in field data; and (3) The inability to observe and measure the values of dynamical variables everywhere in and throughout the sample. Although serious questions can be raised about how and whether the numerical simulations we have used in developing our ideas accurately represent reality, the same questions can be raised about laboratory experiments.

Numerical simulations have a number of important advantages. It is possible, for example, to have "perfect" knowledge of the values of all dynamical variables, at all locations and times, whereas such knowledge is not even remotely possible for field data, and is still not possible for even well-characterized laboratory experiments. Standard simulation technology allows numerical experiments to be repeated exactly, with different random number seeds. As a result, the influence of unknown and stochastically unknowable factors such as noise and disorder can be quantitatively examined. Examples of the usefulness of numerical simulations include both the discovery of deterministic chaos (LORENZ, 1963) and the universality of the period doubling route to chaos (FEIGENBAUM, 1977), both of which were seen in the computer first, and later confirmed in observations of natural phenomena and laboratory experiments.

3. Fluctuations, Seismic Clustering and Statistical Mechanics

Theories to explain the physical processes operating on a fault must have a statistical mechanical character. The presence of seismic clustering clearly implies the need to consider fluctuations around a time averaged state of stress accumulation and release on a fault. The elementary physical object of interest is a spatially course-grained segment of a fault, whose size is chosen either on some physical basis (such as the inelastic thickness of a fault zone, or minimum size necessary for instability to develop, or some small-scale characterizing heterogeneity), or which arises because of the nonlinear dynamics. RICE (1993) argues that instability will be suppressed if the course-grained size is small enough to approximate the "continuum limit" for specialized models, but these conclusions have recently been shown to be inapplicable to more general models (SHAW, 1995). Because the medium external to the fault zone is elastic, the interactions between different patches of fault, and between different faults, is fundamentally elastic on short-time scales. In turn, elasticity leads to "mean field" dynamics (KLEIN *et al.*, 1996) characterized by a diverging second moment of the elastic interaction, which in turn is roughly $\sim 1/|\mathbf{x} - \mathbf{x}'|^3$ between interacting fault patches. In this mean field regime, the long-range elastic forces between defects (dislocations) lead to an averaging over short wavelength details. Fluctuations are increasingly important at longer spatial wavelengths and longer time intervals, and constrain important dynamics and correlations. Other effects can include screening due to the presence of defects, which can lead to cutoffs that effectively reduce the range of the interaction (KLEIN *et al.*, manuscript in preparation, 1998).

For driven mean field threshold systems like earthquake faults, subject to a fluctuating noisy force $VK_L + \eta(\mathbf{x}, t)$, RUNDLE *et al.* (1995) showed that the dynamics lead to metastable, or "punctuated equilibrium" behavior. Here V is plate velocity, K_L is a modulus, and $\eta(\mathbf{x}, t)$ is the component of mechanical noise due to unmodeled effects such as smaller faults, microseisms, wearing and abrasion of the fault plane, nonplanar regions of the fault, inhomogeneous rock properties, pore fluid flow, etc. The system settles into a state in which the elastic energy executes small fluctuations around an "equilibrium" energy E_1 for an interval T, at the end of which there is a major event that causes the system to reorganize and settle in at another energy E_2. During the time interval T, the system's energy distribution is characterized by a density of states function multiplied by an exponential Boltzmann factor (see, e.g., RUNDLE *et al.*, 1995c; KLEIN *et al.*, 1996). This latter property, which has been verified using three entirely different methods (RUNDLE *et al.*, 1995, 1996, 1997), makes possible the use of the huge arsenal of analytical techniques developed for the analysis of equilibrium systems over the last hundred years. In particular, the nucleation of high stress regions and earthquake dynamics should therefore be similar to nucleation processes (especially nonclassical processes) described in the recent literature for equilibrium systems.

In our research, we have considered two end-member planar fault models (RUNDLE *et al.*, 1995, 1996a,b, 1997a,b,c; KLEIN *et al.*, 1996, 1997): (1) Fluctuations of seismicity about a steady state (KAGAN, 1994; KLEIN *et al.*, 1997) in which events of all sizes occur and whose frequency-magnitude statistics define a scaling distribution with a well-defined scaling exponent; and (2) characteristic earthquake models (SCHOLZ, 1990; RUNDLE *et al.*, 1996), in which a distribution of smaller events defines a scaling distribution, but in addition to which there exist infrequent maximal events involving the entire fault plane. The latter model shares important similarities to the idea of "Intermittent Criticality" recently advanced by Sammis and coworkers (SAMMIS *et al.*, 1996; SAMMIS, personal communication, 1997), and may be the mechanism behind the Intermittent Criticality idea. Moreover the mainshock in Model 2 has a physically different mechanism from the smaller events that precede and follow it. In Model 1, all events are self-similar, and the only distinguishing feature of a mainshock is that it is the largest event in a series of events.

4. Fluctuations and Potentials

When considering space-time correlations that may exist in earthquake populations arising in Models 1 and 2, two important quantities of interest are the length ξ_c over which fluctuations are correlated, and the corresponding correlation time τ_c.

These quantities set appropriate length and time scales for the fluctuations. For equilibrium, metastable equilibrium, and some driven dissipative systems (RUNDLE *et al.*, 1995, 1996; FISHER *et al.*, 1997) systems that are characterized by a Free Energy (equilibrium) or Lyapunov (driven dissipative) functional potential U, ξ_c and τ_c can be used to measure the proximity of the system to a thermodynamic singularity (GUNTON and DROZ, 1983). These generally come in two forms: (1) A "Spinodal" at which the first derivative of the thermodynamic potential is discontinuous (first-order transition); and (2) A "Critical Point," at which the second derivative of the potential is discontinuous (second-order transition). Figure 1 illustrates the behavior of a Ginzburg–Landau fourth-order potential U (defined below) near the spinodal, as well as near the critical point. For equilibrium systems, the spinodal is the classical limit of stability of the system, and the first derivative of U is proportional to the entropy. The discontinuity of the entropy is often called the latent heat. The second derivative of U is proportional to the specific heat, so that at a critical point, the specific heat is discontinuous.

Summaries of phenomena related to scaling, nucleation and critical phenomena, for systems residing in an equilibrium state can be found in standard references and

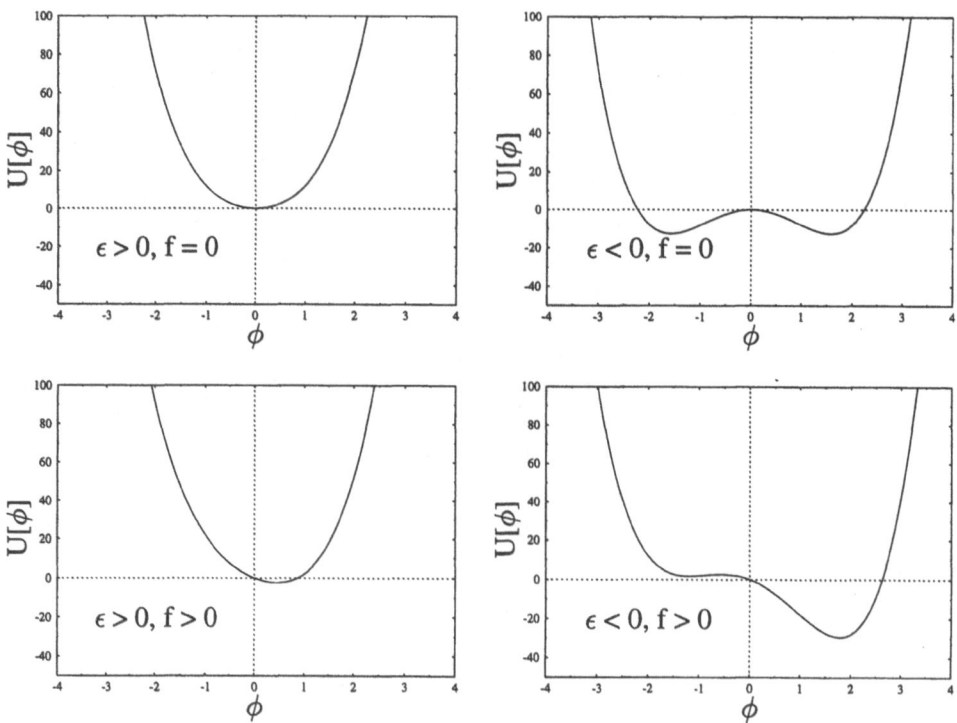

Figure 1
Illustrative plot of the function $U[\phi]$, Equation (3), for various values of ε and f, assuming $U_0 = 0$.

texts (STANLEY, 1971; BINNEY et al., 1993). The significance of these systems lies in the fact that certain types of nonlinear threshold systems, including those that describe earthquakes, have been found to possess a number of the properties of equilibrium systems (RUNDLE et al., 1995, 1996; KLEIN et al., 1997; FISHER et al., 1997), including the fact that they can be characterized by a functional potential. The existence of the potential depends on the long-range nature of the elastic interaction, and the resulting mean field regime in which all but the longest wavelength fluctuations are strongly damped.

5. Example: Velocity-dependent Potential for a Simple Slider Block

As an example of these ideas, consider a simple massless slider block pulled along a frictional surface by a loader spring. This example, although somewhat naive and possibly unrealistic, can be used to illustrate a number of the basic ideas behind modern notions of critical phenomena. The model we present here is a simple variation of more general considerations discussed by FISHER et al. (1997).

Basic Model: For our simple sliding block, we denote the force pulling on the spring by f, the velocity at which the block slides by V, and the stretch of the loader spring by $X - \phi$, where X is the value of the spring stretch at which static friction is first overcome. Note in particular that V_b is the velocity of the sliding block, not the velocity of the load point. We assume that f is prescribed independently of ϕ, and that ϕ will adjust as a result of the dynamics. Thus the work W_P per unit time done by the pulling force in keeping the loader spring stretched by the additional amount $-\phi$ is:

$$W_P = -f\phi. \tag{1}$$

For this example, we also assume that the pulling force is resisted by a velocity-dependent frictional force acting on the sliding block. The stretch of the loader spring is an implicit function of the force f, and an explicit function of the sliding velocity V_b, where $|V_b| > 0$ thus $\phi = \phi(V_b, f)$. Then the energy dissipated per unit time done by the force of friction W_F can be written as:

$$W_F = \varepsilon\phi^2 + \alpha\phi^4. \tag{2}$$

It should be noted that (2) is quite general, in that W_F can only depend on ϕ^2 as a result of symmetry considerations (the physics should be independent of whether the block is pushed or pulled by the spring). By assumption, $\varepsilon = \varepsilon(V_b)$, $\alpha = \alpha(V_b)$. For the remainder of this example, we specialize to the case for which $\alpha = $ constant > 0, so that for large ϕ, total work done per unit time U will always be positive for arbitrarily large ϕ:

$$U = U_0 + W_F + W_P = U_0 + \varepsilon\phi^2 + \alpha\phi^4 - f\phi. \tag{3}$$

U_0 is selected so that $U > 0$. Equation (3) is the standard form for what in the critical phenomena literature is called a "Landau–Ginzburg" model (Fig. 1).

Equilibrium of forces on the sliding block is given by:

$$\frac{\partial U}{\partial \phi} = 0 = 2\varepsilon\phi + 4\alpha\phi^3 - f. \tag{4}$$

For $f \to 0$, there are three solutions to (4):

$$\phi_0 = 0; \quad \phi_\pm = \pm\left(\frac{-\varepsilon}{2\alpha}\right)^{1/2}. \tag{5}$$

The second two solutions, $\phi_\pm$, are only applicable if $\varepsilon < 0$, this is called the "double well" potential. If $\varepsilon > 0$, there is only a single well, and $\phi_\pm = \phi_0 = 0$. Stability of the potential (3) is governed by the second derivative of (3):

$$\frac{\partial^2 U}{\partial \phi^2} = 2\varepsilon + 12\alpha\phi^2. \tag{6}$$

The block is stable if:

$$\frac{\partial^2 U}{\partial \phi^2} > 0 \tag{7}$$

and unstable if:

$$\frac{\partial^2 U}{\partial \phi^2} < 0. \tag{8}$$

The status that is of interest to our work is when $\varepsilon < 0$ and f is large enough so that both (4) and:

$$\frac{\partial^2 U}{\partial \phi^2} = 0 \tag{9}$$

are satisfied. In that case:

$$\phi_1 = -\left(\frac{-\varepsilon}{6\alpha}\right)^{1/2} \qquad \phi_r = +\left(\frac{-\varepsilon}{6\alpha}\right)^{1/2}. \tag{10}$$

Also, to satisfy both (4) and (9), f has the precise value:

$$f_S = \pm\left(\frac{4\varepsilon}{3}\right)\left(\frac{-\varepsilon}{6\alpha}\right)^{1/2}. \tag{11}$$

In the event that (4) and (9) are both satisfied for $\varepsilon < 0$ and the value (11) for $f = f_S$, the potential U is said to have a "spinodal" (GUNTON and DROZ, 1983; see also Fig. 1). The spinodal value $\phi_S = \phi_1$ for $f_S > 0$, and $\phi_S = \phi_r$ for $f_S < 0$. The spinodal value ϕ_S represents a coalescence of the extremum of U corresponding to the local metastable minimum, and the extremum of U corresponding to the local maximum. The other value of ϕ, which we call ϕ_M, represents the globally stable energy minimum. For $f > 0$, $\phi_M = \phi_r$, while for $f < 0$, $\phi_M = \phi_1$. The spinodal is the classical limit of stability for the system.

Physically the basic idea behind (1)–(11) is the following. It has been found experimentally (DIETERICH, 1994; RUINA, 1983) that for small $V_b \to 0$, sliding surfaces are velocity weakening, in the sense that as V_b increases, the force f needed to maintain sliding decreases. Thus one has the property:

$$V_b \uparrow \text{ corresponds to } f \downarrow. \tag{12}$$

It has also been found that as V_b increases through a characteristic velocity value V_{b0} (BLANPIED et al., 1987), there is a transition from velocity weakening to velocity strengthening.

To investigate these effects, let us take:

$$\varepsilon \sim \varepsilon_0 (V_b - V_{b0}) \tag{13}$$

for V_b near V_{b0}. The form of (13) for ε is clearly not unique. We could, for example, have taken:

$$\varepsilon \sim \varepsilon_0 \log\left(\frac{V_b}{V_{b0}}\right) \tag{14}$$

which might be more appropriate in light of the observational evidence (DIETERICH, 1994; RUINA, 1983).

Time Dependence: Finally, time dependence can be introduced into the model by assuming that if any changes in $\phi(V_b, f)$ take place, such changes are consistent with minimizing the potential U (HAKEN, 1983):

$$0 > \frac{\partial U}{\partial t} = \left(\frac{\partial U}{\partial \phi}\right) \frac{\partial \phi}{\partial t} \quad \text{implies} \quad \frac{\partial \phi}{\partial t} \propto -\frac{\partial U}{\partial \phi}. \tag{15}$$

Thus:

$$\frac{\partial U}{\partial t} = M\{f - 2\varepsilon\phi - 4\alpha\phi^3\} \tag{16}$$

where M is a new constant called the "mobility."

To understand how a potential such as (3) with ε defined by (13) or (14) relates to the laboratory experiments, we note that for fixed ϕ, the interval between $[\phi_1, \phi_r]$ is a velocity weakening region in which the condition (12) is satisfied. Fixed ϕ would represent the limit of an infinitely stiff loader spring. Conversely, outside the interval of interest, $[\phi_1, \phi_r]$, U becomes velocity strengthening. There is thus a transition in U that is associated with the value V_{b0}. Although these arguments do not as yet demonstrate that the model (3) with conditions (13), (14) adequately represents the observed laboratory frictional process, we have at least made it plausible. Further details of a more realistic version of the model can be found in FISHER et al. (1997).

Field Theory: Now consider two media in contact at an irregular sliding surface, having bumps, divots, and asperities, in which $\phi(\mathbf{x}, t)$ represents the difference between the displacement at point $\mathbf{x}$ and time t and the load point displacement. In that case, we must replace the potential function (3) by a functional potential (BINNEY *et al.*, 1993):

$$U = \int \{u_0 + R^2|\nabla\phi|^2 + \varepsilon\phi^2 + \alpha\phi^4 - f\phi\}\, d\mathbf{x}. \tag{17}$$

The new term, $R^2|\nabla\phi|^2$, represents an energy associated with the mutual spatial interactions of various sites (see for example RUNDLE *et al.*, 1995). Each site is assumed to interact with $q = (2R)^2 - 1$ other sites by means of a simple exchange potential. In fact, $R^2|\nabla\phi|^2$ represents only the lowest order term of more general and realistic interactions that can in principle have considerably more complex forms, including dipolar $1/|\mathbf{x} - \mathbf{x}'|^3$ interactions with cutoffs, or perhaps an infinite series of terms involving products of gradients, and other higher derivatives of ϕ.

Once (17) has been defined, results similar to the simpler case outlined above are obtained, along with new results. Note that partial derivatives are replaced by functional derivatives, producing an Euler–Lagrange equation (BINNEY *et al.*, 1993):

$$\text{Equilibrium of forces:} \quad \frac{\partial U}{\partial \phi} = 0 \quad \Rightarrow \quad \frac{\delta U}{\delta \phi} = 0$$

Thus:

$$-R^2\nabla^2\phi + 2\varepsilon\phi + 4\alpha\phi^3 - f = 0. \tag{18}$$

The spinodal equation remains the same as (6), even under functional differentiation. However, we are now in a position to obtain some new and interesting results.

As we have noticed above, the character of the equations changes depending on whether ε is positive or negative. Also recall that the two independent variables are V_b (or equivalently ε) and f. Thus we first set $f = 0$, and scale (or nondimensionalize) Equation (18) in terms of the parameter $\varepsilon \to 0$:

$$\mathbf{x} = \xi_c \mathbf{x}'$$

$$\phi = \varepsilon^{1/2}\phi'(\mathbf{x}')$$

$$\xi_c = R\varepsilon^{-1/2}. \tag{19}$$

Consequently Equation (18) can be written in the universally valid form:

$$-\nabla'^2\phi' + 2\phi' + 4\alpha\phi'^3 = 0 \tag{20}$$

at $f = 0$, and for $V_b \to V_{b0}$ or equivalently $\varepsilon \to 0$.

Alternatively, we could have set $\varepsilon = 0$ and defined quantities in terms of the other independent forcing variable in the limit as $f \to 0$:

$$\mathbf{x} = \xi_c \mathbf{x}'$$

$$\phi = f^{1/3} \phi'(\mathbf{x}')$$

$$\xi_c = R f^{-1/3} \tag{21}$$

producing the universally valid equation:

$$-\nabla'^2 \phi' + 4\alpha \phi'^3 - 1 = 0 \tag{22}$$

at $\varepsilon = 0$ and as $f \to 0$.

Both (19) and (21) define a set of scaling laws that are valid near the particular values in parameter space:

$$V_b \to V_{b0} \quad \text{and} \quad f = 0. \tag{23}$$

These values for (V_b, f), i.e., $(V_{b0}, 0)$ define the Landau–Ginzburg "critical point." The control parameters $(V_b - V_{b0}, f)$, which measure proximity of the system to the critical point, are called "scaling fields." A unique property of such a critical point is that, independent of whether V_b or f is the control parameter, one has a scaling relation for the length ξ in terms of one of the scaling fields. ξ_c is the correlation length, because it sets the scale for the spatial fluctuations arising from the Laplacian (STANLEY, 1971; GUNTON and DROZ, 1983; BINNEY et al., 1993).

Although it is reasonable that sliding velocity V_b could be adjusted to the value V_{b0}, it is not clear that the value $f = 0$ has physical meaning, since this implies zero driving force. However, the model (18) can be put into the form (20) by using the idea that the driving force at a site is applied through an "effective loader spring" which has spring constant K_L acting on that site (see for example, FISHER et al., 1997, for justification). In this case, the force f can no longer be set independently of ϕ, rather, V_b is now assumed to be the control variable. Therefore, we set:

$$f = -\frac{K_L}{2} \phi \tag{24}$$

in (17) and rewrite (18) as:

$$-R^2 \nabla^2 \phi + 2\varepsilon^* \phi + 4\alpha \phi^3 = 0 \tag{25}$$

where

$$\varepsilon^* = \varepsilon + \frac{K_L}{2}. \tag{26}$$

Using (13), we could alternately define a new scaling field V_b^*:

$$\varepsilon^* = \varepsilon_0(V_b^* - V_{b0}), \quad V_b^* = V_b + \frac{K_L}{2\varepsilon_0}. \tag{27}$$

Then the nondimensional form of (25) is the same as (20) but with all of the scaling relations written in terms of ε^*.

Note that the scaling relations (19) and (21) are special cases of more general relations in the literature. A series of exponents β, v, δ, are commonly defined for scaling relations such as these (STANLEY, 1971). For example:

$$\phi \sim \varepsilon^\beta \quad (\text{if } f = 0)$$

$$\phi \sim f^{1/\delta} \quad (\text{if } \varepsilon = 0)$$

$$\xi_c \sim \varepsilon^{-v} \tag{28}$$

and so forth. Clearly, $\beta = 1/2$, $v = 1/2$, $\delta = 3$. These are called "Gaussian" exponents.

Two more scaling exponents are also of interest. The first is a dynamical scaling exponent that arises when the model (18) is made time-dependent in the same way as (16). Thus:

$$\frac{\partial \phi}{\delta t} = M\{f + R^2 \nabla^2 \phi - 2\varepsilon \phi - 4\alpha \phi^3\}. \tag{29}$$

When (29) is nondimensionalized, time t turns out to be nondimensionalized in terms of a correlation time $\tau_c = \varepsilon^{-1}$. Thus

$$\tau_c \propto \xi_c^z \tag{30}$$

in terms of the dynamical scaling exponent z. In this model, we find that $z = 2$.

The second scaling exponent of interest describes the frequency at which fluctuations occur in the neighborhood of a critical point. At a critical point, fluctuations are self-similar, and the frequency with which they occur is described by a power law similar to (19) and (21) which has an exponent in the standard notation of $\tau_f - 1$ (the exponent is $\tau_f - 1$ and not τ_f for clusters of failed sites grown outward from an initial failed site). In our case, areas on the sliding surface can be thought of as either "stuck" or "just slipped." Then the number $n(A)$ of "just slipped" regions of area A can be described in the neighborhood of the critical point by the Fisher–Stauffer model:

$$n(A) = \left(\frac{n(0)}{A^{\tau-1}}\right) \exp[-k\varepsilon^{1/\sigma_f}A]. \tag{31}$$

Here σ_f is another exponent called the surface exponent, and k is a constant. More generally, (31) is a Gamma distribution. Kagan argues that such distributions describe the frequency moment relation of earthquakes. Notice that (31) could also have been written:

$$n(A) = \left(\frac{n(0)}{A^{\tau-1}}\right) \exp[-k'A\xi_c^{-1/\nu\sigma_f}]. \tag{32}$$

On a plot of $\text{Log}\{n(A)\}$ against $\text{Log}\,A$, the location of the exponential cutoff will therefore depend on the size of the correlation length ξ_c.

Finally, it is important to point out that the scaling exponents β, ν, δ, z, τ_f, σ_f are generally not all independent. They are often linked by scaling relations (STANLEY, 1971), which are derived from scaling analyses such as the Renormalization Group (BINNEY et al., 1993). In the models that are discussed below, scaling relations have been calculated, and more detailed discussions can be found elsewhere (RUNDLE et al., 1997).

6. Driven Dissipative Models for Earthquakes

The model just discussed is not meant to represent the kinds of physics that may be present on real faults, rather it is instead intended to be an illustrative model that shows how scaling and critical phenomena can arise. As we have discussed, scaling relations have been seen in earthquake data dating from the time the Gutenberg–Richter and Omori Laws on earthquake frequencies were first observed. We now consider several models for slip on earthquake faults that are more physically motivated than the simple model described above. The first model describes phenomena associated with faults that have a distinct cycle, along with a limited scaling region in event sizes and what are often called "characteristic earthquakes." The second model is more descriptive of a steady-state condition, and shows a continuous range of event sizes that fall along a scaling curve. Both models are characterized by mean field dynamics, and as such can be shown to possess a Lyapunov functional potential that encodes the dynamics (RUNDLE et al., 1996a, 1997a,b; KLEIN et al., 1997).

Model 1: Intermittent Criticality Model for Earthquakes. In models of this type, there is a gradual approach to a critical state, followed by the main event, then a retreat from the critical state. Such a model has been called an "Intermittent Criticality Model" (SAMMIS et al., 1996; SAMMIS, personal communication, 1997). We therefore expect the Lyapunov functional $U[\phi]$ to have a cyclic character. Here as above, $\phi(\mathbf{x}, t) = s(\mathbf{x}, t) - Vt$ is the slip deficit field, written in terms of the slip $s(\mathbf{x}, t)$, the plate velocity V and time t. Notice that now V is the load point velocity, not the velocity of sliding on the surface.

The association of earthquakes with the existence of a critical state motivates the development of more general models for system dynamics based upon the construction of new, and general, Lyapunov functionals for frictional sliding. A particular example of such models is the Traveling Density Wave model (RUNDLE et al., 1996a, 1997a,b), which is based upon the kind of slip weakening frictional physics described in the experimental literature over the past half century (e.g., see

BOWDEN and TABOR, 1950; and RABINOWICZ, 1995; for summaries). In this model, the system evolves persistently toward a state of lowest free energy. The system dynamics are given by:

$$\frac{\partial \phi}{\partial t} = -M \frac{\delta U}{\delta \phi} + \eta(\mathbf{x}, t) \tag{33}$$

similar to (29), but in which a new term $\eta(\mathbf{x}, t)$ represents mechanical noise that arises from small unmodeled faults, pore fluid flow, and other physical processes. We adopt a system of units, or time scale, such that $M = 1$.

In the simplest case, $U[\phi]$ is given by (RUNDLE *et al.*, 1997):

$$U[\phi] = \int \int \{-(1/2)[T(\mathbf{x} - \mathbf{x}')\phi(\mathbf{x}, t)\phi(\mathbf{x}', t) \, d^2\mathbf{x}']$$

$$- 2\gamma_c \cos[\kappa\{\phi(\mathbf{x}, t) + Vt + \varepsilon_c\}] + h\phi(\mathbf{x}, t)\} \, d^2\mathbf{x}. \tag{34}$$

Here, $T(\mathbf{x} - \mathbf{x}')$ is a stress Green's function, κ describes the wave number of cohesive patches or bumps, $2\gamma_c$ and h are surface energies, and $\varepsilon_c(\mathbf{x}, t)$ is a (possibly random) or correlated phase. Both $2\gamma_c$ and h depend generally on the normal stress σ_N and on the driving velocity V. The dynamical equation corresponding to (33), or Itô-Langevin equation, is (RUNDLE *et al.*, 1997):

$$\frac{\partial \phi(\mathbf{x}, t)}{\partial t} = \left\{ \int T(\mathbf{x} - \mathbf{x}')\phi(\mathbf{x}', t) \, d^2\mathbf{x}' - 2\gamma_c\kappa \sin\{\kappa\{\phi(\mathbf{x}, t) + Vt + \varepsilon_c(\mathbf{x}, t)\} - h \right\}$$

$$+ \eta(\mathbf{x}, t). \tag{35}$$

Under the driving action of the tectonic plate stresses, the system finds itself in a progressively higher energy state. At intervals of roughly $P = 2\pi/(\kappa V)$, a recurring energy barrier appears that separates a new metastable state of the system from a new, lower energy state that has come into existence. The height of the barrier diminishes with time as a result of the tectonic stress applied. Eventually, the barrier height becomes low enough so that the effects of the noise $\eta(\mathbf{x}, t)$ allow the system to escape from the metastable well and decay into the lower energy state. The point at which the barrier height vanishes is a spinodal, exactly as described previously, at which the system is unstable to sudden changes. An example of these dynamics is shown in Figure 2 for the mean field situation in which $\phi(\mathbf{x}, t) = \Phi(t)$ is assumed to be independent of $\mathbf{x}$. Note that metastability can only occur in this model if the parameter $\Lambda \equiv 2\gamma_c\kappa^2/K_L > 1$, a value that is a consequence of the condition (8).

It is possible to expand the static version of Equation (35) about the spinodal, and to demonstrate that scaling solutions exist. If the time of appearance of the spinodal is t_{sp}, an expansion about t_{sp} yields:

$$\phi(\mathbf{x}, t) = \Phi(t_{sp}) + \psi(\mathbf{x}, t)$$

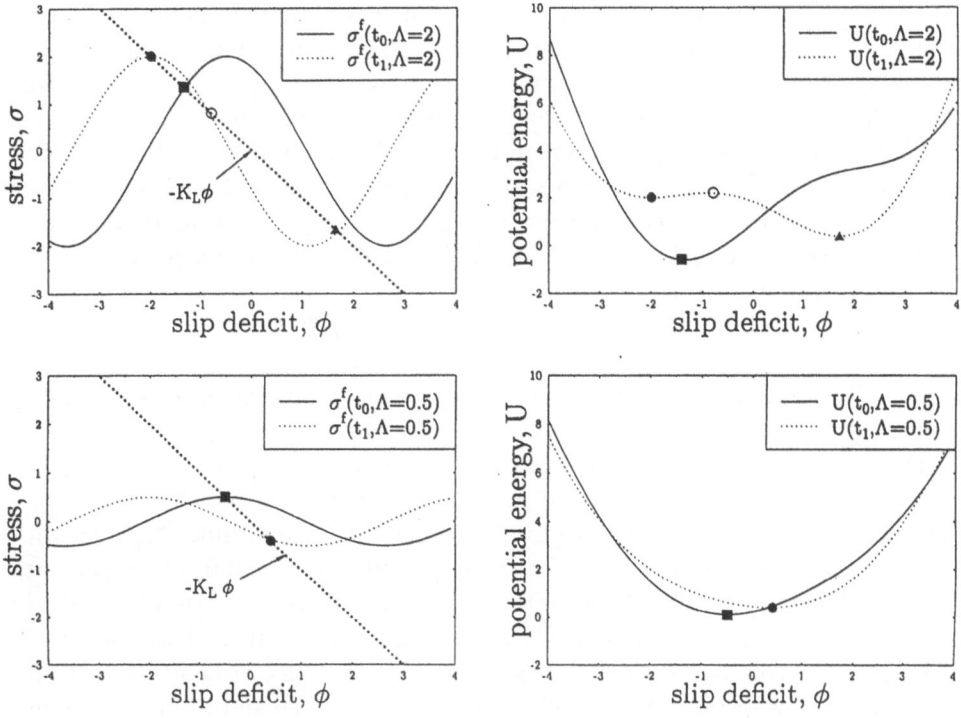

Figure 2

Plot of stress $\sigma(t)$ and potential energy $U(t)$ against ϕ in mean field limit. Values of parameter Λ determine whether no metastable states do ($\Lambda > 1$) or do not ($\Lambda < 1$) exist. Plots are shown for both $\Lambda = 2$ (top) and $\Lambda = 0.5$ (bottom), and for two distinct times, $t_1 > t_0$. For plots of stress $\sigma(t)$, possible solutions exist at the intersection of the sine curve with the dashed unloading line. The same solutions correspond to extrema of the potential $U(t)$. Top: At time t_0, only one globally stable state exists (square); at time t_1, metastable (filled circle), unstable (open circle), and globally stable (triangle) states now exist. Bottom: At both times t_0 and t_1, only one globally stable state exists (square at t_0 and circle at t_1).

$$t = t_{sp} - \delta t \qquad (36)$$

where

$$|\psi(\mathbf{x}, t)| \ll |\Phi|$$

$$\kappa V \delta t \ll 2\pi. \qquad (37)$$

Using the notation of (29), one finally obtains the spinodal equation (RUNDLE *et al.*, 1997):

$$-R^2 \nabla^2 \psi + K_L V \delta t - \alpha \psi^2 = 0. \qquad (38)$$

Equation (38) yields scale-invariant solutions $\tilde{\psi}(|\mathbf{x}|/\xi_c)$ of the form:

$$\psi(\mathbf{x}, t) \sim (K_L V \delta t)^{1/2} \tilde{\psi}(|\mathbf{x}|/\xi_c). \qquad (39)$$

The process of decay can be understood by an analysis of the solutions (39), and in particular, these solutions have a number of the scaling properties observed in nature, including Gutenberg–Richter scaling, Omori-type scaling, and so forth (RUNDLE et al., 1996a, 1997a,b). Note that in (38), the time interval δt acts as a scaling field because it brings the system towards or away from the spinodal, which is acting as a critical point at which scaling occurs. Moreover, the Fisher–Stauffer model (31) indicates that the frequency of spinodal fluctuations $n(A, \delta t)$ of area A at time δt, which are realized here as clusters of failed sites, is given by:

$$n(A, \delta t) = \frac{n_0}{A^{\tau - 1}} \exp\{-k[K_L V \delta t]^{1/\sigma_f} A\}. \tag{40}$$

It can be shown that the scaling property (40) applies to pre-mainshock as well as post-mainshock fluctuations (RUNDLE et al., 1997).

The pre- and post-event fluctuations, and the process of decay from metastability may be related to the underlying physical mechanisms behind the foreshock–mainshock–aftershock sequence. Examples of our simulation results, together with the real data from cycles of activity in the Mammoth Lakes, California, region, are shown in Figure 3. Similar such cycles can be seen in other areas of the world (SCHOLZ, 1990). The simulation results were obtained on a lattice of size $64 \times 64 = 4096$ sites. The correlation length ξ_c (KLEIN et al., 1996; RUNDLE et al., 1997a,b) varies inversely with the barrier height separating the stable and metastable states:

$$\xi_c \sim (K_L V |\delta t|)^{-1/4}. \tag{41}$$

Equation (41) states that as the mainshock time is approached, the correlation length grows. At the spinodal, the correlation length diverges as the inverse-fourth power of the barrier height, rather than the inverse-half power encountered in (28). The description of this physical process closely resembles the "Intermittent Criticality" model proposed by C. Sammis and coworkers (personal communication, 1997; see also, SAMMIS et al., 1996) to explain the development and growth of correlations on an earthquake fault as the time of the mainshock approaches. We will thus be motivated to explore these similarities in detail, to ascertain what observable properties should be reflected in seismicity distributions and clustering. Depending on the physical details of the configuration of the metastable well, this model may be associated either with *quiescence* (depressed foreshock activity), or with *elevated precursory activity* (enhanced foreshock activity). The closer to the spinodal at which system decay occurs, the larger the mainshock can be, although decay of the system is influenced by the amount and type of disorder present, the existence of lower lying states that are still metastable, and other factors.

Model 2: Steady-state (SS) Model with Disorder. In these models (KLEIN et al., 1997; FISHER et al., 1997; FERGUSON et al., 1998), there is sufficient disorder such that the system fluctuates around a steady state in which the correlation length stays nearly constant with time. The nucleation processes leading up to the

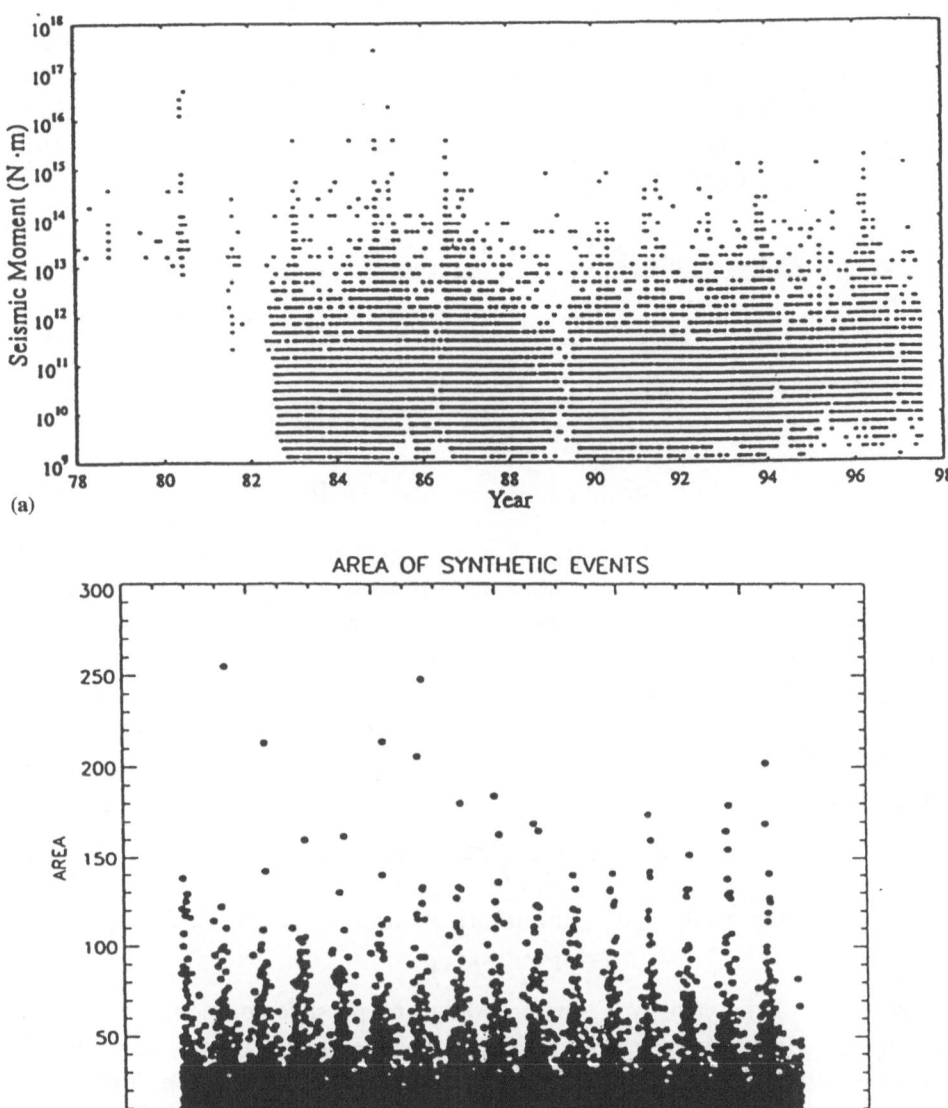

Figure 3
(a) This scatter plot shows seismic moment released in the Mammoth Lakes, CA, area since regional networks began monitoring. Seismicity is limited to events with hypocentral location quality letter A, and focal depths greater than 2 km. (b) Area vs. time for simulations using the Traveling Density Wave model for earthquakes (RUNDLE *et al.*, 1996, 1997). Foreshock–mainshock–aftershock sequences are shown on a 64 × 64 lattice of sites.

formation of a high stress region whose decay comprises the mainshock can be described using equations and analysis developed by KLEIN *et al.* (1997).

Again, the basic idea is that for mean field models, in which R^2 becomes large, the system can be characterized by a Lyapunov functional U. The mean field nature of the interaction induces a spatial and temporal averaging over regions of size $(2R)^2 - 1$ centered at $\mathbf{x}'$, and averaging times τ. Thus we are interested only in a spatially and temporally coarse-grained version of the functional U, and its highly nonlinear functional derivative $F[\sigma(\mathbf{x}', \tau)] + K_L V$:

$$
\begin{aligned}
F[\sigma(\mathbf{x}', \tau)] + K_L V = K_L V + &\frac{(R^2 \nabla^2 - K_L)}{K_T} \frac{(\sigma^F - \sigma^R)}{2} \\
&\times \{\text{erf}\{-\sqrt{\beta_n}[\sigma^F - \sigma(\mathbf{x}', \tau)]\} - \text{erf}\{-\sqrt{\beta_n}[\sigma_0 - \sigma(\mathbf{x}', \tau)]\}\} \\
&- \frac{\beta^{-1}}{\sigma^F - \sigma^R}\left[\text{Log}\left(\frac{\sigma(\mathbf{x}', \tau) - \sigma^R}{\sigma^F - \sigma(\mathbf{x}', \tau)}\right) - \sqrt{\frac{\beta_n}{\pi}}\right. \\
&\left. \times \int_{\sigma^R}^{\sigma^F} d\sigma \, \text{Log}\left(\frac{\sigma - \sigma^R}{\sigma^F - \sigma}\right) \exp\{-\beta_n[\sigma - \sigma(\mathbf{x}', \tau)]\}\right].
\end{aligned}
\tag{42}
$$

Here σ^F is a failure threshold stress, σ^R is a residual stress to which the stress on a block falls following slip, β_n is a noise scale (inverse "temperature"), and erf$\{z\}$ is the error function. The stress $\sigma(\mathbf{x}', \tau)$ is the spatially and temporally coarse-grained stress. Figure 4 plots $F[\sigma(\mathbf{x}', \tau)] = K_L V$.

The active agent adding stress to the system is the tectonic loading $\partial/\partial t\{\sigma_{\text{Load}}\} = VK_L$, which is proportional to the plate velocity V (GOLDSTEIN *et al.*, 1998). Neglecting fluctuations, the stress in the system is then given by solutions of the equation:

$$
\frac{\partial \sigma}{\partial t} = \frac{\partial \sigma_{\text{Load}}}{\partial t} - \frac{\partial \Delta \sigma_{\text{Seis}}}{\partial t} = VK_L - F[\sigma].
\tag{43}
$$

For this model, the steady-state solutions are obtained by setting $\partial \sigma/\partial t = 0$ so that:

$$
VK_L = F[\sigma].
\tag{44}
$$

At low values of the velocity V, there is only one possible solution, a low stress solution σ_{low}. However, there is a lower critical value of velocity $V_c^{(1)}$ above which a second, metastable, high stress solution becomes possible, σ_{high}, as well as an unstable solution, σ_{us}. For elastic systems, we find that this critical velocity $V_c^{(1)} \to 0$ as the range of interaction between sites is increased, so that the mean field regime is approached. Therefore, if the velocity were initially low, the stress would be low. If the velocity were then raised above the critical value, nucleation into the high stress state becomes possible. If the system was then in the high stress state, and the velocity was lowered, decay from the high stress solution is inevitable. Note that there is also an upper critical value of velocity $V_c^{(u)}$ at which only one solution again exists. $V_c^{(u)}$ defines a spinodal velocity, in the same manner as described earlier.

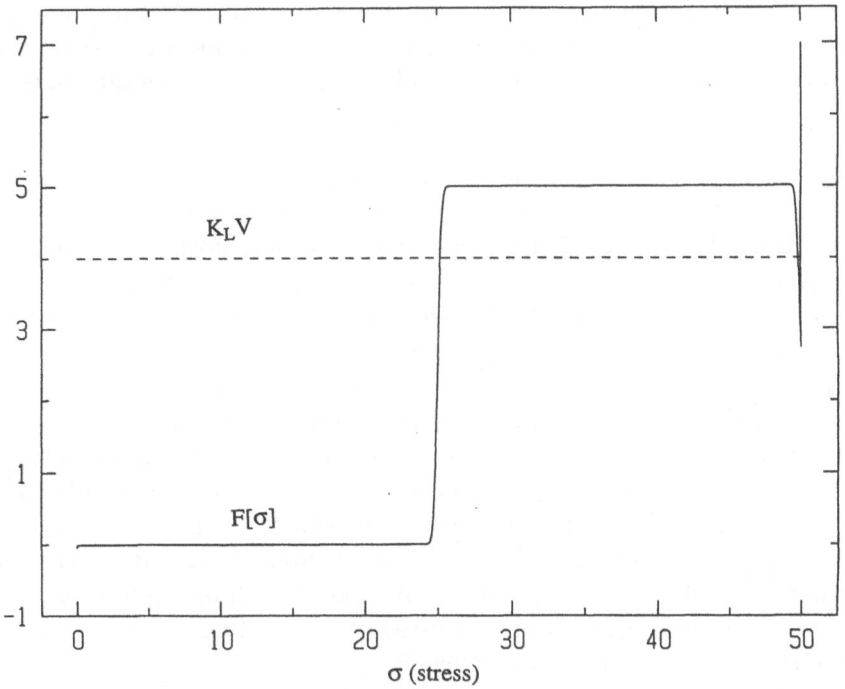

Figure 4
Plot of $F[\sigma(\mathbf{x}', \tau)]$ from Equation (42), solid line, and $K_L V$, dashed line, for the case $\beta = 10$. Intersection of solid and dashed lines represents the steady-state solution to the equation $F[\sigma(\mathbf{x}', \tau)] = K_L V$.

In analyzing the dynamics described by (43), we have found that fluctuation-induced corrections $\Delta\sigma_{\text{Seis}}$ to (43), due to the time variation of seismicity, cause the velocity to vary, producing a correction term. Thus instead of a constant loading stress rate $\partial/\partial t\{\sigma_{\text{Load}}\} = VK_L$, one has an effective loading rate modified by the stress lost due to seismicity fluctuations, $\partial/\partial t\{\sigma_{\text{Load}}\} = VK_L - \partial/\partial t\{\Delta\sigma_{\text{Seis}}\}$. Such a correction to Equation (43) may then be replaced by:

$$\frac{\partial\sigma}{\partial t} = VK_L - \frac{\partial\Delta\sigma_{\text{Seis}}}{\partial t} - F[\sigma]. \tag{45}$$

During long periods of quiescence, the effective velocity is high enough so that the system tends to nucleate into the high stress state, forming a two-dimensional region or "droplet" of high stress. Because of the elevated stress level, the seismicity rate (foreshock activity) begins to increase, meaning that $\partial/\partial t\{\Delta\sigma_{\text{Seis}}\}$ increases and $\partial/\partial t\{\sigma_{\text{Load}}\}$ decreases. As foreshock activity increases, $\partial/\partial t\{\sigma_{\text{Load}}\}$ drops below the value necessary to sustain the system in the high stress state, and the high stress droplet decays in the mainshock. Thus, in this picture, the foreshock activity is causally related to the mainshock, and in fact is responsible for the decay of the mainshock. This model is therefore associated with *elevated* foreshock activity. In this model, there is essentially no temporal change in the correlation length ξ_c and time τ_c with time, thus the size of the mainshock is determined by the size of the high stress droplet and the statistical nature of the stress field outside the droplet.

7. Growth and Arrest

It is frequently observed that the mainshock in a foreshock–mainshock–aftershock sequence has a magnitude substantially greater than that of the largest foreshock or aftershock (see Figs. 2a,b; also, see e.g., SCHOLZ, 1990; RUNDLE *et al.*, 1997c). Why does this magnitude "band gap" exist? Why does the foreshock–mainshock–aftershock sequence not have a smooth continuum of magnitudes, with the mainshock being only the largest event of the continuum? Our models (RUNDLE *et al.*, 1997c), which need to be tested with seismicity data, imply that the mainshock represents a high stress region that grows explosively beyond its original boundaries during decay to lower stress. These rare "breakaway" high stress regions or droplets can only expand because of statistical conditions in the stress field beyond the droplet boundaries. The implications of crack growth on rupture, arrest, and scaling laws have also been considered by MAIN *et al.* (1993, 1994); MAIN (1996) and FUKUYAMA and MADARIAGA (1998).

A simple application of ideas from Griffith theory for spatially compact tensile cracks indicates that as an earthquake grows, it tends to concentrate stress at its edges. According to this idea, if a stress field $p_e > p_{ec}$ is applied that is larger than a critical size p_{ec}, the fracture radius λ runs away to "infinite" size, $\lambda \to \infty$. This property is a result of the structure of the Griffith energy F_G

$$F_G = -|\Delta E_{el}|\lambda^2 + a_0\lambda \tag{46}$$

which is a sum of a (negative) elastic energy release term involving an elastic energy decrease $\Delta E_{el} < 0$ multiplying the squared radius λ^2, and a term representing a (positive) surface energy density a_0 multiplying λ. It can easily be seen that F_G has a local maximum at λ_{crit}:

$$\lambda_{\mathrm{crit}} = \sqrt{\frac{a_0}{2|\Delta E_{el}|}}. \tag{47}$$

The idea is that the crack evolves so as to minimize the Griffith free energy F_G. If $\lambda < \lambda_{\mathrm{crit}}$, the crack remains static (but there are actually isolated examples of brittle cracks healing at small λ: see KANINEN and POPELAR, 1985). However, if $\lambda > \lambda_{\mathrm{crit}}$ the crack will grow. As it grows larger as a self-similar compact object, it increasingly concentrates stress at the edges, enabling it more easily to overcome the surface energy a_0 tending to retard growth. While this may be a reasonable prediction for tensile cracks in a homogeneous stress field, where the surface energy $a_0 = 2\gamma$ is approximately constant, it is not reasonable for shear cracks in a heterogeneous stress field.

In most models of shear rupture used to understand earthquake phenomena, sites are characterized by a stress threshold $\sigma^F(\mathbf{x}, t)$ which can change with time as a result of wearing and abrasion, or other time-dependent processes. The asperity and barrier ideas proposed more than a decade ago (LAY et al., 1982; AKI, 1979; BEN-ZION and RICE, 1993) asserted that spatial variations in $\sigma^F(\mathbf{x}, t)$ were responsible for whether or not an earthquake grows or arrests. However, instead of focusing just on the threshold $\sigma^F(\mathbf{x}, t)$, or the elastic stress level $\sigma(\mathbf{x}, t)$, it is more appropriate to focus on the space-time field $\Sigma(\mathbf{x}, t) = \sigma^F(\mathbf{x}, t) - \sigma(\mathbf{x}, t)$. We retain the idea that variability is important in determining whether an earthquake becomes a breakaway event, but we think it more reasonable that spatial variability in $\Sigma(\mathbf{x}, t)$ is the important factor.

For growth of an earthquake rupture in a heterogeneous environment, we must allow the surface energy $a_0 \rightarrow a(\mathbf{x}, \mathbf{x}_0)$, a stochastic function of position. Moreover, if variations in $\Sigma(\mathbf{x}, t)$ are to lead to arrest of the crack, the field $a(\mathbf{x}, \mathbf{x}_0)$ will be related to the square of the field $\Sigma(\mathbf{x}, t)$ divided by a modulus, taking account of physical units. The field $a(\mathbf{x}, \mathbf{x}_0)$ must depend on the spatial mean $\Sigma_m(t)$ of $\Sigma(\mathbf{x}, t)$, because if $\Sigma_m(t) = 0$, every site would be at failure. As the crack extends from $\mathbf{x}_1$ to $\mathbf{x}_2$ to $\cdots$ to $\mathbf{x}_n$, it must overcome the stress difference $\Sigma(\mathbf{x}_1, t) + \Sigma(\mathbf{x}_2, t) + \cdots + \Sigma(\mathbf{x}_n, t)$, therefore the process of crack propagation can be viewed as a process of "integrating" the Σ-noise field. As a consequence, the quantity $a(\mathbf{x}, \mathbf{x}_0)$ should depend on stochastic fluctuations about $\Sigma_m(t)$ through an integral of $\Sigma(\mathbf{x}, t)$ that we call $W(r, t)$, where $r = |\mathbf{x} - \mathbf{x}_0|$.

Using the example of a circular crack extending in a heterogeneous $\Sigma(\mathbf{x}, t)$-field, one can construct a Stochastic Griffith Energy F_S (RUNDLE et al., 1998):

$$F_S = -\Delta E_{el}\lambda^2 + 2\omega\lambda + 2\rho\lambda^{2H+1} \qquad (48)$$

where

$$\omega = \frac{\pi}{K_L}[\Sigma_m(t)]^2, \quad \rho = \frac{\pi}{(2H+1)K_L}\text{Var}\{\Sigma(\mathbf{x}, t)\}. \qquad (49)$$

The first term is the elastic energy release, $\Delta E_{el} \propto \Delta\sigma^2/\pi$, where $\Delta\sigma$ is the average stress drop. The second term arises from the spatial mean $\Sigma_m(t)$ of $\Sigma(\mathbf{x}, t)$, and the third term originates from $W(r, t)$. In the second term, $\omega > 0$ and ω is proportional to the squared mean of $\Sigma(\mathbf{x}, t)$. In the third term, $\rho > 0$ and ρ is proportional to the variance of $\Sigma(\mathbf{x}, t)$. If the assumption of circularity of the crack is relaxed, allowing the occurrence of irregularly shaped, but still compact cracks, one has instead:

$$F_S = -\Delta E_{el}\lambda^2 + 2\omega\lambda + 2\rho\lambda^{2dH} \qquad (50)$$

where d is the dimension of space (usually $d = 2$ for cracks).

The first two terms in (48) or (50) describe the same kind of classical nucleation process as in tensile fracture, but the third term is new and provides a means of arresting the growth of the shear fracture. For example, regardless of the magnitude of ρ, arrest will eventually occur if $H > 0.5$, the value characteristic of a Brownian Walk. Thus if one regards the $\Sigma(\mathbf{x}, t)$ field as a noise, the spectrum of the noise must be red if the crack is to eventually arrest, so that there must be proportionately more power at longer wavelengths. A white $\Sigma(\mathbf{x}, t)$ noise field represents the boundary between $\Sigma(\mathbf{x}, t)$ fields that can lead to arrest, and those that cannot. Using these ideas, we have found that foreshocks and aftershocks generally represent high stress regions that cannot grow beyond their boundaries. However, mainshock events that display a large magnitude band gap relative to the associated foreshock and aftershock magnitudes represent breakaway events that can lower F_S by growing beyond their original, high stress region boundaries.

Figure 5 shows an example in which we measure the surface energy terms in the Stochastic Griffith Energy from simulations based on the Traveling Density Wave model for earthquakes (RUNDLE *et al.*, 1996). Here the events are not circular,

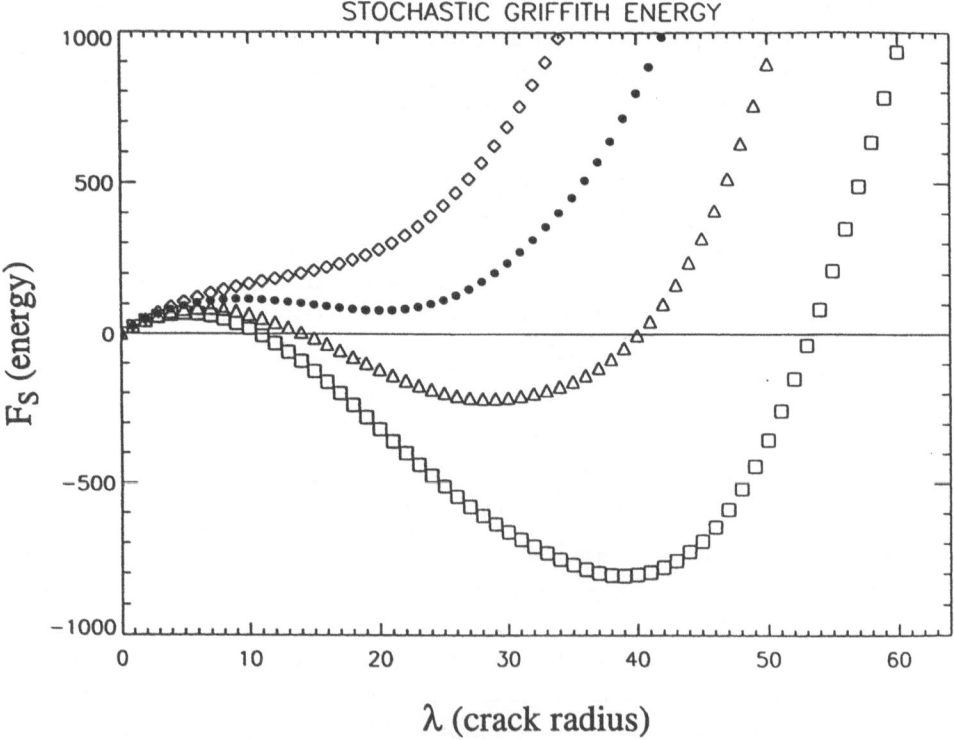

Figure 5

Complete Stochastic Griffith Energy curves. The entire F_s function is constructed by direct integration of simulation data to obtain terms equivalent to the second and third terms in (5), but making no assumptions about Gaussian statistics or crack shape. In these plots, we use $|\Delta E_{el}|$ as a parameter to construct an elastic energy release term in which the area of the failed sites for each event in the simulation multiplies $-|\Delta E_{el}|$. From the top, these curves have the values $|\Delta E_{el}| = 17, 18, 19, 20$. For comparison, measured values of $|\Delta E_{el}|$ for the largest events in the situation ($>$ size 50) have $|\Delta E_{el}| \sim 18$.

rather instead can have arbitrary, although still compact, shapes. For this figure, we integrated the stochastic surface energy density from the observed $\Sigma(\mathbf{x}, t)$ field to obtain F_S directly. Assuming a range of values for ΔE_{el}, we show that F_S can either be monotonically increasing with event radius λ, or it can have a local maximum ("stochastic spinodal"). In the former case, high stress regions that nucleate and then decay cannot grow beyond their original boundaries, because they will always be in a region where F_S increases with λ. In the latter case, if high stress regions nucleate on a part of the $F_S - \lambda$ curve that is decreasing with λ, their decay will produce events that can grow explosively beyond their original boundaries, thus minimizing F_S.

8. Consequences, Predictions, and Possible Observational Tests

The physical picture outlined above contains a number of consequences, and makes a number of predictions for the dynamics of the mainshock, and for seismicity and seismic clustering, that can be tested with observations. These consequences, all of which imply observable characteristics for space-time clustering, including the following.

8.1. Gutenberg–Richter Relation—b-value

For both Intermittent Criticality model (IC) and Steady State Model (SS), there are two possible observables: (1) the exponent for the scaling (power law) region $\tau - 1$, and (2) the size of the maximum magnitude cutoff ($= k[K_L V \delta t]^{1/\sigma_f}$), in the regional GR relation. Both IC and SS models demonstrate the family of scaling exponents appropriate to a spinodal (see RUNDLE et al., 1997a, or STANLEY, 1971 for definitions):

$$\sigma_f = 1$$

$$v = 1/4$$

$$\beta = 1/2$$

$$z = 2 \tag{51}$$

In addition, one can write the usual *cumulative* Gutenberg–Richter magnitude frequency as

$$\text{Log } N_{GR}(m) = -bm + a. \tag{52}$$

We use the Fisher–Stauffer relation (40) to obtain a scaling relationship for b and another exponent c, which defines the scale dependence of the seismic moment M_0 on the event area A

$$M_0 \propto A^c. \tag{53}$$

Together with the moment-magnitude relation (SCHOLZ, 1990):

$$m = \frac{2}{3} \text{Log } M_0 - 6.0 \tag{54}$$

we find by integrating (40) over areas $[A, \infty)$, using (53)–(54), and equating powers of area that

$$b = \frac{3(\tau - 2)}{2c}. \tag{55}$$

In both the IC and SS models, we find that (a) $\tau = 2.5$ or (b) $\tau = 3.0$, depending on whether the system is either (a) "not near" the spinodal, or (b) "near" the spinodal (KLEIN *et al.*, 1997). The extra factor of 0.5 in the value of τ is due to the presence of "critical slowing," when the correlation time scale τ_c defined in (30) diverges very near the spinodal. Notice that the value of b depends not only on τ, but also upon c from (53), which in turn is determined by whether or not healing occurs during the sliding process. With no healing, crack-like behavior is expected, slip should be proportional to the square root of event area, and $c = 1.5$. If healing takes place, pulse-like behavior is expected, slip should be independent of event area, and $c \sim 1$ (see the discussion in HEATON, 1990). For earthquakes which display a crack-like displacement profile and occur very near the spinodal, $c \sim 1.5$, $\tau \sim 3$, and $b \sim 1$. Since in any case, the total number of earthquakes is always dominated by the foreshocks and aftershocks occurring near the time of the mainshock (spinodal), $\tau \sim 3$, and b can be expected to vary between 0.75 and 1. Removing the foreshocks and aftershocks ("declustering") is equivalent to setting $\tau \sim 2.5$, consequently b will vary between 0.5 and 0.75. Thus the declustering process results in substantially lower b values, as is observed (JONES, 1995).

8.2. Gutenberg–Richter Relation. Maximum Earthquake Size and Correlation Length ξ_c

In addition to b value, there is an exponential cutoff, or maximum earthquake size, that appears in the Fisher–Stauffer relation (40). Using (53) and (54), we find that

$$A \sim M_0^{1/c} \sim [0^{(3m/2 + 9.0)/c}. \tag{56}$$

Note that if $c = 1$, and $\tau = 2.5$, one finds the same Gamma distribution for the Fisher–Stauffer relation as the observations of KAGAN (1994) using the Harvard catalog:

$$n(A, \delta t) = \frac{n_0}{M_0 \zeta} \exp\{-M_0/M_{0_{max}}\} \tag{57}$$

where ζ is a number found empirically to be near $3/2$. For the case of worldwide seismicity as discussed by KAGAN (1994), $M_{0_{\max}}$ is a fixed cutoff moment multiplying M_0. A fixed cutoff would be appropriate to the SS model, which involves small fluctuations around a constant value.

However, the situation is more interesting for the IC model, which displays a strongly near-cyclic, or recurrent, behavior. As can be deduced from (40) and (41), the factor $k[K_L V \delta t]^{1/\sigma_f}$ multiplying M_0 in the argument of the exponential of (40) is actually a function of the correlation length ζ_c. For the IC model, $\zeta_c = \gamma(K_L V|\delta t|)^{-1/4}$ where γ is a constant. Using the value $\sigma_f = 1$, which is appropriate for a spinodal, one finds that

$$k[K_L V \delta t]^{1/\sigma_f} = \frac{k'}{\zeta_c^{4/\sigma_f}} \equiv \frac{1}{A^*} \tag{58}$$

where A^* is a "cutoff area" and $k' = \frac{l}{\gamma^{4/\sigma_f}}$. Equation (40) therefore becomes

$$n(A, \delta t) = \frac{n_0}{A^{\tau_f-1}} \exp\{-A/A^*\}. \tag{59}$$

Since ζ_c is inversely proportional to the fourth power of $|\delta t| = |t_{sp} - t|$, it can be seen that the correlation length ζ_c, or equivalently the cutoff area A^* will grow in magnitude as the time of the spinodal t_{sp} (i.e., mainshock) approaches. After the mainshock, both ζ_c and A^* will then diminish as time proceeds. Note in particular that the exponent τ which characterizes the scaling region is not a function of $|\delta t|$ and is thus independent of time relative to the mainshock. We note that observations of this type have been reported by Jaumé (this issue).

Finally, note also that ζ_c varies inversely with the inverse fourth power of plate velocity V and loading modulus K_L. This implies that faults characterized by small V will have small ζ_c and thus the maximum size event on these faults will tend to be small, all other factors being equal. Larger events may still occur, although they will be extremely improbable. Note that this conclusion holds for both the IC and SS models. Observational evidence (RUFF and KANAMORI, 1980) indicates that the maximum size event possible on a fault system depends both on the plate velocity V, and upon a couple parameter. In our models, "coupling" is determined by the size of $\Lambda = 2\gamma_c \kappa^2/K_L$ and the amount of disorder present, that is by the variance in $\varepsilon_c(\mathbf{x}, t)$ in (34). ζ_c tends to increase with decreasing $|\Lambda - \Lambda_c| = |\Lambda - 1|$ and decreasing variance in $\varepsilon_c(\mathbf{x}, t)$.

8.3. Time Scale for Aftershocks

We remark that for the IC model, the expansion about the spinodal (36)–(39) that transforms the general dynamical equation (35) into a scaling form (38) valid near the spinodal is only appropriate when (RUNDLE et al., 1997):

$$\kappa V |\delta t| \ll 2\pi \tag{60}$$

where κ is the wave number for the frictional cohesion, and V is the driving (plate motion) velocity. The wavelength of the frictional cohesion is proportional to the nominal recurrence interval T_R through the plate velocity:

$$T_R = \frac{2\pi}{\kappa V}. \tag{61}$$

Therefore (60) expresses the condition that the scaling forms for foreshocks and aftershocks (i.e., Omori Law) will only be valid during time intervals around the mainshock that are short compared to the nominal recurrence interval

$$|\delta t| \ll T_R. \tag{62}$$

Observations of this type have recently been reported by DIETERICH (1994).

8.4. Omori Relation

Omori's Law is derived from observational evidence on the rate at which aftershocks decay following a mainshock. It has the form (SCHOLZ, 1990):

$$n_{as} = \frac{K_{as}}{(c_{as} + \Delta t_{ms})^p}. \tag{63}$$

Here n_{as} is the rate at which aftershocks occur at a time $\Delta t_{ms} = t - t_{ms}$ following the mainshock, and K_{as}, c_{as}, and p are all empirical constants. The remarkable fact about the Omori relation is that p is mostly observed to have the value $p \approx 1$. Equation (63) is thus a scaling form like the Gutenberg–Richter relation, and may be characteristic of an essential part of the earthquake-related critical phenomena. Equation (63) has also recently been found to characterize foreshocks (JONES *et al.*, 1995).

We now summarize the derivation of a relation very similar to (63), which differs only for times very close to the mainshock. We emphasize that both (63) and the relation below are both finite at $\Delta t_{ms} = 0$, which is the reason why the empirical constant c_{as} was originally inserted into (63). We first identify the observationally defined interval $\Delta t_{eq} = t - t_{ms}$, the time since the mainshock, with the time interval $-\delta t$ since the spinodal (at which scaling is observed), thus $\Delta t_{eq} \approx -\delta t$. Consider first the problem of foreshocks, in which Δt_{eq} is replaced by $-\Delta t_{eq}$. To calculate the frequency of events at time δt in terms of the frequency at time $\delta t = 0$, we use the event frequency relation (59) to integrate over a band of events lying between area (A_{min}, A_{max}):

$$\int_{A_{min}}^{A_{max}} \frac{n(A, \delta t)}{n(A, 0)} \, dA = A^* \{ (\exp[-A_{min}/A^*] - \exp[-A_{max}/A^*]) \}. \tag{64}$$

Recalling that A^* is a function of δt, and from (51) that $\sigma_f = 1$, we set

$$A^* = B^* \{\delta t\}^{-1/\sigma_f} = B^* \{\delta t\}^{-1} \tag{65}$$

and write (64) as

$$\int_{A_{\min}}^{A_{\max}} \frac{n(A, \delta t)}{n(A, 0)} \, dA = B^* \left\{ \frac{\exp[-\delta t \, A_{\min}/B^*] - \exp[-\delta t \, A_{\max}/B^*]}{\delta t} \right\}.$$

Thus one finds that the rate of occurrence for the band within $(A_{\min}, A_{\max})$ is

$$n(A_{\min}, A_{\max}, \delta t) = n(A_{\min}, A_{\max}, 0)$$
$$\times B^* \left\{ \frac{\exp[-\delta t \, A_{\min}/B^*] - \exp[-\delta t \, A_{\max}/B^*]}{\delta t} \right\}. \tag{67}$$

We see that for times $\delta t > 0$, the frequency varies with time according to $(\delta t)^{-1}$, modified by an exponential decay. Thus an Omori p' value of $p' = 1$ is predicted, in agreement with observation (SCHOLZ, 1990). Equation (66) clearly predicts a "run-up" of activity to the mainshock. Calculation of the aftershock rate is carried out in the same way, with δt replaced by $-\delta t$. The value of p is the same, and $p = 1$ is predicted as well, in agreement with observations of real faults.

Integrating over all events up to the size of the mainshock, one finds a foreshock (aftershock) frequency:

$$n_{\text{tot}}(\delta t) = n_{\text{tot}}(0)B^* \left\{ \frac{1 - \exp[-\delta t \, A_{\text{main}}/B^*]}{\delta t} \right\} \tag{68}$$

where A_{main} is the area of the mainshock.

This relationship behaves asymptotically as δt^{-1} for "large" δt, where the approach to the asymptotic form is controlled by the size of $A_{\max}$, the mainshock area. Again $p' = p = 1/\sigma_f = 1$ is found. Note that while the short time cutoff is exponential in (68), it is inverse-linear in (63).

Equations (64)–(68) also predict something new—that the rate at which aftershocks of area A decay depends on A. Thus the band of aftershocks centered on a large area will decay more rapidly than aftershocks within a band on a smaller area. While data on this effect are at present limited, there is at least some observational evidence in support of this prediction (Fig. 6).

8.5. Growth, Arrest, Stress Heterogeneity

Our results indicate that growth and arrest of earthquakes are determined by the heterogeneity of the stress field, and are governed by equations such as (48)–(50). For models in which a "stochastic spinodal" or limit of stability against unstable growth exists (Fig. 5), there will be (a) no magnitude "band gap" if the $F_S - \lambda$ curve is monotonic with increasing λ, and is therefore stable for all λ; or (b)

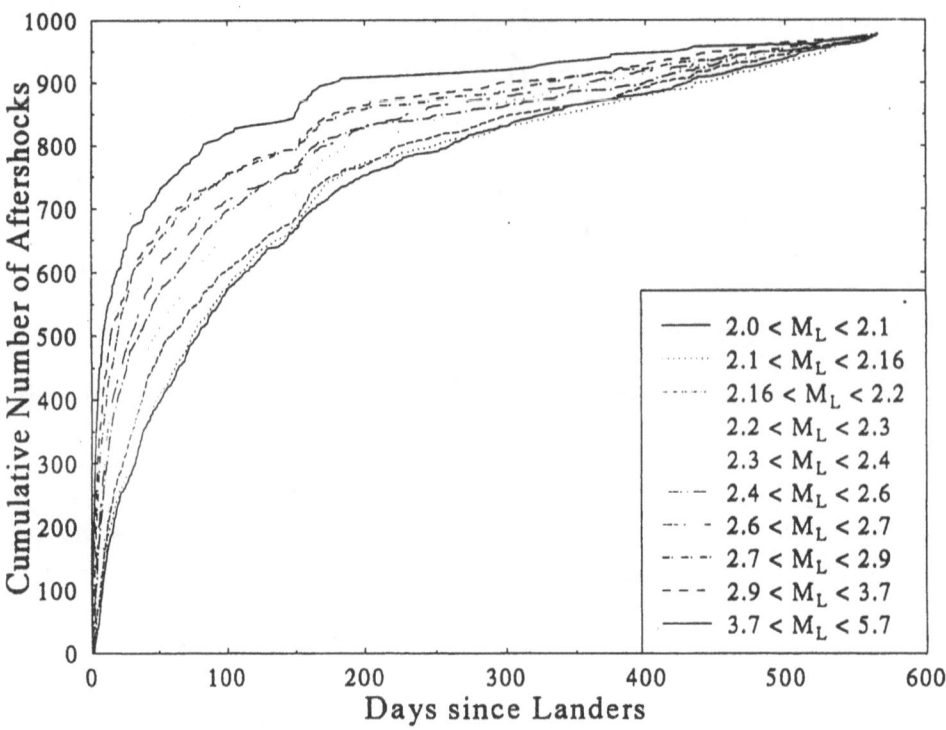

Figure 6

Cumulative Landers aftershocks in days since mainshock. All magnitude bins have ~970 events. The largest event bins clearly relax the most rapidly, as predicted by Equation (67).

a gap if the $F_S - \lambda$ has an unstable region in which F_S decreases as λ increases. We consider predictions for cases (A) and (B) for both the IC model and the SS model:

Intermittent Criticality (IC) Model: This model generally corresponds to the case of small disorder in $U[\phi]$, thus $\Sigma(\mathbf{x}, t)$ has a red spectrum, and case (b) applies. The larger the magnitude band gap, the fewer the number of foreshocks and aftershocks for a given size mainshock will be. Correlation lengths should increase throughout the earthquake cycle, becoming largest ("diverging") just prior to the mainshock, implying critical slowing down, $\tau_c \to \infty$. The associated H value is ≤ 0.5 if $|B/\Delta E_{el}|$ is $\gg 1$, or $H > 0.5$ if $|B/\Delta E_{el}| \ll 1$. For these cases, moment release rate and nucleation zone size should be uncorrelated with mainshock size. The mainshock can be triggered by events having a wide range of stress drops.

Steady State (SS) Model: This model corresponds to the case of large disorder in $U[\phi]$, thus $\Sigma(\mathbf{x}, t)$ will tend to have a blue spectrum. The magnitude band gap is small or nonexistent, with comparatively more foreshocks and aftershocks for a given size mainshock. Correlations lengths are relatively constant in size and do not diverge at any time. Values for parameters and exponents are $H > 0.5$ and

$|B/\Delta E_{el}| \sim 1$. For these cases, the analysis of ELLSWORTH and BEROZA (1995) indicates that moment release rate, nucleation size, and other indicators should be correlated with the size of the mainshock. Breakaway events (mainshock) can only be triggered by high stress drop foreshocks.

We have found that simulations of intervals between repeating earthquakes, which are events registering the same magnitude and the same location, indicate that these events can be used as discriminants (GROSS, 1998) for the spatial heterogeneity of the fault failure threshold $\sigma^F(\mathbf{x}, t)$. Models with spatially heterogeneous thresholds indicate that repeating events occur with higher probability and at shorter time intervals than in models with spatially homogeneous thresholds. Furthermore, high stress regions, particularly in the SS model, should nucleate via a nonclassical mechanism in which the initial mode of growth is a densification, or infilling mode (KLEIN et al., 1997). High stress sites grow by a coalescence mechanism. Testing this idea will involve looking at the growth and decay of "clusters of quiescent regions," rather than seismically active regions. In these faults, any observed "run-up" to failure represents loading by events on nearby faults or neighboring regions of the same fault.

9. Final Remarks

Our previous work has shown that clustering is not a random process (RUNDLE et al., 1996b). Using aftershocks from the June 28, 1992 Landers earthquake, we looked for examples of "anomalous" clustering, the clustering that is over and above that expected from the Omori decay. We then compared the anomalous Landers clustering to that obtained from the simplest possible minimal simulations (defined and described, in, e.g., RUNDLE and KLEIN, 1995b). We constructed distributions of interevent times that have the expected temporally decaying Omori clustering removed, and which are normalized to the Poisson distribution expected for a random process. These data manifested a behavior characteristic of the "anti-gap" forecast model (KAGAN and JACKSON, 1991), in which an event is more likely to take place soon after a previous event, rather than later. Small values of interevent time are more probable, and intermediate values are less probable than the Poisson process would predict, implying the existence of clustering. Longer interevent times are also more probable than the Poisson point process, due to the fact that most of the activity is temporally concentrated within the clusters.

There are two major candidate mechanisms for *spatial* clustering in seismicity, (1) clustering of faults, and (2) dynamical interaction between faults. Fault geometries have been shown to be fractal (e.g., BROWN and SCHOLZ, 1985; AVILES et al., 1987; TURCOTTE, 1992; SAHIMI et al., 1993) so the network of structures upon which earthquakes occur occupies space in an irregular and spatially clustered fashion. This mechanism for earthquake clustering is not present in the minimal

Table 1

Definition of major parameters (in order of appearance)

Parameter	Definition
V	Loading (plate) velocity
K_L	Loading spring constant
$\eta(\mathbf{x}, t)$	Fluctuating noise term (space $\mathbf{x}$ and time t dependent)
$s(\mathbf{x}, t)$	Slip on fault
$\phi(\mathbf{x}, t)$	Slip deficit $= s(\mathbf{x}, t) - Vt$
$U[\phi]$	Lyapunov functional potential
ξ_c	Correlation length
τ_c	Correlation time
V_b	Sliding velocity of block
V_{b0}	Critical value of V_b
f	Loading force in excess of static friction
W_P	Work per unit time (power) done by loading force
W_F	Work per unit time (power) dissipated by friction
ε	Velocity-dependent coefficient in W_F
α	Velocity-dependent coefficient in W_F
ϕ_S	Spinodal value of ϕ
ϕ_M	Globally stable value of ϕ
R	Range of interaction sites
β, δ, ν, z	Critical exponents
M	Mobility
A	Area of earthquake
$n(A)$	Frequency of earthquakes having area A
τ_f	Fisher–Stauffer critical exponent for frequency-area relation
σ_f	Surface critical exponent in frequency-area relation
κ	Wave number of cohesion term in $U[\phi]$
h	Cohesive energy in $U[\phi]$
$\varepsilon_c(\mathbf{x}, t)$	Random phase (noise) term in $U[\phi]$
$\Phi(t_{sp})$	Spinodal value of $\phi(\mathbf{x}, t)$
$\psi(\mathbf{x}, t)$	Fluctuations about $\Phi(t_{sp})$
t_{sp}	Time at which spinodal occurs
σ^F	Stress failure threshold
σ^R	Residual stress
τ	Coarse-graining time
$\sigma(\mathbf{x}', \tau)$	Coarse-grained stress
σ_{Load}	Tectonic loading stress
$\Delta\sigma_{\text{Seis}}$	Stress lost due to seismicity
λ	Fracture radius
ΔE_{el}	Elastic energy change due to fracture
a_0	Griffith surface energy $\{= 2\gamma\}$
p_e	Externally applied stress field
F_G, F_S	Griffith free energy, Stochastic Griffith free energy
$\Sigma(\mathbf{x}, t)$	Stress difference field $\{= \sigma^F(\mathbf{x}, t) - \sigma(\mathbf{x}, t)\}$
$\Delta\sigma$	Static stress drop
H	Hausdorff dimension
m	Moment magnitude
M_0	Seismic moment
b	b value
ζ	Scaling exponent in frequency-moment relation
T_R	Recurrence time interval
p	Omori p value
K_{as}, c_{as}	Constants in observed Omori Law
Δt_{ms}	Time since mainshock $\{= t - t_{ms}\}$

simulation models, because they represent a single planar fault. The second primary cause of clustering is stress transfer, in which a fault alters the stress state near it, and brings neighboring faults closer to failure. Stress transfer between faults on a large scale is an active area of research (e.g., STEIN et al., 1992; JAUMÉ and SYKES, 1992; HARRIS and SIMPSON, 1992). Clearly, *spatial* clustering of faults cannot automatically explain *temporal* clustering of activity, without an associated plausible mechanism for linking spatial and temporal behavior via a nonlinear process.

Increasingly convincing evidence has been compiled to demonstrate that changes in the static stress field are an important triggering mechanism beyond the usual aftershock zone, implying the existence of long-range correlations (see, for example JAUMÉ, 1999, this issue). A possible example of the existence of these long-range correlations has been given by HILL et al. (1993), who document observations of the Landers earthquake triggering seismicity at very great distances, a conclusion that remains controversial. It has been argued that these long-range interactions must be due to the transient stresses that result from the passage of seismic waves, because static stress changes at great distances are so small. Recent modeling and observations also support the idea that the triggering amplitude in the far field is inversely proportional to the frequency of the waves (see, e.g., GOMBERG et al., 1997). An alternative explanation for far field triggering is that seismicity is correlated over considerably longer distances than previously thought, the so-called "action at a distance" idea. How such long-range correlations can physically exist in nature for complex fault systems has been something of a mystery, but we note that models such as the Intermittent Criticality Model exhibit these properties near the spinodal.

Acknowledgements

Research by JBR has been supported by NSF grant EAR-9526814 (theory), and by USDOE grant DE-FG03-95ER14499 (simulations), to Univ. of Colorado. Work by WK has been supported by USDOE grant DE-FG02-95ER14498 to Boston University. The authors would also like to acknowledge support from the Colorado Center for Chaos and Complexity; and conversations with D.L. Turcotte, S. Gross, J. Goldstein, and V.K. Gupta.

REFERENCES

AKI, K. (1979), *Characterization of Barriers on an Earthquake Fault*, J. Geophys. Res. *84*, 6140–6148.
AVILES, C. A., SCHOLZ, C. H., and BOATWRIGHT, J. (1987), *Fractal Analysis Applied to Characteristic Segments of the San Andreas Fault*, J. Geophys. Res. *92*, 331–340.
BEN-ZION, Y., and RICE, J. R. (1993), *Earthquake Failure Sequences along a Cellular Fault Zone in a Three-dimensional Elastic Solid Containing Asperity and Nonasperity Regions*, J. Geophys. Res. *98*, B8, 14,109–14,131.

BINNEY, J. J., DOWRICK, N. J., FISHER, A. J., and NEWMAN, M. E. J., *The Theory of Critical Phenomena* (Oxford University Press, New York 1993).

BLANPIED, M. L., TULLIS, T. E., and WEEKS, J. D. (1987), *Frictional Behavior of Granite at Low and High Sliding Velocity*, Geophys. Res. Lett. *14*, 554–557.

BOWDEN, F. P., and TABOR, D., *The Friction and Lubrication of Solids* (Oxford University Press, Oxford, UK 1950).

BROWN, S. R., and SCHOLZ, C. H. (1985), *Broad Bandwidth Study of the Topography of Natural Surfaces*, J. Geophys. Res. *90*, 12,575–12,584.

CASTI, J., *Would-be Worlds, How Simulation is Changing in the Frontiers of Science* (John Wiley, New York 1997).

DIETERICH, J. (1994), *A Constitutive Law for Rate of Earthquake Production and its Application to Earthquake Clustering*, J. Geophys. Res. *99*, 2601–2618.

ELLISWORTH, W., and BEROZA, G. (1995), *Seismic Evidence for an Earthquake Nucleation Phase*, Science *268*, 861–864.

FISHER, D. S., DAHMEN, K., RAMANATHAN, S., and BEN-ZION, Y. (1997), *Statistics of Earthquakes in Simple Models of Heterogeneous Faults*, Phys. Rev. Lett. *78*, 4885–4888.

FERGUSON, C. F., KLEIN, W., and RUNDLE, J. B. (1998), *Spinodals, Scaling and Ergodicity in a Model of an Earthquake Fault with Long-range Stress Transfer*, Phys. Rev. E., submitted.

FEIGENBAUM, M. J. (1978), *Quantitative Universality for a Class of Nonlinear Transforms*, J. Stat. Phys. *19*, 25–52.

FUKAYAMA, E., and MADARIAGA, R. (1998), *Rupture Dynamics of a Planar Fault in a 3-D Elastic Medium: Rate and State Weakening Friction*, Bull. Seismol. Soc. Am. *88*, 1–17.

GOLDSTEIN, J., KLEIN, W., GOULD, H., RUNDLE, J. B., and FERGUSON, C. F. (1998), *Arrested Nucleation in Slider Block Models of Earthquake Faults*, Phys. Rev. E., submitted.

GOMBERG, J., BLANPIED, M. L., and BEELER, N. M. (1997), *Transient Triggering of Near and Distant Earthquakes*, Bull. Seismol. Soc. Am. *87*, 294–309.

GROSS, S. J. (1998), *Repeating Earthquakes on Heterogeneous Faults*, Bull. Seismol. Soc. Am., submitted.

GUNTON, J. D., and DROZ, M. (1983), *Introduction to the Theory of Metastable and Unstable States*, Lecture Notes in Physics 183, Springer-Verlag, Berlin.

HABERMANN, T. (1991), *Seismicity Rate Variations and Systematic Changes in Magnitudes in Teleseismic Catalogs*, Tectonophysics *193*, 277–334.

HAKEN, H. (1983), *Synergetics, An Introduction*, 3rd ed. (Springer-Verlag, Berlin 1983).

HARRIS, R. A., and SIMPSON, R. W. (1992), *Changes in Static Stress on Southern California Faults after the 1992 Landers Earthquake*, Nature (London) *360*, 251–254.

HEATON, T. H. (1990), *Evidence for and Implications for Self-healing Pulses of Slip in Earthquake Rupture*, Phys. Earth Planet. Int. *64*, 1–20.

HILL, D. P., and many others (1993), *Seismicity Remotely Triggered by the Magnitude 7.3 Landers, California, Earthquake*, Science *260*, 1617–1623.

JAUMÉ, S. C., and SYKES, L. R. (1992), *Changes in State of Stress on the Southern San Andreas Fault Resulting from the California Earthquake Sequence of April to June 1992*, Science *258*, 1325–1328.

JOHNSTON, M. J. S., LINDE, A. T., GLADWIN, M. T., and BORCHERDT, R. D. (1987), *Fault Failure with Moderate Earthquakes*, Tectonophysics *144*, 189–206.

JONES, L. M., CONSOLE, R., DI LUCCIO, F., and MURRU, M. (1995), *Are Foreshocks Mainshocks Whose Aftershocks Happen to be Big?*, EOS Trans. Am. Geophys. Un. Suppl. *76*, F388.

KAGAN, Y. Y., and JACKSON, D. D. (1991), *Seismic Gap Hypothesis: Ten Years After*, J. Geophys. Res. *96*, 21,419–21,431.

KAGAN, Y. Y. (1994), *Observational Evidence for Earthquakes as a Nonlinear Dynamic Process*, Physica D. *77*, 160–192.

KANAMORI, H., and KIKUCHI, M. (1993), *The 1992 Nicaragua Earthquake: A Slow Tsunami Earthquake Associated with Subducted Sediments*, Nature *361*, 714–716.

KANNINEN, M. F., and POPELAR, C. H., *Advanced Fracture Mechanics* (Oxford University Press, New York 1985).

KLEIN, W., FERGUSON, C., and RUNDLE, J. B., *Spinodals and Scaling in Slider Block Models. In Reduction and Predictability of Natural Hazards* (Rundle, J. B., Turcotte, D. L., and Rundle, J. B., eds.) (Santa Fe Inst. Ser. Sci. Complexity, XXV, Addison Wesley 1996) pp. 223–242.

KLEIN, W., RUNDLE, J. B., and FERGUSON, C. (1997), *Scaling and Nucleation in Models of Earthquake Faults*, Phys. Rev. Lett. *78*, 3793–3796.

LAY, T., KANAMORI, H., and RUFF, L. (1982), *The Asperity Model and the Nature of Large Subduction Zone Earthquakes*, Earthquake Pred. Res. *1*, 3–71.

LORENZ, E. N. (1993), *Deterministic Nonperiodic Flow*, J. Atmos. Sci. *20*, 130–141.

MAIN, I. G., SAMMONDS, P. R., and MEREDITH, P. G. (1993), *Application of a Modified Griffith Criterion to the Evolution of Fractal Damage During Compressional Rock Failure*, Geophys. J. Int. *115*, 367–380.

MAIN, I. G., HENDERSON, J. R., MEREDITH, P. G., and SAMMONDS, P. R. (1994), *Self-organized Criticality and Fluid-rock Interactions in the Brittle Field*, Pure appl. geophys. *142*, 529–543.

MAIN, I. G. (1996), *Statistical Physics, Seismogenesis and Seismic Hazard*, Rev. Geophys. *34*, 433–462.

PACHECO, J. F., SCHOLZ, C. H., and SYKES, L. R. (1992), *Changes in Frequency-size Relationship from Small to Large Earthquakes*, Nature (London) *355*, 171–173.

RABINOWICZ, E., *Friction and Wear of Materials* (John Wiley, New York 1995) (second edition).

RICE, J. R. (1993), *Spatio-temporal Complexity of Slip on a Fault*, J. Geophys. Res. *98*, 9885–9907.

RICHTER, C. F.. *Elementary Seismology* (W. H. Freeman and Co., San Francisco 1958).

RUFF, L., and KANAMORI, H. (1980), *Seismicity and the Subduction Process*, Phys. Earth Planet. Int. *23*, 240–252.

RUINA, A. (1983), *Slip Instability and State Variable Friction Laws*, J. Geophys. Res. *88*, 10,359–10,370.

RUNDLE, J. B. (1988), *A Physical Model for Earthquakes, Application to Southern California*, J. Geophys. Res. *93*, 6255–6274.

RUNDLE, J. B., and KLEIN, W. (1993), *Scaling and Critical Phenomena in a Cellular Automaton Slider Block Model for Earthquakes*, J. Stat. Phys. *72*, 405–412.

RUNDLE, J. B., and KLEIN, W. (1995a), *New Ideas about the Physics of Earthquakes*, AGU Quadrennial Report to the IUGG (invited), and Rev. Geophys. Space Phys. suppl., 283–286.

RUNDLE, J. B., and KLEIN, W. (1995b), *Dynamical Segmentation and Rupture Patterns in a "Toy" Slider-block Model for Earthquakes*, J. Nonlin. Proc. Geophys. *2*, 61–81.

RUNDLE, J. B., KLEIN, W., GROSS, S., and TURCOTTE, D. L. (1995), *Boltzmann Fluctuations in Numerical Simulations of Nonequilibrium Threshold Systems*, Phys. Rev. Lett. *75*, 1658–1661.

RUNDLE, J. B., KLEIN, W., and GROSS, S. (1996a), *Dynamics of a Traveling Density Wave Model for Earthquakes*, Phys. Rev. Lett. *76*, 4285–4288.

RUNDLE, J. B., TURCOTTE, D. L., and KLEIN, W., editors, *Reduction and Predictability of Natural Disasters*, Santa Fe Institute Studies in the Sciences of Complexity (Addison-Wesley 1996b).

RUNDLE, J. B., KLEIN, W., GROSS, S., and FERGUSON, C. D. (1997a), *The Traveling Density Wave Model for Earthquakes and Driven Threshold Systems*, Phys. Rev. E. *56*, 293–302.

RUNDLE, J. B., GROSS, S., KLEIN, W., FERGUSON, C. D., and TURCOTTE, D. L. (1997b), *The Statistical Mechanics of Earthquakes*, Tectonophysics *277*, 147–164.

RUNDLE, J. B., KLEIN, W., GROSS, S., and TURCOTTE, D. L. (1997c), *Reply to a Comment*, Phys. Rev. Lett. *78*, 3798.

RUNDLE, J. B., PRESTON, E., McGINNIS, S., and KLEIN, W. (1998a), *Why Earthquakes Stop: Growth and Arrest in Stochastic Field*, Phys. Rev. Lett. *80*, 5698–5701.

RUNDLE, J. B., KLEIN, W., TIAMPO, K., GROSS, S. (1998b), *Linear Pattern Dynamics in Nonlinear Threshold Systems*, Phys. Rev. E, submitted.

SAHIMI, M., ROBERTSON, M. C., and SAMMIS, C. G. (1993), *Fractal Distribution of Earthquake Hypocenters and its Relation to Fault Patterns and Percolation*, Phys. Rev. Lett. *70*, 2186–2189.

SAMMIS, C. G., SORNETTE, D., and SALEUR, H., *Complexity and Earthquake Forecasting*. In *Reduction and Predictability of Natural Hazards* (D. L. Turcotte and J. B. Rundle eds.) (Santa Fe Inst. Ser. Sci. Complexity, XXV, Addison-Wesley 1996).

SCHOLZ, C. H., *The Mechanics of Earthquakes and Faulting* (Cambridge, Cambridge, UK 1990).

SHAW, B. E. (1995), *Frictional Weakening and Slip Complexity on Faults*, J. Geophys. Res. *100*, 18,239–18,252.

STANLEY, H. E., *Introduction to Phase Transitions and Critical Phenomena* (Oxford, New York 1971).

STEIN, R. S., KING, G. C. P., and LIN, J. (1992), *Changes in Failure Stress on the Southern San Andreas Fault System Caused by the 1992 Magnitude = 7.4 Landers Earthquake*, Science *258*, 1328–1332.

TURCOTTE, D. L., *Fractals and Chaos in Geology and Geophysics* (Cambridge, Cambridge, UK 1992).

(Received August 12, 1998, revised/accepted November 17, 1998)

Pure appl. geophys. 155 (1999) 609–624
0033–4553/99/040609–16 $ 1.50 + 0.20/0

⌐Pure and Applied Geophysics

Use of Statistical Models to Analyze Periodic Seismicity Observed for Clusters in the Kanto Region, Central Japan

MASAJIRO IMOTO[1], KENJI MAEDA[2] and AKIO YOSHIDA[3]

Abstract—A periodic pattern of seismicity has been reported for the Kinugawa cluster in the Kanto region, where several earthquake clusters are observed at depths between 40 and 90 km. To analyze this periodicity, statistical studies are performed for the Kinugawa cluster together with eight other clusters. Hypocentral parameters of the earthquakes with magnitudes 4.5 and larger for the period between 1950 and 1995 are taken from the JMA catalogue. The simple sinusoidal function, the exponential of sinusoidal function and the stress release model are applied as the intensity function. Model parameters are determined by the maximum likelihood method and the best model for each cluster is selected by using the Akaike Information Criterion (AIC). In six cases the sinusoidal model or the exponential of the sinusoidal model is selected as the best option and achieves AIC reductions of values between 2.4 and 13.2 units from the simple Poisson model. The stress release model is selected for two clusters. The three clusters, the Kinugawa, Kasumigaura, and Choshi clusters, have a similar optimal period of about 10 years, and align in the northwest–southeast direction at a similar depth range of 40 to 70 km. A model modified from the stress release model is applied to the three clusters so to analyze the relationship among them. In the modified model, an earthquake occurrence in one zone increases the stress in the other zone, which is different from the original stress release model which assumes a linear increase with time. Applying the modified model to the Kinugawa cluster, an AIC reduction from the Poisson model is significantly larger than the value obtained with the sinusoidal model. This suggests that the periodic seismicity observed for the Kinugawa cluster can be explained with the more comprehensive model than the sinusoidal model.

Key words: Statistical model, periodic seismicity, Kanto, stress release model, cluster.

1. Introduction

A study of seismicity patterns is taking advantage of the database for seismicity, which exists for all active seismic regions in the world and is considerably larger than that for any other class of precursor that has been suggested. A large number of phenomena proposed as promising candidates have been reported, based on the

---

[1] National Research Institute for Earth Science and Disaster Prevention, Tsukuba, Japan.
[2] Meteorological Research Institute, Tsukuba, Japan.
[3] Japan Meteorological Agency, Tokyo, Japan.

database of this advantage, such as foreshocks, seismic gaps, seismic quiescences, changes in the *b* value of magnitude-frequency relation and others. Some research has concentrated on this field, since a promising precursor might play an important role in making a reliable prediction. Notwithstanding these efforts, to date a reliable precursor which precedes the main event has not been found. Consequently, the performance of a proposed precursor will be assessed as an earthquake prediction tool in terms of statistical parameters such as reliability, validity, and probability gain (REASENBERG and MATTHEWS, 1988; AKI, 1981). These statistical terms lead earthquake prediction research to probability forecasting. From the viewpoint of probability forecasting, an appropriate model based on each proposed precursor could state an earthquake probability of target events within the entire space-time limits for the study. A model fitting better statistically to observed seismicity could estimate more accurate probabilities for future events.

An example of probability forecasting (RIKITAKE, 1976) can be taken from studies of earthquake recurrence models, where earthquake sequences are considered to follow a renewal process. In this case, a statistical model for the target events can be constructed, based only on their own parameters (time intervals between successive target events). In a similar case, YOSHIDA (1995) reported periodic seismicity of moderate size earthquakes observed in the Kinugawa cluster, in the Kanto region, Central Japan. In his work he fitted a sinusoidal curve to yearly numbers of earthquakes obtained by a three-year moving average method. As the fitting was made in this manner, the statistical significance of the result remains uncertain. Although the fitted curve is a sinusoidal function and drastic increases in earthquake probabilities could not be expected, if it is the case, the sinusoidal model for this earthquake cluster could play a more important role in risk assessment than the simple Poisson model. From this perspective, the present study has been conducted to examine the statistical significance of the periodic seismicity observed in the Kinugawa cluster, using a more exact method of the point process in statistics. Taking into consideration the possibility that a more drastic change in seismicity appears than is indicated by the sinusoidal function, the exponential of the sinusoidal function is also studied as an intensity rate function for statistical models in addition to the sinusoidal function. In addition to these two models, the stress release model is attempted for fitting, as a model which provides a physically plausible explanation for at least quasi-periodic effects (ZHENG and VERE-JONES, 1994). On the other hand, for the purpose of clarifying the features of periodic seismicity, which may help physical interpretation, in addition to the Kinugawa cluster eight other clusters in the Kanto region are also studied.

In the present study, we will focus on the periodic seismicity observed in the Kinugawa cluster and attempt to make a statistical model with a plausible physical interpretation. Three basic models will be applied to the nine clusters, including the Kinugawa cluster, and model fitting is examined by the Akaike Information Criterion procedure (AKAIKE, 1977). After examining these basic results, we will attempt to develop a new class of models for the Kinugawa cluster, which we call the modified version of the stress release model. These modified versions allow for stress transfer or the interaction between regions (ZHENG and VERE-JONES, 1991) and are considered more appropriate for a physical interpretation of seismicity patterns than a model based purely on mathematical functions.

2. Data

Earthquakes with a magnitude of 4.5 and larger are used for the present study, which occurred during the period from 1950 to 1995 in Kanto, central Japan. Hypocentral parameters of the earthquakes are taken from the catalogue prepared by the Japan Meteorological Agency. With this cutoff magnitude, earthquakes are considered to be uniformly detected in the time space window for the study (ISHIKAWA, 1987). Figure 1 shows an epicenter map of those earthquakes in the depth range between 30 km and 90 km within the study area. Several clusters can be seen in this map. These clustering features are demonstrated more clearly by an epicenter map of microearthquakes in the same region (Fig. 2). Tectonics in this area is quite complex with three plates, the Pacific, the Philippine Sea, and the North American plates, converging. The configurations of these plates are defined differently by different authors (SHIMAZAKI et al., 1982; NOGUCHI, 1985; KASA-HARA, 1985; ISHIDA, 1992), and are summarized by OKADA (1990). The clusters are interpreted as being the result of relative movement between either the Pacific and Philippine Sea plates or the Philippine Sea and North American plates. This resulted in a depth range for the active seismic zone, of approximately 30 km to 90 km, and a varying depth range from cluster to cluster. Taking a recent hypocentral distribution of microearthquakes into consideration, we select nine polyhedrons enclosing active zones. The polygons in Figure 1 indicate the results of the selection; all the earthquakes inside the polyhedron are considered to be the members of the respective cluster. The depth range of each polyhedron and the number of earthquakes for each cluster are listed in Table 1. Cluster A (named Kinugawa cluster) is the same one for which periodic seismicity was reported on by YOSHIDA (1995). The other eight clusters are also studied here in comparison with cluster A.

3. Models and Analysis

We assume three statistical models to apply to the time series of events taken from the nine clusters. We consider an intensity function λ, corresponding to each statistical model. The most simple function of the present study is the simple sinusoidal function taken from the idea of YOSHIDA (1995), which is defined as

$$\lambda_s = \alpha_s + \beta_s \sin(2\pi/T_0 t + \gamma_s) \tag{3.1}$$

where α_s, β_s, T_0, and γ_s are free parameters which are obtained through the optimization of the likelihood function. The likelihood for the point process with the rate function λ takes the form,

$$L(t_1, t_2, \ldots, t_n; \theta) = \exp\left\{ -\int \lambda \, dt \right\} \prod_{i=1}^{n} \lambda(t_i). \tag{3.2}$$

Here, $t_1, t_2, \ldots, t_n$ are the times of earthquake occurrences in each series. For the second model we consider the exponential of the sinusoidal function, which is defined as

$$\lambda_e = \exp\{\alpha_e + \beta_e \sin(2\pi/T_0 t + \gamma_e)\}. \tag{3.3}$$

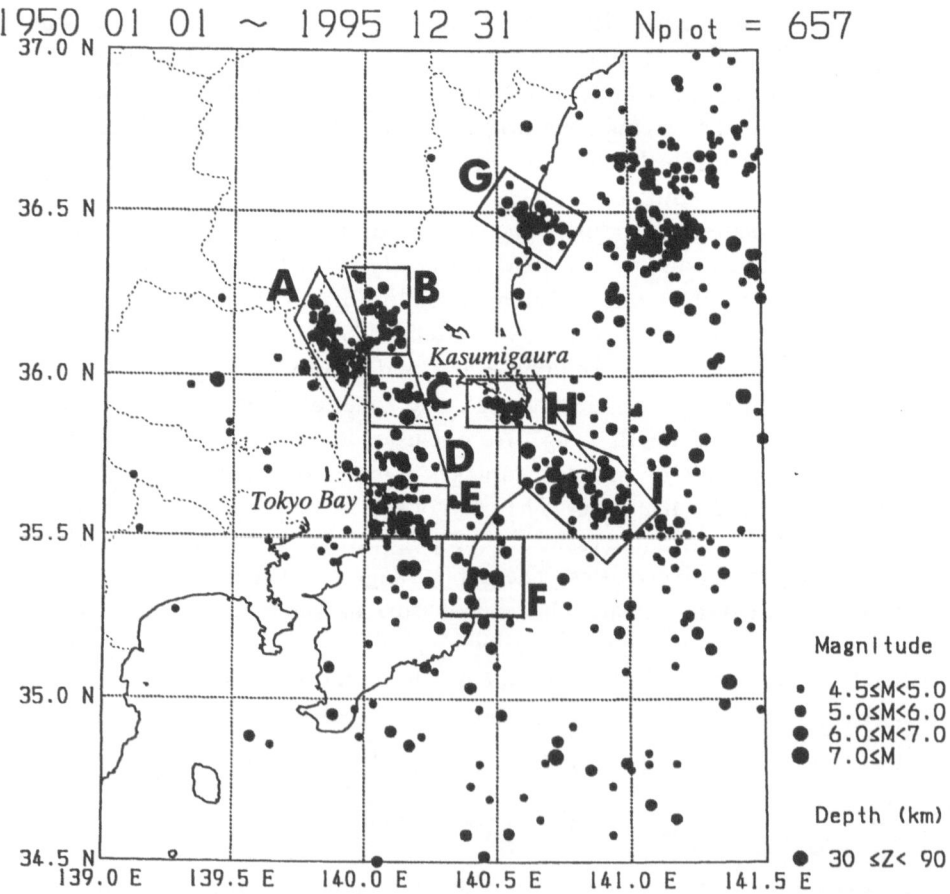

Figure 1
Epicenters of earthquakes with a magnitude of 4.5 and larger after JMA for the period 1950 to 1995. Nine clusters are indicated by polygons labelled A to H.

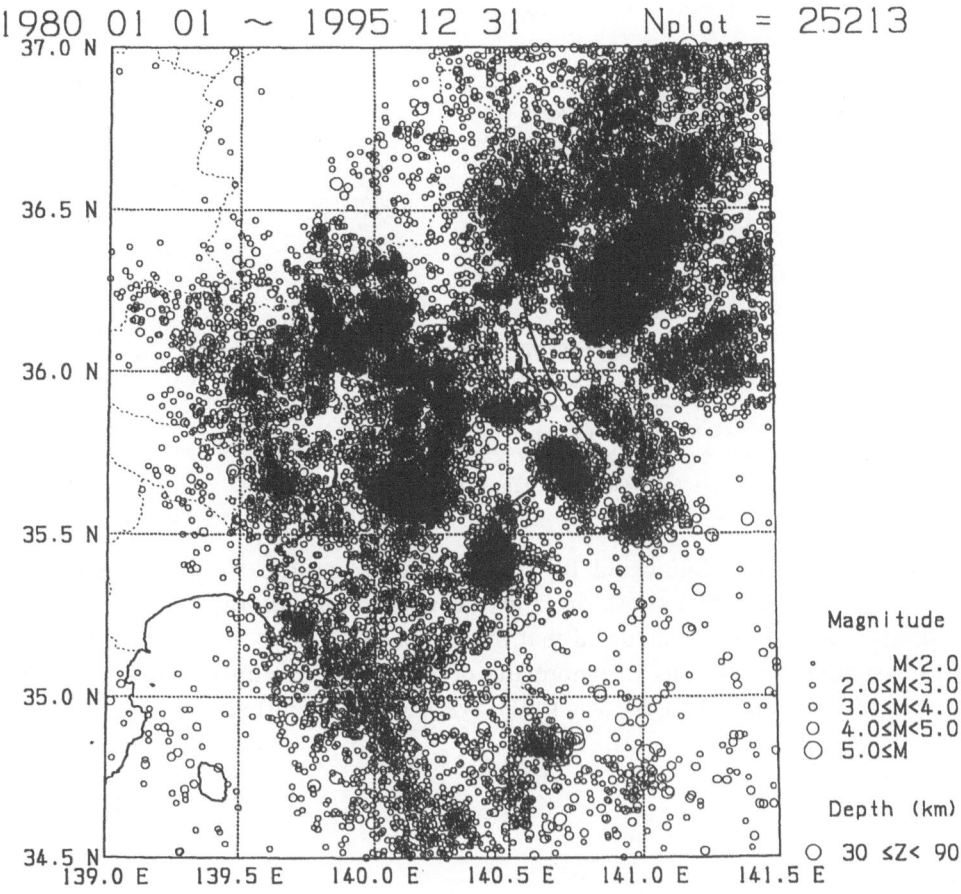

Figure 2
Epicenters of microearthquakes located by NIED for the period 1986 to 1995.

These two functions basically behave in similar ways, except for a few points. The parameters of the simple sinusoidal function α_s, β_s are limited in their ranges with the condition of $\alpha_s > \beta_s \geq 0$. The optimal values for T_0 of both functions are surveyed in the range from 5 to 15 years by a grid search technique.

The third model we consider is the simple stress release model, which was developed by Vere-Jones et al. (VERE-JONES and DENG, 1988; ZHENG and VERE-JONES, 1991, 1994). The model attempts to provide a general formulation of the idea of the elastic rebound theory of major earthquakes. Referring to the previous articles by Vere-Jones et al., the stress release model is formulated as follows. The general assumption of the model is that the probabilities of events occurring within the region are determined by a state variable. It is presumed that this variable can be represented by a scalar quantity $X(t)$ which increases linearly with time and decreases instantaneously just after events. This quantity could be tentatively

Table 1

Terms of clusters

Cluster	Depth (km)	N
A	40–70	66
B	50–80	34
C	60–80	12
D	60–90	25
E	60–90	45
F	30–60	14
G	40–70	32
H	30–70	13
I	40–70	53

A number of earthquakes with a magnitude of 4.5 and larger is given in the last column. Clusters A, H and I are named Kinugawa, Kasumigaura and Choshi, respectively.

interpreted as stress, but could be a general characteristic. It is reasonable to assume that the average stress drop in the region depends on the size of each respective event. It is also assumed that the size of the event is independent of $X(t)$. This assumption allows us to ignore the detail of the size distribution of earthquakes (ZHENG and VERE-JONES, 1991). We obtain the current value of $X(t)$ as,

$$X(t) = a_0 + b_0 t - c_0 S(t) \tag{3.4}$$

where

$$S(t) = \sum_{i:0 \le t_i \le t} S_i. \tag{3.5}$$

The average stress drop of the ith event, S_i is related to its magnitude, M_i by the formula,

$$S_i = 10^{\eta M_i}. \tag{3.6}$$

Combining the relationship between stress and strain energy with the Gutenberg–Richter relation $\log_{10} E = 1.5 M + $ constant, η takes the value 0.75. According to the previous study (ZHENG and VERE-JONES, 1994), the fit of the model to data was not strongly dependent on η in a range of 0.5–1.0. The fixed value $\eta = 0.75$ has been used throughout the study. In the present study, we shall choose for λ_0 the exponential function:

$$\lambda_0 = \exp\{X(t)\}. \tag{3.7}$$

The choice of the exponential function was justified in the previous study done by ZHENG and VERE-JONES (1991, 1994). Thus we obtain the intensity rate function for the third model.

By using the AIC (AKAIKE, 1977) procedure, we can make comparisons between the models. The definition of AIC is given by:

$$\text{AIC} = -2 \ln \hat{L} + 2k \tag{3.8}$$

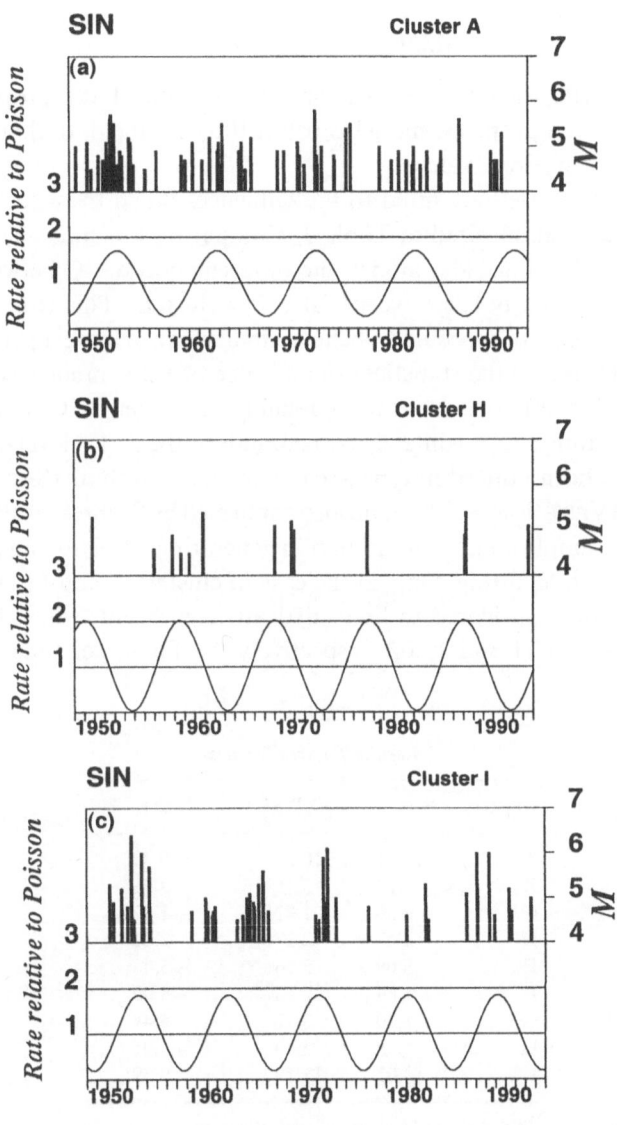

Figure 3

(a) The sinusoidal function fitted to data obtained from cluster A; (b) Same as (a) except cluster H; (c) Same as (a) except cluster I. Time magnitude plots are shown at the top and the optimal intensity rate function is drawn at the bottom. The intensity rate is measured in units of the stationary Poisson rate.

where $\hat{L}$ denotes the maximum likelihood for a given model, and k is the number of fitted parameters in the model. The better model attains to the smaller AIC. In the present study, we take the stationary Poisson process as a standard model for comparisons, and the difference in AICs between a given model and the Poisson model (denoted as dAIC) is considered as the measure for model selections, which is defined by:

$$dAIC = AIC_0 - AIC_p \tag{3.9}$$

where AIC_0 and AIC_p denote AICs of the Poisson and of the proposed models, respectively. When the proposed model is better fitted to the data than the Poisson, dAIC takes a large positive value.

The above three models are fitted to the sequences taken from each cluster. The results obtained are summarized in Table 2. Comparing the sinusoidal model with the exponential of the sinusoidal model, the differences in dAIC between these two models are not particularly large except for a few clusters. This is not unexpected, since these two models are expected to behave similarly providing a periodicity is not too prominent. Therefore the statistical significance of these models compared with the Poisson model is not considered to be reliable since their AIC reductions are as much as 2–5. This range approximately corresponds to the 5–1% level of significance, which has already been noticed in typical cases by many authors (SAKAMOTO *et al.*, 1983; ZHENG and VERE-JONES, 1994) and also confirmed by the present authors (IMOTO *et al.*, 1996) with a simple numerical simulation generating a Poissonian random series for the sinusoidal model fitting. Only in the case of clusters A and I (Choshi cluster), are their periodicities considered to be statistically significant, where the maximum dAICs attained are 11.11 and 13.02, respectively, by far larger than 5.

Table 2

Results for tree models

Cluster	SIN	EXP(SIN)	SRM
A	11.11	10.33	0.89
B	3.74	5.15	2.31
C	2.39	1.82	−1.47
D	−2.68	−2.37	−2.24
E	3.68	2.76	−3.46
F	1.84	2.98	−4.00
G	−1.10	3.52	5.48
H	4.78	5.60	8.05
I	13.02	13.19	0.69

The AIC reductions of the three models from the Poisson model are given. SIN, EXP(SIN), SRM denote the sinusoidal, the exponential of the sinusoidal and the stress release models, respectively.

Table 3

Optimal values for the sinusoidal model

Cluster	dAIC	α_s	β_s	γ_s	T_0
A	11.11	0.00385	0.00285	281	3652
B	3.74	0.00202	0.00158	4	2739
C	2.39	0.00073	0.00071	191	2191
D	−2.68	0.00149	0.00080	338	1826
E	3.68	0.00268	0.00177	47	2921
F	1.84	0.00083	0.00082	338	4200
G	−1.10	0.00194	0.00109	202	2009
H	4.78	0.00079	0.00077	50	3469
I	13.02	0.00322	0.00264	241	3286

For units, α_s and β_s: number of earthquakes per day, γ_s: degree, T_0: day.

The optimal values of the parameters for each cluster are listed in Table 3. Cluster A takes 10 years as the optimal value of T_0 and cluster I has a similar value. YOSHIDA (1995) pointed out the periodic seismicity of cluster A with a period of 10 years. In his study, he fitted the sinusoidal curve to a three-year moving average of earthquakes by a graphical fitting method. This result is confirmed by the present method based on the point process procedure in statistics. It is a notable result that three of the clusters A, H (Kasumigaura cluster), and I take not only a similar optimal value at around 10 years for the period, but also rather large dAIC values compared to those of the other clusters.

Figures 3a,b,c show time magnitude plots for the clusters, A, H, and I, respectively (top), and the optimal sinusoidal curve as the intensity rate function relative to the Poisson constant rate (bottom). It is quite obvious at a glance that the seismicity pattern of cluster A is well coincident with that of cluster I and inversely coincident with that of cluster H. This finding is directly confirmed in Table 3 if we examine the phase delay term in the penultimate column. These three clusters correspond in depth range from 30–40 km down to 70 km (see Table 1) and align in the northwest–southeast direction. The Philippine Sea plate is going down beneath the North American plate to the northwest, the plate boundary between these two plates is located at a depth range of 30 to 70 km in a region covering clusters A, H and I. These tectonic circumstances and the remarkable features of seismicity patterns may imply a certain mechanical interaction among these three clusters associated with the movement of the Philippine Sea plate. Regarding the exponential sinusoidal model, the AIC reductions are only slightly better than those of the sinusoidal model (see Table 2). The results presented in Table 4 are consistent with those of the sinusoidal model (Table 3) with similar optimal values of the phase delay and the period. These confirm the periodic features of seismicity patterns for clusters A, H and I.

Table 4

Optimal values for the exponential sinusoidal function model

Cluster	dAIC	α_e	β_e	γ_e (deg)	T_0 (day)
A	10.33	−5.70	0.745	267	3542
B	5.15	−6.41	0.876	347	2666
C	1.82	−7.58	1.280	204	2227
D	−2.37	−6.58	0.554	350	1862
E	2.76	−6.02	0.643	19	2848
F	2.98	−7.51	1.319	349	4346
G	3.52	−6.41	0.820	242	1899
H	5.6	−7.75	1.636	12	3360
I	13.19	−5.98	0.929	267	3433

The result obtained for the stress release model in Table 2 shows dAIC values (AIC reductions from the stationary Poisson model) less than those for the other two models except in a few cases. The dAIC for cluster H estimates at more than 8, which betters that of either the sinusoidal model or the exponential sinusoidal model, and is the third largest value of dAIC following clusters A and H. Figure 4 shows time magnitude plots (top) and the optimal intensity rate function relative to the Poisson constant rate (bottom) for cluster H.

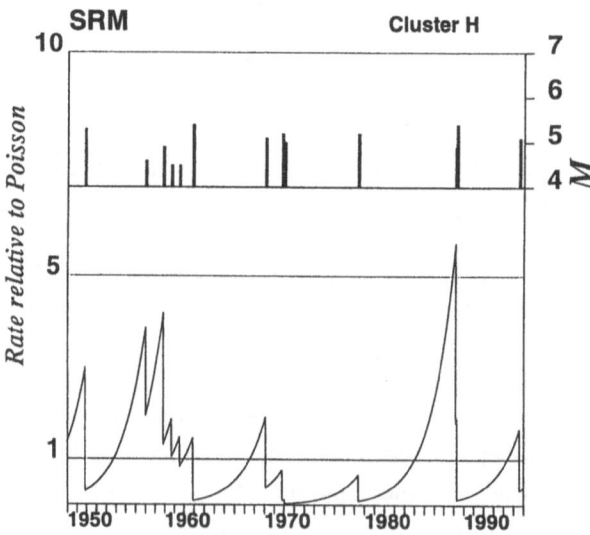

Figure 4

The stress release model fitted to data obtained from cluster H. Time magnitude plots are shown at the top and the optimal intensity rate function is drawn at the bottom. The intensity rate is measured in units of the stationary Poisson rate.

4. Modified Models

The results obtained from the three basic models suggest that the sinusoidal or the exponential sinusoidal model is better fitted to the data than the stress release model for clusters A and I and the stress release model is for cluster H. All these models are statistically significant, with AIC reductions exceeding 5 from the Poisson model. It might be possible to make physical interpretations starting from the results that the sinusoidal model (or the exponential sinusoidal model) is better fitted to both clusters A and I than the stress release model, and that the three sinusoidal models fitted to clusters A, H and I all take the similar optimal values of about 10 years for the period. In this case, we should speculate on the physical reasons for the approximately 10-year period. Alternatively, it may also be possible to accomplish this from the results that the stress release model is best fitted to cluster H among the three basic models, and that the seismicity pattern of cluster H is inversely coincident with those of clusters A and I. In this case, we must speculate on a reasonable mechanical interaction between either clusters H and A or H and I, and develop a statistical model consistent with this speculation. Various modified versions of the stress release model are worth examining, in which an interaction between regions is realized in the model through either the term of stress increase (the second term in (3.4)) or the term of stress decrease (the third term) or both. Many versions of the stress release model have been developed, some of which allow for stress transfer between regions (ZHENG and VERE-JONES, 1991; LIU et al., 1998). In this section, we will follow the second scheme, in which the stress release model for cluster H in Table 2 is concluded as the final one and modified models will be considered for clusters A or I under the influence of cluster H.

Taking into account the results of the difference in phase (see Table 3) between cluster H and cluster A (or I), we assume that effects of earthquakes in cluster H are transferred to cluster A with a certain time delay. Referring to (3.4), two simple modifications are considered, where the second and third terms of the equation are replaced by functions of stress release in cluster H, respectively. As for the first modification (delay model), the third term, which corresponds to stress release calculated from each magnitude of earthquakes in cluster A (or I), is replaced by stress release in cluster H with some delay. Thereby, $S(t)$ in (3.4) is given by

$$S(t) = \sum_{j: 0 \le t_j \le t - t_d} S_j, \tag{4.1}$$

where j refers to earthquakes in cluster H, and t_d denotes a time delay of effects from cluster H to A. In the optimization process, twenty different t_d are applied, encompassing 0 to 3640 days at every 182 days step. With this model, a similar intensity rate function to that of cluster H can be applied to cluster A (or I) (except for a delay). The AIC reductions obtained by this model are listed in Table 5 as in

Table 5

The AIC reductions for modified stress release models

Cluster	Delay model dAIC	t_d (day)	Coupled model dAIC	t_d (day)
A	−1.6	1274	16.03	728
I	5.8	1456	10.70	546

the second column. It should be noted here that a time span for the likelihood calculation becomes short by t_d due to the shift of times. It may cause a bias of dAIC from that in Table 2, which varies from case to case. However, assuming that the bias is expected to be a maximum of about 20% and underestimated from that of the full span (1950 to 1995), the bias caused by the introduction of a delay is ignored here. These assumptions are reasonable for the first approximation since the AIC reduction may become proportional to the length of analysis if both an intensity rate function of the model in question and seismicity are considered to repeat similar patterns, respectively. Comparing these values with those obtained by the sinusoidal model, the AIC reductions are not large enough to select this version instead of the sinusoidal model.

For the second modification (coupled model), the second term of (3.4) (stress increase linearly with time), is replaced by a function of stress release in cluster H with a delay. In this case, stress release in cluster H results in stress increase in cluster A with a delay. Thereby (3.4) is replaced by:

$$X(t) = a_2 + b_2 S1(t) - c_2 S2(t) \qquad (4.2)$$

where $S1(t)$ and $S2(t)$ are given by

$$S1(t) = \sum_{j:0 \le t_j \le t - t_d} S_j, \qquad (4.3)$$

and

$$S2(t) = \sum_{i:0 \le t_i \le t} S_i. \qquad (4.4)$$

The subscripts i and j refer to events in clusters A and H, respectively. Similar to that of the previous case, the optimization of the parameters is performed. The reductions of AIC are listed in the last column of Table 5. The reduction for cluster A estimates 16.03 and by about 5 units larger than that of the sinusoidal model. This difference in dAIC suggests that the coupled model is significantly better than the sinusoidal model. The intensity rate function of this model is shown at the bottom of Figure 5 similar to the previous case. As is indicated in Table 5, the reduction for cluster I takes a slightly smaller value than that of the sinusoidal case, which means that the coupled model is not conclusively better than the sinusoidal model, rather it is one possible alternative to the sinusoidal model.

5. Discussion

The sinusoidal model fits significantly better than the Poisson model to the clusters A, H, and I for earthquakes with magnitude 4.5 and larger. This cutoff magnitude is adopted taking the detection capability into consideration. IMOTO *et al.* (1996) have reported that a similar result is obtained with the lower cutoff magnitudes down to 4.0. To confirm the present result, a simple simulation experiment has been carried out, in which a random sequence of events has been generated, and has been fitted to the sinusoidal model in the same way as in an earlier section. The number of events in the sequence is set to be equivalent to that of the cluster for test. One thousand cases have been attempted for each of the clusters A, H, and I. This experiment demonstrates that dAICs larger than those in Table 2 (11.11, 4.76, and 13.02) are obtained in 5, 63, and 2 cases out of 1000 trials for the clusters A, H and I, respectively. Accordingly, the probabilities of a dAIC value larger than that in Table 2 by chance are estimated approximately 0.5%, 7% and 0.2%. The probability that dAICs larger than 10 are obtained at least for two clusters is less than 0.1% (approximately given by $_9C_2 0.005 \times 0.005 = 0.0009$). This

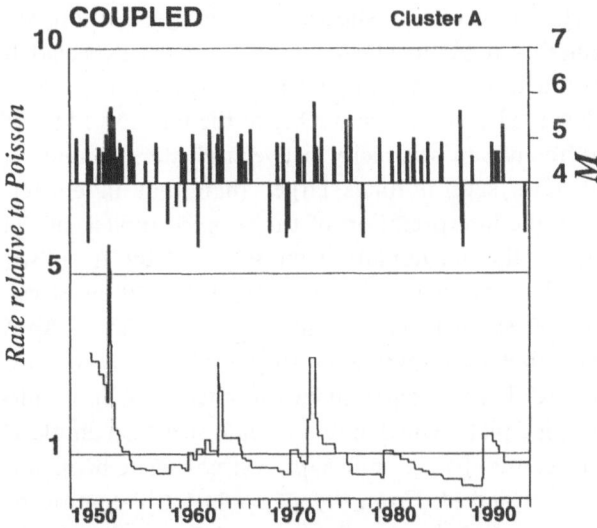

Figure 5
The intensity rate function for the coupled model fitted to cluster A. Time magnitude plots for cluster A are shown at the top, plots for cluster H are also shown in reverse sense as cluster A in the center, and the optimal intensity rate function is drawn at the bottom (see Fig. 3 for unit). Stress increase (step) has taken place a few years after an earthquake in cluster H.

implies that the periodic pattern of seismic activities in clusters A and I is not observed by chance but is substantial. In general, the larger the number of samples on which a statistical analysis is based, the more reliable the result obtained, and it is preferred in our case to adopt a magnitude cutoff as small as possible. However, a lower cutoff magnitude would cause another problem in that a large number of aftershocks with smaller magnitudes sampled for the analysis might distort the results from those obtained without aftershocks. In order to estimate the effects of aftershocks, IMOTO (1997) has attempted to fit the epidemic type of aftershock sequence model (Etas) proposed by OGATA (1988) to the data sets employed in the present study. No reduction in AIC value is obtained for each cluster except for the three clusters E, F and I. This suggests that the results obtained for clusters A and H should not be distorted by aftershocks although the result for cluster I may be somewhat affected by them. Accordingly, we have not thoroughly inspected the reductions of AIC for cluster I in fitting the modified versions of the stress release model. A mixed type of the stress release model or its modified version with the Etas model should be considered in making a detailed analysis for cluster I. Its study will be deferred.

YOSHIDA (1995) discussed the periodic seismicity of cluster A from the perspective of stress accumulation by the relative motion between two plates; the Philippine Sea plate and the subducted plate, and stress release by earthquakes of moderate size. His interpretation is essentially consistent with the elastic rebound theory. The stress release model is considered to provide a physically plausible explanation for quasi-periodic effects. However, as shown in an earlier section, the stress release model is not as suitably fitted to cluster A as could be expected from the above consideration. Introduction of the coupled model results in success from the dual perspective of both statistics and physics. One of the unresolved problems, however, is related to the difference in seismicity between clusters H and A, where energy released in cluster A is several times larger than that in cluster H. This may engender difficulty in the interpretation of the coupled model, in that stress release in cluster H results in the accumulation stress in cluster A. If we consider only elastic stress changes by earthquakes in cluster H, the resulting stress increase could be estimated to be too small to cause earthquakes in cluster A. Alternatively, it could be interpreted that earthquakes in cluster H play a role in regulating the seismicity in cluster A. That is, earthquakes in cluster H may unlock the relative motion of the Philippine plate, which could initiate stress accumulation in cluster A. Time delay might be caused by aseismic slip or visco-elastic behavior in a less active zone between clusters A and H. From the viewpoint of plate movement, clusters B, C, D, E, and G are interpreted basically as being the results of relative movement between the Pacific and Philippine Sea plates, and clearly separated from cluster A which corresponds to the relative movement between the Philippine Sea and North American plates. Although clusters H, I and F are not well interpreted, cluster H appears to contain such activity as those in cluster A from a consideration of its depth range. Hence, an interaction between A and H is attempted in the present paper. The coupled model proposed here may not be the final model however, as a trial model which is better fitted to cluster A, it provides a more plausible explanation than the sinusoidal model.

In summary, the periodic patterns in clusters A (Kinugawa) and I (Choshi) are of statistical significance with the use of the simple sinusoidal model and the exponential of sinusoidal function. The stress release model is well fitted only to cluster H (Kasumigaura). Considering the coincidence of the periods among the three clusters, a modified version of the stress release model is proposed to interpret interaction between the clusters. This model is better fitted to cluster A than the sinusoidal model. These analyses imply that the periodic seismicity observed in clusters A and H is modeled by the stress release model and its modified model, which are consistent with the elastic rebound theory.

Acknowledgements

The authors would like to express their gratitude to Professor David Vere-Jones and his colleagues for providing the source code of the stress release model and for their helpful discussions. We also thank Professor Max Wyss and the anonymous reviewers who provided valuable critical remarks. The source code of the DALL algorithm used in this study was supplied by the Institute of Mathematical Statistics in Tokyo.

REFERENCES

AKAIKE, H., *On entropy maximisation principle*. In *Applications of Statistics* (P. R. Krishnaiah, ed.) (North Holland, Amsterdam 1977) pp. 27–41.

AKI, K., *A probabilistic synthesis of precursory phenomena*. In *Earthquake Predictions* (D. W. Simpson and P. G. Richards, eds.) (AGU 1980) pp. 566–574.

IMOTO, M., MAEDA, K., and YOSHIDA, A. (1996), *Periodic Change of Seismic Activity—Clusters in the Kanto Area*, Programme and Abstracts, SSJ *2* (in Japanese).

IMOTO, M. (1997), *Periodic Change of Seismic Activity in Earthquake Clusters, Kanto, Central Japan*, EOS, AGU *78*, 46, F490.

ISHIDA, M. (1992), *Geometry and Relative Motion of the Philippine Sea Plate and Pacific Plate beneath the Kanto–Tokai District, Japan*, J. Geophys. Res. *97*, 489–513.

ISHIKAWA, Y. (1987), *Changes of JMA Hypocenter Data and Some Problems*, Quarterly J. Seismol. *51*, 47–56.

KASAHARA, K. (1985), *Patterns of Crustal Activity Associated with the Covergence of Three Plates in the Kanto–Tokai Area, Central Japan (in Japanese)*, Rep. of Nat. Res. Cent. for Disaster Prev. *35*, 33–137.

LIU, J., CHEN, Y., SHI, Y., and VERE-JONES, D. (1998), *Coupled Stress Release Model for Time Dependent Seismicity*, Abstract, the Univ. of Alaska Workshop on Seismicity Patterns, their Statistical Significance and Physical Meaning.

NOGUCHI, S. (1985), *Configuration of the Philippine Sea Plate and Seismic Activities beneath Ibaraki Prefecture* (in Japanese), Earth Mon. *7*, 97–104.

OGATA, Y. (1988), *Statistical Models for Earthquake Occurrences and Residual Analysis for Point Processes*, J. Am. Stat. Assoc. *83*, 9–27.

OKADA, Y. (1990), *Seismotectonics in the Southern Kanto District, Central Japan*, Zisin *2*, 43, 153–175 (in Japanese).

REASENBERG, P. A., and MATTHEWS, M. V. (1988), *Precursory Seismic Quiescence: A Preliminary Assessment of the Hypothesis*, Pure appl. geophys. *126*, 373–406.

RIKITAKE, T. (1976), *Recurrence of Great Earthquake at Subduction Zones*, Tectonophysics *35*, 335–362.

SAKAMOTO, Y., ISHIGURO, M., and KITAGAWA, G., *Akaike Information Criterion Statistics* (Reidel, Dordrecht 1983) 298 pp.

SHIMAZAKI, K., NAKAMURA, K., and YOSHII, T. (1982), *Complicated Pattern of the Seismicity beneath Metropolitan Area Japan: Proposed Explanation by the Interactions among the Superficial Eurasian Plate and the Subducted Philippine Sea and Pacific Slabs*, Mathematical Geophysics, Chateau de Bonas, France, 20–25 June 1982, Terra Cognita, *2*, 403.

VERE-JONES, D., and DENG, Y. L. (1988), *A Point Process Analysis of Historical Earthquakes from North China*, English translation in Earthquake Research in China 2 (2), 165–181.

YOSHIDA, A. (1995), *Periodic Change of Seismic Activity Observed in the Kinugawa Cluster Area, Southwestern Ibaraki Prefecture, Central Japan*, Zisin *2*, 48, 51–56 (in Japanese).

ZHENG, X., and VERE-JONES, D. (1991), *Application of Stress Release Models to Historical Earthquakes from North China*, Pure appl. geophys. *135* (4), 559–576.

ZHENG, X., and VERE-JONES, D. (1994), *Further Applications of the Stochastic Stress Release Model to Historical Earthquake Data*, Tectonophysics *229*, 101–121.

(Received July 17, 1998, revised December 8, 1998, accepted December 11, 1998)

 To access this journal online:
http://www.birkhauser.ch

Pure appl. geophys. 155 (1999) 625–647
0033–4553/99/040625–23 $ 1.50 + 0.20/0

Pure and Applied Geophysics

Pore Creation due to Fault Slip in a Fluid-permeated Fault Zone and its Effect on Seismicity: Generation Mechanism of Earthquake Swarm

TERUO YAMASHITA[1]

Abstract—Spatio-temporal variation of rupture activity is modeled assuming fluid migration in a narrow porous fault zone formed along a vertical strike-slip fault in a semi-infinite elastic medium. Pores are assumed to be created in the fault zone by fault slip. The effective stress principle coupled to the Coulomb failure criterion introduces mechanical coupling between fault slip and pore fluid. The fluid is assumed to flow out of a localized high-pressure fluid compartment in the fault with the onset of earthquake rupture. The duration of the earthquake sequence is assumed to be considerably shorter than the recurrence period of characteristic events on the fault. The rupture process is shown to be significantly dependent on the rate of pore creation. If the rate is large enough, a foreshock–mainshock sequence is never observed. When an inhomogeneity is introduced in the spatial distribution of permeability, high complexity is observed in the spatio-temporal variation of rupture activity. For example, frequency-magnitude statistics of intermediate-size events are shown to obey the Gutenberg–Richter relation. Rupture sequences with features of earthquake swarms can be simulated when the rate of pore creation is relatively large. Such sequences generally start and end gradually with no single event dominating in the sequence. In addition, the *b* values are shown to be unusually large. These are consistent with seismological observations on earthquake swarms.

Key words: Faulting, fractures, earthquake swarm, permeability, porosity, seismicity.

Introduction

It is now widely believed that fluids exert significant mechanical effects on earthquake faulting. As a typical effect, we can mention the reduction of the effective confining stress by zones of high pore fluid pressure, which facilitates frictional slip at low fault shear stress (RALEIGH *et al.*, 1976). Hence the transient elevation of pore fluid pressure is believed to trigger earthquake faulting. SLEEP and BLANPIED (1992, 1994) proposed a mechanism for elevating pore fluid pressure that ductile creep compacts the pore spaces. The compaction leads to fluid pressure

[1] Earthquake Research Institute, University of Tokyo, 1-1-1 Yayoi, Bunkyo-ku, Tokyo 113-0032, Japan. Fax: 81-3-5841-5693, Tel: 81-3-5841-5699, E-mail: tyama@eri.u-tokyo.ac.jp

increase when fluid flow into the country rock is slow. Some field observation of exhumed fault zones suggest that low-permeability seals hydraulically isolate a fault zone from the country rocks (e.g., SLEEP and BLANPIED, 1992; CHESTER et al., 1993). Laboratory experiment clearly showed that impermeable seals can be self-generated along a fault zone boundary due to redistribution of materials in solution (BLANPIED et al., 1992). On the other hand, SIBSON (1990, 1992) proposed a conceptually different model for the mechanism of local elevation of fluid pressure. He considered that the transition from hydrostatic to suprahydrostatic fluid regime occurs at some depth across a discrete permeability barrier on the basis of geological evidence. High-pressure fluids develop beneath the barrier from aquathermal pressuring, progressive metamorphic dehydration, and etc. (SIBSON, 1990).

Recently, considerable attention is directed to the effects of fluid migration in a fault zone on the spatio-temporal complexity of earthquake occurrence (MILLER et al., 1996; HENDERSON and MAILLOT, 1997; YAMASHITA, 1997, 1998). Mechanical interactions between fluid-conduit properties and rupture occurrence may play an important role in the generation of complexity. Two mechanisms are well known for these interactions. The first is the fault zone expansion due to fault slip. In fact, laboratory and field observations suggest that fault zone thickness generally scales with accumulated slip (e.g., ENGELDER, 1974; HULL, 1988; CHESTER et al., 1993) although the data scatter is rather large (EVANS, 1990). If such fault zone expansion occurs, the pore fluid pressure is reduced locally. The reduction in the fluid pressure tends to retard the fluid migration, so that the rupture occurrence also tends to be delayed. YAMASHITA (1997, 1998) studied the effect of fault zone expansion on the rupture occurrence and observed a high complexity in the sequence of ruptures in his numerical simulations. Second, the creation of pores (dilatancy) in a fault zone due to fault slip is another important mechanism of interactions between fluid flow and earthquake ruptures (TEUFEL, 1981; MORROW and BYERLEE, 1989; MARONE et al., 1990). TEUFEL (1981) investigated dilatancy along sliding surfaces in drained, triaxial compression, pore pressure laboratory experiments with precut specimens of sandstone. He suggested that dilatancy during frictional sliding is a direct consequence of an increase in porosity. MARONE et al. (1990), on the other hand, assumed water-saturated layers of sand sheared between 45° surfaces and measured porosity changes under nominally drained conditions. Their experiments also suggest a monotonic increase in porosity with increasing slip. Such pore creation can reduce the pore fluid pressure, which tends to retard the rupture occurrence.

The aim of this paper is to study the effect of pore creation due to fault slip on the spatio-temporal complexity of earthquake occurrence, and to compare the results with those obtained by YAMASHITA (1998), in which only the fault zone expansion is assumed. The paper of YAMASHITA (1998) is hereafter referred to as Paper I. We assume the same fault model as in Paper I (Fig. 1), which enables us to investigate the differences in the effects of pore creation and fault zone expansion. In this fault model, a localized high-pressure fluid source near the bottom of a sealed fault is assumed before the nucleation of ruptures. The high-pressure fluid begins to flow out with the onset of rupture and migrates in the fault zone.

Fluid Migration and Fault Model

The fault zone model used here is the same as discussed in Paper I, therefore it is only briefly described here. We consider a vertical rectangular strike-slip fault S in a semi-infinite isotropic homogeneous elastic medium (Fig. 1). A narrow porous fault zone sealed from the country rock is assumed along the fault, which behaves as a fluid conduit. The remotely applied stress is assumed to be kept constant, which implies that the assumed time range is negligibly small compared to the recurrence period of characteristic events on the fault.

The rupture occurrence is assumed to be dependent on the pore fluid pressure p through the Coulomb fracture criterion coupled to the principle of effective stress

$$\tau_s = a_0 + m_s(\sigma_n - p), \tag{1}$$

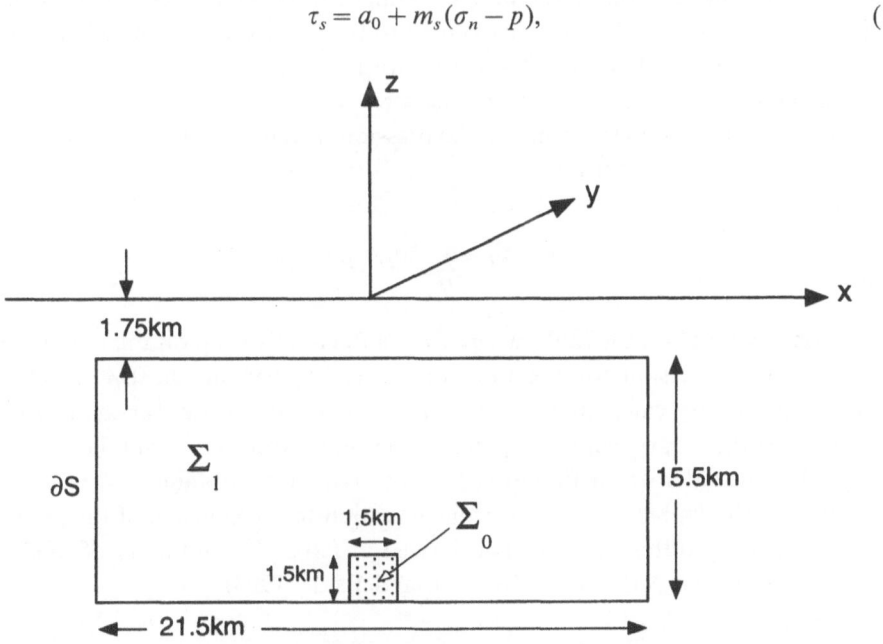

Figure 1
Elastic half-space with vertical strike-slip fault S $(=\Sigma_0 + \Sigma_1)$. The plane $z = 0$ corresponds to the free surface. High-pressure fluid is initially localized at Σ_0; the rest of the fault, Σ_1, is under considerably lower fluid pressure.

where τ_s is the static shear traction at fracture, m_s is the coefficient of static friction, σ_n is the total normal traction on the fault, and a_0 is a parameter independent of σ_n and p. Our assumption of the strike-slip fault suggests that the rupture is initiated if the stress component p_{yx} exceeds the threshold stress τ_s; the stress suddenly drops to the residual level τ_f with the onset of the rupture. The dependence of τ_f on σ_n and p can also be written in the same form as in Equation (1) (WONG, 1986; YAMASHITA and OHNAKA, 1992).

The fault is assumed to comprise a computational grid in which space and time evolution of stress, relative slip and fluid pressure fields are calculated. All the physical quantities are assumed to be constant on each fault segment; each segment has the same size 500 m × 500 m. These fault segments are assumed to rupture independently. The relative slip Δu, fluid pressure p, and shear stress p_{yx} are calculated at the center of each segment. The fault length and width are assumed to be 21.5 km and 15.5 km, respectively, and the top edge of the fault is at a depth of 1.75 km (Fig. 1); the rigidity and Poisson's ratio of the medium are fixed at 4×10^4 MPa and 0.25, respectively. According to RICE (1993), our fault model is classified as an inherently discrete fault model, which contrasts with a continuous model; see Paper I for the particulars.

The fluid flow alone will cause negligible deformation in the country rock because of the assumption of the narrow straight fault zone. Hence the change in the stress p_{yx} in the country rock is assumed to be described in terms of the relative slip alone as in Paper I. Catastrophic rupture is assumed to be instantaneous, so that quasistatic treatment is allowed. The stress due to relative slips on the fault segments can be obtained, using an expression given by OKADA (1992). It is generally expressed in a form

$$p_{yx}(\mathbf{x}) = \sum_{i,j} \Delta u_{i,j} f_{i,j}(\mathbf{x}) \tag{2}$$

at location $\mathbf{x} = (x, z)$ on the fault, where $\Delta u_{i,j}$ is the relative slip on the fault segment (i, j), and the expression for the function $f_{i,j}(\mathbf{x})$ is given in OKADA (1992). The relative slip can be calculated by the inversion of Equation (2), assuming the residual stress $p_{yx} = \tau_f$ on slipped segments. The stress transfer from failed segments can also be calculated from Equation (2) once the slip is obtained.

As will be shown below, the governing equation for the change of the pore fluid pressure is slightly different from that derived in Paper I. Continuity of fluid mass yields (WALDER and NUR, 1984; SEGALL and RICE, 1995)

$$\nabla \cdot \mathbf{q} + \rho \frac{\partial}{\partial t} \phi_p + \rho \phi \beta \frac{\partial}{\partial t} p = 0 \quad (\beta = \beta_f + \beta_\phi), \tag{3a}$$

and

$$\frac{\partial}{\partial t} \phi = \phi \beta_\phi \frac{\partial}{\partial t} p + \frac{\partial}{\partial t} \phi_p, \tag{3b}$$

where $\mathbf{q} = (q_x, q_y, q_z)$ is the fluid flux, ρ is the fluid density, and ϕ_p is the plastic component of the porosity ϕ. Here the plastic component of the porosity is assumed to be formed by the fault slip (see Equation (9) below), which was neglected in Paper I. The fluid compressibility and elastic pore compressibility are given by β_f and β_ϕ, respectively. Darcy's law relates the fluid flux to pore fluid gradient through

$$\mathbf{q} = -\rho \frac{\kappa}{\eta} \nabla p, \tag{4}$$

where η and κ are the viscosity and permeability of the fluid, respectively.

As in Paper I, we simplify Equation (3a) by assuming the fluid flow in the direction normal to the fault plane is small compared to the flow along the fault plane. This will be allowed when the fault extent substantially exceeds the thickness of the fluid conduit. The result is a reduction in the dimensionality of the equation, which is obtained by the integration of Equation (3a) in the direction normal to the fault. As stated before, the fault zone is assumed to be sealed from the country rock, so that no fluid communication occurs across the fault zone boundaries at $y = r(\mathbf{x})$ and $s(\mathbf{x})$, that is,

$$q_x(x, z, s, t) \frac{\partial s}{\partial x} + q_z(x, z, s, t) \frac{\partial s}{\partial z} = q_y(x, z, s, t),$$

$$q_x(x, z, r, t) \frac{\partial r}{\partial x} + q_z(x, z, r, t) \frac{\partial r}{\partial z} = q_y(x, z, r, t). \tag{5}$$

Hence we finally have

$$\frac{\partial}{\partial x}\left(b(\mathbf{x})\kappa(\mathbf{x})\frac{\partial p}{\partial x}\right) + \frac{\partial}{\partial z}\left(b(\mathbf{x})\kappa(\mathbf{x})\frac{\partial p}{\partial z}\right) = \eta b(\mathbf{x})\left(\phi\beta \frac{\partial p}{\partial t} + \frac{\partial \phi_p}{\partial t}\right) \tag{6}$$

from Equations (3a) and (4), where $b(\mathbf{x}) = r(\mathbf{x}) - s(\mathbf{x})$ is the fault zone thickness at $\mathbf{x}$. The permeability κ is assumed to be a function of $\mathbf{x}$ in Equation (6). The pore fluid pressure p in Equation (6) is regarded as the pressure averaged over the fault zone thickness. It should be noted that a major difference in the governing equation for p between that in the present case and that in Paper I is the introduction of the sink term $\partial \phi_p / \partial t$ in Equation (6).

As in Paper I, we assume that shear-induced pore compaction in a sealed fault (SLEEP and BLANPIED, 1992, 1994; BYERLEE, 1993) is a mechanism for locally elevated fluid pressure. BYERLEE (1993) proposed that the formation of sealed fluid compartments of various sizes in a fault zone and their compaction can lead to locally elevated fluid pressure. The seal formation is considered to be due to the redistribution of materials in solution as discussed by SLEEP and BLANPIED (1992,

1994). The rupture of an impermeable seal separating one of the high-pressure fluid compartments and the surrounding low-pressure one in a fault zone can trigger a sequence of earthquakes. On the basis of these studies, we assume

$$p(\mathbf{x}) = p_0 \quad \mathbf{x} \in \Sigma_0$$

$$= p_1 \quad \mathbf{x} \in \Sigma_1 \tag{7}$$

for the distribution of the pore fluid pressure at $t = 0$ on the fault S ($= \Sigma_0 + \Sigma_1$), where $p_0 > p_1$, and Σ_0 is a localized high-pressure fluid source at the bottom of the fault (Fig. 1); the rest of the fault Σ_1 is assumed to be under lower fluid pressure. An impermeable seal is assumed at the boundary between Σ_0 and Σ_1 for $t < 0$; the failure of the seal at $t = +0$ nucleates a rupture sequence. The region Σ_0 can be regarded as a sealed high-pressure fluid compartment in a fault zone in the framework of BYERLEE's model (1993). The periphery of the fault ∂S is assumed to be impermeable during the rupture sequence, hence the boundary condition there is given by

$$\partial p / \partial n = 0 \quad \text{on} \quad \partial S \quad \text{for} \quad t > 0, \tag{8}$$

where n is the normal to the periphery. Equations (3b) and (6), coupled to each other, are solved with the conditions (7) and (8) in a finite difference scheme.

The rupture of the impermeable barrier at the boundary between Σ_0 and Σ_1 may also be regarded as the initiation of valve action if it is allowed to assume a sealed fault and a localized overpressure zone in the framework of the model proposed by SIBSON (1990, 1992). The rupture of the permeability barrier between the hydro-static and suprahydrostatic regime acts as a valve promoting upward discharge of fluids from the overpressured zone in his model.

Evolution Equations for Porosity and Permeability

The experimental data discussed above suggest that the plastic component of porosity increases with increasing slip. A tendency is also observed for the increase rate of the porosity to be reduced for larger slip (SEGALL and RICE, 1995). Thus, we consider the simple evolution equation for the plastic porosity

$$\phi_p = \phi_0 + (\phi_i - \phi_0) \exp(-\Delta u_a / \Delta u_c), \tag{9}$$

where Δu_a is the accumulated slip, ϕ_i is the porosity when the slip Δu_a is zero, and the plastic porosity is assumed to evolve toward the upper bound ϕ_0 over the critical slip distance Δu_c.

Simple formulas have been proposed for relationships between porosity ϕ and permeability κ although all of them are not necessarily conclusive because of the geometrical complexity of the earth's materials. BRACE (1977) evaluated the formula

$$\kappa = \kappa_0 \phi^3, \tag{10}$$

and found that the calculated values of permeability compared well with measurements of a number of rock types, where κ_0 is a quantity proportional to the hydraulic radius. In this work we also assume that the permeability is proportional to ϕ^3. However, as noted by HENDERSON and MAILLOT (1997), the uncertainty in the geometry of pore space and the wide range of permeabilities found in rocks (BRACE, 1980, 1984) demand that a wide range of κ_0 be investigated.

Numerical Values of the Model Parameters

As in Paper I, we assume one of the simplest assumptions for the spatial distributions of p_{yx}, σ_n and p, that is, they are assumed to be constant over the fault at $t = 0$ except at the high pressure fluid compartment Σ_0. As noted in Paper I, this is largely because there is no reliable information on the depth variation of p_{yx}. In addition, although many researchers assume a hydrostatic pore fluid pressure gradient in the seismogenic layer (e.g., SIBSON, 1974, 1982, 1984; YAMASHITA and OHNAKA, 1992), it may be far from realistic (GOLD and SOTER, 1985). We assume the simple fault model in the present paper and in Paper I considering these complicated circumstances, which suggests that our understanding is rather qualitative.

It should be noted that the contribution from the fault slip to the normal traction on the fault is negligible because of the assumptions of the narrow straight fault zone and due to the occurrence of shear slip only. Hence Equation (1) is rewritten as

$$\tau_s = \alpha_s - m_s p \quad (\alpha_s > 0), \tag{11}$$

where α_s is constant. The residual stress can be assumed in the form

$$\tau_f = \alpha_f - m_f p \quad (\alpha_f > 0), \tag{12}$$

where α_f is constant, and m_f is the coefficient of sliding friction. We assume $m_s = m_f = 0.7$ and the swinging back of the relative slip is prohibited in all the following calculations as in Paper I. The fault slip is coupled to the fluid migration through the conditions (9), (10), (11), and (12).

Mechanical interactions are assumed to occur between some model parameters in our modeling. For example, the fluid pressure p and the porosity ϕ are coupled through Equations (3b) and (6). The interactions between such model parameters are schematically illustrated in Figure 2 for readers' convenience. The arrows denote the direction of influence and the numerals do the equations or conditions in the text to express the interactions.

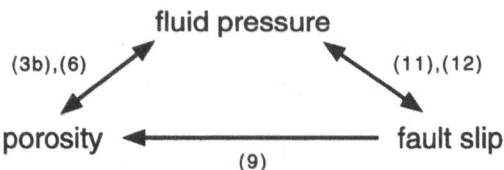

Figure 2
Schematic illustration of interactions between model parameters. See the text on the arrows and the numerals.

One of the difficulties in estimating the time scale of fluid migration is associated with the fact that permeability of crustal rocks varies over a wide range; such a large variation of κ is represented by the variation of κ_0 as mentioned above. BRACE (1980) compiled laboratory, *in situ*, and inferred values of permeability of crystalline and argillaceous rocks. He indicated that 4 to 6 orders of magnitude variation of κ is typical at a particular site for *in situ* measurement; the data at depth from 0.5 to 3 km show the range of the value $10^{-18} \sim 10^{-12}\,\mathrm{m}^2$. FORSTER *et al.* (1994) examined the permeability of exhumed fault rock reported to have been formed at depths from 4 to 7 km. They showed that the fault damaged zone has a consistently higher permeability than the adjacent rocks. Their estimate is 10^{-16} to $10^{-14}\,\mathrm{m}^2$ in the damaged zone. DAVISON and KOZAK (1988) gave the range 10^{-17} to $10^{-11}\,\mathrm{m}^2$ for *in situ* values of damage and gouge zone permeability of a thrust fault in crystalline rocks. It will therefore be required to consider the range $10^{-18} \sim 10^{-11}\,\mathrm{m}^2$ for κ_0. For the other poroelastic parameters, we assume $\eta = 2 \times 10^{-4}\,\mathrm{Pa \cdot s}$, $\beta_\phi = 1 \times 10^{-8}\,\mathrm{Pa}^{-1}$ and $\beta_f = 2 \times 10^{-10}\,\mathrm{Pa}^{-1}$ (e.g., DAVID *et al.*, 1994; SEGALL and RICE, 1995; WONG *et al.*, 1997). The initial value of the porosity ϕ is rather arbitrarily assumed to take the value 0.05. We also assume $\phi_0 = 0$ and $\phi_i = 0.3$. It will be shown in the following calculations that the final value of ϕ_p is about 0.01. This value is much smaller than 0.09, which is considered to be the saturation porosity for sandstone and granite (SLEEP, 1995).

Since our concern is in the effect of pore creation, the fault zone thickness is assumed to be constant in all the following calculations; note that the governing equation for the fluid pressure change (6) is independent of the fault zone thickness when it is constant.

Effect of Pore Creation

We first investigate how the critical slip distance for pore creation Δu_c in Equation (9) affects the simulation results; note that the pore creation is more facilitated by a smaller value of Δu_c. The coefficient κ_0 is now assumed to be constant ($\kappa_0 = \kappa_c$) over the fault to elicit the effect of Δu_c. Some model parameters are rather arbitrarily assumed here (see Table 1). It is characteristic in the model shown in Table 1 that the tectonic shear stress τ_0 is larger than the residual stress τ_f at $t = 0$ on Σ_1; higher activity is expected in this case than in the case $\tau_0 < \tau_f$ (Paper I). Table 1 implies $\tau_s = \tau_0$ on the high-pressure fluid compartment Σ_0 at $t = 0$, so that the rupture verges on nucleation there. As stated in the preceding section, the impermeable seal at the boundary between regions Σ_0 and Σ_1 is broken at $t = +0$ simultaneously with the rupture of the region Σ_0.

We now assume two models concerning the magnitude of Δu_c; in one model we assume $\Delta u_c = \infty$, so that no pores are created by any slip in this model. In the other model, we assume $\Delta u_c = 12.5$ m rather arbitrarily. Calculated results of the temporal variation of rupture activity are displayed in Figure 3 for both models; simulations are carried out for different values of the strength α_s on Σ_1. Magnitudes of events are given by the moment magnitude in all the calculations in this paper. The moment magnitude M is calculated from the relation $M = 2/3 \log_{10} M_0 - 6.0$, in which the rigidity is assumed to be 4×10^4 MPa in the calculation of the scalar seismic moment M_0. The nondimensional time T is defined as $T = \kappa_c t / \eta \beta \phi_e h^2$, where h ($= 500$ m) is the side length of the fault segment. The definition of T implies that the value of κ_c affects the calculation results only through the time scale. For example, the nondimensional time $T = 1000$ corresponds to $t = 28.9$ days and 7.93×10^5 years for $\kappa_c = 10^{-11}$ m^2 and 10^{-18} m^2, respectively. Such a wide variation of κ_0 makes a reliable estimate of the duration of the rupture sequence impossible.

It is commonly observed in both models that the entire fault is ruptured at $T = +0$ simultaneously with the onset of rupture at Σ_0 for sufficiently small values of α_s. We also observe in both models a swarm-like activity with no single predominant principal event for α_s in a certain range. Creating a striking contrast between the two models, is that sequences which have a feature of foreshock–main-shock sequences are observed only in the model with $\Delta u_c = \infty$. For example, such sequences are observed for $\alpha_s = 14.0$ to 15.25 MPa in Figure 3b; the largest event in each sequence ruptures most of the segments that have remained unbroken. This will occur because rupture can grow spontaneously ahead of the migration front of the fluid after the dimension of the ruptured region attains some critical level: the rupture tip stress generally increases with increasing rupture size when $\tau_0 > \tau_f$ (Paper I). The largest event in a sequence that ruptures most of the fault is referred

Table 1

Parameters assumed on the regions Σ_0 and Σ_1; p denotes the pore fluid pressure at time $t = 0$. See the text for the other parameters.

	p (in MPa)	α_s (in MPa)	α_f (in MPa)	τ_0 (in MPa)
Σ_0	100.0	80.0	70.0	10.0
Σ_1	2.0	variable	8.4	10.0

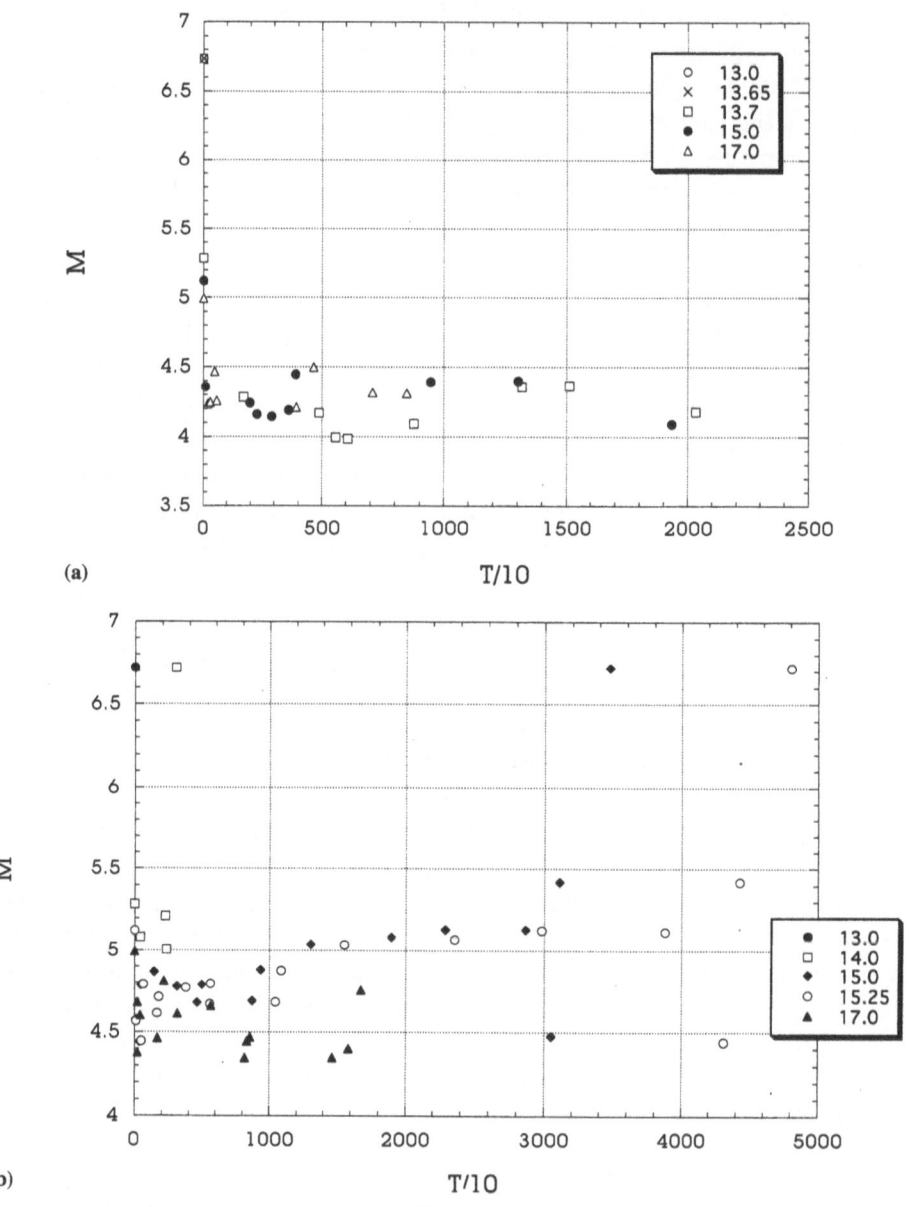

Figure 3
Temporal variation of rupture activity for the models (a) with $\Delta u_c = 12.5$ m and (b) with $\Delta u_c = \infty$. The numerals in the figure denote the values of α_s (in MPa) on Σ_1, and T and M are the nondimensional time and magnitude, respectively.

to as a mainshock in this paper; smaller events preceding it are cited as foreshocks. The above consideration implies that the magnitude of a mainshock would be larger if a larger fault were assumed; however, the magnitudes of smaller events would be quite similar. It is noteworthy in Figure 3a that the entire fault is suddenly ruptured at $T = +0$ for $\alpha_s = 13.65$ MPa, while swarm-like activity appears if α_s is only increased by 0.05 MPa.

We now investigate causes of the differences in the rupture pattern in the above two models, from the examination of spatio-temporal changes of the rupture front and fluid pressure. We first study the model with $\Delta u_c = 12.5$ m. The spatio-temporal change of pore fluid pressure is illustrated in Figure 4 for the sequence with $\alpha_s = 13.7$ MPa; we clearly observe low fluid pressure zone along the advancing rupture front. This is related to the creation of pores at the newly ruptured zone. As schematically illustrated in Figure 5, numerous pores are created at the newly ruptured zone. Since such pores are not filled with fluids at the time of creation, fluids flow into such zone from the surroundings. This causes a temporal decrease of fluid pressure in the regions near the rupture front. Since the fluid pressure decreases result in the increase in the fracture strength τ_s, the rupture extension is temporally suppressed after a large number of pores is created. A large part of the fluids initially accumulated in Σ_0 is used to fill newly created pores in this way when Δu_c is small enough. Hence there will be insufficient fluids to drive the rupture to grow to a critical size for the excitation of a mainshock. This will be the reason that foreshock–mainshock sequences cannot be observed in the model with $\Delta u_c = 12.5$ m.

Since new pores are not created during rupture sequence in the model with $\Delta u_c = \infty$, fluids are not consumed to fill such pores. Hence the fluids can efficiently contribute to driving the rupture growth when the strength α_s is not very large. This will cause foreshock–mainshock sequences for values of α_s within a certain range.

Effect of Inhomogeneities on the Stochastic Properties of Rupture Activity

As stated above, a wide range of κ_0 should be investigated for more realistic simulations. The coefficient κ_0 is now assumed to be statistically distributed within a specified range on Σ_1, and only the effect of the extent of the distribution is considered. We specifically assume two models for its distribution, model N ($0.5 < \kappa_0/\kappa_c < 1.5$) and model W ($0 < \kappa_0/\kappa_c < 2.0$); the probability density is assumed to be homogeneous in each range. All other model parameters are assumed to be the same as in the case $\alpha_s = 13.7$ MPa in Figure 3a, so that swarm-like activity is expected in each model. Model W is characterized by its larger extent of the distribution, while the mean value of the distribution is fixed at $\kappa_0 = \kappa_c$ in both models.

Presently it seems to be a prevailing view that high diversity in earthquake ruptures is related to the wide variation of fracture strength τ_s (e.g., DAS and AKI, 1977). In fact, the effect of widely variable strength α_s was investigated in Paper I. It will therefore be meaningful to investigate the effects of distributed α_s for the comparison with models N and W; we assume that the value of α_s is homoge-

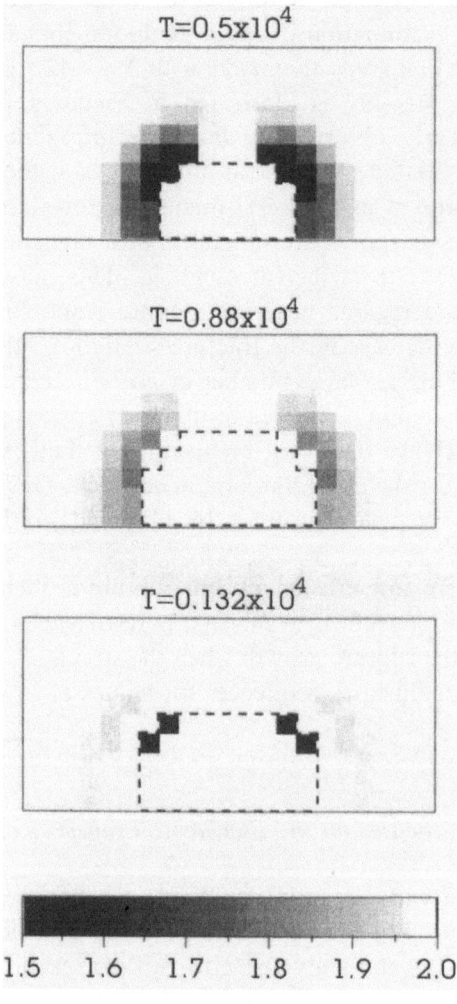

P (in MPa)

Figure 4

Spatio-temporal change of pore-fluid pressure near the rupture front for the example $\alpha = 13.7$ MPa shown in Figure 3a. Low-pressure fluid zone is recognized along the advancing rupture front. Only the segments on which pore-fluid pressure is smaller than the initial value (2 MPa) on Σ_1 are illustrated. The darker area stands for lower pressure zone and the broken line denotes the location of rupture front.

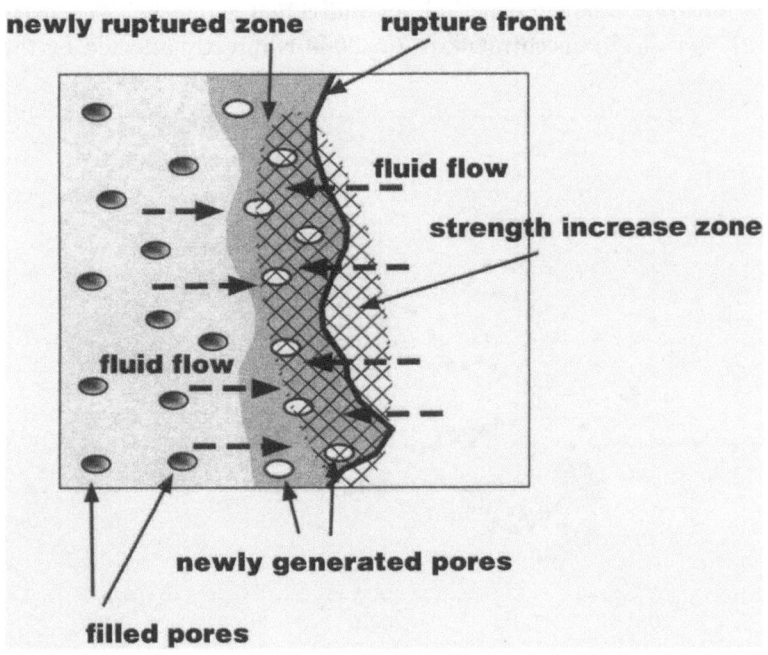

Figure 5
Schematic illustration of fluid flow near the advancing rupture front.

neously distributed in the range $14.0 \, \text{MPa} < \alpha_s < 16.0 \, \text{MPa}$, which is referred to as model F. We assume $\kappa_0 = \kappa_c$ over the fault in model F. We can expect a foreshock–mainshock sequence in model F in contrast to the cases in models N and W since foreshock–mainshock sequences are observed for the values $\alpha_s = 14.0$ and $16.0 \, \text{MPa}$; the case for $\alpha_s = 14.0 \, \text{MPa}$ is shown in Figure 3b.

Several simulations are carried out in each model with a different set of values of κ_0 in models N and W and of α_s in model F; each numerical experiment is the simulation of one rupture sequence. The number of simulations is 20 in models N and W, and 10 in model F; the activity is relatively low in each sequence of models N and W compared to that in model F. The calculation is carried out in each simulation of models N and W until the fluid pressure is reduced enough so as not to further excite catastrophic ruptures over the fault. The calculation is stopped in model F once a mainshock occurs.

It is ascertained in models N and W that no single predominant principal event is observed in each sequence except the initial event to rupture the entire region of Σ_0. Twenty simulations are superimposed in Figure 6 for each model to see average properties; two typical sequences are also indicated by different symbols in each model to illustrate characteristics of an individual sequence. The occurrence of events with $M \simeq 4.25$ concentrated at $T \simeq 2000$ is directly affected by the assump-

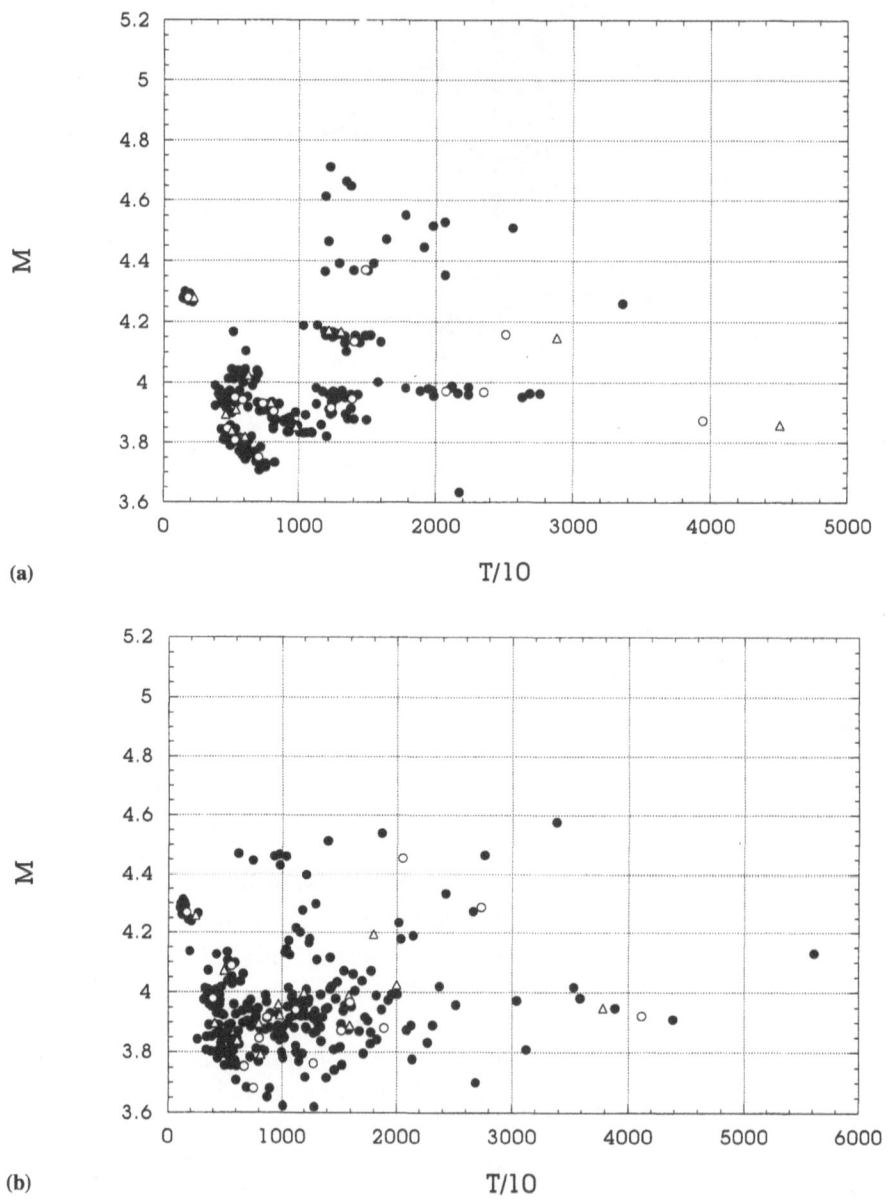

(a)

(b)

Figure 6
Superimposed sequences of (a) model N and (b) model W for the temporal variation of rupture activity; twenty sequences are superimposed in each model. Two typical individual sequences are also designated by different symbols. The nucleation event with $M \simeq 5.1$ is omitted in the figure.

tion of high pressure fluid at Σ_0; such an event occurs in every individual sequence. Hence these events may be independent of the dynamics of the system. Figure 6 shows for models N and W that the magnitude of the event tends to increase with time at the beginning of the superimposed sequence except the initial event to rupture Σ_0, while it decreases near the termination of the sequence. This occurs because the driving effect of fluid on ruptures decreases with time near the termination of a sequence because of the consumption of fluids in newly created pores. The temporal variation of rupture activity found for models N and W, shown in Figure 6, closely resembles that of activity in earthquake swarms. In fact, earthquake swarms are sequences of events that often start and end gradually and in which no single event dominates in size (e.g., MOGI, 1963). It should be noted that these features are somewhat observable even when κ_0 is constant over the fault: see the examples with $\alpha_s = 13.7$, and 15.0 MPa in Figure 3a. Although swarm-like activity is apparently simulated in some sequences for $\Delta u_c = \infty$ as well (e.g., see the example with $\alpha = 17.0$ MPa) in Figure 3b, these sequences do not necessarily explain actual earthquake swarms. In other words, the rupture activity tends to stop abruptly in these sequences, which contradicts the seismological observation on earthquake swarm. The above consideration indicates that features of actual earthquake swarms can be simulated well when the rate of pore creation is relatively high.

The largest event occurs at the beginning of each sequence in Figures 3a and 6, which is rarely observed in actual earthquake swarms. This can be interpreted as due to the simplicity in our assumption regarding the fault model. That is, we assume that the mechanical properties are homogeneous over the high-pressure fluid source Σ_0. Hence all the segments in Σ_0 and some nearby segments rupture simultaneously at $T = +0$; the rupture of the nearby segments is excited by that of Σ_0. This gives rise to a relatively large event at $T = +0$. However, such homogeneous conditions will rarely be satisfied over a large area in actual earthquakes.

There is a tendency that the length of sequences is larger when κ_0 varies in a wider range. The average lengths of the individual sequences are $T = 28{,}967$ and $22{,}818$ for models W and N, respectively, while the length of the sequence is only $T = 20{,}350$ when $\kappa_0 = \kappa_c$ and $\alpha_s = 13.7$ MPa over the fault (see Fig. 3a). This may occur because fault segments which have locally small values of κ_0 retard the fluid flow, which makes a rupture sequence longer in Model W. In contrast to models N and W, all the sequences have a feature of foreshock–mainshock sequences in model F (Fig. 7).

It has been established in a number of simulation studies that the observation of power-law frequency distribution for the event sizes is closely related to the mechanical coupling between fault segments (e.g., BAK and TANG, 1989; RICE, 1993; YAMASHITA, 1993, 1997). We now investigate the frequency distribution of event sizes for models N, W, and F. As exemplified in Figure 3, smaller size events are not necessarily more abundant in the models in which all the model parameters

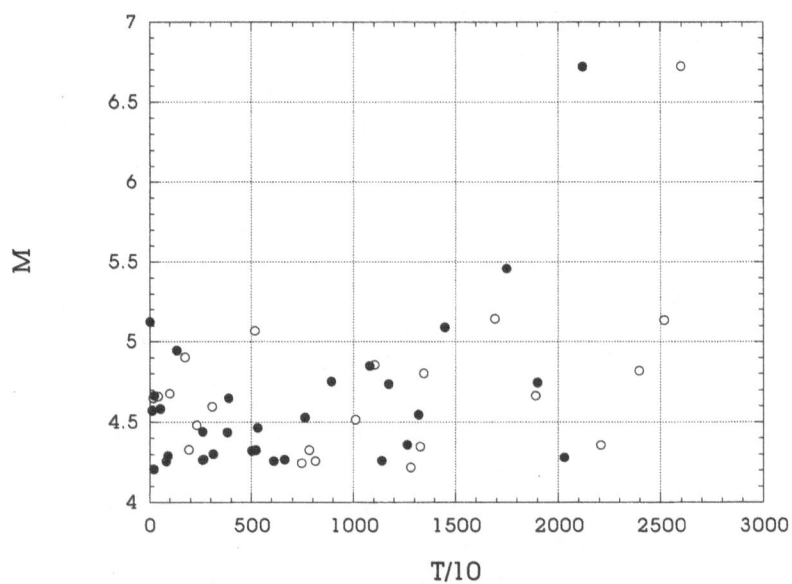

Figure 7
Two typical examples of the temporal variation of rupture activity for model F.

take constant values over the region Σ_1, so that the Gutenberg–Richter (GR) relation cannot be expected in such models. In contrast, the number of smaller size events appears to be larger in each of the models N, W, and F as typically observed in Figures 6 and 7. In fact, the cumulative frequency-magnitude distribution satisfies the power-law distribution for intermediate size events (Figs. 8 and 9). The b values are approximately equal to 2.0 in models N and W, while it is close to 1.0 in model F. The relation $b \simeq 1.0$ was also obtained in the simulations of foreshock–mainshock sequences on the assumption of fault zone expansion and $\Delta u_c = \infty$ in Paper I. Larger b values found for models N and W, in which the temporal variation of rupture occurrence was shown to closely resemble that in earthquake swarms, are consistent with the seismological observation that the b value is often unusually large in earthquake swarms (SYKES, 1970). Earthquake swarms occurring at volcanoes and mid-ocean ridges sometimes take the b values close to 2.0 (TOMODA, 1954; MIYAMURA, 1962). Hence it is inferred that an earthquake swarm with a large b value is due to the contribution of fluids to ruptures and the high rate in pore creation by fault slips.

It is reported in experimental and seismological observations that the b value changes during rupture sequence. For example, the decrease is observed during the evolution of a foreshock process (VON SEGGERN, 1980; SAMMONDS et al., 1992). The same tendency is clearly observed in the simulation of model F as illustrated in

Figure 10. Figure 10 shows the frequency distribution of event sizes at two successive time periods before the occurrence of the mainshock; the result of 10 simulations are superimposed in each time period. The decrease in the b value occurs because a larger area tends to be ruptured at a later stage of sequence. The decrease in the b value was also observed in Paper I during the evolution of foreshock sequences, while the decrease was shown to be markedly smaller when the extent of the distribution of α_s is larger.

The temporal change in the b value is also expected for models N and W; Figure 6 indicates that small size events are more abundant at the earlier part of the superimposed sequence than at the later part. Figure 11 illustrates the frequency distribution of events sizes in the time ranges $0 < T \le 15,000$ and $15,000 < T < \infty$; the results of 20 individual sequences are superimposed. Only the result for model W is shown here since larger number of events are observed in model W. A slight tendency for the b value to decrease with time is observed in Figure 11 although it is not so obvious as in model F.

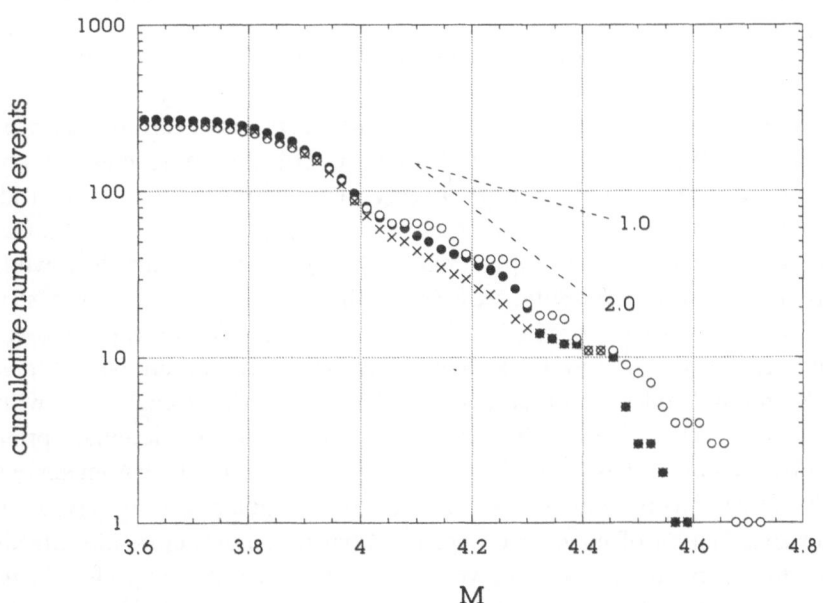

Figure 8

Cumulative frequency-magnitude distribution for models N (open circle) and W (closed circle); all the simulated results are superimposed in each model. The bump near $M \simeq 4.25$ corresponds to the events at $T \simeq 2000$ in Figures 6a and 6b. As noted in the text, these events seem to be independent of the dynamics of the system. If these events are removed, the curves become much smoother; an example for model W is shown by the crosses.

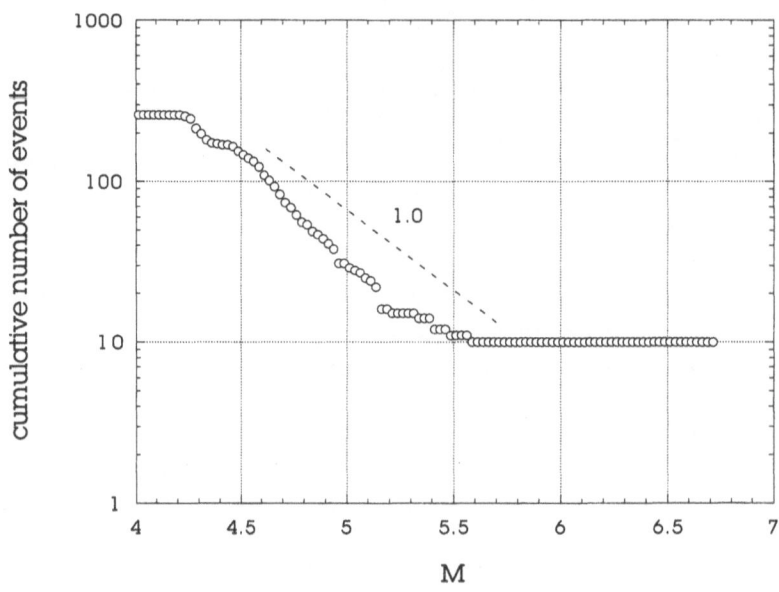

Figure 9
Cumulative frequency-magnitude distribution for model F; all the simulated results are superimposed.

Discussion and Conclusions

We studied the effect of mechanical coupling between rupture occurrence and fluid migration. We particularly investigated the effect of pore creation on the spatio-temporal change of rupture occurrence. Two extreme cases are considered for that purpose and the simulation results are compared; in one case (models N and W), pores are assumed to be created in a fault zone by fault slips, while no pores are formed during a rupture sequence in the other case (model F). A striking difference in the simulation results between the two cases is that a rupture sequence which has features of a foreshock–mainshock sequence cannot be observed in models N and W, while it is observed in model F when the strength α_s is within a certain range. In other words, only a sequence with no single predominant principal event is observed in models N and W for any values of the model parameters unless the entire fault is ruptured at once. A statistical regularity is observed in the frequency distribution of magnitudes when spatial inhomogeneities are introduced in the permeability (models N and W) and in the strength α_s (model F), which agrees well with the GR relation; the b values are shown to be larger in models N and W than in model F. Our calculation also shows the temporal variation of the b value during rupture sequences.

It was shown in the present paper that rupture characteristics are significantly dependent on the value of the critical slip Δu_c. This suggests that ruptures are also dependent on the upper threshold value of the porosity ϕ_0 in some cases, while it was fixed in our calculations. Knowledge of these values taken in the earth's crust is therefore crucial to predict characteristics of earthquake rupture. Porosity increase has been measured as a function of fault slip in some laboratory experiments (e.g., MORROW and BYERLEE, 1989), nonetheless the values of Δu_c and ϕ_0 cannot easily be estimated from their results because of the limited range of fault slip in their experiments.

It was shown in the simulations that the increase in the plastic component of the porosity is only 0.01 at most. While it is difficult to judge whether this change is large or small, the plastic pore creation was demonstrated to give rise to a significant effect on the rupture sequence. Such a substantial effect may occur mainly because the change in the fluid pressure is affected by the rate of temporal variation of the plastic component of the porosity: see Equation (6). Principally the sudden creation of new plastic pores was shown to considerably influence the rupture occurrence as illustrated in Figure 5.

It is observed that the GR relation is satisfied over ranges $M = 3.8$ to 4.4 and 4.2 to 5.5 in models N and W and model F, respectively. It is also seen that the GR relation is satisfied in narrower ranges in models N and W than in model F. The difference in the upper limit of the magnitude range over which the GR relation is satisfied occurs because ruptures do not grow so large in models N and W as in

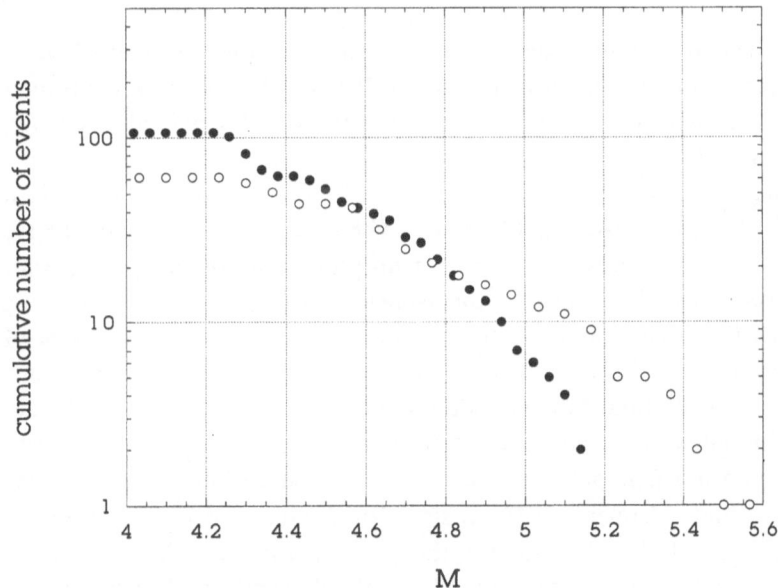

Figure 10
Cumulative frequency-magnitude distribution in two successive time periods $T_m - 20,000 \leq T < T_m - 10,000$ (closed circle) and $T_m - 10,000 \leq T < T_m$ (open circle) for model F, where T_m is the occurrence time of the mainshock in each individual sequence. For the calculation, all simulation results are superimposed in each time period.

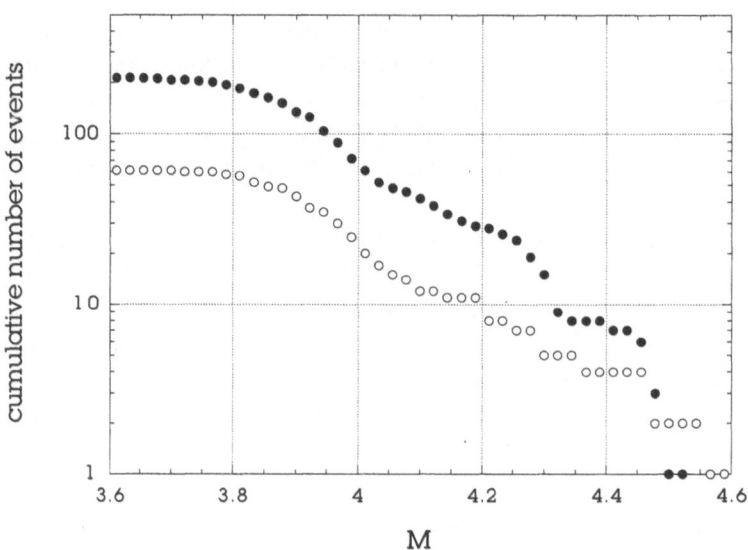

Figure 11
Cumulative frequency-magnitude distribution in two successive time periods $0 < T \leq 15,000$ (closed circle) and $15,000 < T < \infty$ (open circle) for model W. For the calculation, all simulation results are superimposed in each time period. Note that the activity appears to be highest at $T \simeq 15,000$ in the superimposed sequence (Fig. 6b).

model F. The lower limit value of the magnitudes is, however, mainly dependent on the assumed fault element size as noted in Paper I. If a lesser value is assumed for the fault element size, small size events with $M < 4.0$ will also satisfy the GR relation in each model.

Earthquake swarm is generally characterized as a sequence of earthquakes in which no single event dominates in size. The occurrence of earthquake swarm has been observed to be closely associated with pore fluid flow in many examples. One of the well-known examples is the Matsushiro, Japan, earthquake swarm that began in 1964; a large amount of fluid outflow was observed at a peak of seismic activity (OHTAKE, 1976). Sequences featuring earthquake swarm were simulated in Paper I, assuming fluid flow in a fault zone when the strength α_s is relatively low and the initial tectonic stress is lower than the residual stress. Another possible mechanism for earthquake swarm was advanced in the present paper. It was established that a sequence with no single predominant principal event is obtained when the critical slip distance for the pore creation is relatively small. Such a sequence was shown to start and end gradually, and the b value was unusually large, which is consistent with seismological observation of earthquake swarms (TOMODA, 1954; MIYAMURA, 1962; MOGI, 1963; SYKES, 1970). It seems that the mechanism obtained in the present paper more accords with seismological observation. MOGI (1963) pointed out that earthquake swarm is likely to occur in highly fractured regions such as volcanic regions and mid-ocean ridges; it is noteworthy that the existence of substantial fluid is suggested in such regions. Hence it will be reasonable to assume that the critical slip distance for pore creation is less for highly fractured regions since pores will be more easily created in highly fracture regions; note that earthquake swarm is shown to be facilitated when the critical slip distance for pore creation is minor. If the extent of permeability variations is greater in regions with a higher degree of fracturing, the duration of seismic activity is significantly long in such regions, according to our simulation results. Lastly, as remarked in Paper I, it should be noted that it will be inevitable to consider the effects of 3-D fluid flow path for quantitative understanding of earthquake swarm: swarm activity such as the Matsushiro swarm tends to spread throughout a substantial rock volume rather than being restricted to a planar zone (OHTAKE, 1976).

Acknowledgments

Helpful comments by two anonymous reviewers are greatly appreciated. I have used the computer systems of the Earthquake Information Center of the Earthquake Research Institute, University of Tokyo.

REFERENCES

BAK, P., and TANG, C. (1989), *Earthquakes as a Self-organized Critical Phenomenon*, J. Geophys. Res. *94*, 15,635–15,637.

BLANPIED, M. L., LOCKNER, D. A., and BYERLEE, J. D. (1992), *An Earthquake Mechanism Based on Rapid Sealing of Faults*, Nature *358*, 574–576.

BRACE, W. F. (1977), *Permeability from Resistivity and Pore Shape*, J. Geophys. Res. *82*, 3343–3349.

BRACE, W. F. (1980), *Permeability of Crystalline and Argillaceous Rocks*, Int. J. Rock Mech. Min. Sci. and Geomech. Abstr. *17*, 241–251.

BRACE, W. F. (1984), *Permeability of Crystalline Rocks: New in situ Measurements*, J. Geophys. Res. *89*, 4327–4330.

BYERLEE, W. F. (1993), *Model for Episodic Flow of High-pressure Water in Fault Zones before Earthquakes*, Geology *21*, 303–306.

CHESTER, F. M., EVANS, J. P., and BIEGEL, R. L. (1993), *Internal Structure and Weakening Mechanisms of the San Andreas Fault*, J. Geophys. Res. *98*, 771–786.

DAS, S., and AKI, K. (1977), *Fault Plane with Barriers: A Versatile Earthquake Model*, J. Geophys. Res. *82*, 5658–5670.

DAVID, C., WONG, T.-F., ZHU, W., and ZHANG, J. (1994), *Laboratory Measurement of Compaction-induced Permeability Change in Porous Rocks: Implications for the Generation and Maintenance of Pore Pressure Excess in the Crust*, Pure appl. geophys. *143*, 425–456.

DAVISON, C. C., and KOZAK, E. T. (1988), *Hydrogeological Characteristics of Major Fracture Zones in a Large Granite Batholith of the Canadian Shield*, Proc. 4th Can./Am. Conf. Hydrogeol., 52–60.

ENGELDER, J. T. (1974), *Cataclasis and the Generation of Fault Gouge*, Geol. Soc. Amer. Bull. *85*, 1515–1522.

EVANS, J. P. (1990), *Thickness-displacement Relationships for Fault Zones*, J. Struct. Geol. *12*, 1061–1065.

FORSTER, C. B., GODDARD, J. V., and EVANS, J. P., *Permeability structure of a thrust fault*. In *The Mechanical Involvement of Fluids in Faulting* (eds. Hickman, S., Bruhn, R. L., and Sibson, R.) USGS Open File Report 94–228, pp. 216–223 (U.S. Geological Survey, Menlo Park, Calif. 1994).

GOLD, T., and SOTER, S. (1985), *Fluid Ascent through the Solid Lithosphere and its Relation to Earthquakes*, Pure appl. geophys. *122*, 492–530.

HENDERSON, J. R., and MAILLOT, B. (1997), *The Influence of Fluid Flow in Fault Zones on Patterns of Seismicity: A Numerical Investigation*, J. Geophys. Res. *102*, 2915–2924.

HULL, J. (1988), *Thickness-displacement Relationship for Deformation Zones*, J. Struct. Geol. *10*, 431–435.

MARONE, C., RALEIGH, C. B., and SCHOLZ, C. H. (1990), *Frictional Behavior and Constitutive Modeling of Simulated Fault Gouge*, J. Geophys. Res. *95*, 7007–7025.

MILLER, S. A., NUR, A., and OLGAARD, D. L. (1996), *Earthquakes as a Coupled Shear Stress—High Pore Pressure Dynamical System*, Geophys. Res. Lett. *23*, 197–200.

MIYAMURA, S. (1962), *Seismicity and Geotectonics*, Zisin, Ser. 2, *15*, 23–52 (in Japanese).

MOGI, K. (1963), *Some Discussions of Aftershocks, Foreshocks and Earthquake Swarms*, Bull. Earthq. Res. Inst. *41*, 615–658.

MORROW, C. A., and BYERLEE, J. D. (1989), *Experimental Studies of Compaction and Dilatancy during Frictional Sliding on Faults Containing Gouge*, J. Struct. Geology *11*, 815–825.

OHTAKE, M. (1976), *A Review of the Matsushiro Earthquake Swarm*, Kagaku *46*, 306–313 (in Japanese).

OKADA, Y. (1992), *Internal Deformation due to Shear and Tensile Faults in a Half-space*, Bull. Seismol. Soc. Am. *82*, 1018–1040.

RALEIGH, C. B., HEALY, J. H., and BREDEHOEFT, J. D. (1976), *An Experiment in Earthquake Control at Rangely, Colorado*, Science *191*, 1230–1237.

RICE, J. R. (1993), *Spatio-temporal Complexity of Slip on a Fault*, J. Geophys. Res. *98*, 9885–9907.

SAMMONDS, P. R., MEREDITH, P. G., and MAIN, I. G. (1992), *Role of Pore Fluids in the Generation of Seismic Precursors to Shear Fracture*, Nature *359*, 228–230.

SEGALL, P., and RICE, J. R. (1995), *Dilatancy, Compaction, and Slip Instability of a Fluid-infiltrated Fault*, J. Geophys. Res. *100*, 22,155–22,171.

SIBSON, R. H. (1974), *Frictional Constraints on Thrust, Wrench and Normal Faults*, Nature *249*, 542–544.

SIBSON, R. H. (1982), *Fault Zone Models, Heat Flows, and the Depth Distribution of Earthquakes in the Continental Crust of the United States*, Bull. Seismol. Soc. Am. *72*, 151–163.

SIBSON, R. H. (1984), *Roughness of the Base of the Seismogenic Zone: Contributing Factors*, J. Geophys. Res. *89*, 5791–5799.

SIBSON, R. H. (1990), *Rupture Nucleation on Unfavorably Oriented Faults*, Bull. Seismol. Soc. Am. *80*, 1580–1604.

SIBSON, R. H. (1992), *Implications of Fault-valve Behaviour for Rupture Nucleation and Recurrence*, Tectonophysics *211*, 283–293.

SLEEP, N. H. (1995), *Ductile Creep, Compaction, and the Rate and State-dependent Friction within Major Fault Zones*, J. Geophys. Res. *100*, 13,065–13,080.

SLEEP, N. H., and BLANPIED, M. (1992), *Creep, Compaction, and the Weak Rheology of Major Faults*, Nature *359*, 687–692.

SLEEP, N. H., and BLANPIED, M. (1994), *Ductile Creep and Compaction: A Mechanism for Transiently Increasing Fluid Pressure in Mostly Sealed Fault Zones*, Pure appl. geophys. *143*, 9–40.

SYKES, L. R. (1970), *Earthquake Swarms and Sea-floor Spreading*, J. Geophys. Res. *75*, 6598–6611.

TEUFEL, L. W., *Pore volume changes during frictional sliding of simulated faults*. In *Mechanical Behavior of Crustal Rocks* (eds. Carter, N. L., Friedman, M., Logan, J. M., and Stearns, D. W.) (AGU, Washington D.C. 1981) pp. 135–145.

TOMODA, Y. (1954), *Statistical Description of the Time Interval Distribution of Earthquakes and on its Relations to the Distribution of Maximum Amplitude*, Zisin, Ser. 2, *7*, 155–169 (in Japanese).

VON SEGGERN, D. (1980), *Evidence for Precursory Changes in the Frequency-magnitude b Values*, Geophys. J. R. Astr. Soc. *86*, 815–838.

WALDER, J., and NUR, A. (1984), *Porosity Reduction and Crustal Pore Pressure Development*, J. Geophys. Res. *89*, 11,539–11,548.

WONG, T.-F., *On the normal stress dependence of the shear fracture energy*. In *Earthquake Source Mechanics* (eds. Das, S., Boatwright, J., and Scholz, C. H.) (AGU, Washington D.C. 1986) pp. 1–11.

WONG, T.-F., KO, S. C., and OLGAARD, D. L. (1997), *Generation and Maintenance of Pore Pressure Excess in a Dehydrating System 2. Theoretical Analysis*, J. Geophys. Res. *102*, 841–852.

YAMASHITA, T. (1993), *Application of Fracture Mechanics to the Simulation of Seismicity and Recurrence of Characteristic Earthquakes on a Fault*, J. Geophys. Res. *98*, 12,019–12,032.

YAMASHITA, T. (1997), *Mechanical Effect of Fluid Migration on the Complexity of Seismicity*, J. Geophys. Res. *102*, 17,797–17,806.

YAMASHITA, T. (1998), *Simulation of Seismicity due to Fluid Migration in a Fault Zone*, Geophys. J. Int. *132*, 674–686.

YAMASHITA, T., and OHNAKA, M. (1992), *Precursory Surface Deformation Expected from a Strike-slip Fault Model into which Rheological Properties of the Lithosphere are Incorporated*, Tectonophysics *211*, 179–199.

(Received June 19, 1998, revised/accepted December 17, 1998)

 To access this journal online:
http://www.birkhauser.ch

Pure appl. geophys. 155 (1999) 649–667
0033–4553/99/040649–19 $ 1.50 + 0.20/0

© Birkhäuser Verlag, Basel, 1999

Coupled Stress Release Model for Time-dependent Seismicity

JIE LIU,[1] YONG CHEN,[1] YAOLIN SHI[2] and DAVID VERE-JONES[3]

Abstract—Based on the original stress release model of seismicity proposed by VERE-JONES (1978), this paper has developed a stochastic coupled stress release model of time-dependent seismicity, which considers the earthquake interaction and stress transfer between different seismic subregions. As an example, the model is applied to a statistical analysis of the historical earthquake catalog with magnitude $M \geq 6.0$ during the period from 1480 to 1996 in North China. According to the Akaike information criterion (AIC), the results show that the coupled stress release model is better than the original model, which demonstrates the existence of long-range correlations between different seismic subregions. We also apply both the stochastic (original and developed coupled) models to analyze the synthetic catalog produced by a cellular automata model, which is based on mechanics of a slide-spring-damper system to model the fault network. The stress release model provides a good fit to the synthetic regional stress, and the coupled stress release model provides an improvement in fit to the synthetic catalog over the original model.

Key words: Coupled stress release model, AIC criterion, historical earthquake catalog, synthetic catalog.

1. Introduction

The stress release model was first proposed by Vere-Jones in 1978 (VERE-JONES, 1978), which is used to study the statistical regularity of seismicity. The physical essence of this model is elastic rebound theory. The classical elastic rebound model of earthquake occurrence considers that the stress in a region is accumulated slowly over a prolonged time, afterward, the stress would be suddenly released and an earthquake would occur when the stress exceeds the strength of the medium. In the field of stochastic process, this conception can be simulated using a leap-type Markov process (KNOPOFF, 1971). The stress release model is a development of Knopoff's Markov model. It uses a notional regional stress level as the key variable, or state, that controls the probabilities of earthquakes within a region.

[1] Center for Analysis and Prediction, China Seismological Bureau, Beijing 100036, China.
[2] Graduate School, University of Science and Technology of China, Beijing 100039, China.
[3] ISOR, Victoria University of Wellington, New Zealand.

VERE-JONES and DENG (1988) applied this model to the historical earthquake catalog in North China and obtained some definite results. ZHENG and VERE-JONES (1991, 1994) studied this model in detail, developed a complete and concrete computation method, and applied this model to the historical earthquake catalogs in China, Japan, Iran, etc. They also conducted further discussion on their results. However, they only considered the condition of one region. Although ZHENG and VERE-JONES (1991) divided North China into four seismic belts and combined them to calculate the parameters of the stress release model, they did not consider the spatial interaction of earthquake occurrence through stress transfer between seismic belts (or different parts) in the model. In this paper, a coupled stress release model is proposed by adding stress interaction between different parts to estimate the effect on the seismic hazard in each part, based on the original stress release model, and the coupled model is applied to analyze the historical earthquake catalog in North China. Based on the Akaike information criterion, a comparison study is done by studying the results from various models.

Due to the time-length limitation of the historical earthquake catalog, we also apply the stochastic models to analyze the synthetic catalog produced by a cellular automata model. A cellular automata model was first proposed by BAK and TANG in 1989, which considered earthquake occurrence as a typical self-organized critical phenomena (BAK and TANG, 1989; ITO and MATSUZAKI, 1990; MAIN, 1996). An essential assumption in their models is that stress increases in each cell purely randomly. This assumption is not appropriate for a real earthquake. Tectonic stress field varies according to mechanical laws, not randomly. A slide-spring-damper mechanical model of a continental earthquake produced by a network of faults was proposed by SHI *et al.* (1994). Based on the mechanical model, the global rules of stress adjustment in the system can be summarized, and correspondingly a similar cellular automata model can be constructed (LIU *et al.*, 1995). Both the fault network model and corresponding cellular automata model cannot only produce chaotic synthetic catalogs, but also simulate the stress evolution of the system. In this paper, we analyze the synthetic earthquake catalogs by applying the stochastic models. The results indicate that even if the model and parameters are unknown, some characteristics of the system, such as the system stress, can be predicted from the synthetic catalog alone. The simplified stress is useful for prediction of the earthquake in the synthetic catalog.

2. Coupled Stress Release Model

2.1. Simple Stress Release Model

Stress level can be simplified as a scalar function $X(t)$ in the studied region

$$X(t) = X(0) + \rho t - S(t) \qquad (1)$$

where ρ is a constant loading rate from external tectonic force. $S(t)$ is the accumulated stress release from events within the region over the period $(0, t)$

$$S(t) = \sum_{0 < t_i < t} S_i \tag{2}$$

assume that stress drop is only related to magnitude

$$S_i = 10^{0.75(m_i - m_0)} \tag{3}$$

where m_i is earthquake magnitude. m_0 is considered as normalized magnitude, which is a constant.

The stochastic behavior of the model is controlled by a "risk function" $\Psi(x)$. It represents the conditional probability intensity on condition that $\Psi(x)$ is equal to x (x is a stress value), or $\Psi(x) \, dt$ is the occurrence rate in the expected time interval $(t, t + dt)$. In the Poisson model, seismic risk is always considered as a constant regardless the stress value, but, in the stress release model, the seismic risk function is assumed to exponentially increase with stress, i.e.,

$$\Psi(x) = A \, \exp(d + bx). \tag{4}$$

In the above assumption, the key element of the stochastic point process analysis is to find conditional probability intensity function λ, which is also the basis of statistical inference, simulation and prediction

$$\lambda(t, s) \, dt \, ds = \text{Prob}\{\text{event in time interval } (t, t + dt)$$

$$\text{and stress drop within } (s, s + ds)$$

$$\text{given the past history up to time } t\}. \tag{5}$$

How to get λ? There is a set of strict theories and methods in statistics. ZHENG and VERE-JONES (1991, 1994) have presented detailed discussion in their papers. The conclusion is that the effect for λ by time and stress can be separated to calculate. The change of λ with time can be written:

$$\lambda(t) = \Psi(X(t)) \tag{6}$$

substitute Equations (1), (2) and (3) into Equations (4) and (5), and obtain Equation (7)

$$\lambda(t) = \exp\{a + b[t - cS(t)]\} \tag{7}$$

where a, b and c are parameters, they can be obtained by achieving $\log L$ maximum using a likelihood method

$$\log L = \sum_{i=1}^{N(T)} \log \lambda(t_i) - \int_0^T \lambda(u) \, du \tag{8}$$

where T is the length of the observation interval (conventionally presumed to start time $t = 0$) and $N(T)$ is the total number of events observed in that interval, t_i is the origin time of each earthquake.

2.2. AIC Criterion

There are many statistical models to be selected for describing a natural phenomenon. If just comparing their log-likelihood values, the conclusion will be that the more parameters, the better the results. However, that is not correct in a practical sense. In order to solve this problem, AKAIKE (1977) proposed the AIC (Akaike Information Criterion) method to select models. The method is: let $\log \hat{L}$ denote the log-likelihood for a given model, so

$$\text{AIC} = -2 \log \hat{L} + 2k \tag{9}$$

where k is the number of fitted parameters in the given regional model. The AIC value can be obtained for different models. The model with the smallest AIC value is considered as the best model to be selected. Generally, if model difference reaches a 5% credible level, these two models can be selected, which corresponds to a difference in AIC values of around $1.5 \sim 2.0$.

2.3. Coupled Stress Release Model

From the above description, it can be seen that a simple stress release model only considers the stress-level effect of a region by the seismicity in the same region, yet, in practical seismicity, the seismicity around this region can also produce a definite effect on stress level in the region. Conversely, its seismicity can cause a definite effect on seismic hazard around this region too. Therefore, a coupled stress release model is proposed in this paper. For simplicity, we only consider coupled interaction in two subregions. Assuming that $\lambda_1(t)$ and $\lambda_2(t)$ are conditional probability intensity functions in the first subregion and the second subregion respectively, they can be described as

$$\begin{aligned}\lambda_1(t) &= \exp\{a_1 + b_1[t - c_{11}S_1(t) - c_{12}S_2(t)]\} \\ \lambda_2(t) &= \exp\{a_2 + b_2[t - c_{21}S_1(t) - c_{22}S_2(t)]\}\end{aligned} \tag{10}$$

where $a_1, b_1, c_{11}, c_{12}, a_2, b_2, c_{21}, c_{22}$ are model parameters. The total number is 8. $S_1(t)$ and $S_2(t)$ are the accumulated stress release in the first and second subregions separately. The computation method is the same as Equation (2). Comparing Equation (7) with Equation (10), coupled interaction is realized by introducing the stress release effect of the other subregion by parameters c_{12} and c_{21}. This definition is also simplified as that of stress function. The model parameters can be obtained by using the log-likelihood method

$$\log L = \sum_{i=1}^{N_1(T)} \log \lambda_1(t_i) + \sum_{j=1}^{N_2(T)} \log \lambda_2(t_j) - \int_0^T (\lambda_1(u) + \lambda_2(u))\, du \qquad (11)$$

where $N_1(t)$ and $N_2(t)$ are the total earthquake number in the first and second subregions in the time interval T respectively; t_i and t_j are occurrence times of earthquakes in those two subregions separately.

3. Computation Models and Results of Historical Earthquakes

When the statistical model is applied to the historical earthquake catalog in North China, we select only $M_S \geq 6.0$ earthquakes from 1480 to 1996, considering the completeness of the earthquake catalog and referring to the division method for the North China region in Zheng's papers (ZHENG et al., 1991). The subregion is divided, and the boundary line is the Taihangshan Mountains (Fig. 1). The earthquakes in the western subregion are mainly events around the Ordos Plateau (not including earthquakes in the Qilianshan seismic belt and southeast of Gansu Province, and the boundary line is delimited along the edge of North China tectonic plate). The earthquakes in the eastern subregion mainly include the events in the Hebei plain seismic belt and the Tanlu seismic belt. Because each earthquake is considered as an independent event in stochastic point process, aftershocks are deleted from the catalog using space-time window criteria suggested by KEILIS-BOROK and KNOPOFF (1980). Although a satisfactory definition of such a scheme is very difficult, aftershocks are not present in large numbers due to the magnitude cut-off of $M \geq 6.0$ fortunately. After the removal of aftershocks, there remains a total of 65 earthquakes in the entire region, of which there are 33 earthquakes in the eastern subregion and 32 in the western subregion. The resulting data set is provided in the Appendix. The following three models are used in computation.

3.1. Simple Stress Release Model

Simple stress release model is used in the entire North China region, and the random allocation method is used to distribute the earthquakes in the region. Their computation formulas are as follows:

$$\lambda(a, b, c) = \exp\{a + b[t - cS(t)]\}$$

$$\log L_1 = \sum_{i=1}^{N} \lambda(t_i) - \int_0^T \lambda(u)\, du$$

$$\log L_2 = N_1 \log \frac{N_1}{N_1 + N_2} + N_2 \log \frac{N_2}{N_1 + N_2} \qquad (12)$$

$$\text{AIC} = -2(\log L_1 + \log L_2) + 2 * (3 + 1)$$

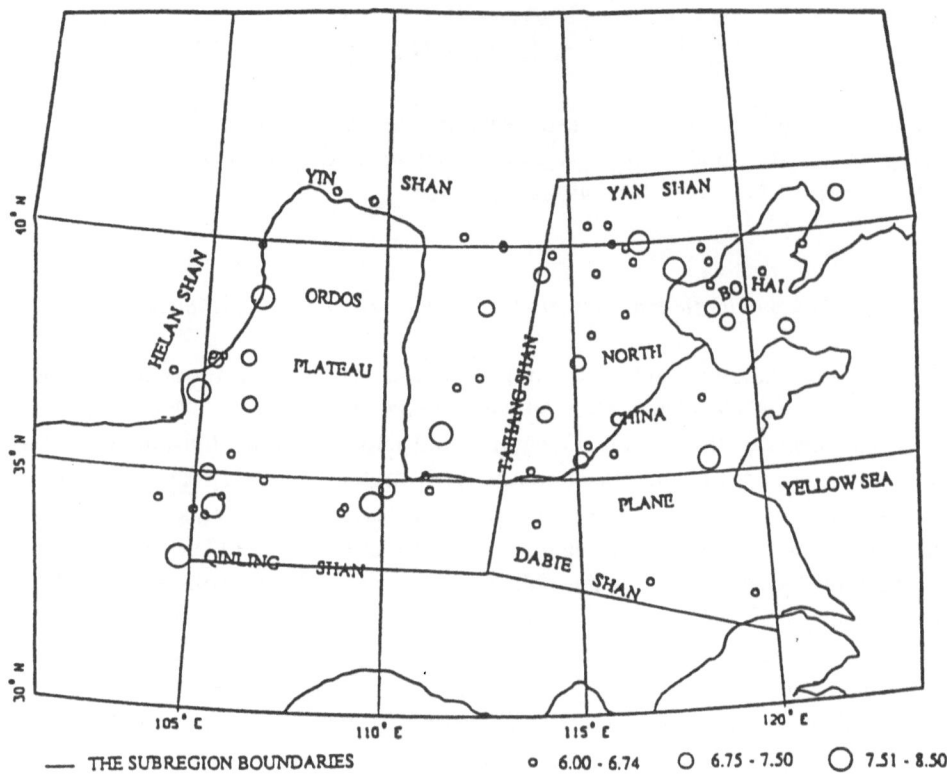

Figure 1
Map of North China showing historical records and subregion boundaries.

where N, N_1 and N_2 is the total earthquake number, the earthquake number in the first and second subregion, respectively. There are three parameters a, b and c in stress release model. Because the total number of earthquakes in the entire region is a fixed number, it needs only one parameter to allocate earthquakes randomly to two subregions. Consequently, the total parameters are 4.

Table 1

Simple stress release model

a	B	c	$-\log L_1$	$-\log L_2$	AIC
-2.462	0.01128	0.1513	195.87	45.05	489.83

Table 2

Independent stress release model

a_1	b_1	c_1	$-\log L_1$	a_2	b_2	c_2	$-\log L_2$	AIC
-3.46	0.0132	0.329	118.13	-3.15	0.0155	0.284	117.86	483.97

3.2. Independent Stress Release Model

In each subregion, simple stress release model is computed independently to obtain the conditional probability intensity function. The conditional probability intensity in the entire region can be gained by simple addition.

$$\lambda_1(a_1, b_1, c_1) = \exp\{a_1 + b_1[t - c_1 S_1(t)]\}$$

$$\log L_1 = \sum_{i=1}^{N_1} \log \lambda_1(t_i) - \int_0^T \lambda_1(u)\, du$$

$$\lambda_2(a_2, b_2, c_2) = \exp\{a_2 + b_2[t - c_2 S_2(t)]\}$$

$$\log L_2 = \sum_{j=1}^{N_2} \log \lambda_2(t_j) - \int_0^T \lambda_2(u)\, du \tag{13}$$

$$\lambda(t) = \lambda_1(t) + \lambda_2(t)$$

$$\text{AIC} = -2 * (\log L_1 + \log L_2) + 2 * (3 + 3).$$

There are three parameters in each subregion, thus the total number of parameters is 6.

3.3. Coupled Stress Release Model (CSRM)

For coupled stress release model, its computation method is shown in Equations (10) and (11), considering the interaction between two parts. In addition, we also use a simplified model in this paper which is called a symmetrical coupled stress release model, i.e., assume that interaction within the two subregions is the same, consequently, it is in the equation of conditional probability intensity

$$c_{12} = c_{21}. \tag{14}$$

Thus there are only seven parameters in symmetrical CSRM. The equations to calculate AIC values of the two kinds of CSRM are separately as follows:

$$\text{AIC} = -2 * \log L + 2 * (4 + 4) \quad \text{(for common CSRM)}$$

$$\text{AIC} = -2 * \log L + 2 * 7 \quad \text{(for symmetrical CSRM).} \tag{15}$$

In all above computations, normalized magnitude $m_0 = 5.0$. The results are given in Tables 1–3 and Figures 2–4.

When comparing obtained AIC values in the three models, the difference between them is larger than 2. This demonstrates that the difference between models is obvious. Coupled stress release model is better than independent stress release model, or better than the stress release model in which the stress is calculated in each subregion independently and the interaction is not considered. Furthermore, independent stress release model is better than the simple stress release model, in which the whole region is set up as only one stress parameter. This indicates that the result, considering region division is better than the results in the entire region when applying stress release model. Zheng considered region division in his paper (ZHENG *et al.*, 1991), and he remarked mainly on the comparison of various models (Poisson model, Poisson model with trend and simple stress release model) after region division, however he did not reveal how to compare their AIC values between the subregions and the entire region (for example, we introduce the random-distribution parameter in simple stress release model).

When comparing symmetrical CSRM and common CSRM, and AIC value in symmetrical CSRM is smaller than that of common CSRM although the minus log-likelihood value of symmetrical CSRM is larger than that of common CSRM. Because the difference of AIC values is smaller (less than 2), we cannot determine which model is better by the AIC criterion method despite the fact that there is one more parameter in common CSRM than in symmetrical CSRM.

In order to study coupled stress release model further, the North China region is divided into four subregions. In the eastern part, the Hebei plain seismic belt is a subregion (E1), and the other is the Tanlu seismic belt (E2). The total numbers are 21 and 12, respectively. In the western part, the Shanxi seismic belt is the subregion (W1), and the other is the western boundary area of Ordos plateau (W2). The total numbers are 20 and 12 separately (see Appendix). The above three kinds of models are applied to both these two parts. Comparing their difference of AIC values, the same conclusion can be reached (Table 4). In addition, because the tectonic environment is similar in the eastern and western parts of North

Table 3

Coupled stress release model

	a_1, a_2	b_1, b_2	c_{11}, c_{21}	c_{12}, c_{22}	$-\log L$	AIC
Symmetrical	−3.949	0.02578	0.1580	0.1390	232.603	479.21
CSRM	−3.293	0.01680	0.1390	0.1686		
Common	−3.939	0.02632	0.1448	0.1506	232.464	480.93
CSRM	−3.322	0.01804	0.1030	0.1971		

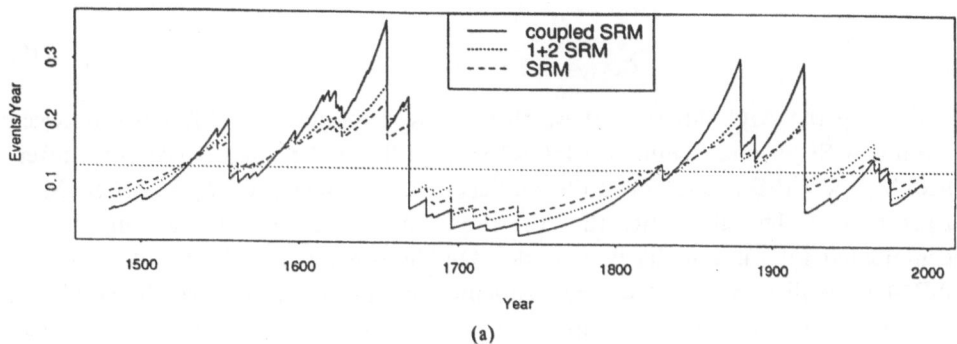

(a)

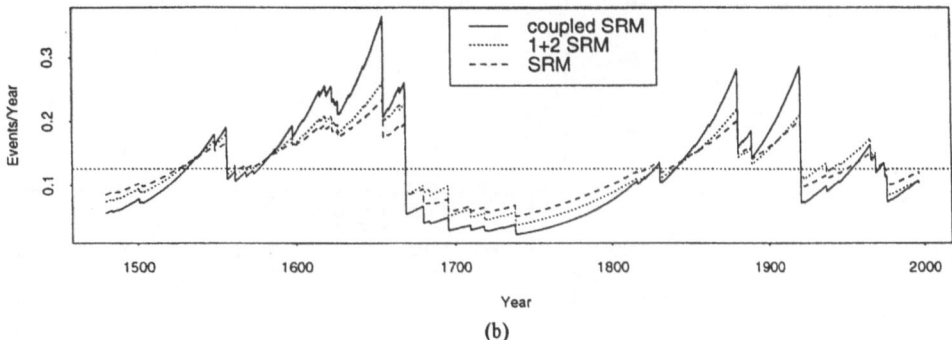

(b)

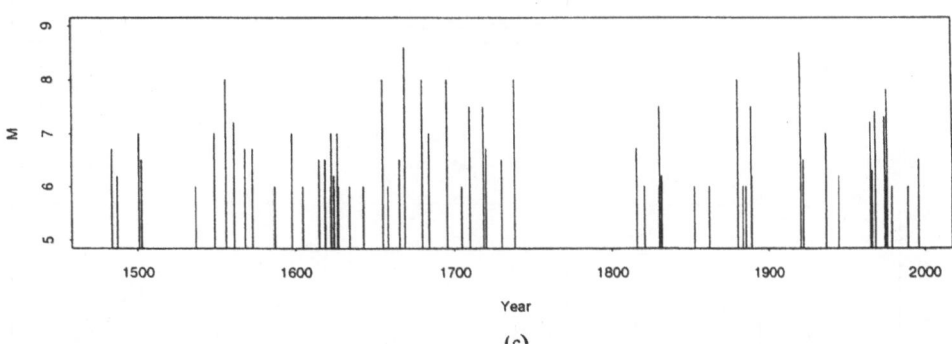

(c)

Figure 2

The conditional probability intensity plot of stress release model and M-T plot in North China $(1 + 2$ SRM means independent stress release model). (a) Coupled stress release model, (b) symmetrical coupled stress release model, (c) magnitude of event versus time (M-t plot).

China (ZHENG *et al.*, 1991), the stress-accumulated condition may be assumed as the same, therefore, another computation method could be proposed for CSRM, i.e.,

$$b_1 = b_2$$
$$\text{AIC} = -2 * \log L + 2 * 7. \tag{16}$$

Comparing the AIC values of these three CSRM models, the difference between common CSRM and symmetrical CSRM is still smaller than 2, yet the difference between the results of both models and the result of $b_1 = b_2$ CSRM is larger than 2. This illustrates that the model assuming there is the same stress-accumulated rate, is not the best model. On the other hand, the result of $b_1 = b_2$ CSRM is smaller than that of independent stress release mode, which establishes that coupled interaction is an important factor in stress release models. However, because there is one subregion in which the earthquake number is small (only 12) for both regions, its credible level is slightly lower from the point of statistics.

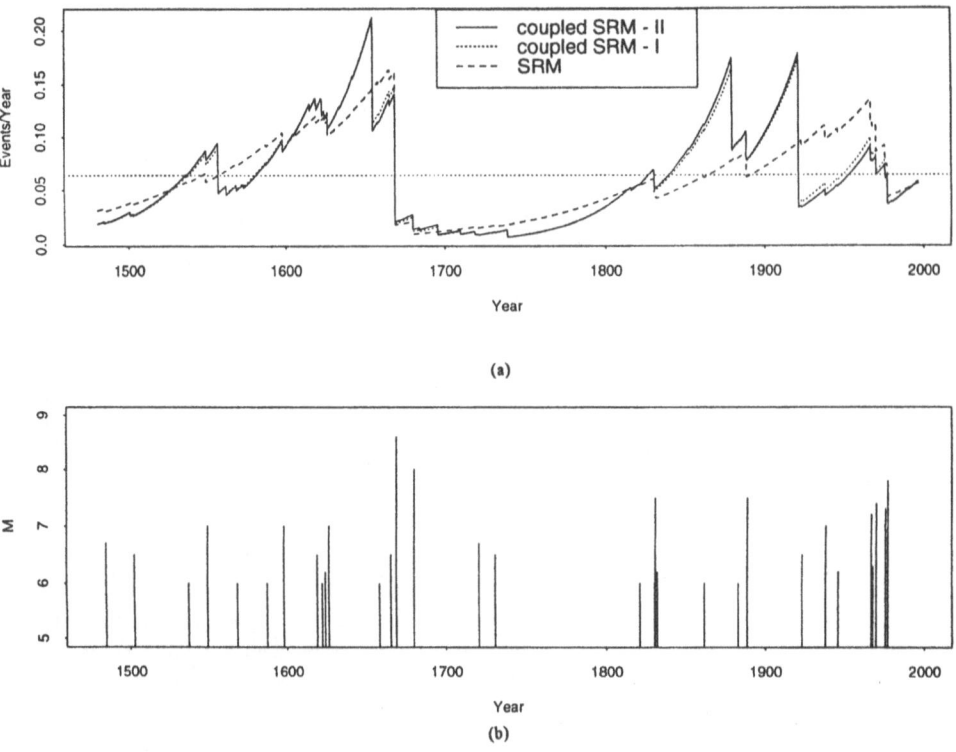

(a)

(b)

Figure 3

The conditional probability intensity plot of stress release model (a) and M-T plot (b) in the eastern part (the first subregion) of North China (coupled SRM-I is common coupled stress release model, coupled SRM-II is symmetrical coupled stress release model).

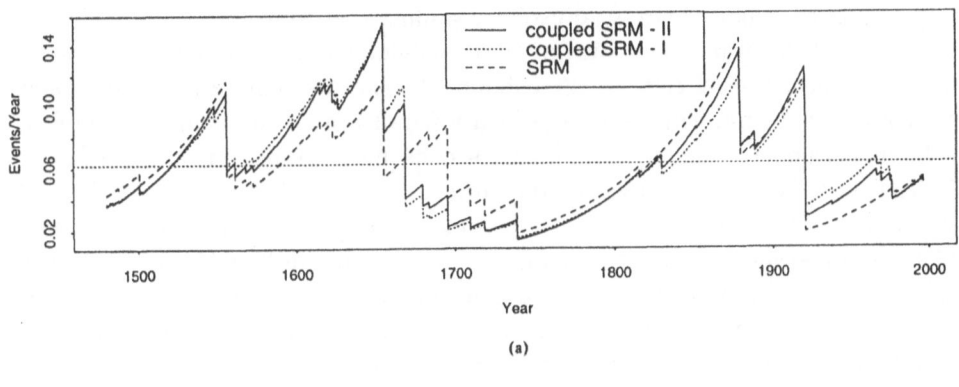

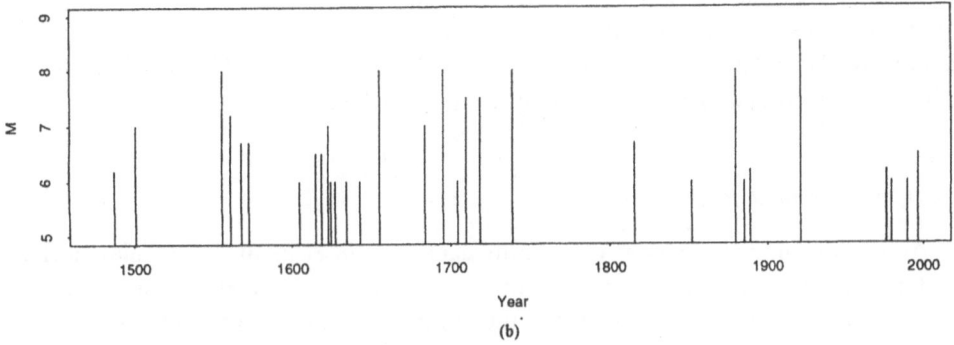

Figure 4
The conditional probability intensity plot of stress release model (a) and M-T plot (b) in the western part (the second subregion) of North China (coupled SRM-I is common coupled stress release model, coupled SRM-II is symmetrical coupled stress release model).

4. Application of the Stochastic Models to Synthetic Earthquakes

4.1. Mechanical Model for Earthquake in a Fault Network

SHI et al. (1994) proposed a spring-slide-damper mechanical model to simulate a fault-network earthquake. By solving a set of differential equations under uniform boundary strain rate conditions, stresses of each element and the entire system can be calculated. An element will break as its stress exceeds its static strength, and the stress will begin to drop until it reduces to its dynamic strength. Potential energy released during the earthquake defines the magnitude of the earthquake. Stress adjustment after an earthquake can be calculated according to the equilibrium condition of the system. A large earthquake may occur if the adjustment of stress induces further break adjacent to the broken element. A synthetic catalog simulates many aspects of continental earthquakes produced by such computation (ZHANG et al., 1995).

Based on the rules of stress adjustment summarized from the network model,
LIU *et al.* (1995) proposed a corresponding cellular automata model. The model
consists of an array of element. A column of elements represents a fault, and
different columns represent a group of nearly parallel faults mechanically coupled.
In the model under boundary conditions of constant deformation rate, stresses in
all elements as well as the stress of the entire system can be calculated. They all
increase at a different rate linearly with time until break of an element occurs.
Break of an element produces significant stress drop in the element. Nearby
elements in the same fault undergo stress increase because of the stress concentration produced by the broken element. The stress increase decays with the distance
to the broken element. As concerns the entire fault, the summation of stresses in all
elements of the fault is still reduced relative to that before the event. Stresses on the
other faults sustain the same stress drop, and the amount of stress drop for each
element is also related to its distance to the broken element. Continuous breaks
induced by a broken element represent a large earthquake.

4.2. Computational Results

A cellular automata model of 2 faults each has 78 elements is calculated. The
model contained 620 independent parameters, such as element statistic friction,
dynamic friction, stress increase rate, and initial stress, etc. The synthetic catalog
produced by the model is similar to real seismicity in many aspects.

Figure 5 displays the catalog and the stresses calculated from the cellular
automata model, and the calculated conditional intensity from 141 events spanning
the time period 20,000 to 21,000 time unit in the synthetic catalog. The logarithm
of conditional intensity can be compared with the stress. It is noted that the two
curves look quite similar. The stress is calculated from a cellular automata model
with 620 parameters, while the stress release model uses only the catalog to fit 3

Table 4

The results of stress release model in the eastern and western parts of North China

Division	Stress release model		$-\log L$	AIC
The eastern part	Coupled model	$b_1 = b_2$	126.830	267.660
		Symmetrical	116.196	246.392
		Common	114.955	245.909
	Independent model		136.476	284.952
	Simple model		139.758	287.516
The western part	Coupled model	$b_1 = b_2$	122.741	259.482
		Symmetrical	113.550	241.100
		Common	113.365	242.730
	Independent model		135.792	283.584
	Simple model		139.027	286.054

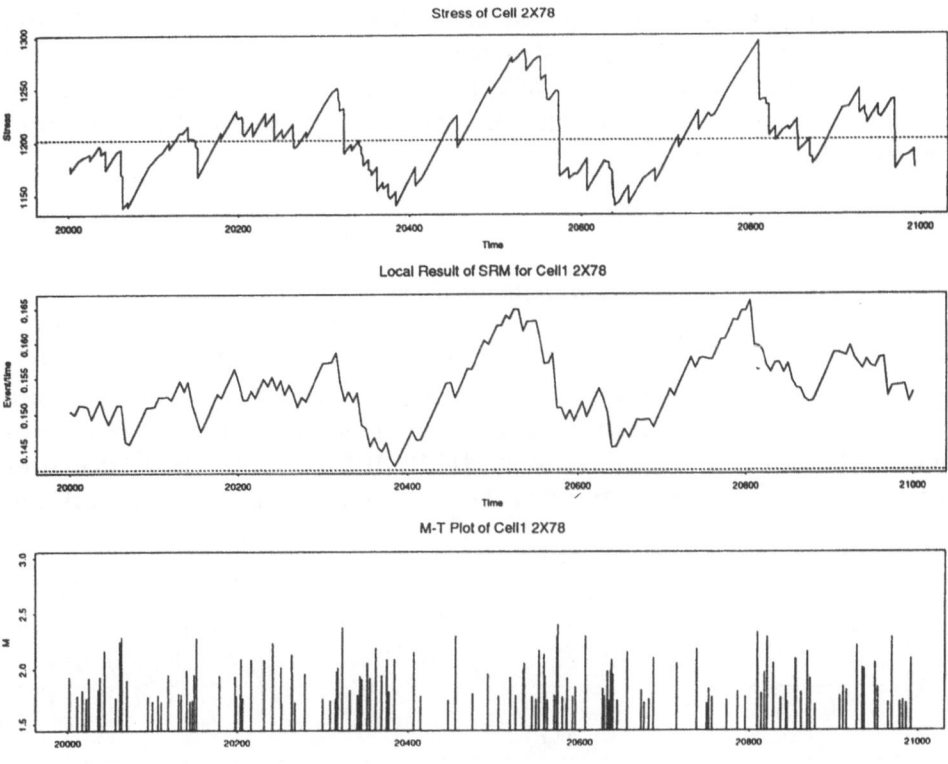

Figure 5

Computational results by using 141 events from relative time 20000 to 21000 in cellular automata model. (a) System stress in cellular automata model; (b) conditional probability intensity of earthquake occurrence calculated from stress release model (logarithm of the intensity is proportional to system stress); (c) M-t plot.

parameters, but, successfully simulates the stress variation of the system. AIC value of stress release model is smaller than that of Poisson model by 3.6, therefore, stress release model is superior to Poisson model for the synthetic catalog.

In practice, small earthquakes usually cannot be completely recorded. In order to study the effect of a catalog absent small events, we cut small events ($M < 2.2$) from 2,843 events during time 10,000 to 30,000 in our synthetic catalog, and keep only 262 relatively large events. Although this number is 10% of the total events, stress release model still can produce a satisfactory fit as shown in Figure 6. In this case, AIC value of stress release model is smaller than Poisson model by 7.4, evidently superior to Poisson model.

In actuality, we generally study a local region, instead of the entire system. In order to study the effect of the division of subregions, we cut the region parallel to the fault, and each subregion contains a complete fault. The results are shown in Figure 7. The stress release model now behaves even worse than Poisson model in each subregion. Coupled stress release model has AIC value lower than Poisson model by 222.6, and lower than independent stress release model by 203.9. It still provides the best fit. The results indicate that coupled stress release model can significantly improve the fitting, although the stress release model may not work well in each subregion.

5. Conclusion and Discussion

Poisson model is often used in seismic hazard assessment. Comparing Poisson model with stress release model, the former is a common model with no physical meaning in statistics, but, the latter is a statistical model based on the geophysical model. The papers by ZHENG *et al.* (1991, 1994) have shown clearly that stress release model is obviously better than Poisson model. Based on these works, we

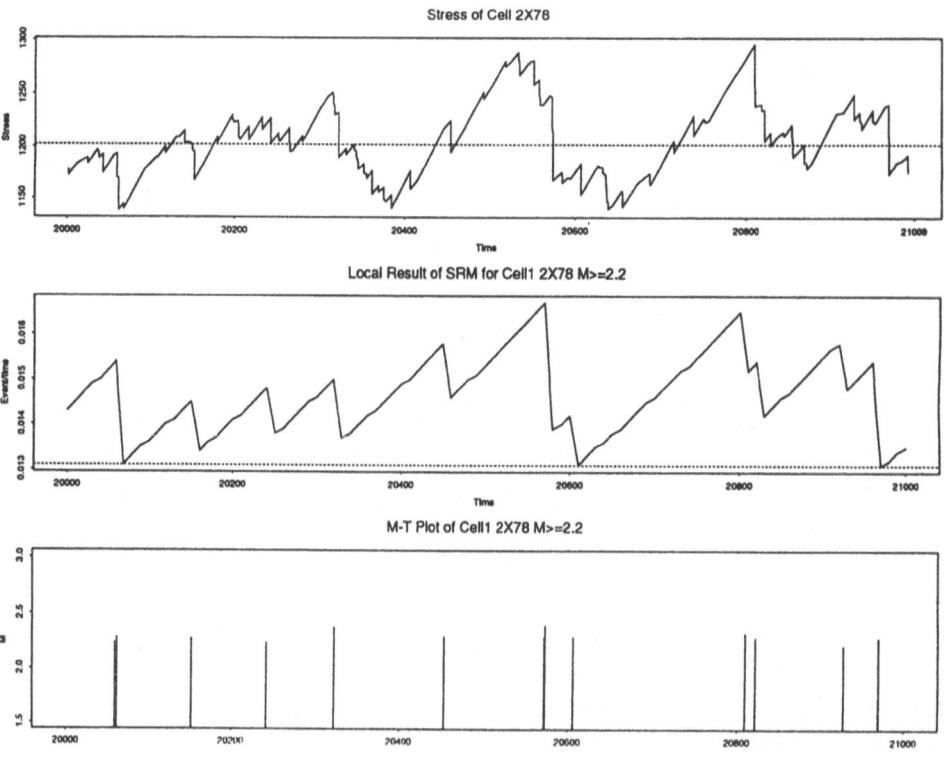

Figure 6
Computational results from 248 large earthquakes ($M > 2.2$) among 2,846 earthquakes in synthetic catalog from time 10000 to 30000. Only the curves within time period 20000–21000 are shown for comparison with Figure 5. (a) System stress in the cellular automata model; (b) conditional probability intensity of earthquake occurrence calculated from the stress release model; (c) M-t plot.

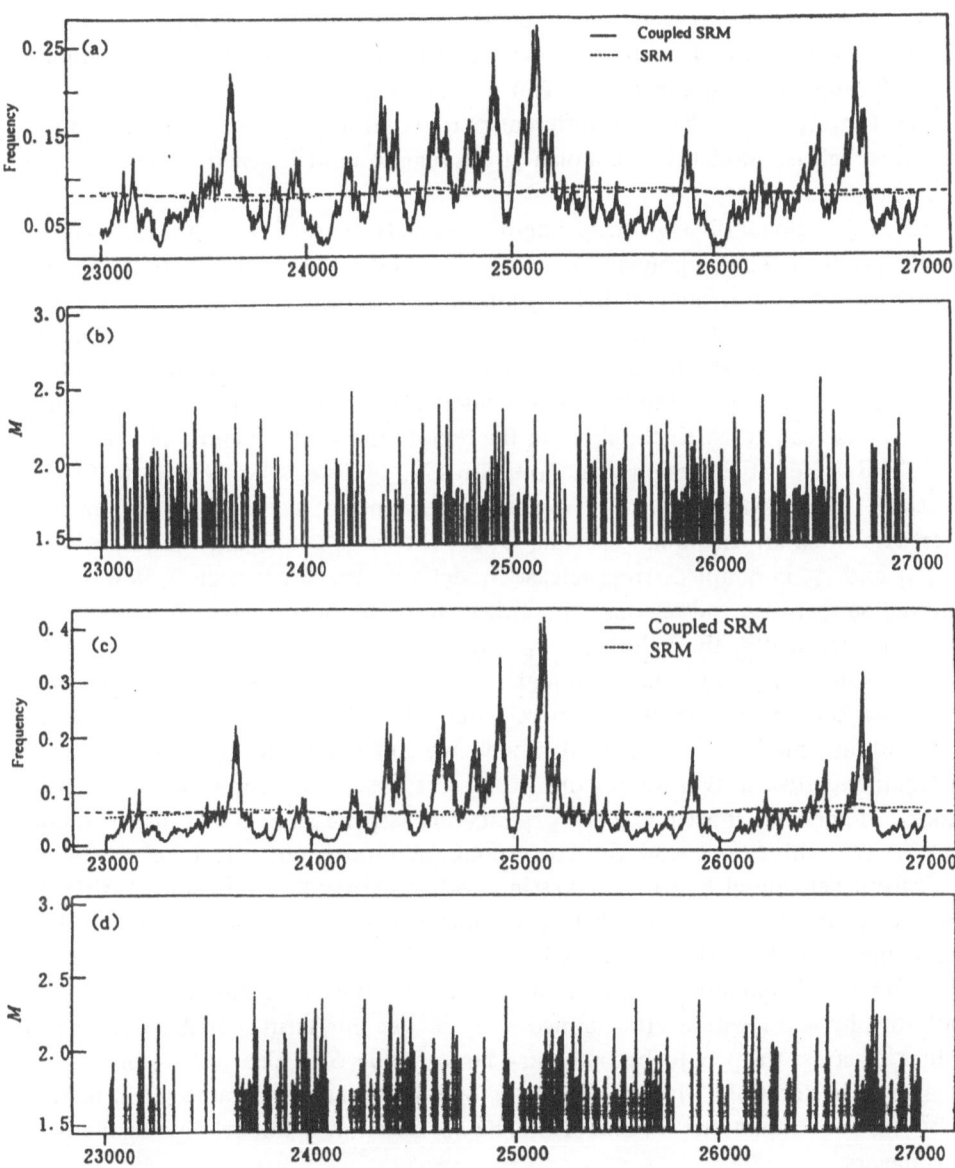

Figure 7

The results for a model with two subregions divided by a line parallel to the faults. (a) Conditional probability intensity of earthquake occurrence in the first subregion, calculated from coupled stress release model (solid line) and simple stress release model (dash line) respectively; (b) M-t plot for the first subregion; (c) conditional probability intensity of earthquake occurrence in the second subregion, calculated from coupled stress release model (solid line) and simple stress release model (dash line), respectively; (d) M-t plot for the second subregion.

have considered the characteristics of practical seismicity (interaction between different seismic subregions) and we have developed a coupled stress release model. The following research conclusions can be obtained:

(1) The interaction between different parts is an important factor when applying stress release model, i.e., coupled stress release model is better than common stress release model.

(2) The regional division according to seismicity characteristics is also a considerable problem when applying stress release model. The computation of stress release model in subregions is better than in the whole region.

(3) The regional stress can be simulated very well by our stress release model, which suggests that the model is a very useful tool in seismic hazard assessment.

Because stress release model only considers inner stress accumulation and inner stress release caused by earthquakes in the region, the values b and c of conditional probability intensity function λ in Equation (7) are always larger than 0. From a physical view, it means that the regional stress increases linearly with time and decreases when an earthquake occurs. Under the same condition, the values of b_1, b_2, c_{11} and c_{22} in coupled stress release model are also larger than 0, however the interaction between subregions can either decrease or increase the subregional stress, consequently, the values of c_{21} and c_{12} may be larger or less than 0. The positive values mean that the earthquakes in a subregion will cause stress release in the other subregion, and the negative values will cause stress increase. Because stress release model is a statistical model, we can determine the whole effect of interaction between two subregions using the results of coupled stress release model. In addition, if assuming $c_{12} = c_{21}$, its physical meaning is that the interaction way in two subregions and the corresponding time length of stress release (or increase) level caused by earthquakes in another subregion are identical. Although there is no definite conclusion between symmetrical CSRM and common CSRM, the symmetrical CSRM is not an adverse model.

Normalized magnitude m_0 is used only for simplifying computation, which is without physical meaning. It is unnecessary to use this term, but it will cause the values of stress drop S to become very large (value S is also just a comparative measure). In addition, when m_0 decreases one magnitude, the value c will increase $10^{0.75}$ times in Equations (7) and (10). The m_0 selection is not relative to the other parameters.

There are many similar assumptions in the stress release model and the cellular automata model of this paper. Both models assume stress linear increase of stresses during earthquake preparation, although the rates are different for each element in the cellular automata model and the entire system is assumed to have the same rate in the stress release model. The cellular automata has definite rules for stress adjustment during earthquake occurrence, and the stress release model also assumes every earthquake exerts a definite influence on system stress depending only on the magnitude of the earthquake. This similarity may explain why the stress release model can provide a very good fit of system stress of the cellular automata model. Although the actual data is surely more complicated than the synthetic data, it is indicated by the fitting of the synthetic data of cellular automata that some system characteristics (the system stress in this study) can be predicted with a limited time, based on the catalog (or an incomplete catalog), even if the details of model and its parameters are not known.

When computing the accumulated stress release $S_j(t)$ in coupled stress release model, we think that the $S_j(t)$ of $\Sigma_{i=1}^{N_1(T)} \log \lambda_1(t_i)$ in Equation (11) (same as for $\Sigma_{j=1}^{N_2(T)} \log \lambda_2(t_j)$) just considers event time in the first subregion, which means that $S_j(t)$ is calculated up to time t_{i-1} for the ith event in the first subregion. The reason is that the log-likelihood method is to maintain balance between stress accumulation $b(t_i - t_{i-1})$ and stress release $\Sigma_{k=1}^{L_i} 10^{0.75(m_k - m_0)} + 10^{0.75(m_i - m_0)}$ during the time interval $[t_{i-1}, t_i]$, where L_i and m_k are separately the total number and magnitude of earthquakes of the second subregion which occurred in time interval $[t_{i-1}, t_i]$, m_i is the earthquake magnitude of the first subregion which occurred in time t_i.

Furthermore, we have also found problems in our work. These are of value for further works:

(1) The results obtained by coupled stress release models clearly raise the seismic occurrence probability of some main earthquakes (or a start earthquake in a cluster of earthquakes) from the conditional probability intensity plot, compared with the results obtained from independent stress release model and simple stress release model. It is very useful to predict these earthquakes using conditional probability intensity. However, on the other hand, the prediction for following earthquakes (not aftershocks) in a cluster of earthquakes by coupled stress release model is not good. In fact, the problem exists in all involved stress release models in the paper. One solution to this problem is to further improve the conditional probability intensity function in stress release model and to consider clustering earthquake activity in the function. The second method is to further divide the region and to allocate clustering earthquakes into each subregion.

(2) The regional division is very important in the stress release model. The region is divided into an area as small as possible, considering the condition of the tectonic environment. However, in practicality, because the reoccurrence time is prolonged, there will be fewer earthquakes in some subregions if they are too small, and fewer events will not satisfy the condition of the statistical calculation. For example, although the AIC value in common CSRM (or symmetrical CSRM) is less than that of other stress release models when the North China region is divided into four seismic belts (four subregions), the result is not in accord with the practical stress accumulation condition from the conditional probability intensity plot. The reason is that the value b is very large, causing stress increase rate to be excessively large (this is another reason that $b_1 = b_2$ CSRM is used in the paper). Thus, it is reasonable to decrease the lower limit of magnitude in order to increase the sample number in future research. Another method is to study stress release model using the synthetic earthquake catalog produced by the physical model.

(3) The stress release model does provide a means to estimate the temporal variation of system stress. It works very well for the synthetic catalog produced by cellular automata model. It is an interesting problem if this model also works very well for real earthquake data. Independent observations on regional stress changes are necessary.

(4) How to understand accumulated stress release $S(t)$ in coupled stress release model is worthy of further endeavor.

Acknowledgements

This project is supported by Asia 2000 foundation of New Zealand and completed in Victoria University of Wellington in New Zealand. We thank Professors Li Ma, Zhengxiang Fu, Dr. Xiaogu Zheng and Dr. David Harte for valuable discussions and assistance. This project is also partly supported by NFSC (Project No. 59574203) and the China Seismological Bureau under the contract number 95-04-03-03-03.

Appendix
North China earthquake data set (1480–1996)

No.	Time	Lat.	Long.	Mag.	Reg.	No.	Time	Lat.	Long.	Mag.	Reg.
1	1484.01.29	40.4	116.1	6.75	E2	34	1704.09.28	34.9	106.8	6.00	W2
2	1487.08.10	34.3	108.9	6.25	W1	35	1709.10.14	37.4	105.3	7.50	W2
3	1501.01.19	34.8	110.1	7.00	W1	36	1718.06.29	35.0	105.2	7.50	W2
4	1502.10.17	35.7	115.3	6.50	E2	37	1720.07.12	40.4	115.5	6.75	E2
5	1536.10.22	39.6	116.8	6.00	E2	38	1730.09.30	40.0	116.2	6.50	E2
6	1548.09.13	38.0	121.0	7.00	E1	39	1739.01.03	38.8	106.5	8.00	W2
7	1556.01.23	34.5	109.7	8.00	W1	40	1815.10.23	34.8	111.2	6.75	W1
8	1561.07.25	37.5	106.2	7.25	W2	41	1820.08.03	34.1	113.9	6.00	E2
9	1568.04.25	39.0	119.0	6.00	E1	42	1829.11.19	36.6	118.5	6.00	E1
10	1568.05.15	34.4	109.0	6.75	W1	43	1830.06.12	36.4	114.2	7.50	E2
11	1573.01.10	34.4	104.1	6.75	W2	44	1831.09.28	32.8	116.8	6.25	E1
12	1587.04.10	35.2	113.8	6.00	E2	45	1852.05.26	37.5	105.2	6.00	W2
13	1597.10.06	38.5	120.0	7.00	E1	46	1861.07.19	39.7	121.7	6.00	E1
14	1604.10.25	34.2	105.0	6.00	W2	47	1879.07.01	33.2	104.7	8.00	W2
15	1614.10.23	37.2	112.5	6.50	W1	48	1882.12.02	38.1	115.5	6.00	E2
16	1618.05.20	37.0	111.9	6.50	W1	49	1885.01.14	34.5	105.7	6.00	W2
17	1618.11.16	39.8	114.5	6.50	E2	50	1888.06.13	38.5	119.0	7.50	E1
18	1622.03.18	35.5	116.0	6.00	E2	51	1888.11.02	37.1	104.2	6.25	W2
19	1622.10.25	36.5	106.3	7.00	W2	52	1920.12.06	36.7	104.9	8.50	W2
20	1624.02.10	32.4	119.5	6.00	E1	53	1922.09.29	39.2	120.5	6.50	E1
21	1624.04.17	39.8	118.8	6.25	E2	54	1937.08.01	35.4	115.1	7.00	E2
22	1624.07.04	35.4	105.9	6.00	W2	55	1945.09.23	39.5	119.0	6.25	E2
23	1626.06.28	39.4	114.2	7.00	E2	56	1966.03.22	37.5	115.1	7.20	E2
24	1627.02.15	37.5	105.5	6.00	W2	57	1967.03.27	38.5	116.5	6.30	E2

Appendix (Continued)

No.	Time	Lat.	Long.	Mag.	Reg.	No.	Time	Lat.	Long.	Mag.	Reg.
25	1634.01.15	34.1	105.3	6.00	W2	58	1969.07.18	38.2	119.4	7.40	E1
26	1642.06.30	35.1	111.1	6.00	W1	59	1975.02.04	40.7	122.8	7.30	E1
27	1654.07.21	34.3	105.5	8.00	W2	60	1976.04.06	40.2	121.1	6.20	W2
28	1658.02.03	39.4	115.7	6.00	E2	61	1976.07.28	39.4	118.0	7.82	E2
29	1665.04.16	39.9	116.6	6.50	E2	62	1976.09.23	39.9	106.4	6.20	W1
30	1668.07.25	35.3	118.6	8.60	E1	62	1979.08.25	41.2	108.1	6.00	W2
31	1679.09.02	40.0	117.0	8.00	E2	64	1989.10.18	40.0	113.7	6.00	W1
32	1683.11.22	38.7	112.7	7.00	W1	65	1996.05.03	40.8	109.6	6.50	W2
33	1695.05.18	36.0	111.5	8.00	W1						

REFERENCES

AKAIKE, H., *On entropy maximization principle*. In *Applications of Statistics* (ed. Krishnaiah, P. R.) (North Holland, Amsterdam 1977) pp. 27–41.

BAK, P., and TANG, C. (1989), *Earthquakes as a Self-organized Critical Phenomena*, J. Geophys. Res. *94*, 15635–15637.

ITO, K., and MATSUZAKI, M. (1990), *Earthquakes as Self-organized Critical Phenomena*, J. Geophys. Res. *95*, 6853–6860.

KEILIS-BOROK, V. I., and KNOPOFF, L. (1980), *Bursts of Aftershock of Strong Earthquakes*, Nature *283*, 5744259–5744263.

KNOPOFF, L. (1971), *A Stochastic Model for Occurrence of Main-sequence Earthquakes*, Rev. Geophys. Space Phys. *9*, 175–188.

LIU, G., SHI, Y., and MA, L. (1995), *Seismicity and Cellular Automata Model*, Bull. Northwest Seismol. *17*, 20–25 (in Chinese).

MAIN, I., (1996), *Statistical Physics, Seismogenesis, and Seismic Hazard*, Rev. Geophys. *34*, 433–462.

SAKAMOTO, Y., *Categorical Data Analysis by AIC* (KTK Scientific Publishers, Tokyo 1991).

SHI, Y., GENG, L., and ZHANG, G. (1994), *Non-linear dynamic modeling of earthquake prediction in Nonlinear Dynamics and Predictability of Geophysical Phenomena*. In *Geophysical Monograph 83 IUGG volume 18* (eds. Newman, W. I., Gabrielov, A., and Turcotte, D. L.) pp. 81–89.

VERE-JONES, D. (1976), *A Branching Model for Crack Propagation*, Pure appl. geophys. *114*, 711–726.

VERE-JONES, D. (1978), *Earthquake Prediction—A Statistician View*, J. Phys. Earth. *26*, 129–146.

VERE-JONES, D. (1988), *On the Variance Properties of Stress Release Models*, J. Statist. *30A*, 123–135.

VERE-JONES, D., and DENG, Y. L. (1988), *A Point Process Analysis of Historical Earthquakes from North China*, Earthquake Res. China *4*, 8–19.

ZHANG, G., GENG, L., and SHI, Y. (1995), *Analysis of Grouped Continental Earthquake Model and their Relation to Precursors*, Acta Seismologia Sinica *17*, 1–10.

ZHENG, X., and VERE-JONES, D. (1991), *Application of Stress Release Models to Historical Earthquakes from North China*, Pure appl. geophys. *135*, 559–576.

ZHENG, X., and VERE-JONES, D. (1994), *Further Applications of Stochastic Stress Release Model to Historical Earthquake Data*, Tectonophysics *229*, 101–121.

(Received June 20, 1998, revised/accepted December 17, 1998)

Pure appl. geophys. 155 (1999) 669–687
0033–4553/99/040669–19 $ 1.50 + 0.20/0

┌Pure and Applied Geophysics

Recognition of a Locked State in Plate Subduction from Microearthquake Seismicity

SHOZO MATSUMURA[1] and NAOYUKI KATO[2]

Abstract—A tectonic state of a locked subduction is considered to be a possible source of a future interplate earthquake. Discriminating an actually locked state to verify its extent is therefore essential in constructing an accurate prospect against the forthcoming earthquake. Micorearthquake seismicity is an effective tool for such an analysis because it is considered to be a faithful indicator of the stress state, and is expected to exhibit a characteristic pattern in the area where the locked state in the subduction appears with a certain stress concentration. Focusing on the microearthquake seismicity around the Tokai district in central Japan, where a large interplate earthquake is feared to occur, we tried to identify such an area of locked subduction on the Philippine Sea plate, possibly related to the future earthquake. We investigated the microearthquake seismicity from various perspectives. First, the hypocenter distribution was analyzed to identify the extent of the locked area. The characteristic profile of the distribution was presumed to represent a stress concentrated area induced from the mechanical contact between both plates. The second approach is to interpret stress patterns reflected in focal mechanisms. The locked state was recognized and verified by a comparison of the P-axis distribution pattern with that expected from a model imaging a partially locked subduction. The third approach is to monitor the temporal change of the seismic wave spectrum. Analyzing predominant frequencies of P and S waves and monitoring their changes for a period of 10 years, we found a trend of gradual increase common to both waves. This means an increase of stress drop in microfracturings, and in its turn implies accumulation of stress around the focus area. The rate of the stress change converted from the frequency change was compared with the result derived from a numerical simulation. The simulation, performed on the basis of a constitutive friction law for a stick sliding on the plate interface, computed a changing rate of the maximum shear stress around the locked zone and showed its spatial variation along the subduction axis. Thus the simulated result indicated a certain compatibility with the observed one. Although ambiguities and uncertainties still exist in the study, all the results derived here seem to indicate an identical conclusion that the plate subduction is actually locked in this region at present.

Key words: Locked subduction, microearthquake seismicity, frictional sliding.

1. Introduction

The key factor which essentially reflects seismogenic potential is not strain but stress. Therefore, in order to create some prospects regarding the possibility of future earthquake occurrences, it is required to obtain information on the present

[1] National Research Institute for Earth Science and Disaster Prevention, Tennodai 3-1, Tsukuba-shi, Ibaraki-ken 305-0006, Japan.

[2] Geological Survey of Japan, Higashi 1-1-3, Tsukuba-shi, Ibaraki-ken 305-8567, Japan.

state of stress and its change. However, it is generally difficult to measure stress itself, especially nearly impossible in a deep underground site. Under such circumstances, the only way we can gain access is to observe a microearthquake, although it provides only indirect information regarding the stress state. Microearthquake seismicity, generating steady and characteristic patterns at one locality, can be regarded as a faithful indicator of stress. However, the relationship between seismicity and stress is not always straightforward. For example, in some cases a seismic cluster or a swarm activity reflect stress concentration at the place, while in other cases an aseismic state or a quiescence of the activity mean the same condition. Furthermore, an active swarm does not always indicate a stress accumulating state corresponding to the potential of a large earthquake. In order to reach a significant conclusion from microearthquake observation, it is necessary to examine it from various sides, and to interpret its meaning in the tectonic condition based on a physical model.

In this study, focusing on the microearthquake activity in the Tokai district, we intend to reveal the stress state around this region. The Tokai district, the central part of the Japanese islands (Fig. 1), was previously designated as a seismogenic

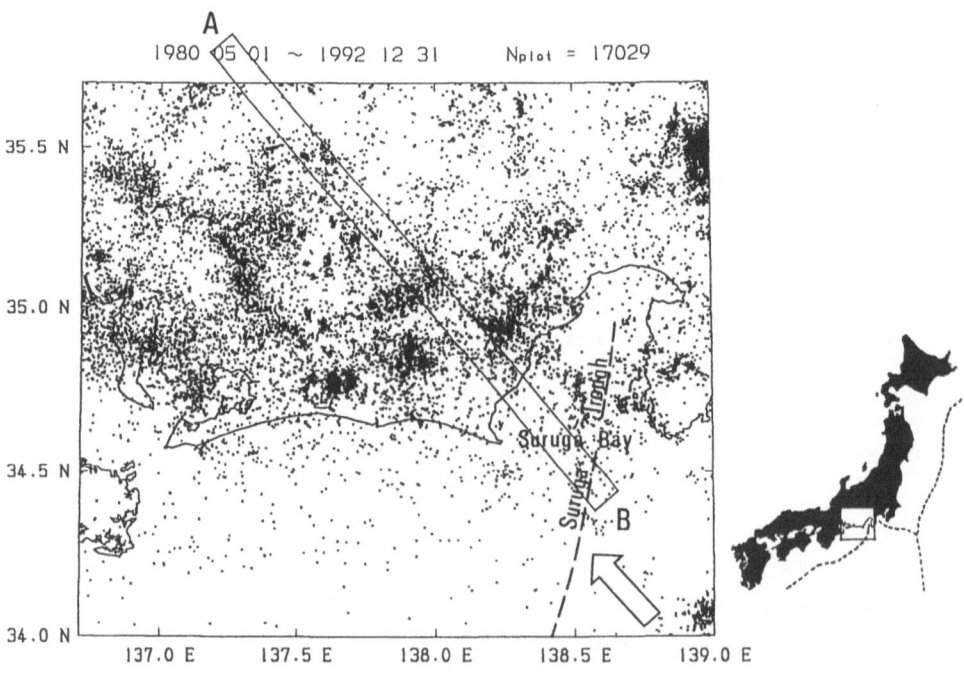

Figure 1

Microearthquake seismicity in the Tokai district, the central region of the Japanese islands. The arrow indicates the convergence direction of the Philippine Sea plate. The *A−B* section parallel to the arrow is sampled to view the structure of the plate subducting from the Suruga trough.

zone of a forthcoming major earthquake by seismologists in the early 1970s. MOGI (1970), based on strain accumulation detected in triangulation survey, pointed out that the Tokai district might be a candidate zone in which a large earthquake was impending. ISHIBASHI (1976, 1981) specified a focal zone to be a thrust fault at the Suruga trough which would be formed by the subduction of the Philippine Sea plate. He examined old documents which described historical earthquakes, tsunami disasters, and geomorphological evidence, and noticed the difference between both fault zones of the previous Tokai earthquakes, the 1854 Ansei Tokai earthquake (M 8.4), and the subsequent one, the 1944 Tonankai earthquake (M 7.9). The former earthquake ruptured the zone ranging over 200 km along the Tokai and Nankai troughs, and intruded deep into Suruga Bay. On the other hand, the latter quake ruptured only the southwestern part of the 1854 failure zone, consequently the bay area was left unbroken at that time. Thus he concluded that the potential for an M 8 earthquake beneath the Suruga Bay still existed. This was the first warning concerning the so-called "Tokai earthquake." However, that which was presented by Ishibashi did not immediately project the existence of the plate subduction, nor the stress accumulation, consequently it left a doubt behind it; is there really the potential for a large earthquake?

Since Ishibashi's warning more than 20 years have passed with no manifestation of the anticipated earthquake. During this period, observation techniques have made rapid progress, and have produced more direct evidence of subduction in the Suruga Bay. For example, the geometrical configuration of the Philippine Sea plate was inferred from the hypocenter distributions of microearthquakes (YAMAZAKI and OOIDA, 1985; ISHIDA, 1992; NOGUCHI, 1996), which confirmed that the subduction actually occurs at the Suruga trough. The next problem is whether or not an interplate earthquake is near failure. In order to solve this problem, it is necessary to produce evidence of a locked state and stress accumulation accompanying the subduction. We try to identify the zone under such a tectonic state by analyzing microearthquake data from the following perspectives. First, the pattern of hypocenter distribution is used to bound the zone of stress concentration. Next, the focal mechanism pattern is analyzed to interpret the mechanical condition, by attributing it to the locked subduction while introducing a partial locking model. Finally, the temporal changes of the seismic wave spectrum are examined to ascertain the stress accumulation due to the locked subduction, in which, a numerical simulation is introduced to explain the observed phenomena.

2. Recognition of Locked Subduction

In this section both the seismicity and the focal mechanism patterns are viewed to recognize the state of locked subduction, and finally to identify the zone of locking. The seismic activity around the Tokai district can be definitively discriminated into two groups; one inside the overriding plate, and the other inside the subducted slab. We focus mainly on the latter because its seismicity appears more uniform than the former in both space and time, which implies simplicity of the mechanical condition that controls the stress state inside the slab.

2.1 Locked Subduction Recognized from Seismicity Pattern

Figure 1 exhibits the seismicity pattern of microearthquakes around the Tokai district, observed by the network of our institute, the National Research Institute for Earth Science and Disaster Prevention (NIED network, e.g., HAMADA et al., 1985). The threshold magnitude that assures uniform detection without exception is about M 2.0 for this entire area, and M 1.5 or less for the inland area. The arrow indicates the motion direction of the approaching Philippine Sea plate, which is subducting beneath the overriding Eurasian plate from the trough axis. Along this direction a cross section $A-B$ is taken, a sectional view of which is shown in Figure 2(a). In this figure the geometry of the subducting plate can easily be identified as an activity down dipping from the right side (trough axis) can be distinguished from another one that distributes the hypocenters within a shallow layer (inside the crust). The latter activity seems to traverse the former one along the zone marked in the figure, where both are separated by a narrow aseismic zone between ('q' in Fig. 2(a)). Surrounding this aseismic zone, we find regions of relatively high seismicity, especially in the dipping side. It is commonly accepted that such a dipping structure of the seismicity reflects the subduction of the Philippine Sea plate, and most of the dipping activity belongs to the subducted slab (YAMAZAKI and OOIDA, 1985; ISHIDA, 1992; NOGUCHI, 1996). Thereafter, the aseismic zone, as first noted by ISHIDA (1995), is thought to locate the position through which the plate boundary passes. Furthermore, those activities surrounding the aseismic zone may be interpreted to reflect the stress concentration due to the locked subduction. As a result, the range marked with the brackets in the figure is regarded to be an approximate locked zone at present. It is problematic as to why the interface zone of locking appears aseismic in spite of the fact that it should be most stressed. Although we have no persuasive answer here, evidence that the zone is not free from shear stress, which is possibly attributed to the locked subduction, is presented in the next section by analyzing the focal mechanism pattern.

2.2 Locked Subduction Recognized from Focal Mechanisms

Figure 2(b) presents a P-axis profile projected on the same section as Figure 2(a). In order to interpret both profiles of Figures 2(a) and (b), we introduce a partial locking model as shown in Figure 3, in which we assume that the relative plate motion accompanying the subduction is locked along the thick line, and unlocked along the rest. The stress concentration due to such locked subduction

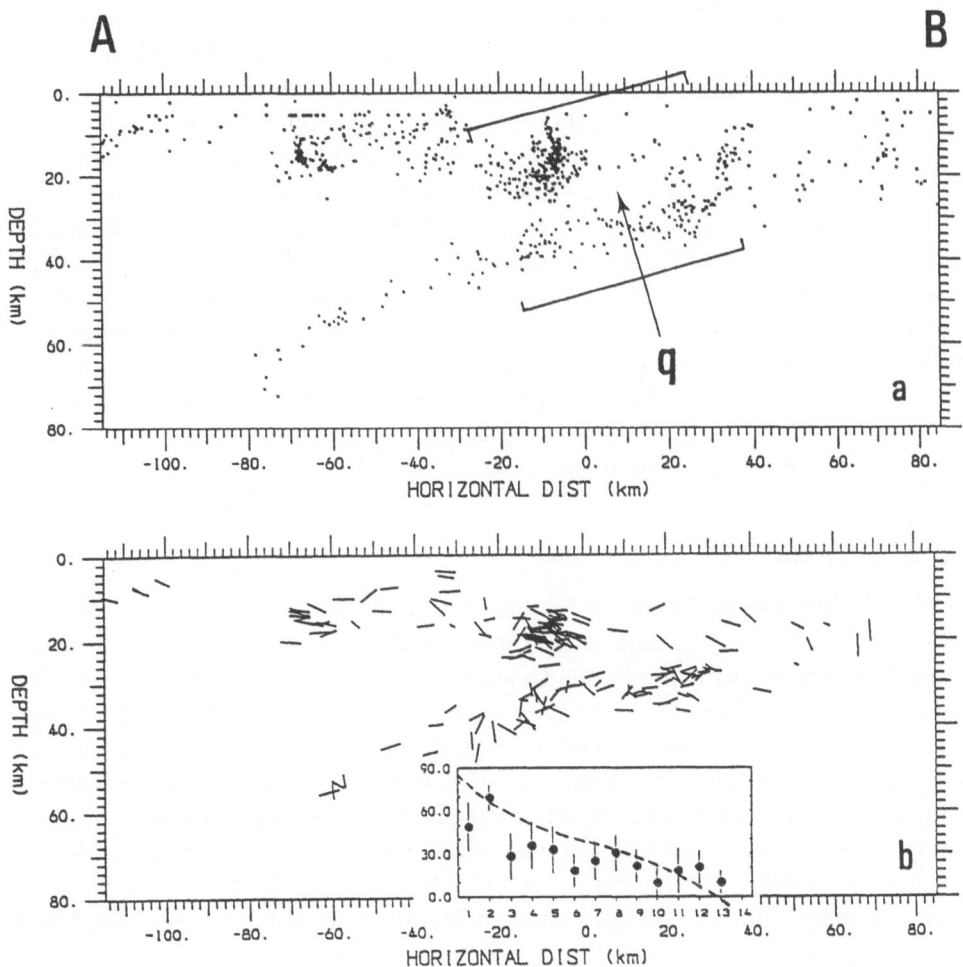

Figure 2

Vertical cross sections (the strip $A-B$ in Fig. 1) of the hypocenter distribution (top), and the P-axis profile projected on this section (bottom). The aseismic zone explained in the text is indicated by 'q'. The couple of thin brackets in the top figure mark the zone where relative plate motion is thought to be locked at present. The enclosure in the bottom figure presents the dip-angle distribution of the P axis indicated by dots with error bars, and the dip-angle variation of the pressure direction in the model of Figure 3 indicated by a broken line.

can be analytically estimated. Those stress patterns displayed in Figure 3 were calculated by an elasticity theory for a uniform half-space given by OKADA (1992) for the circumference area surrounding the locked zone, which corresponded to the active seismicity area in Figure 2(a). Comparing Figure 2 with Figure 3, we equated the zone marked with the brackets in Figure 2 to the thick line part in Figure 3, based on the following standpoint. In the model of Figure 3, we recognize the spatial change of the stress pattern along the locking as the direction of the maximum compression stress in the slab rotates clockwise with the down-dip direction of the plate boundary, and accordingly it is approximately horizontal in a shallow part and is vertical in a deep part. We find a similar variation in the stress pattern obtained from the data. In Figure 2(b), P axes in the subducted slab seem to dip almost horizontally in the right side, changing to vertical in the left. Both variations are compared in a small graph enclosed at the bottom of Figure 2(b), where a broken line indicates the dip-angle variation of the pressure direction predicted by the model, and dots with bars plot averages and uncertainties of the P-axis dip angle measured in this projection. If the comparison is performed in a 3-D space, we find both patterns not entirely to coincide with each other. However, the coincidence of the stress pattern projected in the subduction direction confirms our understanding that the dominant mechanism acting on the focus area would be precisely what was explained in the model as partially locked subduction.

2.3 Identification of the Locked Zone

We divided the study area into sixteen sections parallel to the relative plate motion to examine the seismic activity in the same manner. The above-introduced procedure of recognizing the locked subduction was applied to each of the sections covering the entire Tokai district. After determining the range of locking for each section, and linking them, we could delineate the region of the locked zone as enclosed with the thick line in Figure 4, where only earthquakes within the slab are plotted. This procedure, as a result, corresponds to enclosing the active seismic cluster 'a' in the figure. The depth range of the zone extends from 10 km to 30 km, which seems consistent with our knowledge that both shallower and deeper parts than this range must be unlocked on the plate interface (e.g., SHIMAMOTO, 1985).

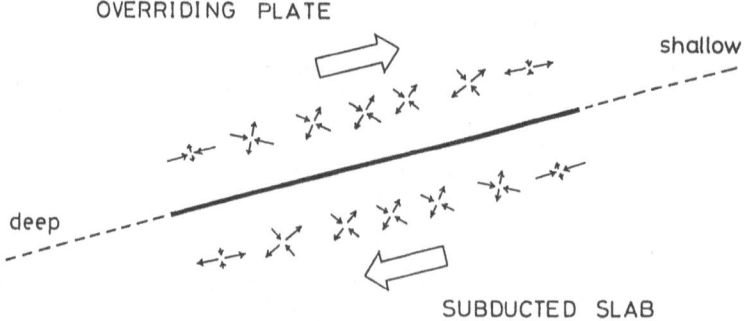

Figure 3

A model imaging a partially locked subduction. The plate boundary is regarded as being locked on the thick part, and unlocked on the thin part. A stress concentration due to the locked plate motion is expected to distribute a characteristic stress pattern surrounding the locked zone, which is drawn by using an analytical estimation proposed by OKADA (1992).

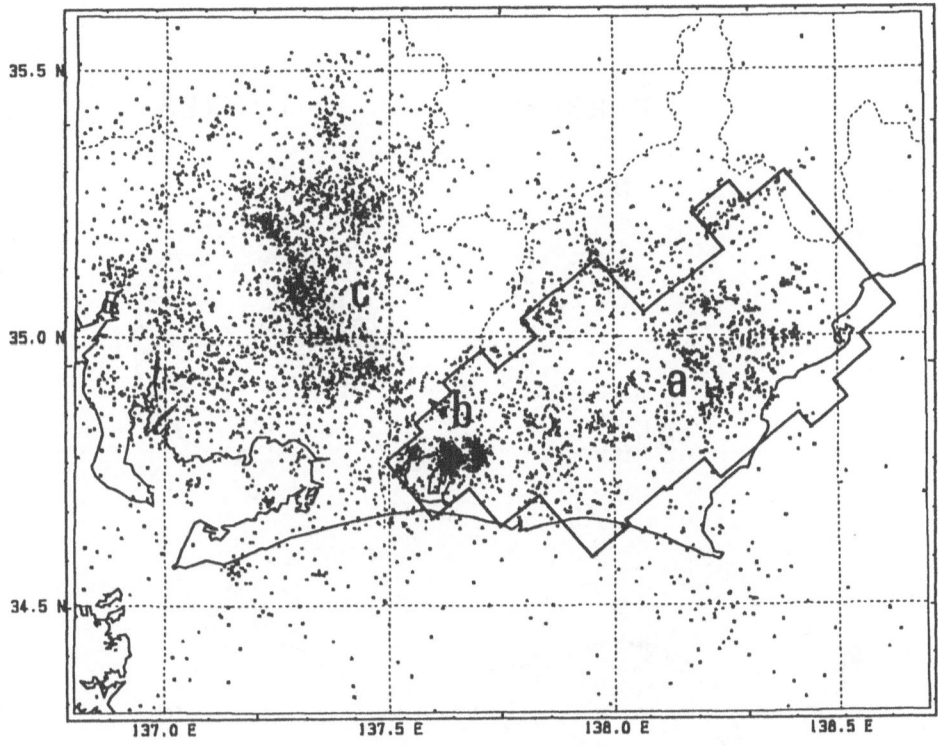

Figure 4
The locked zone identified in this study is indicated by the thick enclosure. The hypocenters plotted are those belonging to the subducted slab, from which three significant clusters 'a', 'b', and 'c' are differentiated and marked.

Besides cluster 'a', two other clusters 'b' and 'c' are marked in the figure. The cluster 'b' indicates the most active seismicity in this area. This was located in the southwestern edge of the specified zone, and its high activity might be attributed to stress concentration due to edge effects in locking. In contrast, another cluster 'c' was not located within the specified zone. Figure 5 shows the focal mechanisms compiled for each cluster, from which we notice that both 'a' and 'b' indicate a common type of strike slip, while 'c' indicates roughly normal faulting. Such a difference in the typical mechanism pattern led us to surmise that shear stresses transmitted across the plate interface act on both the regions of 'a' and 'b', although not on the region of 'c'. For this reason we excluded 'c' from the locked zone. Conversely, the typical pattern of strike-slip for 'a' and 'b' also significantly differs from that expected from the model, the thrust type as shown in Figure 3. In order to explain the stress field estimated from seismological observations, another

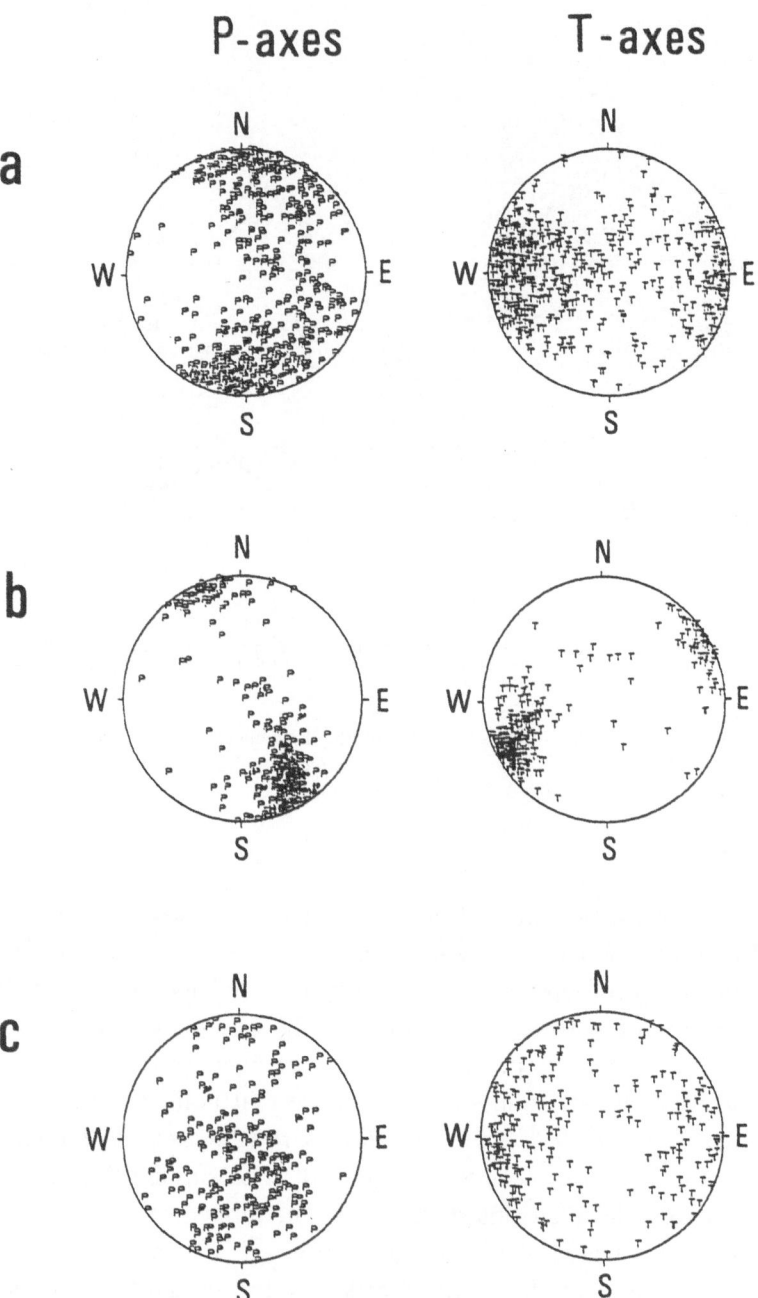

Figure 5
The focal mechanisms compiled for each cluster 'a', 'b', and 'c' in Figure 4.

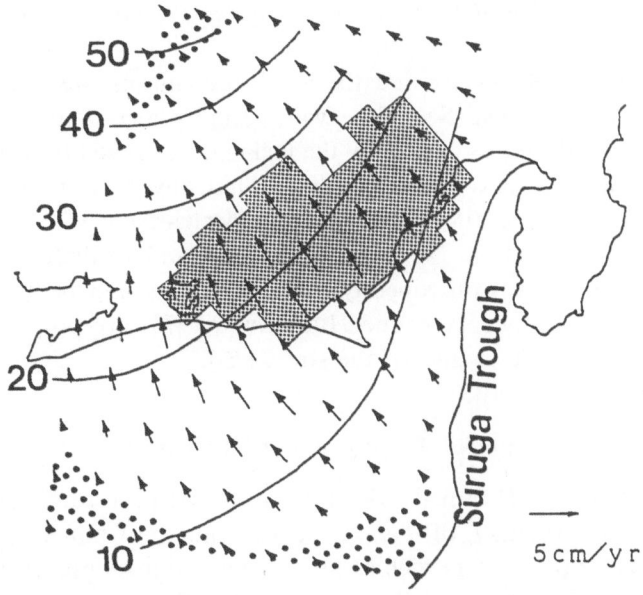

Figure 6

The locked zone we assumed is placed on the YOSHIOKA's result (1993) which presents a back-slip distribution on the plate interface estimated from the geodetic survey data. The curves indicate the depth contours of the upper surface of the Philippine Sea plate given by ISHIDA (1992).

stress source that which can affect the stresses in the slab must be introduced. We presume that the most likely candidate of such stress source is the "lateral stretching force" which should be generated due to the deformation of the plate with subduction, which was proposed by UKAWA (1982) to interpret the characteristic seismicity inside the slab. Details and further explanations should be referred to a previous paper (MATSUMURA, 1997).

The thusly assumed locked zone is verified by comparing it with that obtained from another approach. In Figure 6, the result of this study is designated as the shaded zone. The arrows are the back-slip distribution given by YOSHIOKA et al. (1993) who calculated it by using an inversion technique and data from level surveys. The back slip is defined as an imaginary relative motion between adjoining plates, and it corresponds to motion of the overriding plate dragged by the subduction. An abundance of the back slip is therefore strongly related to the locked subduction. In the figure, the dominant back-slip distribution appears to nearly cover the locked zone estimated in the present study. Such coincidence having been formulated from quite different approaches, makes the assumption more convincing that the locked subduction actually exists in and around the specified zone.

3. Recognition of Stress Accumulation in the Locked Zone

By interpreting the seismicity and the focal mechanisms, we inferred that the subduction of the Philippine Sea plate was partially locked around the Tokai district. In this case, the stress caused by the locking is expected to increase year by year under the condition of a steady plate motion. Conversely speaking, if we can detect such a stress change, that should become persuasive evidence to emphasize the validity of the assumption. However, we have no way of directly measuring *in situ* stresses underground. Consequently, using a relationship between stress and seismic wave spectrum (e.g., SATO and HIRASAWA, 1973), we propose an alternative means to monitor the change of the stress field.

3.1 Changes of the Predominant Frequency of Seismic Waves

In the operation of NIED network observation, we have routinely determined the predominant frequencies f_p of P and S waves of microearthquakes as well as the standard earthquake parameters such as hypocenter and magnitude since 1987. The procedure of determining f_p is as follows (refer to Fig. 7): (1) picking a data window of the seismic signal with a duration of 1.6 seconds (128 samples) just after the arrivals of P and S waves, (2) deriving a power spectrum through FFT, (3) packing the spectrum into 14 columnar bands on a frequency axis, (4) fitting a cubic curve and calculating the frequency which indicates the extreme value of the curve. Such a procedure is performed semi-automatically daily, and millions of data have been stored during the last decade. Based on these data, we examine the temporal changes of the predominant frequency for those earthquakes occurring inside the assumed locked zone. As shown in Figure 8, the examined area is slightly extended beyond the originally specified zone along the subduction direction, while the southwestern edge is excluded in order to omit the anomalous cluster '*b*' in Figure 4. Only the earthquakes inside the slab were sampled again, and the magnitude range was limited between 1.0 and 2.0 in order to weaken the influence of magnitude dependence. Furthermore, 17 stations in and around the focus zone were selected in the analysis, for which every epicentral distance ranged less than 100 km. The values of f_p selected in this manner still contain dependants on magnitude, distance, and site effect. We evaluated them by using an equation:

$$\log(f_p) - \log(f_{p_0}) = -a \cdot (M - M_0) - b \cdot \log(d_j/d_0) + c_j,$$

where M is the magnitude, d_j is the hypocentral distance, c_j a constant respecting a station correction for the j-th station, and those values with the subscript 0 are reference bases for evaluation the dependants.

Coefficients a, b, and c_j are determined for P and S waves by applying the least-squares method once to all the data. By using the coefficients thus obtained, each f_p was reduced at $M_0 = 1.5$ and $d_0 = 40$ km. Furthermore, taking an average value of f_p for at least 5 stations among 17, we could have approximately five hundred data, which were plotted on a time axis in Figure 9. Although the data show considerable scatter, we find a slight tendency for f_p to gradually increase with

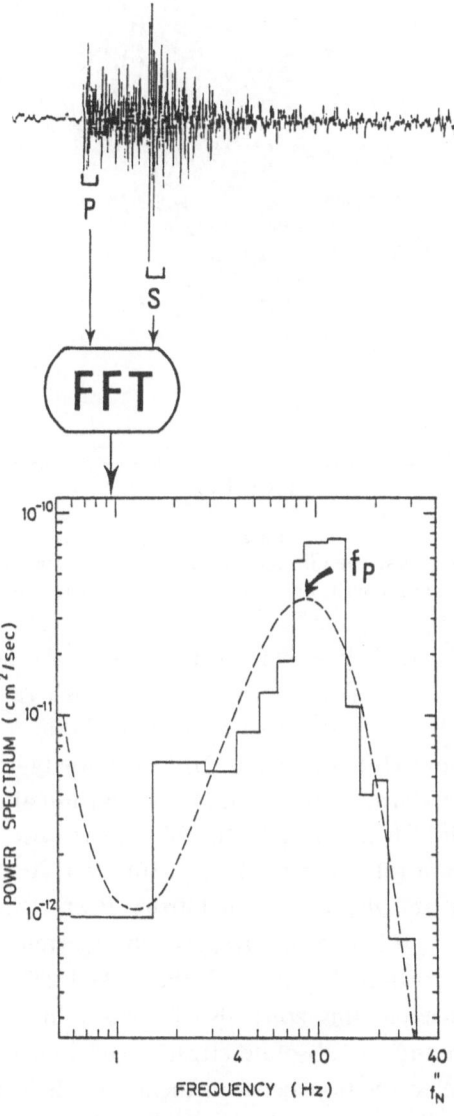

Figure 7

Procedure to determine the predominant frequency f_p of the seismic waves. The initial portion for 1.6 s (128 samples) just after the arrivals of P and S phases are picked and transformed into a power spectrum through FFT. The spectrum, after being packed into 14 columnar bands on a frequency axis, is fitted to a cubic curve function. The predominant frequency f_p is determined as a value which indicates the extreme point of the curve.

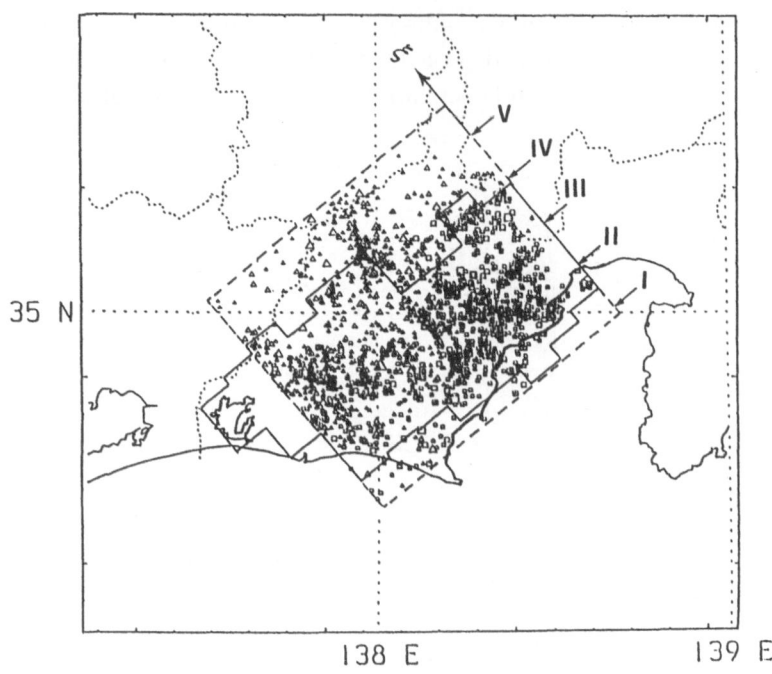

Figure 8
Microearthquakes distribution used for monitoring the stress change. Only earthquakes occurring inside the subducted slab are sampled. Points I to V correspond to those in Figure 10.

time for both cases of P and S waves. The increase rates of f_p are $+(0.34 \pm 0.26)\%/$ year for P, and $+(0.47 \pm 0.19)\%/$year for S, where each rate and its confidence interval were estimated by using the least-squares method.

According to SATO and HIRASAWA (1973), f_p of velocity spectra of body waves can be related to the stress drop, σ by $\sigma \sim f_{p^3}$ for an earthquake with a fixed seismic moment, i.e., magnitude. Then a small rate of f_p increase can be approximately converted into an increase rate of σ by $\delta\sigma/\sigma \sim 3(\delta f_p/f_p)$. As a result, the increase rates of the stress drop $(d\sigma/dt)/\sigma$ estimated from P and S waves are $+(1.0 \pm 0.78)\%/$year, and $+(1.4 \pm 0.57)\%/$year, respectively. Current results indicate that the stress drop of microearthquakes beneath the locked zone increases with time, suggesting tectonic stresses in this zone also increase. Although the relationship between the stress drop and the absolute stress is still indefinite, we have reports which suggest a positive correlation between them. One is a result derived from a laboratory experiment. KUSUNOSE et al. (1980), observing acoustic emissions, revealed that emissions with higher frequencies predominated under higher axial stresses. Another is a report by ISHIDA and OHTAKE (1984) who revealed that the foreshocks radiated waves of higher frequencies than the aftershocks for the case of a small event observed around the Tokai district. These reports imply that earthquakes of a high stress drop would be caused under a high level of absolute tectonic stresses. Then, subsequently surmising that the increase of σ is roughly proportional to the change of the absolute stress level due to the locking, an approximate 1% increase of σ per year at the present stage seems reasonable under the assumption that the stress accumulation commenced around the time of the last release, due to the 1854 Tokai earthquake.

We divided the study area into 5 segments I to V along the subduction as shown in Figure 8, and determined the increase rate of predominant frequency or stress drop for microearthquakes in each segment. In the next section, the result will be compared with those predicted by a numerical simulation for spatiotemporal variation of stresses.

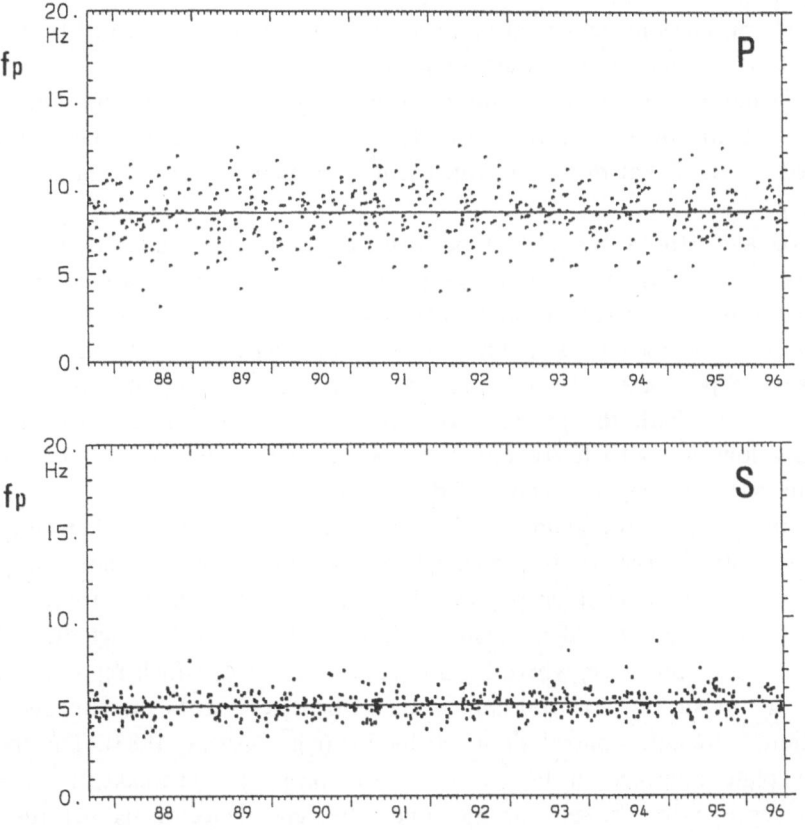

Figure 9
Temporal changes of the predominant frequency f_p for the P wave (top), and for the S wave (bottom).

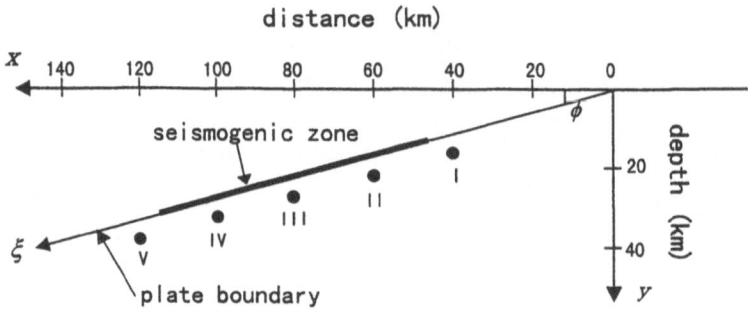

Figure 10

A 2-D model of the plate boundary in the subduction zone in a half-space, introduced to simulate a locked subduction. The plate boundary is regarded as being locked on the thick part, and unlocked on the thin part. Points I to V, placed 5 km beneath the plate boundary, are referred to in Figures 8, 11 and 12 to evaluate and compare the stress change rate.

3.2 Numerical Simulation of the Locked Subduction

Recent progress in numerical models based on laboratory-derived friction laws has made it possible to construct realistic models of entire processes of seismic cycles, including both seismic and aseismic sliding (e.g., TSE and RICE, 1986; STUART, 1988). In this section, we demonstrate that the seismicity variation observed in the Tokai district is consistent with that expected from a numerical modeling.

KATO and HIRASAWA (1996) performed a numerical simulation of seismic cycles of large interplate earthquakes in the Tokai district by assuming that the frictional force acting on a plate interface obeys a rate- and state-dependent friction law developed by DIETERICH (1979) and RUINA (1983) on the basis of laboratory experiments. To compare the modeling results of KATO and HI-RASAWA (1996) with the present observational data, we review the simulation method below. Refer to KATO and HIRASAWA (1997, 1998) in addition to KATO and HIRASAWA (1996) for details of the simulation method.

Figure 10 shows the geometrical scheme of the present model, where a 2-D uniform elastic half-space is presumed. Stable sliding of the sliding rate of 4 cm/year is assumed on a deep part of the plate interface. The frictional force obeying a rate- and state-dependent friction law is assumed to act on a shallow part of the plate interface, where friction parameters $a-b$, which represent the rate dependence of steady-state friction, and L, which represents the slip dependence of frictional strength, control sliding behavior (e.g., RUINA, 1983). The thick-line part of plate interface in Figure 10 is the part of rate-weakening ($a-b < 0$) friction, representing the seismogenic zone. The dependence on temperature of the friction parameters has been examined in laboratories (e.g., BLANPIED et al., 1995).

Furthermore, the simulation results, such as a recurrence interval of large earthquakes and a seismic slip amount are dependent on friction parameters (e.g., KATO and HIRASAWA, 1997). Taking into account the above facts, we selected appropriate values of friction parameters. The simulation result indicates that large earthquakes repeatedly occur at a constant time interval of 119 years on a shallow part of plate interface, being consistent with that for past large earthquakes along the Nankai trough (e.g., ISHIBASHI, 1981).

According to the simulation result, aseismic sliding propagates on the plate interface before the occurrence of a large interplate earthquake, effecting significant variation of stresses in a subducting oceanic plate as discussed by KATO and HIRASAWA (1998). This stress variation may control stress drop and, accordingly, predominant frequencies of small earthquakes. Figure 11 shows simulated histories of stress rate $(d\tau/dt)/\tau$ at points I to V (Fig. 10) in a subducting oceanic plate from 50 years to 1.95 years before a large earthquake, where τ is the maximum shear stress and t is time. In calculating the maximum shear stress τ, we consider only the stresses due to slip on the plate interface, neglecting the effects of other sources of tectonic stresses such as bending of the oceanic plate. This approximation is considered to be adequate because we are concerned with the stress rate, and because the stress rate due to slip predominates over those due to other sources. The depths of points I to V are 5km deeper than the plate interface (Fig. 10), and roughly correspond to

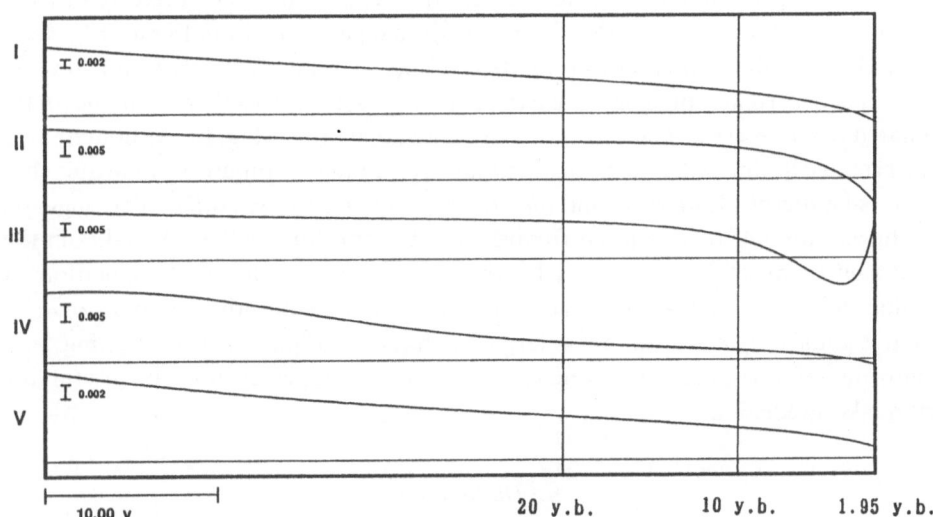

Figure 11

Rate changes of the maximum shear stress τ simulated on the basis of the friction constitution laws at each point I to V in Figure 10. Horizontal lines indicate zero level, and scales are given for each graph on the far left-hand side in the unit of (year)$^{-1}$. The time range is from 50 y.b. to 1.95 y.b. (years before the breakage).

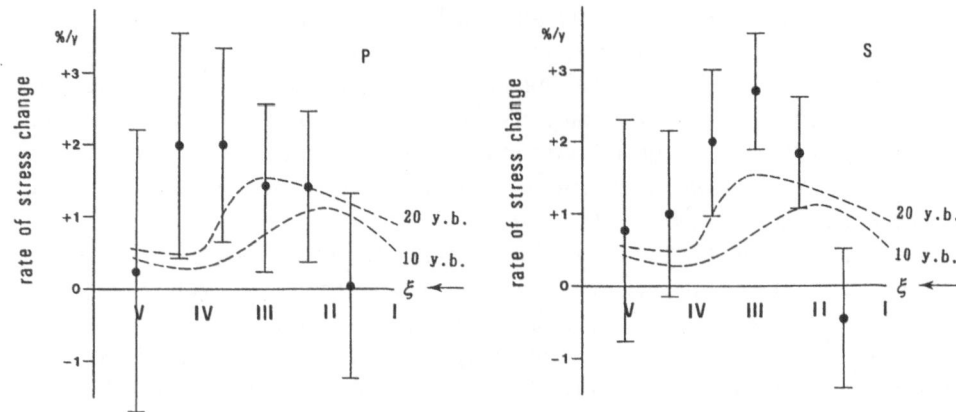

Figure 12

Spatial distribution of the temporal rate change of the stress drop $(d\sigma/dt)/\sigma$ evaluated form the seismic wave spectrum. The broken lines indicate the corresponding values derived from the simulation, that is the rate of the maximum shear stress change $(d\tau/dt)/\tau$ read from Figure 11 at 10 y.b., and 20 y.b. (years before the breakage).

a seismically active zone (Fig. 8). Although the exact relationship between the magnitudes of tectonic stresses and the stress drop of earthquakes is not known, the maximum shear stress seems to be related to the stress drop (e.g., ISHIDA and OHTAKE, 1984).

We compare the rate of maximum shear-stress change $(d\tau/dt)/\tau$ predicted by the simulation with the observed rate of stress drop change $(d\sigma/dt)/\sigma$ in Figure 12, where the simulated values of $(d\tau/dt)/\tau$ at points I to V at the times of 20 years and 10 years before the occurrence of a large earthquake are plotted together with $(d\sigma/dt)/\sigma$ estimated from observed P and S waves in the regions I to V (Fig. 8). Although there is uncertainty in time interval to the next large earthquake, Figure 12 indicates that the spatial patterns of $(d\tau/dt)/\tau$ and $(d\sigma/dt)/\sigma$ are similar to each other. This suggests that the variation of predominant frequencies observed for small earthquakes in the subducting oceanic plate is caused by the stress variation due to propagation of aseismic sliding on the plate interface. The simulation result indicates that $(d\tau/dt)/\tau$ should significantly vary before a large earthquake (Figure 11), suggesting that monitoring of predominant frequencies of microearthquakes may be useful for earthquake prediction.

4. Discussion

In this study, we attempt to recognize the state of stress loading due to the locked subduction based on the microearthquake observation data. By examining the data from various points of view comprising the seismicity pattern, focal mechanism distribution, and the wave spectrum change, we have delineated the zone where the locked state is thought to occur and progress toward a future large interplate earthquake.

However, it must be noted that several problems remain behind the above conclusion, containing a certain ambiguity in each step of the analysis. One of the problems is the uncertainty involved in the data. The microearthquake data are generally quite scattered in the seismicity, especially in the focal mechanism, and also in the spectrum. Therefore it is meaningless to treat the data individually, and it is necessary to view it as a mass in order to extract a general trend hidden in the data. This means that a statistical treatment is required in any step of the analysis. However, the amount of data is apt to be insufficient to sustain a statistical significance.

Secondly, the model introduced in this study is defective in simulating the actual tectonic state. For example, comparison of the stress pattern between the observed data in Figure 2 and the model in Figure 3 is inaccurate when it is viewed in the 3-D space. As explained in Section 2.3, the observed focal mechanisms indicate their principal stress axes existing approximately within a horizontal plane, while those of the model exist in a vertical plane. Such a condition is common to monitoring of the temporal stress changes, where $(d\sigma/dt)/\sigma$ measured in a horizontal plane was compared with a model estimation of $(d\tau/dt)/\tau$ in a vertical plane. Such a contradiction compelled us to introduce an advanced model, suggesting the adoption of another stress source which acts on the whole slab, for example, the "lateral stretching force" (UKAWA, 1982). This is one of the possible answers, although it is not always a unique answer.

As a result, it must be noted that the assumption derived here is not always a unique solution, but only one of the possibilities. However, at the same time it must be stressed that this assumption suggesting the existence of the locked subduction seems most probable to explain all the features extracted from the observation.

Finally, we again discuss the comparison of our result with that obtained from the back-slip analysis. In Figure 6, the back-slip distribution seems to cover a somewhat broader area than our proposing zone, especially in the oceanic region. A similar result is given by using the recent GPS survey (e.g., GSI, 1998), which indicates that the largest back slip is located in the oceanic region, slightly deviating from our result. It is considered that there exists a significant discrepancy between both which should be attributed to the difference in the physical parameter to be analyzed. The back-slip analysis would give a displacement distribution on the overriding plate dragged with the subduction. It is then considered that on the overriding plate, not only the part just above the locked portion but the part extended toward the trough axis may be dragged together because the latter part is mechanically free from the boundary restriction in the trough side. Such a condition may result in extension of the back-slip distribution toward the trough axis, compared with the actual locked zone. It should be thus noted that the back-slip distribution may not always represent the seismogenic focal zone as pertains to the locked subduction.

5. Conclusion

The microearthquake data were analyzed to delineate the locked state in plate subduction beneath the Tokai district. The results obtained are as follows.

1) On the basis of the stress model imaging a partially locked subduction zone, the patterns of the seismicity and the focal mechanisms were investigated to interpret the mechanical condition around the locked zone. As a result, the zone assumed to be locked could be delineated at the present stage.

2) The temporal change of the predominant frequencies of P and S waves was examined, and a gradually increasing trend common to both the waves was found. The increasing rate of the frequency was converted into an increase rate of stress drop, which was about 1% per annum.

3) A numerical experiment to simulate locked subduction was performed to evaluate the temporal changes of the maximum shear stress around the locked zone. The simulated results were compared with the observed ones estimated from the seismic wave spectra. Having obtained a certain consistency between both results, we conclude that the zone proposed here must actually be in the locked state, and assume that the environmental stress in and around this zone increases year after year with a trend which possibly started after the 1854 earthquake.

As discussed in the preceding section, the results obtained in this study contain inconsistencies or uncertainties. Notwithstanding, they present important information with which to plan investigations to make accurate estimates for a future event. Provided a concrete model based on observations and experiment, we will think out a definite strategy for detecting precursory phenomena. For example, the simulation in Figure 11 engenders an expectation that the stress state will be drastically changed at the time approaching the failure. Such a prospect suggests to us what our focus should be, and how we will be able to grasp a meaningful signal for encountering the big one, the forthcoming Tokai earthquake.

Acknowledgements

We express our gratitude to Professors Wyss, Ito, and Shimazaki, who coordinated a Symposium on Seismicity Patterns, and offered us an opportunity to bring this study to a conclusion. We also express our appreciation to Dr. Kusunose for his suggestive discussion on experimental results, and the reviewers for their support revising our manuscripts.

REFERENCES

BLANPIED, M. L., LOCKNER, D. A., and BYERLEE, J. D. (1995), *Frictional Slip of Granite at Hydrothermal Conditions*, J. Geophys. Res. *100*, 13,045–13,064.

DIETERICH, J. H. (1979), *Modeling of Rock Friction 1. Experimental Results and Constitutive Equations*, J. Geophys. Res. *84*, 2161–2175.

GEOGRAPHICAL SURVEY INSTITUTE (1998), *Crustal Movements in the Tokai District*, Report of the Coordinating Committee for Earthquake Prediction *59*, 366–412.

HAMADA, K., OHTAKE, M., OKADA, Y., MATSUMURA, S., and SATO, H. (1985), *A High Quality Network for Microearthquake and Ground Tilt Observations in the Kanto-Tokai Area, Japan*, Earthq. Predict. Res. *3*, 447–469.

ISHIBASHI, K. (1976), *Reexamination of a Great Earthquake Expected to Occur in the Tokai District, Central Japan—the Great Suruga Bay Earthquake*, Proc. Fall Meet. Seismol. Soc. Jpn., pp. 30–34 (in Japanese).

ISHIBASHI, K., *Specification of a soon-to-occur seismic faulting in the Tokai district, central Japan, based upon seismotectonics*, In *Earthquake Prediction: An International Review* (eds. Simpson, D. W., and Richards, P. G.) (American Geophysical Union, Washington D. C. 1981) pp. 297–332.

ISHIDA, M., and OHTAKE, M. (1984), *Seismicity and Waveforms of the Microearthquakes before and after the Shizouka-Seibu Earthquake, Central Japan*, Bull. Seimol. Soc. Am. *74*, 605–620.

ISHIDA, M. (1992), *Geometry and Relative Motion of the Philippine Sea Plate and the Pacific Plate beneath the Kanto-Tokai District, Japan*, J. Geophys. Res. *97* (B1), 489–513.

ISHIDA, M. (1995), *The Seismically Quiescent Boundary between the Philippine Sea Plate and the Eurasian Plate in Central Japan*, Tectonophysics *243*, 241–253.

KATO, N., and HIRASAWA, T. (1996), *Probable Aseismic Sliding and Crustal Deformation Prior to the Hypothetical Tokai Earthquake*, Chikyu Extra Volume *14*, 116–124 (in Japanese).

KATO, N., and HIRASAWA, T. (1997), *A Numerical Study on Seismic Coupling along Subduction Zones Using a Laboratory-derived Friction Law*, Phys. Earth Planet. Inter. *102*, 51–68.

KATO, N., and HIRASAWA, T., (1998), *The Variation of Stresses due to Aseismic Sliding and Its Effect on Seismic Activity*, Pure appl. geophys., this special issue.

KUSUNOSE, K., YAMAMOTO, K., and HIRASAWA, T. (1980), *Source Process of Microfractures in Granite with Reference to Earthquake Prediction*, Sci. Rep. Tohoku Univ. *26*, 111–121.

MATSUMURA, S. (1997), *Focal Zone of a Future Tokai Earthquake Inferred from the Seismicity Pattern around the Plate Interface*, Tectonophysics *273*, 271–291.

MOGI, K. (1970), *Recent Horizontal Deformation of the Earth's crust and Tectonic Activity in Japan (1)*, Bull. Earthq. Res. Inst. *48*, 413–430.

NOGUCHI, S. (1996), *Geometry of the Philippine Sea Slab and the Convergent Tectonics in the Tokai District, Japan*, Zisin 2, *49*, 295–325 (in Japanese).

OKADA, Y. (1992), *Internal Deformation due to Shear and Tensile Faults in a Half-space*, Bull. Seismol. Soc. Am. *82*, 1018–1040.

RUINA, A. (1983), *Slip Instability and State Variable Friction Laws*, J. Geophys. Res. *88*, 10,359–10,370.

SATO, T., and HIRASAWA, T. (1973), *Body Wave Spectra from Propagating Shear Cracks*, J. Phys. Earth *21*, 415–431.

SHIMAMOTO, T. (1985), *The Origin of Large or Great Thrust-type Earthquakes along Subducting Plate Boundaries*, Tectonophysics *119*, 37–65.

STUART, W. D. (1988), *Forecast Model for Great Earthquakes at the Nankai Trough Subduction Zone*, Pure appl. geophys. *126*, 619–641.

TSE, S. T., and RICE, J. R. (1986), *Crustal Earthquake Instability in Relation to the Depth Variation of Frictional Slip Properties*, J. Geophys. Res. *91*, 9452–9472.

UKAWA, M. (1982), *Lateral Stretching of the Philippine Sea Plate Subducting along the Nankai-Suruga Trough*, Tectonics *1*, 543–571.

YAMAZAKI, F., and OOIDA T. (1985), *Configuration of Subducted Philippine Sea Plate beneath the Chubu District, Central Japan*, Zisin 2, *38*, 193–201 (in Japanese).

YOSHIOKA, S., YABUKI, T., SAGIYA, T., TADA, T., and MATSU'URA, M. (1993), *Interplate Coupling and Relative Plate Motion in the Tokai District, Central Japan, Deduced from Geodetic Data Inversion using ABIC*, Geophys. J. Int. *113*, 607–621.

(Recieved July 15, 1998, revised December 23, 1998, accepted December 27, 1998)

Pure appl. geophys. 155 (1999) 689–700
0033–4553/99/040689–12 $ 1.50 + 0.20/0

❘Pure and Applied Geophysics

Seasonality of Great Earthquake Occurrence at the Northwestern Margin of the Philippine Sea Plate

MASAKAZU OHTAKE[1] and HISASHI NAKAHARA

Abstract—A significant seasonality is found in the occurrence time of past great earthquakes ($M \geq 7.9$) in the northwestern margin of the Philippine Sea plate. Among the thirteen earthquakes cataloged, five took place in December, and all the earthquakes are included in the seven months from August to February. The probability that such a skewed distribution occurs by chance is as small as 2.0%. This seasonal concentration of earthquakes suggests that a small stress increase may trigger an earthquake when the future focal zone is at a critical condition to release a large rupture. As a candidate factor of the earthquake triggering, we tested a possibility of stress change caused by annual variation of the atmospheric pressure. However, the incremental stress is no larger than 30 Pa, and it hardly accounts for the seasonality of the great earthquakes.

Key words: Seasonality, great earthquakes, Philippine Sea plate, atmospheric pressure, earthquake triggering.

Introduction

At the northwestern margin of the Philippine Sea plate, great earthquakes of magnitude 8 class have repeatedly ruptured the plate boundary, where the oceanic plate subducts beneath the western part of Japan in the Nankai and Sagami Troughs (see Fig. 1). Those great earthquakes include the 1854 Ansei-Tokai earthquake ($M = 8.4$), the 1923 great Kanto earthquake ($M = 7.9$), and the 1946 Nankai earthquake ($M = 8.0$). The future Tokai earthquake (ISHIBASHI, 1981) would also belong to this group.

We found a significant seasonality in the occurrence time of great earthquakes in this region. Seasonal variation of seismicity has been discussed by many authors since the early age of seismology; e.g., OMORI (1902), DAVISON (1928), MOGI (1969), BOSTROM (1975), KAGAN and KNOPOFF (1976), OKADA (1982). For the Japan region, MOGI (1969) reported a strong seasonal concentration of large

---

[1] Graduate School of Science, Tohoku University, Sendai 980-8578, Japan. Fax: + 81-22-217-6783, E-mail: ohtake@zisin.geophys.tohoku.ac.jp

earthquakes ($M \geq 7.5$) by dividing the whole region into four subregions. The area of present study roughly overlaps the western half of Mogi's subregion B, where earthquake concentration in September to December was reported. As will be shown in the following section, the result of MOGI (1969) is confirmed by extending the data to historical earthquakes back to AD 684.

Most past studies, however, lacked strict statistical tests, and did not provide a physical mechanism for the claimed seasonality. In the present study, we will first show that the seasonality in the northwestern part of the Philippine Sea plate is statistically significant, and hardly occurs by chance. For interpreting this phenomenon, we test a model that seasonal change in the atmospheric pressure may trigger an earthquake in this particular subduction zone.

Data

We select great earthquakes of magnitude 7.9 or larger that took place along the Nankai and Sagami Troughs, based on the earthquake catalogs of UTSU (1982), the JAPAN METEOROLOGICAL AGENCY (1982), HAMADA (1990), and USAMI (1996). The JMA catalog includes hypocenter coordinates and magnitude of major earth-

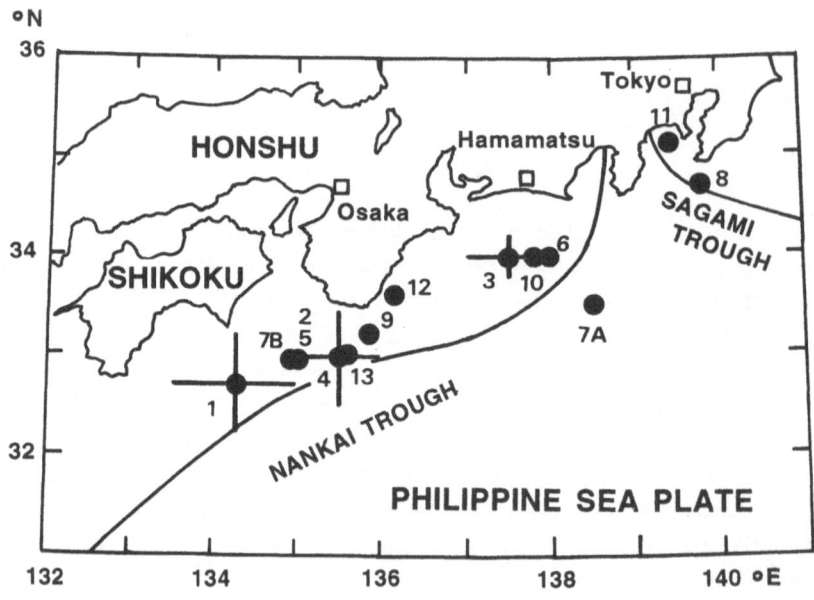

Figure 1

Spatial distribution of the epicenters of great earthquakes ($M \geq 7.9$) that occurred along the Nankai and Sagami Troughs. See text for the earthquake catalogs used. Cross bars indicate the error range of epicenter location. Numericals attached to the epicenters correspond to the event numbers in Table 1.

Table 1

Great earthquakes along the Nankai and Sagami Troughs

| No. | Year | Date | Epicenter | | Magnitude |
			Latitude (°N)	Longitude (°E)	
1	684	Nov. 29	32 1/4–33 1/4	133.5–135.0	8 1/4
2	887	Aug. 26	33.0	135.0	8.0–8.5
3	1096	Dec. 17	33 3/4–34 1/4	137–138	8.0–8.5
4	1099	Feb. 22	32.5–33.5	135–136	8.0–8.3
5	1361	Aug. 03	33.0	135.0	8 1/4–8.5
6	1498	Sep. 20	34.0	138.0	8.2–8.4
*7A	1605	Feb. 03	33.5	138.5	7.9
B	1605	Feb. 03	33.0	134.9	7.9
8	1703	Dec. 31	34.7	139.8	7.9–8.2
9	1707	Oct. 28	33.2	135.9	8.4
10	1854	Dec. 23	34.0	137.8	8.4
11	1923	Sep. 01	35.1	139.5	7.9
12	1944	Dec. 07	33.57	136.18	7.9
13	1946	Dec. 21	33.03	135.62	8.0

* The event of No. 7 is interpreted as a doublet following USAMI (1996).

quakes for the period of 1926 to 1960. For the historical time prior to the instrumental observation, the focal parameters are provided in USAMI (1996) based on the elaborate study of historical documents. For the early period of instrumental observation, UTSU (1982) revised and newly determined the coordinates and magnitude of major earthquakes in and near Japan by using available data from instrumental observation. Considering the characteristics of the earthquake catalogs, we use USAMI (1996) for 1884 and before, UTSU (1982) for 1885–1925, and JMA (1982) for 1926 and after, as a rule.

In Table 1, we list thirteen great earthquakes reported by the catalogs. For event 12 in Table 1, the result of HAMADA's (1990) relocation is adopted. Event No. 7 is interpreted as a doublet of great earthquakes by USAMI (1996). Epicenters of these earthquakes are also shown in Figure 1. All thirteen earthquakes accompanied large tsunamis, and it is most probable that they ruptured the plate boundary of the subducting Philippine Sea plate. For historical time, some great earthquakes may have escaped historical documents, and completeness of the list is not guaranteed. This, however, does not affect the result since our present concern is only the date of earthquake occurrence in a year.

Statistical Test

In Table 1, the month of December appears most frequently; five out of thirteen cases. Other events also distribute in the fall to winter season, for the most part. As a whole, all the events concentrate in the seven months from August to February, and the remaining five months are completely vacant of great earthquakes. In the following, we statistically test whether the seasonality appeared by chance or not. If the seismic seven-month period is fixed at August–February for some reason, the probability of the event concentration is simply

$$(7/12)^{13} = 0.09 \times 10^{-2}, \tag{1}$$

by assuming a uniformly random distribution. However, this estimate is not appropriate for the present case since we do not have a reason to *a priori* specify this particular seven months. If the concentration occurred in another seven-month period, from January to July for instance, one may similarly claim the existence of seasonality. We, therefore, need to calculate the probability that 13 events concentrate in *any* seven-month period in a year. An analytical solution of this unspecified problem was given by one of the authors (M.O., unpublished note) as,

$$F_n(r) = \sum_{i=1}^{[1/(1-r)]} (-1)^{(i-1)} \binom{n}{i} [1 - i(1-r)]^{n-1}, \tag{2}$$

where n is the number of events, and r is the fractional length of the concentration period in a year. The round parentheses indicate the combination.

By substituting $n = 13$, and $r = 7/12$ into Equation (2), we obtain $F_n(r) = 2.0 \times 10^{-2}$ for the present case. The null hypothesis that the seasonal concentration of great earthquakes described above occurs by chance is rejected at the 98% confidence level.

Model Study

The confidence level of 98% is high enough to explore physical mechanisms that bring about the seasonality. As a plausible mechanism, we test a model that the seasonal change in atmospheric pressure may trigger an earthquake.

Figure 2 shows a representative feature of atmospheric pressure change in the central part of Honshu, together with the occurrence time of great earthquakes in Table 1. The pressure plotted is a 30-year average at Hamamatsu (see Fig. 1 for the location). In the winter season, a high pressure appears due to the development of the Siberian high pressure system which covers all of the Japanese Islands. Thus the atmospheric pressure repeats a systematic annual variation, with an amplitude amounting to 10 hPa (10^3 Pa) peak-to-peak. Our model advocates that a small stress perturbation brought about by the surface loading may work as a triggering factor of an earthquake.

We compute the stress change due to the surface load by using a geometry as is illustrated in Figure 3. The area of the Japanese Islands is modeled by a rectangle 1000 km long and 200 km wide. The oceanic plate subducts beneath the islands

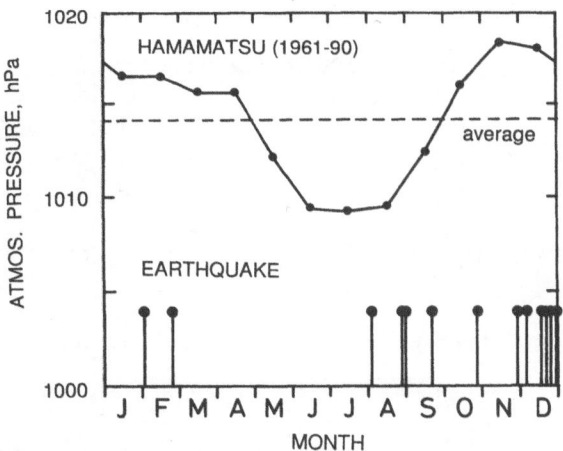

Figure 2
Annual change in the atmospheric pressure at Hamamatsu (folded curve, see Fig. 1 for the location), and occurrence time of great earthquakes along the Nankai and Sagami Troughs (vertical bars). The atmospheric pressure is a 30-year average for 1961–1990.

from the trench axis that is located 100 km offshore. The dip angle is set at 30°. Then we apply a surface load of 10^3 Pa to the rectangle area. We note that the load is applied only to the land area. The pressure applied to the sea surface is compensated by move-out of sea water, resulting in constant pressure at the ocean bottom. This "inverse barometer effect" (e.g., APEL, 1987) holds good as far as a static equilibrium is kept. The stress induced by the surface load is theoretically

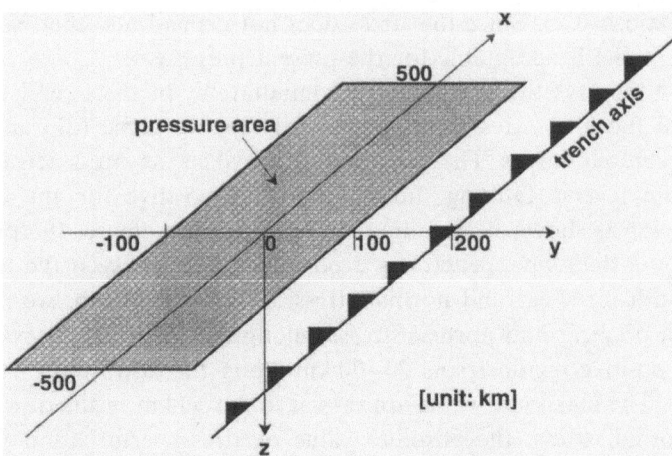

Figure 3
Model for computing the stress change induced by the atmospheric pressure. A surface load of 10^3 Pa is applied to the rectangle area of 200 km × 1000 km, which represents the Japanese Islands. The trench axis is fixed at 100 km off the coast.

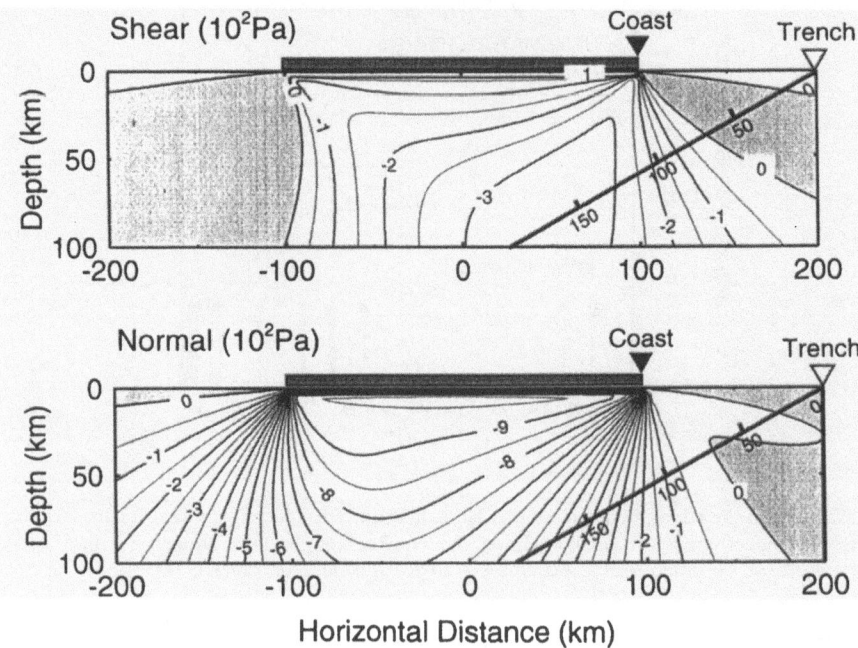

Figure 4
Stress change induced by a surface load of 10^3 Pa. Top and bottom panels show the shear and normal stresses, respectively, on a plane dipping 30° to the left. Shaded are the positive regions of stress change (incremental stress is in the sense to accelerate reverse faulting).

computed by assuming a homogeneous isotropic semi-infinite elastic body, of which Poisson's ratio is 0.25. Since the stress does not depend on other elastic constants, the simple model is acceptable for the present purpose.

Figure 4 displays the result of the computation. In the figure, the shear (top) and normal (bottom) stresses on a leftward dipping plane (dip angle is 30°) are plotted by contour curves. The value is positive when the stress acts in the direction to accelerate reverse faulting; normal stress is positive for the extension. The positive region is shown by shading. As is seen in the figure, the plate boundary, indicated by a thick line, penetrates a positive region between the trench axis and coastline both for shear and normal stresses. In Figures 5a,b, we plot the spatial distribution of shear and normal stresses along the plate boundary. For the shear stress, the positive region spans 20–70 km along the down-dip distance from the trench axis. The maximum value appears at about 50 km in the down-dip distance. For the normal stress, the absolute value of stress perturbation is considerably smaller as compared with the shear stress.

Figure 5c plots the spatial variation of the Coulomb failure function:

$$CFF = \tau + \mu\sigma_n, \tag{3}$$

where τ is the shear stress, μ is the friction coefficient, and σ_n is the normal stress (positive for extension). The incremental value of CFF is computed for $\mu = 0.3$, 0.5 and 0.7. Due to the relatively small change of normal stress, CFF is not sensitive to the friction coefficient for the present case, and the CFF curves show nearly the same features as the shear stress. According to the Coulomb's criterion, a shear rupture occurs when CFF exceeds the shear strength so that an increase in CFF accelerates a rupture onset.

The result of the computation indicates that the high pressure in the winter season brings about an incremental stress, amounting to about 30 Pa at the

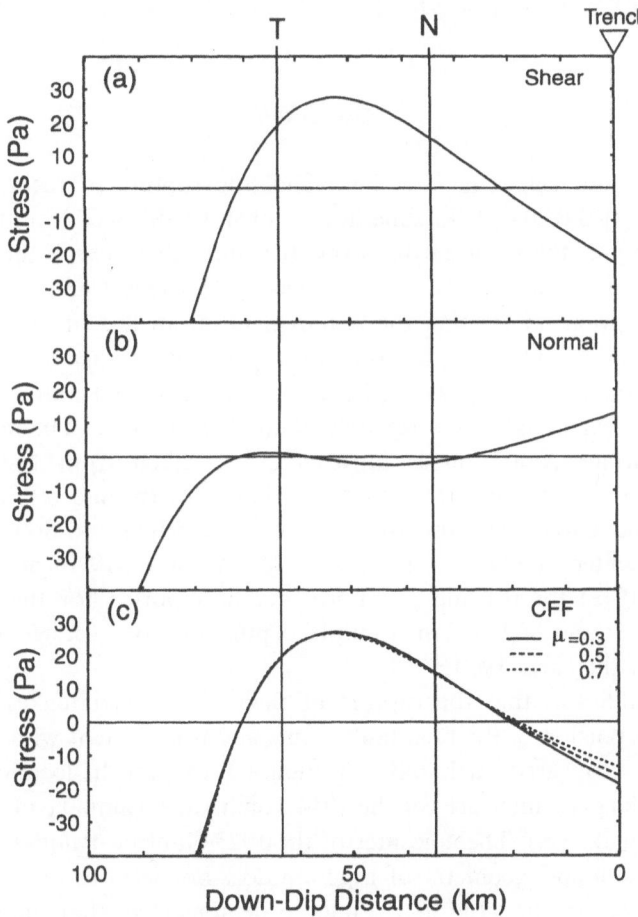

Figure 5

Stress change induced by a surface load of 10^3 Pa for (a) the shear, (b) normal, and (c) CFF components on the plate interface. Abscissa is the down-dip distance from the trench axis along the plate interface (see Fig. 4 for geometry). CFF is shown for three different values of friction coefficient μ. The vertical lines indicated by T and N indicate the location of epicenters for the 1944 Tonankai and the 1946 Nankai earthquakes, respectively.

maximum, on the plate interface. The region of stress increase does not appear beneath the land area, and is confined at a 50-km portion between the trench axis and coastline. This incremental stress may play a role in triggering a large rupture.

Figure 5c exhibits negative *CFF* of a large absolute value for the region deeper than 80 km in the down-dip distance. This negative region changes to a positive region in the summer season, and potentially accelerates a reverse faulting. This effect, however, will not work to trigger an earthquake due to weak coupling of the plate boundary. For the focal area of the pending Tokai earthquake, MATSUMURA (1997) demonstrates that the locked region of the plate interface is restricted in a shallow part, and does not extend deeper than 40 km, which corresponds to 80 km in the down-dip distance (see Fig. 4).

Discussion

If the pressure triggering model is valid, the rupture of the plate boundary should initiate at the 20–70 km zone in the down-dip distance from the trench axis for the majority of the great earthquakes. It is difficult to check this for historical earthquakes, but instrumentally located epicenters for recent great earthquakes will provide information on whether the requirement is satisfied or not.

In Figure 6, we plot the location of the epicenter together with the fault geometry for three recent great earthquakes; (a) the 1923 Kanto earthquake, (b) 1944 Tonankai earthquake, and (c) 1946 Nankai earthquake. The fault planes, for which the surface projection is illustrated, are taken from MATSU'URA and IWASAKI (1983) for (a), and from SATAKE (1993) for (b) and (c). The epicenters, shown by solid circles, are the same as those in Figure 1. Solid triangles are relocated epicenters by KANAMORI and MIYAMURA (1970) for (a), and by KANAMORI (1972) for (b) and (c). Focal depths reported for these earthquakes range between 0 and 40 km, but accurate depths are not expected from the poor observation of the early 1900s.

Figure 6 indicates that the rupture of those three earthquakes initiated at a rather shallow portion of the final fault plane, and is consistent with the theoretical prediction for a triggered earthquake. In Figure 5, we plot the location of epicenter projected on the plate interface for the 1944 Tonankai earthquake (T), and the 1946 Nankai earthquake (N). The epicenter of the 1923 Kanto earthquake is not shown since the simple model geometry of Figure 3 does not seem to represent the actual trench-land geometry (see Fig. 6a). Figure 5 demonstrates that the rupture of the Tonankai and Nankai earthquakes started at a positive region of incremental stress. This observation suggests that the atmospheric pressure change might have triggered those great earthquakes.

In order to further check the validity of the pressure trigger model, statistical study of small earthquakes is of particular interest. However, it is not easy to

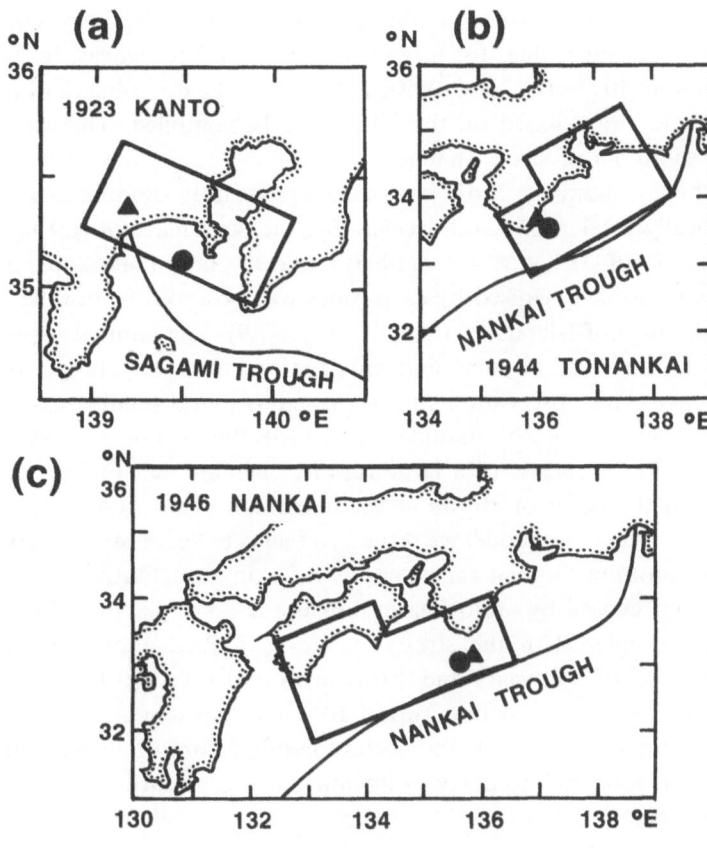

Figure 6
Rupture zones (horizontal projection) and the epicenters (solid circles) of recent great earthquakes; (a) 1923 Kanto earthquake, (b) 1944 Tonankai earthquake, and (c) 1946 Nankai earthquake. The fault models are taken from MATSU'URA and IWASAKI (1983) for (a), and from SATAKE (1993) for (b) and (c). Triangles are relocated epicenters by KANAMORI and MIYAMURA (1970) for (a), and by KANAMORI (1972) for (b) and (c).

extend the present study to small earthquakes since the interplate seismicity in this region is very weak, even at the microearthquake level (e.g., NOGUCHI, 1996). For constructing a reliable data set of interplate earthquakes, we need to discriminate them from relatively high seismicity inside the oceanic and continental plates by using precise hypocenter location and focal mechanism. The analysis of small earthquakes is a subject of future study.

Recent studies have provided strong evidence that a very small stress change can trigger earthquakes, and brings about a significant modulation of seismicity. At the time of the 1992 Landers, California earthquake ($M_s = 7.6$), remote triggering of local earthquakes was observed in a wide area, up to 1200 km in epicentral distance (e.g., HILL et al., 1993; ANDERSON et al., 1994). ANDERSON et al. (1994), and

GOMBERG and DAVIS (1996), attributing the remote triggering effect to the dynamic strain change due to large-amplitude surface waves, estimated the threshold strain at 10^{-6}–10^{-5}. FUJIWARA (1998), on the other hand, obtained 10^{-7} for the threshold strain based on the 70 reports he compiled. This strain roughly corresponds to 10^3 Pa of stress change.

Stress change caused by earth tides also significantly modulates seismicity by quasi-periodically accelerating and decelerating the occurrence of earthquakes (e.g., TSURUOKA et al., 1995). TSURUOKA (1995) reported that a remarkable modulation of background seismicity caused by earth tides was observed before the occurrence of the 1986 Andreanof Islands earthquake ($M_w = 7.9$). The anomaly appeared only in and near the future focal zone, and only for one to two years just prior to the great earthquake. This observation strongly suggests that a small stress change may control the occurrence of earthquakes provided the region is under a critical condition prior to the release of a large rupture. The amount of stress change due to earth tides is the order of 10^3 Pa at the maximum (TSURUOKA et al., 1995).

The pressure triggering model we discussed seems to successfully account for the seasonal concentration of great thrust earthquakes in a qualitative sense. However, the stress change caused by atmospheric pressure is no larger than 30 Pa, and one to two orders smaller than the stress changes described above. By assuming a recurrence interval of 100 years and stress drop of 10^6 Pa (10 bars) for the great earthquakes, an average stress build-up of 10^4 Pa/year is expected for the interseismic period. A stress change of 30 Pa, corresponding to one day part of the tectonic stress increase, is too small to delay or advance the occurrence time of earthquakes by several months.

Conclusion

We found a remarkable seasonality in the occurrence of past great earthquakes ($M \geq 7.9$) in the northwest margin of the Philippine Sea plate. All thirteen earthquakes listed took place in the seven months between August to February. The probability that such a deviated distribution occurs by chance is as small as 2.0%.

In order to interpret the seasonality, we tested a model that annual variation of atmospheric pressure may trigger an earthquake. The result of the computation evidences that the increase in the atmospheric pressure in the winter season by 10^3 Pa induces a positive stress of up to 30 Pa on the subducting plate boundary between the trench axis and coastline. Location of the epicenters of the 1944 Tonankai and 1946 Nankai earthquakes indicates that the rupture of those great earthquakes started at this portion of the plate boundary, in harmony with the pressure triggering model.

However, the small stress change of 30 Pa hardly accounts for the seasonality of earthquakes we observed. It is necessary to explore another model to physically understand the significant seasonal concentration of great earthquakes in the northwestern margin of the Philippine Sea plate.

Acknowledgement

We greatly appreciate helpful comments by anonymous reviewers.

REFERENCES

ANDERSON, J. G., BRUNE, J. N., LOUIE, J. N., ZENG, Y., SAVAGE, M., YU, G., CHEN, Q., and DEPOLO, D. (1994), *Seismicity in the Western Great Basin Apparently Triggered by the Landers, California, Earthquake, 28 June 1992*, Bull. Seismol. Soc. Am. *84*, 863–891.
APEL, J. R., *Principles of Ocean Physics*, International Geophysics Series, vol. 38 (Academic Press, London 1987), 634 pp.
BOSTROM, R. C. (1975), *Modulation of Seismicity by Atmospheric Torques*, Earthq. Notes *46*, 28.
DAVISON, C. (1928), *The Annual Periodicity of Earthquakes*, Bull. Seismol. Soc. Am. *18*, 246–266.
FUJIWARA, K. (1998), *Microearthquake Activity Induced by Large Earthquakes*, Master Thesis, Tohoku University, 139 pp. (in Japanese).
GOMBERG, J., and DAVIS, S. (1996), *Stress/Strain Changes and Triggered Seismicity at the Geysers, California*, J. Geophys. Res. *101*, 733–749.
HAMADA, N. (1990), *Some Amendments for the Earthquake Catalogue in the Special Issue No. 6 of the Seismological Bulletin of the Japan Meteorological Agency*, Zisin (J. Seismol. Soc. Japan) 2, *49*, 307–310 (in Japanese).
HILL, D. P., REASENBERG, P. A., MICHAEL, A., ARABAZ, W. J., BEROZA, G., BRUMBAUGH, D., BRUNE, J. N., CASTRO, R., DAVIS, S., DEPOLO, D., ELLSWORTH, W. L., GOMBERG, J., HARMSEN, S., HOUSE, L., JACKSON, S. M., JOHNSTON, J. S., JONES, L., KELLER, R., MALONE, S., MUNGUIA, L., NAVA, S., PECHMANN, J. C., SANFORD, A., SIMPSON, R. W., SMITH, R. B., STARK, M., STICKNEY, M., VIDAL, A., WALTER, S., WONG, V., and ZOLLWEG, J. (1993), *Seismicity Remotely Triggered by the Magnitude 7.3 Landers, California, Earthquake*, Science *260*, 1617–1623.
ISHIBASHI, K., *Specification of soon-to-occur seismic faulting in the Tokai District, Central Japan, based upon seismotectonics*. In *Earthquake Prediction*, An International Review (eds. Simpson, D. W., and Richards, P. G.) (American Geophysical Union, Washington, DC 1981) pp. 297–332.
JAPAN METEOROLOGICAL AGENCY (1982), *Catalogue of Relocated Major Earthquakes in and near Japan*, Seismol. Bull. Japan Meteor. Agency, Special Issue, no. 6, 109 pp. (in Japanese).
KAGAN, Y. Y., and KNOPOFF, L. (1976), *Statistical Search for Non-random Features of the Seismicity of Strong Earthquakes*, Phys. Earth Planet. Inter. *12*, 291–318.
KANAMORI, H., and MIYAMURA, S. (1970), *Seismometrical Re-evaluation of the Great Kanto Earthquake of September 1, 1923*, Bull. Earthq. Res. Inst. *48*, 115–125.
KANAMORI, H. (1972), *Tectonic Implication of the 1944 Tonankai and the 1946 Nankaido Earthquakes*, Phys. Earth Planet. Inter. *5*, 129–139.
MATSUMURA, S. (1997), *Focal Zone of a Future Tokai Earthquake Inferred from the Seismicity Pattern around the Plate Interface*, Tectonophysics *273*, 271–291.
MATSU'URA, M., and IWASAKI, T. (1983), *Study on Coseismic and Postseismic Crustal Movements Associated with the 1923 Kanto Earthquake*, Tectonophysics *97*, 201–215.
MOGI, K. (1969), *Monthly Distribution of Large Earthquakes in Japan*, Bull. Earthq. Res. Inst. *47*, 419–427.
NOGUCHI, S. (1996), *Geometry of the Philippine Sea Slab and the Convergent Tectonics in the Tokai District, Japan*, Zisin (J. Seismol. Soc. Japan) 2, *49*, 295–325 (in Japanese).

OKADA, M. (1982), *Seasonal Variation in the Occurrence Rate of Large Earthquakes in and near Japan and Its Regional Differences*, Zisin (J. Seismol. Soc. Japan) 2, *35*, 53–64 (in Japanese).

OMORI, F. (1902), *Annual and Diurnal Variations of Seismicity in Japan*, Pub. Earthq. Inv. Comm. *8*, 1–94.

SATAKE, K. (1993), *Depth Distribution of Coseismic Slip along the Nankai Trough, Japan, from Joint Inversion of Geodetic and Tsunami Data*, J. Geophys. Res. *98*, 4553–4565.

TSURUOKA, H. (1995), *Effects of the Earth Tide on Earthquake Occurrence and Its Mechanical Interpretation*, Ph.D. Thesis, Tohoku University, 148 pp. (in Japanese).

TSURUOKA, H., OHTAKE, M., and SATO, H. (1995), *Statistical Test of the Tidal Triggering of Earthquakes: Contribution of the Ocean Tide Loading Effect*, Geophys. J. Int. *122*, 183–194.

USAMI, T., *Summary of Earthquakes with Damage in Japan*, New Edition: Enlarged and Revised Version for 416-1995 (Tokyo University Press, Tokyo 1996) 483 pp. (in Japanese).

UTSU, T. (1982), *Catalog of Large Earthquakes in the Region of Japan from 1885 through 1980*, Bull. Earthq. Res. Inst. *57*, 401–463 (in Japanese).

(Received July 30, 1998, revised/accepted November 18, 1998)

 To access this journal online:
http://www.birkhauser.ch

Pure appl. geophys. 155 (1999) 701–712
0033–4553/99/040701–12 $ 1.50 + 0.20/0

⌈ Pure and Applied Geophysics

Eruptions of Pavlof Volcano, Alaska, and their Possible Modulation by Ocean Load and Tectonic Stresses: Re-evaluation of the Hypothesis Based on New Data from 1984–1998

S. R. McNutt[1]

Abstract—Thirteen of sixteen magmatic eruptions of Pavlof Volcano in nine of the years from 1973 to 1998 have occurred between September 9 and December 29. Volumes of erupted material range from 0.3 to 16×10^6 m^3 (dense rock equivalent). A significant correlation exists between the eruptions and yearly nontidal variations in sea level and may result from ocean loading. Calculated volume changes beneath the volcano due to ocean loading are from 0.02 to 0.6 times eruption volumes, and it is postulated that the volcano acts as a long-period (several months) volume strainmeter, with lava being preferentially erupted when strain beneath the volcano is compressive. Previous observations of a tilt reversal, and new observations of tectonic activity and eruptions in the spring and summer of 1986, also suggest tectonic modulation of eruptions. The volcano appears to be responsive to small, slow changes in ambient stresses or strains, and these changes may modify or trigger eruptions.

Key words: Volcanoes, eruption, triggering periodic eruptions.

Introduction

Pavlof Volcano (latitude 55°24′N, longitude 161°54′W) is a 2518-m-high stratovolcano which sits on the Alaska Peninsula roughly in the middle of the Shumagin seismic gap (Fig. 1). It is the most persistently active volcano in North America, with over 50 reports of activity and 40 documented eruptions since it was first sighted by Russian explorers in the 1760s. An earlier paper by McNutt and Beavan (1987) discussed the eruptions that occurred from 1973–1984, and modeled the effects of possible modulation by ocean load and tectonic stresses. This paper re-evaluates these effects based on new data from the years 1984–1998.

[1] Geophysical Institute, University of Alaska Fairbanks, 903 Koyukuk Drive, P.O. Box 757320, Fairbanks, AK 99775-7320. Tel.: 907-474-7131; Fax 907-474-5618; E-mail: steve@giseis.alaska.edu

Earlier Work

All nine magmatic eruptions at Pavlof from 1973 to 1984 occurred in the fall, between September 9 and November 20 (Fig. 2). Four of them, in four different years, occurred in the same 4-day period between November 11 and 15. Volumes of erupted material range from 0.3 to 16×10^6 m^3 (dense rock equivalent) at an average rate of about 3×10^6 m^3 per year (Table 1). The volumes are estimated from eyewitness reports and photographs for several eruptions; the others are estimated from a locally derived relationship between eruption volume and volcanic tremor duration and amplitude (McNutt and Beavan, 1987). A significant correlation exists between the eruptions and yearly nontidal variations in sea level and may result from ocean loading. Sea levels are corrected for atmospheric pressure (Ingram *et al.*, 1976; Reed and Schumacher, 1981). Calculated volume changes beneath the volcano due to ocean loading are about 10 percent of eruption

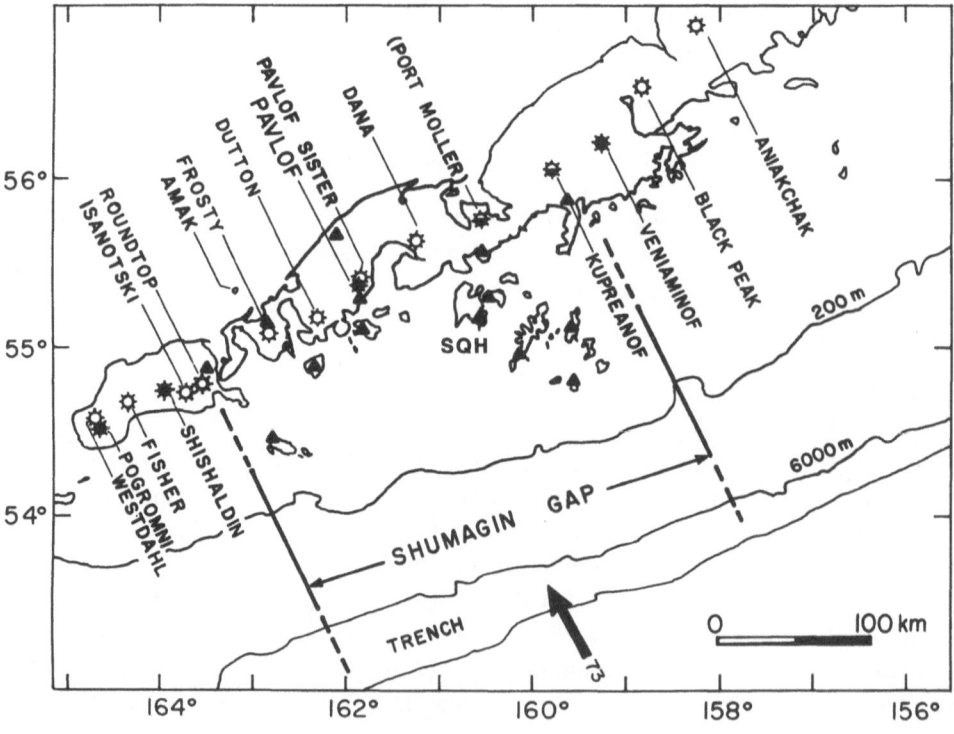

Figure 1
Index map of the study area. Holocene volcanoes are shown as sunbursts; filled symbols indicate eruptions, while half-filled symbols indicate earthquake swarms or geothermal activity. Triangles denote seismic stations. SQH is the levelling line near Squaw Harbor. Shumagin seismic gap boundaries are shown as solid lines for the seismogenic portions of the rupture zone and as dashed lines for nonseismic portions. Arrow shows plate convergence rate in mm/yr.

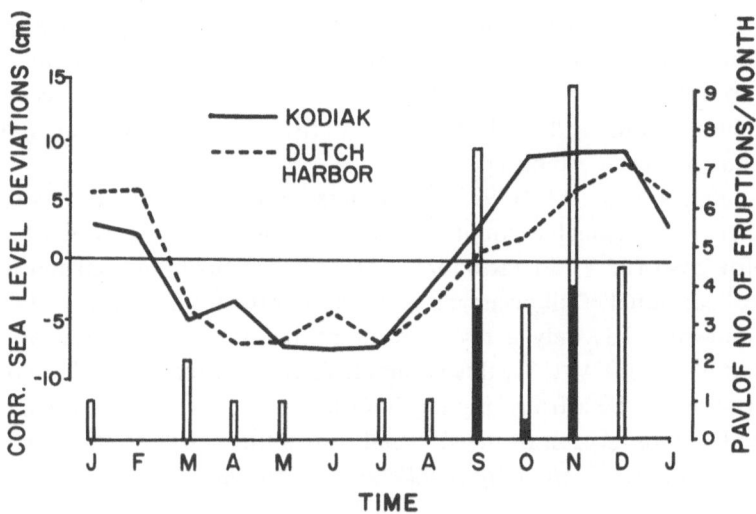

Figure 2

Deviations in monthly mean-corrected (for atmospheric pressure) sea level from long-term mean
(1950–1974) and number of eruptions within each month at Pavlof Volcano (1973–1984). Kodiak is 500
km ENE of Pavlof, while Dutch Harbor is 350 km WSW. Histogram bars are solid for magmatic
eruptions and open for explosive eruptions. Seismicity for the magmatic eruptions (solid bars) is on
average a factor of 10 greater than for the explosive eruptions (open bars). The two smallest magmatic
eruptions, September and October 1976, are shown at half the height of the others. Note the correlation
   between eruptions and increased sea level in fall/winter. (Plot from McNUTT and BEAVAN, 1987.)

volumes, and it is postulated that the volcano acts as a long-period (several month)
volume strainmeter, with lava being preferentially erupted when strain beneath the
volcano is compressive (McNUTT and BEAVAN, 1987). The amplitude of the
sea-level deviation is highest in November, and 4 of the 5 largest volume eruptions
from 1973 to 1984 occurred in November. The volcano did not erupt during the
period 1978–1980, when tilt, seismic, and sea-level data indicate that deep aseismic
slip may have occurred (BEAVAN et al., 1984). Models of this event predict a
volume strain extension beneath the volcano that might have compensated strain
from magma injection. These observations indicate that Pavlof Volcano is responsive to small, slow changes in ambient stresses or strains and that these changes
may modify or trigger eruptions (McNUTT and BEAVAN, 1987).

Other volcanoes and earthquakes have shown apparent seasonal periodicities.
Etna terminal and flank eruptions from 1323 to 1980 show significant seasonal
clustering with a main peak in November and subsidiary peaks in March and May
(CASETTI et al., 1981). Eruptions of Kilauea from 1832 to 1979 clustered in May
(data given in DZURISIN, 1980). In Iceland, Hekla's eruptions from 1104 to 1970
and all known Icelandic eruptions from 1550 to 1978 clustered in May, but not
significantly (THORARINSSON and SIGVALDSSON, 1972; GUDMUNDSSON and SAE-
MUNDSSON, 1980). Earthquakes in the Torfajokull-Myrdahlsjokull area of Iceland,

on the other hand, clustered in the fall, and were interpreted to be triggered by glacial loading (TRYGGVASSON, 1973). Shallow earthquakes at Rabaul caldera from 1967 to 1984 showed an annual periodicity with a peak in December (P. Lowenstein, writt. com., 1985). The formal statistical significance for several of these cases has not been established.

OHTAKE and NAKAHARA (1999, this volume) summarized the observations of earlier workers on seasonal earthquakes, and demonstrate a convincing case of periodicity for great ($M \geq 7.9$) earthquakes along the Nankai and Sagami troughs. The historical earthquakes all occur in the seven months from August to February, a distribution which has only a two percent probability of occurring by chance (OHTAKE and NAKAHARA, 1999, this volume). They interpret seasonal atmospheric pressure variations to be a triggering mechanism, and model the stress changes on the shallow plate interface. Both the seasonal distribution of these earthquakes and the phenomena modeled are remarkably similar to the observations at Pavlof volcano.

New Data

Two new series of eruptions occurred from April 1986 to May 1988 (Fig. 3 and Table 1) and later from September to December 1996 (Table 2). The 1986 eruptions were stronger and of longer duration than any others known (MCNUTT et al.,

Table 1

Magmatic eruptions of Pavlof Volcano, 1973 to 1998

Year	Start date	Volume $\times 10^6$ m^3	Comment
1973	Nov. 13	6.4	Photographs
1975	Sept. 13	4.5	
1975	Sept. 23	1.2	Photographs
1976	Sept. 9	0.3	Ground observations
1976	Oct. 18	0.5	
1976	Nov. 10	7.6	Pilot reports
1980	Nov. 11	6.1	Pilot reports, photographs
1981	Sept. 26	10.8	Pilot reports, photographs, ash sample
1983	Nov. 14	12.5	Pilot reports
1986	Apr. 19	–*	Pilot reports
1986	May 28–Sept. 1	16	Unusually long duration; observed by geologists
1986	Nov. 5	2	Pilot reports
1986	Dec. 8	3	Pilot reports
1987	May 23	2	Pilot reports
1987	Oct. 14	1	Pilot reports and ground observations
1996	Sept. 15	7	Pilot reports, photographs

* Volume of April 19, 1986 eruption is included in estimate for May 28 to Sept. 1, 1986 eruption.

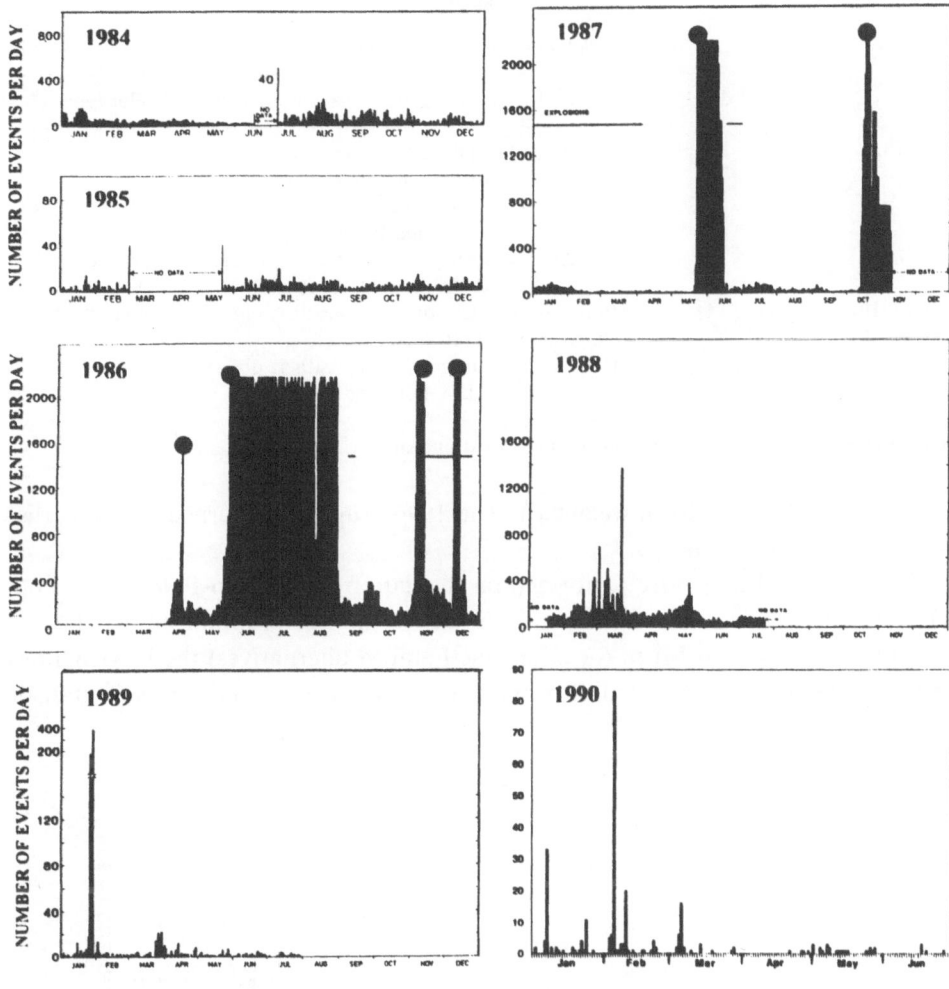

Figure 3
Seismicity plots for Pavlof Volcano for the time period 1984 to 1990. Plots show the number of B-type and explosion events per day. Periods of volcanic tremor are normalized to number of events using a previously published relationship (MCNUTT, 1987). Round dots indicate eruptions accompanied by lava fountains; horizontal bars above the data indicate time periods of explosions. Note the different vertical scales for mid-1984 through 1985, 1989, and 1990. The 1989 data also show a scale break for the swarm in late January. Horizontal (time) scales are the same for all years except 1990.

1991). They produced a possible pyroclastic flow for the first time, and modified the structure of the vent area. They also occurred from April through August, thus changing the temporal pattern of activity that had persisted from 1973–1984 (Fig. 4). Note that although the May 28 to September 1, 1986 portion of the eruption is a single long-lasting eruption, it contributes to the three months June, July and August in Figure 4 (the minor contributions from only three days in May and one

Table 2

Pavlof Volcano 1996 eruptions

Start date	End date	Activity, based on seismicity, satellite images, and pilot reports*
Sep. 15	Dec. 4	Continuous Strombolian eruption with stronger pulses: Oct. 18 — strong volcanic tremor; poor weather Nov. 4 — strong volcanic tremor; ash column to 25,000 ft a.s.1.; ash plume 100 miles long Nov. 22 — strong volcanic tremor; steam plume between 20,000 and 30,000 ft a.s.1.
Dec. 10	Dec. 14	Strong volcanic tremor; poor weather; one observation of ash column of 15,000 ft a.s.1.
Dec. 26	Dec. 29	Strong volcanic tremor; poor weather; observations of hot spot (satellite image), lava flow, and ash plume several tens of miles long

* Observations are reported in the original units of ft and miles.

day in September are down-weighted). The 1996 eruptions returned to the earlier pattern of fall eruptions.

Two statistical tests were performed on the data from 1973 to 1998 (Table 1) to determine the significance of the temporal distribution of eruption onsets. To test for uniform versus von Mises (or other one-humped alternatives) the Rayleigh test (MARDIA, 1972) was conducted. For number of cases $n = 16$, the regression statistic

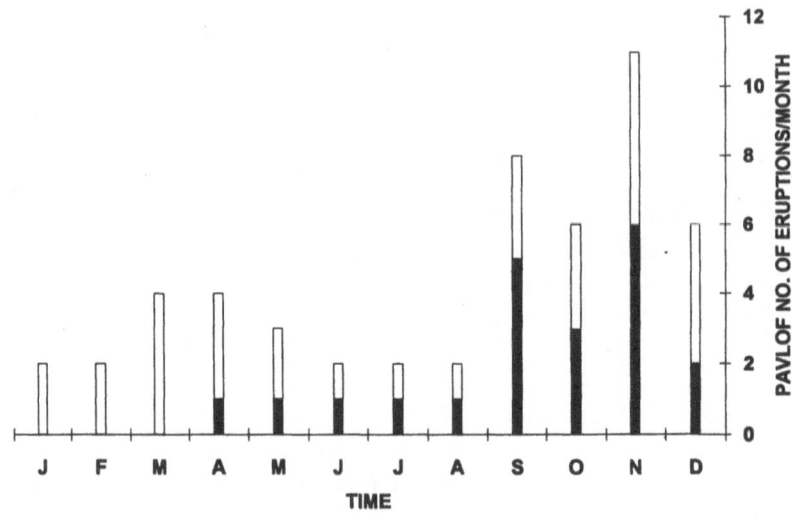

Figure 4
Histogram of the number of eruptions within each month of Pavlof Volcano from 1973–1998. Histogram bars are solid for magmatic eruptions and open for explosive eruptions. Here each eruption is plotted at the same height regardless of volume erupted. Note that despite the increase in the number of eruptions from January to August (compare with Fig. 2), the eruptions are still concentrated in the fall/winter months from September to November.

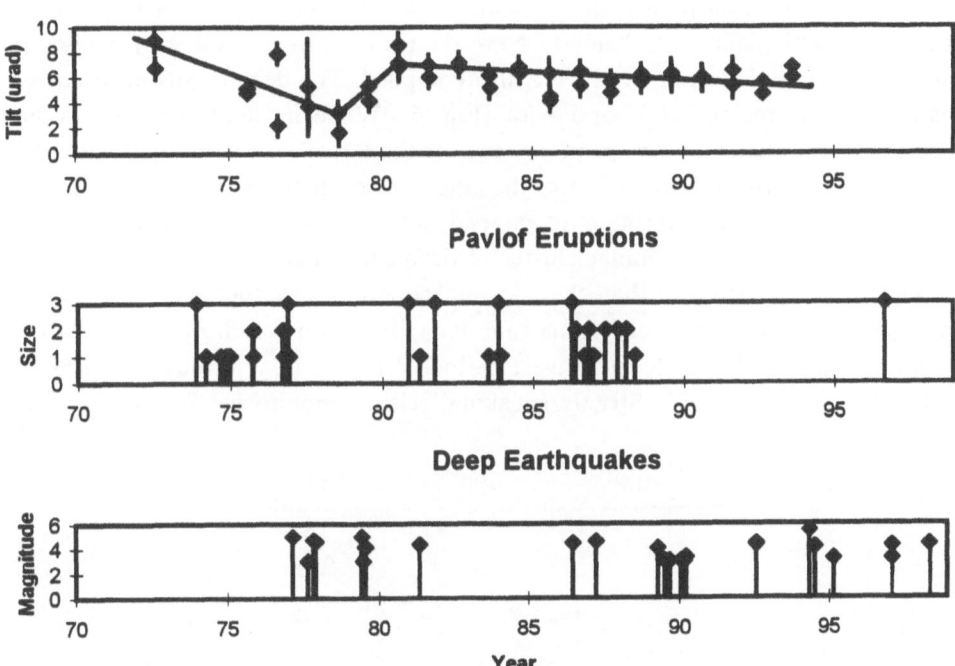

Figure 5

Tilt (microradians), Pavlof eruptions (VEI; Volcanic Explosivity Index), and deep earthquakes (>160 km) in the Shumagin region for the period 1970–1998. An apparent tilt reversal took plate on three level lines between 1978 and 1980. The tilt change is significant at the 98% level on line SQH, which is plotted here (see Fig. 1 for location). The line drawn through the data is our interpretation of the tilt history in the inner Shumagin Islands. During the tilt reversal, the seismicity rate for microearthquakes was higher (not shown; see McNutt and Beavan, 1987), Pavlof Volcano was not erupting and exhibited very low seismicity, and most of the teleseismically recorded deep earthquakes from 1970–1985 occurred in a cluster NNE of Pavlof (see Fig. 6). The estimated magnitude for complete recording is 4.5; quakes plotted as $M = 3$ (smallest events on plot) had no magnitude in the PDE catalog. New tilt data are from Beavan (1994).

$\overline{R} = 0.5602$, thus the probability or p value is approximately 0.005; we can reject a uniform distribution at the 99.5 percent confidence level. A second test conducted was Kuiper's test (Mardia, 1972) of uniformity versus nonuniformity. For $n = 16$, $V_n = 0.5692$, yielding a p value of approximately 0.0004, or a 99.96 percent probability that the distribution is nonuniform. Therefore, both tests indicate that the Pavlof eruption onsets are highly non-random. They appear to be clustered in the fall, as is evident from Figures 2 and 4.

New tilt data from 1984 to 1993 were compiled (Beavan, 1994) and are plotted in Figure 5. These data show continuous slow tilt down towards the trench for site SQH in the inner Shumagins (Fig. 1). No evidence is found for another tilt reversal similar to the one that occurred from 1978 to 1980, however a slight change is noted for the last data point, which occurred just after the May 13, 1993 $M = 6.9$ Shumagin earthquake (see below). New data on deep (>160 km) moderate earthquakes as reported in PDE were also compiled. The deep events all occurred spatially in the region NNE of Pavlof (Fig. 6). While most of the deep events occurred during times of no eruptions, two of the larger events in 1986 and 1987 were associated with eruptions. Also, the cluster of deep events from 1977 to 1979 occurred just before and during a tilt reversal, yet no apparent tilt reversal occurred in 1989 or 1990 during a similar cluster of deep earthquakes.

The new data suggest that the tectonic conditions in the Shumagin region changed in 1986 then returned some time later. It is also worth noting that from 1988 to 1996 Pavlof was in the longest period of repose (8 yr 4 mo) since seismic monitoring began in 1973. Strictly speaking, seismic monitoring by the Lamont-

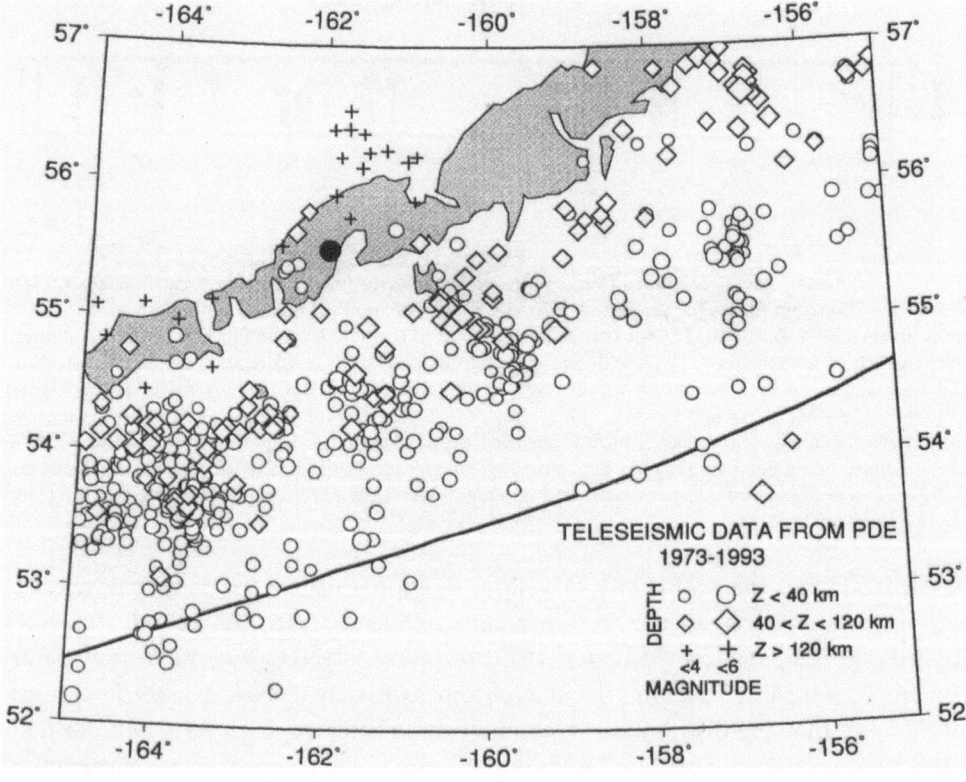

Figure 6

Epicenter map of teleseismic data from PDE for the Shumagin region. All data from 1973–1993 are shown. Note that the only cluster of deep events is located in the back arc region NNE of Pavlof Volcano (solid dot). This spatial pattern was persistent from 1994 to 1998 as well. A time series for these events is displayed in Figure 5.

Doherty Earth Observatory ended in summer 1990. However, stations operated by the University of Alaska Fairbanks at nearby Dutton volcano (35 km SW) remained in operation, but did not record any eruptions at Pavlof from 1988 until fall 1996. Further, there were no pilot reports or any other reports of eruptive activity from 1988 to 1996. In July 1996, Alaska Volcano Observatory installed a new seismic network at Pavlof, which recorded an eruption sequence from September 15 to December 29, 1996 (Table 2; McNutt, 1997). It is salient that the 1996 eruptions again occurred in the fall, similar to the activity from 1973 to 1984.

Discussion

The original work (McNutt and Beavan, 1987) suggested that Pavlof erupted in the fall because eruptions were triggered by compressional stresses or strains due to ocean loading. For modeling, a feeder zone was defined beneath the volcano. This was 40 by 40 km in area extent (average volcano spacing is 40 km), and extended from the base of the crust (30 km) to the top of the Wadati-Benioff zone (100 km). The true magma generation and storage complex probably includes structures with many different rheologies and geometries, which could have a substantial effect on the distribution of stresses or strains from sources such as ocean loading (McNutt and Beavan, 1987). Using the same feeder zone, a 3.8-year period of no eruptions from 1978 to 1980 coincided with an episode of aseismic slip that produced extensional strains at depth. Both effects demonstrated that the volcano was very sensitive and responsive to small changes in stress or strain. It is, however, necessary to re-evaluate the new data in terms of both yearly and longer time scales.

The new data include both additional fall eruptions and several in the spring and summer. However, statistical tests show that the distribution is still highly non-random. Thus, the ocean loading hypothesis still appears to be valid, although other mechanical or tectonic effects may also be occurring. It is speculated that compressional strain may have occurred in 1986 at depth, forcing magma to erupt prematurely starting in April. While no direct data exist to confirm this hypothesis, it is noted that widely separated segments of the Alaska/Aleutian arc were tectonically active during a 6-week period in spring 1986: Mount Augustine erupted 670 km to the east of Pavlof from 27 March–28 April (Kienle et al., 1986) Pavlof erupted 14–26 April (McNutt et al., 1991); and a great earthquake of $M_W = 8.0$ occurred in the Andreanof Islands region 1110 km to the west of Pavlof on 7 May 1986 (Kisslinger, 1988). There is a remote possibility that an arc-wide tectonic strain pulse may have triggered all three of these phenomena.

There is no evidence of another tilt reversal, indicative of aseismic slip (Fig. 5), so it is not possible to evaluate or compare long-term behavior directly. However, the largest earthquake in the Shumagin region in 45 years occurred on May 13,

1993 (Lu *et al.*, 1994; ABERS *et al.*, 1995). This $M = 6.9$ event occurred on the
main thrust zone, had a 600 km^2 rupture area and 1 m of slip (Lu *et al.*, 1994).
The sense of motion for the 1993 earthquake is the same as that for the 1978–
1980 aseismic slip event, leading to a net volume strain extension beneath the
volcano. Levelling data from summer 1993, after the earthquake, show a slight
change in polarity (Fig. 5), consistent with the occurrence of the earthquake
(BEAVAN, 1994). The occurrence of the 1993 earthquake may thus have contributed to the long repose of 8 yr 4 mo from 1988 to 1996. In addition, the
deep earthquakes continued to occur mainly in clusters during times when the
volcano was not erupting (Fig. 5). This suggests a mechanical connection between deep slab processes and shallow eruptive processes. Most earthquakes at
depths greater than 160 km in the Shumagins had focal mechanisms indicating
downdip compression (REYNERS and COLES, 1982; HAUKSSON *et al.*, 1984). All
available first motion data were examined for the newer deep earthquakes, and
all showed downdip compression.

Although the suggested modulating mechanisms, ocean loading, aseismic slip
and related tectonic activity, may have the correct polarities and magnitudes to
produce the observed effects, it must be kept in mind that correlation does not
necessarily imply cause and effect. Therefore, other mechanisms need to be considered. The annual sea-level variations are adding a small amount of strain,
which may be in phase with other larger strains produced by processes for which
we lack observations. For example, there may be annual modulations in stresses
and strains within the solid earth caused by astronomical variations, or rota-
tional changes caused by atmospheric angular momentum variations (e.g.,
LOWRIE, 1997, pp. 54–55). Such effects may be needed to explain similar periodicities of eruptions and earthquake swarms in places as widely separated as
Alaska (Pavlof), Hawaii (Kilauea), Italy (Etna), and Papua New Guinea (Rabaul).

The apparent sensitivity of Pavlof to strains and its location in the Shumagin
seismic gap suggests that its eruptive activity will change systematically if a large
earthquake occurs in the gap. Specifically, based on the above observations, it is
forecast that the volcano will be relatively more active before the earthquake and
less active afterward, possibly for many years. Other studies have shown that
volcanoes tend to erupt more often around the time of great earthquakes (e.g.,
CARR, 1977; KIMURA, 1978), and, in general, volcanoes appear to be affected
by, or sensitive to, regional stresses. A recent example is the increase in earthquake activity at Long Valley, California, and elsewhere in the western U.S.,
triggered by transient stresses generated by the 28 June 1992 Landers, California
earthquake (HILL *et al.*, 1993). Thus, seismic recording at volcanoes offers the
opportunity to use them as "barometers" to monitor stress conditions prior to,
during, and after large earthquakes and other significant tectonic events.

Conclusions

The new computations clearly show that Pavlof's eruptions are non-random; the likelihood of the distribution occurring by chance alone is less than 0.5 percent. New data include a departure from the fall pattern in 1986, then a return later and continuing in 1996. The available data on sea-level variations, aseismic and seismic slip have the correct polarities in terms of compressional and extensional strains, and reasonable magnitudes to modulate the volcanic eruptions. Therefore, all the data are consistent with the hypothesis that Pavlof is responsive to small, slow changes in ambient strains, and these changes modify or trigger eruptions.

REFERENCES

ABERS, G. A., BEAVAN, J., HORTON, S., JAUMÉ, S., and TRIEP, E. (1995), *Large Accelerations and Tectonic Setting of the May 1993 Shumagin Islands Earthquake Sequence*, Bull. Seismol. Soc. Am. *85*, 1730–1738.

BEAVAN, R. J. (1994), *Crustal Deformation Measurements in the Shumagin Seismic Gap, Alaska*, U.S. Geological Survey Open-File Report 94–176, 195–205.

BEAVAN, R. J., BILHAM, R., and HURST, K. (1984), *Coherent Tilt Signals Observed in the Shumagin Seismic Gap: Detection of Time-dependent Subduction at Depth*, J. Geophys. Res. *89*, 4478–4492.

CARR, M. J. (1977), *Volcanic Activity and Great Earthquakes at Convergent Plate Margins*, Science *197*, 655–657.

CASETTI, G., FRAZZETTA, G., and ROMANO, R. (1981), *A Statistical Analysis in Time of the Eruptive Events on Mount Etna (Italy) from 1323 to 1980*, Bull. Volcanol. *44-3*, 283–294.

DZURISIN, D. (1980), *Influence of Fortnightly Earth Tides at Kilauea Volcano, Hawaii*, Geophys. Res. Lett. *7*, 925–928.

GUDMUNDSSON, G., and SAEMUNDSSON, K. (1980), *Statistical Analysis of Damaging Earthquakes and Volcanic Eruptions in Iceland from 1550–1978*, J. Geophys. *47*, 99–109.

HAUKSSON, E., ARMBRUSTER, J., and DOBBS, S. (1984), *Seismicity Patterns (1963–1983) as Stress Indicators in the Shumagin Seismic Gap, Alaska*, Bull. Seismol. Soc. Am. *74*, 2541–2558.

HILL, D. P., REASENBERG, P. A., MICHAEL, A., ARABAZ, W. J., BEROZA, G., BRUMBAUGH, D., BRUNE, J. N., CASTRO, R,. DAVIS, S., dePOLO, D., ELLSWORTH, W. L., GOMBERG, J., HARMSEN, S., HOUSE, L., JACKSON, S. M., JOHNSTON, M. J. S., JONES, L., KELLER, R., MALONE, S., MUNGUIA, L. NAVA, M., PECHMANN, J. C., SANFORD, A., SIMPSON, R. W., SMITH, R. B., STARK, M., STICKNEY, M., VIDAL, A., WALTER, S., WONG, V., and ZOLLWEG, J. (1993), *Seismicity Remotely Triggered by the Magnitude 7.3 Landers, California, Earthquake*. Science *260*, 1617–1623.

INGRAM, W. J., JR., BAKUN, A., and FAVORITE, F. (1976), *Physical Oceanography of the Gulf of Alaska*, Environmental Assessment of the Alaskan Continental Shelf, vol. 2, Principal Investigator's Reports, April–June 1976, pp. 491–624, Environ. Res. Lab., Boulder, Colorado.

KIENLE, J., DAVIES, J. N., MILLER, T. P., and YOUNT, M. E. (1986), *1986 Eruption of Augustine Volcano: Public Safety Response by Alaskan Volcanologists*, EOS, Trans. Am. Geophys. Union *67*, 580–582.

KIMURA, M. (1978), *Significant Eruptive Activities Related to Large Interplate Earthquakes in the Northwestern Pacific Margin*, J. Phys. Earth S557–S570.

KISSLINGER, C. (1988), *An Experiment in Earthquake Prediction and the 7 May 1986 Andreanof Islands Earthquake*, Bull. Seismol. Soc. Am. *78*, 218–229.

LU, Z., WYSS, M., TYTGAT, G., and McNUTT, S. (1994), *Aftershocks of the 13 May 1993 Shumagin Alaska Earthquake*, Geophys. Res. Lett. *21* (6), 497–500.

LOWRIE, W., *Fundamentals of Geophysics* (Cambridge Univ. Press, Cambridge 1997) 354 pp.

MARDIA, K. V., *Statistics of Directional Data. Probability and Mathematical Statistics: A Series of Monographs and Textbooks*, vol. 13 (Academic Press, San Diego 1972).

MCNUTT, S. R. (1987), *Eruption Characteristics and Cycles at Pavlof Volcano, Alaska, and their Relation to Regional Earthquake Activity*, J. Volcanol. Geotherm. Res. *31*, 239–267.

MCNUTT, S. R. (1997), *Seismic Monitoring of the September–December 1996 Eruptions at Pavlof Volcano by the Alaska Volcano Observatory*, EOS, Trans. AGU *78* (17), Spring Meeting Supplement, S47.

MCNUTT, S. R., and BEAVAN, R. J. (1987), *Eruptions of Pavlof Volcano and their Possible Modulation by Ocean Load and Tectonic Stresses*, J. Geophys. Res. *92*, 11,509–11,523.

MCNUTT, S. R., MILLER, T. P., and TABER, J. J. (1991), *Geological and Seismological Evidence of Increased Explosivity during the 1986 Eruptions of Pavlof Volcano, Alaska*, Bull. Volcanol. *53*, 86–98.

OHTAKE, M., and NAKAHARA, H. (1999), *Seasonality of Great Earthquake Occurrence at the Northeastern Margin of the Philippine Sea Plate*, Pure appl. geophys., *155*, 689–700.

REED, R. K., and SCHUMACHER, J. D. (1981), *Sea Level Variations in Relation to Coastal Flow Around the Gulf of Alaska*, J. Geophys. Res. *86*, 6543–6546.

REYNERS, M., and COLES, K. S. (1981), *Fine Structure of the Dipping Seismic Zone and Subduction Mechanics in the Shumagin Islands, Alaska*, J. Geophys. Res. *87*, 356–366.

THORARINSSON, S., and SIGVALDSSON, G. E. (1972), *The Hekla Eruption of 1970*, Bull. Volcanol. *36*, 269–288.

TRYGGVASSON, E. (1973), *Seismicity, Earthquake Swarms, and Plate Boundaries in the Iceland Region*, Bull. Seismol. Soc. Am. *63*, 1327–1348.

(Received October 29, 1998, revised/accepted January 22, 1999)

 To access this journal online:
http://www.birkhauser.ch

Pure appl. geophys. 155 (1999) 713–726
0033–4553/99/040713–14 $ 1.50 + 0.20/0

Pure and Applied Geophysics

Seismicity Patterns: Are they Always Related to Natural Causes?

F. Ramón Zúñiga[1] and Stefan Wiemer[2]

Abstract—By analyzing data from two catalogs of seismicity, one local and one regional, examples are drawn that demonstrate artificial seismicity patterns which can arise when changes in magnitude reporting or in detection ability are introduced by changes in operating procedures of a seismic network. In the first case, a catalog which comprises seismic data from Guerrero, Mexico, is revised and a correction is applied to magnitudes suspected of being affected by different operative practices. Next, several time windows are analyzed which compare the seismicity rate within the window to that of the total record mapped as a function of space. The same procedure is applied to the original (i.e., uncorrected) catalog. A significant seismic quiescence apparent in the original data set in the center of the seismic network all but disappears in the corrected data, indicating that this anomaly is not naturally induced. In the second case, we investigate the homogeneity of seismicity reporting for the Interior of Alaska. We computed the standard deviate z as a function of time by comparing the overall seismicity rate with the rate in a 3-year window. The maps z-values were inspected for all times. The most outstanding rate change is found around 1992.5. Since the b-value remained unchanged, the most reasonable explanation for the observed rate change around mid-1992 is a decrease in the detection ability of the network in the Interior of Alaska. Both case studies demonstrate the usefulness of systematic comparisons of the cumulative and noncumulative frequency-magnitude distribution, and of spatial and temporal mapping of the seismicity rates as a tool to investigate the homogeneity of earthquake reporting.

Introduction

Seismicity patterns have through the years attracted the attention of researchers as a means to investigate the tectonic stress behavior of a region. Changes in the rate of earthquake production are believed to be closely related to changes in stress in a particular volume (DIETERICH, 1994; DIETERICH and OKUBO, 1996; KATO *et al.*, 1997). Precursory phenomena have been proposed to cause changes in seismicity rates, such as localized precursory seismic quiescence (REASENBERG and MATTHEWS, 1988; WYSS, 1997), regional activation (HABERMANN and CREAMER, 1994; KEILIS-BOROK and KOSSOBOKOV, 1990; KOSSOBOKOV and CARLSON, 1995) and accelerated moment release (BUFE *et al.*, 1994; VARNES, 1989). Any study of

[1] Unidad de Investigación en Ciencias de la Tierra, Instituto de Geofísica, UNAM, Campus Juriquilla, Querétaro, Qro., C.P. 76230, México. E-mail: ramon@conin.unicit.unam.mx
[2] Eidgenössische Technische Hochschule Zürich, Institute of Geophysics, ETH-Hönggerberg, 8093 Zürich, Switzerland. E-mail: stefan@seismo.ifg.ethz.ch

seismicity patterns whose goal is to detect variations in the background stress state must start with the identification of a pattern as significant and then proceed to attempt inferring about the possible physical causes of the pattern. One, however, must be certain that the pattern in question is caused by natural activity rather than related to data acquisition or processing, before any conclusions can be drawn on its significance or physical meaning. For this reason, it is of the utmost importance to know the seismic record from which the pattern comes to the finest detail possible and in addition provide a method of testing whether a pattern in question has indeed been generated by natural causes.

Several studies have been carried out pertinent to the problem of artificial or man-made seismicity anomalies (HABERMANN, 1991, 1982; ZÚÑIGA, 1989; WYSS, 1991). Some of the most common sources of errors, as well as techniques to detect them, can be found in ZÚÑIGA and WYSS (1995). Other sources of error are continuously added to the list as we progress in finding new techniques of identifying artificial seismicity variations. Different techniques have been proposed to identify artificial rate changes. Some methods rely on comparisons of seismicity rates, defined as the number of events per unit time, taken at different time intervals and for varying magnitude bands (i.e., Magnitude Signatures and GENAS Algorithm, HABERMANN, 1983, 1987) as well as on its spatial mapping (i.e., ZMAPS, WIEMER and WYSS, 1994). Others have made use of b-value relations (FROHLICH and DAVIS, 1993; ZÚÑIGA and WYSS, 1995) or magnitude distributions (PEREZ and SCHOLZ, 1984) in order to distinguish between natural and man-made seismicity anomalies.

The objective of this study is to highlight the extent of the problem of artificial seismicity rate changes (ASR) by providing examples which demonstrate how spurious patterns can be produced by them. We also show ways in which we can discriminate these spurious patterns from the naturally induced ones. We specifically demonstrate that spatial and temporal mapping of seismicity rates can provide important clues to the existence of ASR and aid in identifying their causes. By doing so, we hope to emphasize the serious repercussions of treating seismicity catalogs in an improper manner.

First Case Study: A Local Network Record in the Guerrero Gap, Mexico

We will discuss the catalog of Guerrero, Mexico, seismicity obtained from a telemetric network of short-period seismographs which has operated from 1988 until 1996 in a basically continuous fashion. This network is still run by the Institute of Geophysics of the National University of Mexico (UNAM) and has been subjected, as have most networks, to various changes in operative procedures throughout its history. The network covers a segment of the Mexican subduction thrust which has not experienced a large earthquake for over 90 years, the so-called Guerrero seismic gap.

Figure 1 shows the seismic activity of the region coupled with the location of the stations. Not all stations have been in operation continuously, although all were installed during 1987. It is immediately apparent that the seismicity is not distributed homogeneously, displaying bands of dense activity parallel to the trench. These bands have been attributed to tensional and compression stresses within the subducting slab (SUÁREZ et al., 1990) due to a varying dip.

We first study the overall characteristics of the seismicity record by means of the cumulative number of events per time unit (hereafter referred to as seismicity rate). In this case we used a time unit equal to one month for our calculations. Figure 2 shows the aforementioned rate for events with a magnitude $M \geq 2.0$ (magnitude is determined from coda duration), which is somewhat lower than the minimum magnitude of completeness (~ 2.5 from the b-value distribution). Various trends in the data can be seen, of which three stand out: the beginning of the time series, from 1987 to 1989; the mid-time interval, from 1989 to 1993; and the most recent trend, from 1993 to 1996. The first trend is related to the start of cataloguing and operation of the network, and was subjected to many adjustments, so we excluded

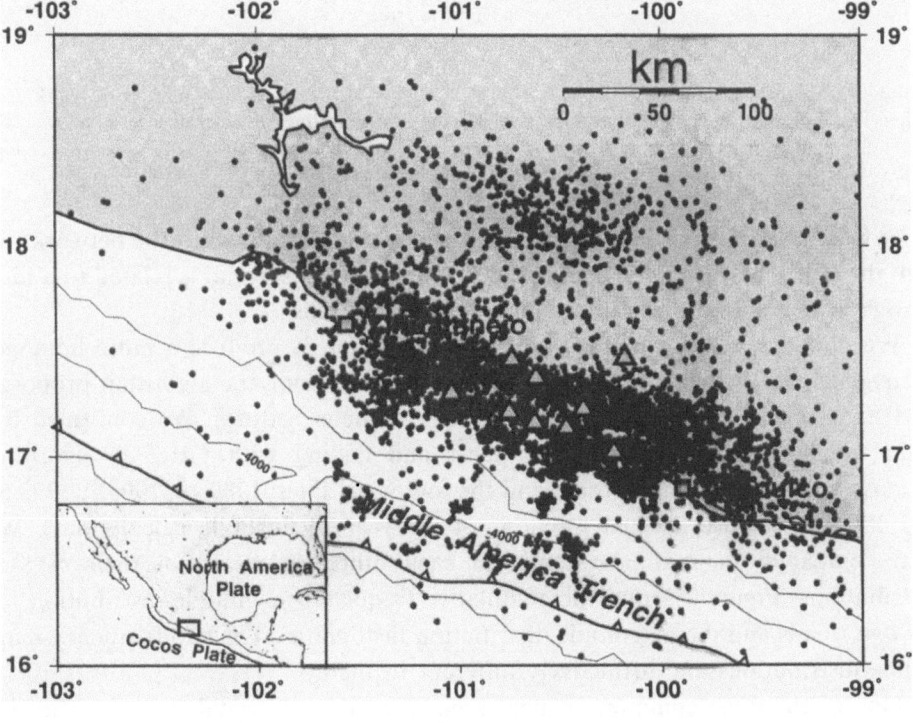

Figure 1

Epicenter map of the Guerrero region. Epicenters located with the local seismic network during the period 1988–1995 are shown as black dots. Triangles mark the location of the seismic station, black lines indicate the bathymetry and the location of the Mid-America trench.

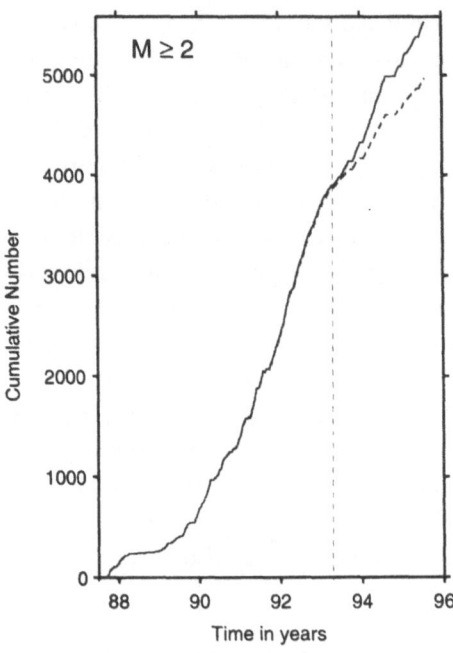

Figure 2

Cumulative number of detected events ($M \geq 2$) as a function of time from the Guerrero network. The dashed line indicates the original data after declustering, the solid line the corrected data set using the magnitude transformation given in the text. The vertical dashed line indicated the time when the change in reporting rate occurred.

it out from our analysis. The second trend is a period during which the network ran with the least changes and in a consistent manner. The last trend is related to a final change in operating procedures, which is discussed later.

We now proceed to "clean" the catalog in order to produce a more homogeneous data set. We start by declustering the catalog, using the algorithm proposed by REASENBERG (1985), using the original parameter settings. We confirmed the main trends and determined their onset and ending with GENAS algorithm (HABERMANN, 1983) which resulted in the following dates: 1989.18, 1993.2, 1993.53 and 1995.48 (we use decimal fractions of the year to facilitate calculations). We then compared the two last trends to each other by means of their b-value distributions. Figure 3 shows the cumulative frequency-magnitude distribution for the two trends and the magnitude distribution histograms. The two frequency-magnitude distributions are distinctively different to the eye.

One of the most common changes seen in earthquake catalogs is an effective additive variation in magnitudes measured after some time (magnitude values increase or decrease by a fixed amount), which has been referred to as a "magnitude shift" (HABERMANN, 1987). Such a change might produce the horizontal translation of the histogram observed in Figure 3c. However, a magnitude shift would not affect the b-value (slope of linear trend in Fig. 3a, ZÚÑIGA and WYSS, 1995) as is the case here.

We investigated whether a more general linear change in magnitude, which is sometimes known as a "stretch" in magnitude, might suit the data better, employing the technique proposed in ZÚÑIGA and WYSS (1995) to that effect. Figures 3b, d display the results of this exercise. The best fit in a least-squares sense between the earlier and later period is achieved by the following transformation: $M_{new} = 0.8$ $M_{old} + 1.07$ for events after 1993.53. We can see that a good match is obtained between the b-value curves for both intervals. The resulting cumulative number versus time curve is shown in Figure 2 as a solid line. A statistically significant change in rate for $M \geq 2.0$, which was detected by comparing a running average rate with the previous rate by means of $AS(t)$ function (HABERMANN, 1983; WIEMER and WYSS, 1994) for time 1993.2, is no longer detected after the correction.

To study the effect of the magnitude correction on seismicity patterns, we analyzed the spatial dependency of seismicity rate changes by comparing seismicity patterns before and after the correction. The patterns we will discuss are those obtained through the ZMAP (WIEMER and WYSS, 1994; WYSS and WIEMER, 1997) technique. In short, this method compares the rate within a specified time window to the seismicity rate of the entire seismic record, excluding the analysis window.

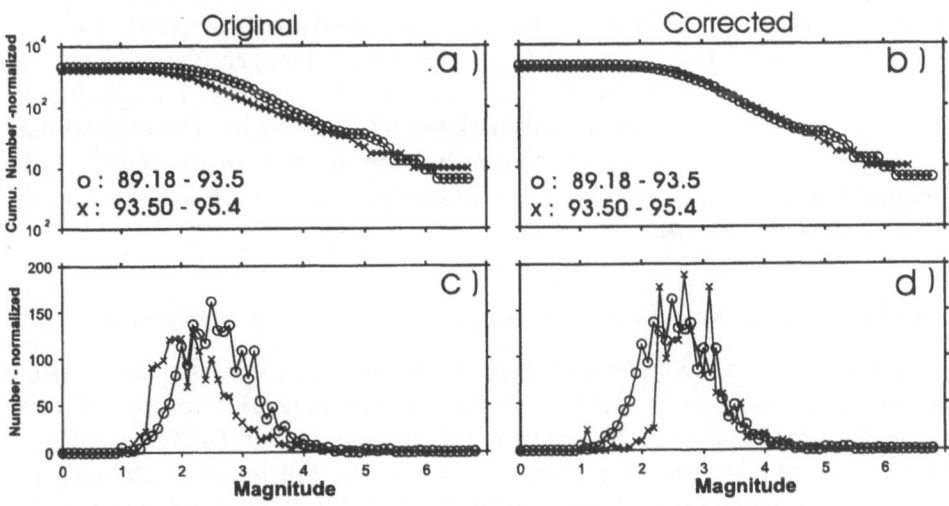

Figure 3
Cumulative number of events (frames a and b) and non-cumulative number of events (frames c and d) as a function of magnitude in Guerrero. The frames a and c correspond to the original data set, frames b and d to the corrected data set. Two periods are compared in each frame: 1989.18–1993.5 (○) and 1993.5–1995.4 (×). The numbers have been normalized by the duration of each period.

The time window is selected such that its length is not significant with respect to the duration of the record. The area of study is subdivided by a densely spaced grid, and for every grid node a fixed number of nearest earthquakes (here: 300) is selected to represent the seismicity around that particular point (that is, we form a subcatalog for every node in the grid which is taken as the seismicity record for that point in space). The same procedure could be performed using events which are located within a fixed distance from the node, however variations in the seismicity density make the former approach of constant sampling sizes statistically more robust. A statistical z-test using the $LTA(t)$ function (HABERMANN, 1987; WIEMER and WYSS, 1994) is carried out for each time series at each node. The resulting three-dimensional matrix of z-values (Latitude, Longitude and Time) can be sliced at any time.

We analyzed time windows with a duration of one year, starting at the beginning of each year. Figures 4a,b depict the results of comparing a rate variation map for a time window (t_w) which starts at 1990.0. We can see that an apparent anomalous decrease (in all Figures, positive z-values indicate seismicity decreases while negative values are indicative of increases in seismicity) located to the center of the study area is enhanced by the correction. z-values larger than 2.57 are significant at the 99% level.

It is important to mention that although we corrected magnitudes only after 1993.5, the rate analysis is carried out considering the entire span of the catalog as a basis for comparison. Therefore, we observe an increase in the z-value because the variance of the seismicity rates decreases due to the subsequent correction. Few changes between the original and corrected catalog are visible for the 1991 and 1992 slices (Figs. 4c–f). The most significant differences between the original and corrected catalog can be identified when comparing the seismicity rates during the period 1993.5–1994.5 with the overall background (Figs. 4g,h). The original data set reveals a statistically important seismicity rate decrease in the center of the Guerrero gap (Fig. 4g). However, this rate change is no longer significant in the corrected data set (Fig. 4h).

Second Case Study: Seismicity Rate Change in the Interior of Alaska in mid-1992

As a second case study we investigate the homogeneity of seismicity reporting for crustal seismicity in the Interior of Alaska. The data used in our study are extracted from the Alaska Earthquake Information Center (AEIC) catalog of digital data. AEIC started recording digitally in the fall of 1988, although the earliest usable data are from January 1989. Earthquakes are routinely located with the computer program HYPOELLIPSE (LAHR, 1989). We investigated crustal seismicity (depths < 40 km) for the period 1989–1998. The catalog is declustered using (REASENBERG, 1985) algorithm with the standard parameter setting. However, the results presented here are largely independent of the declustering applied.

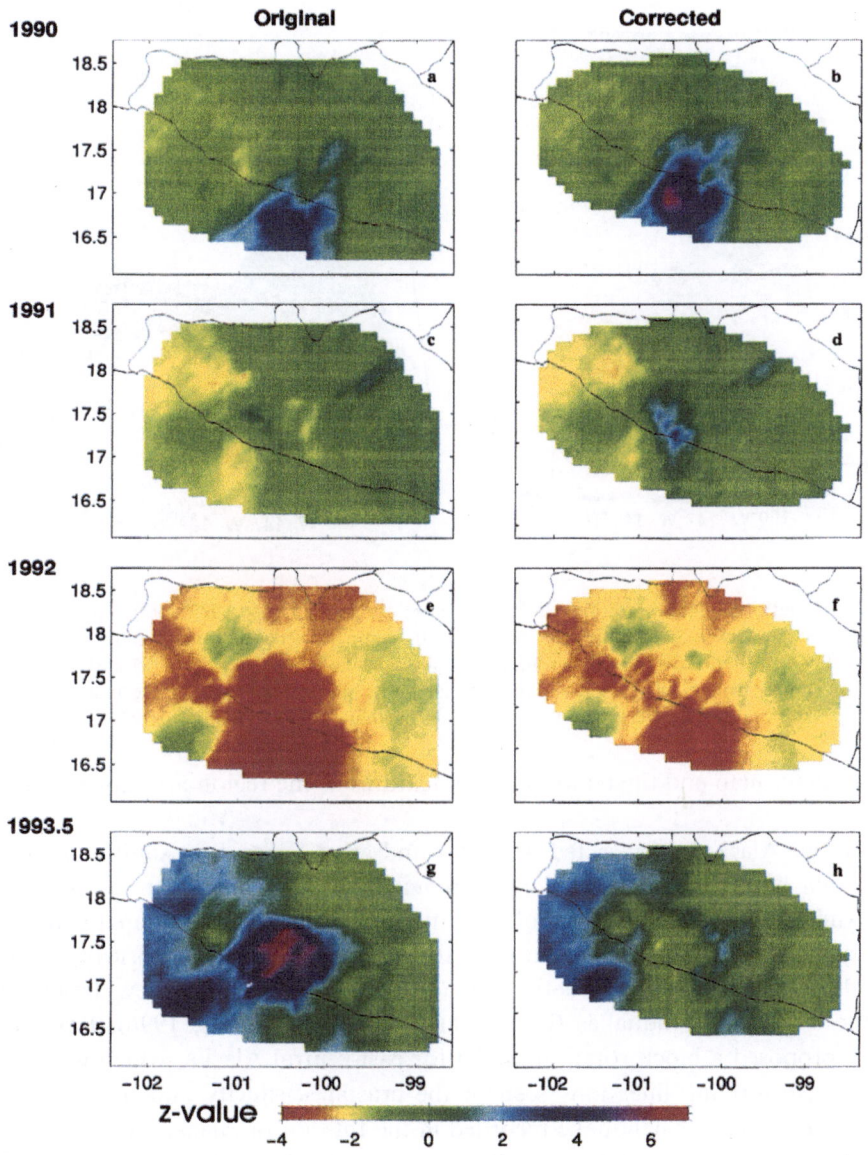

Figure 4

Maps of the standard deviate z for four different times. The left frames correspond to the original data set, the right frames to the corrected data set. At each node of a densely spaced grid, the 300 nearest earthquakes have been sampled and investigated for seismicity rate changes. The color-coding corresponds to the z-value at this grid node, positive z-values indicate a rate decrease. z-values are computed for a window length of one year and compared to the seismicity rate in the one-year window, starting at the time indicated to the left with the overall seismicity rate, excluding the one-year period.

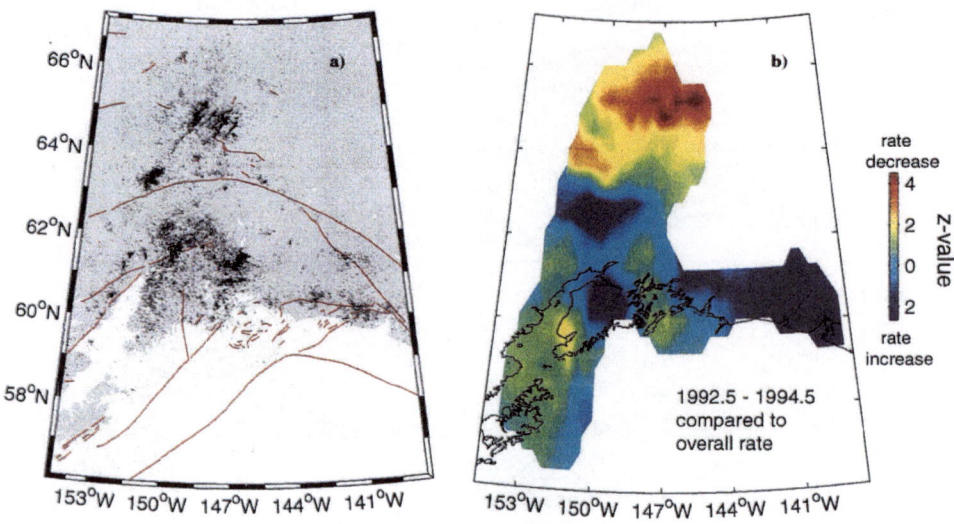

Figure 5

(a) Map of south-central and Interior Alaska. Epicenter during the period 1989–1997 is plotted as black dots, red lines mark major faults. (b) z-value map comparing the seismicity rates in the period 1992.5–1994.5 with the overall background rate. At each node, the nearest 1000 earthquakes were sampled. The red colors (high z-values) indicate an artificially introduced decrease in the number of events detected in the Interior of Alaska that started around mid-1992.

An epicenter map and the major tectonic features of the region are shown in Figure 5a.

Interior Alaska is transected by major and great faults, for example the Denali fault, that have dimensions and morphological expressions similar to those of the San Andreas fault system. Some of these faults are thought to be capable of $M > 8$ earthquakes, although they have not ruptured during the short historically recorded period in Alaska. Several faults that have ruptured during the last few decades have produced $M > 7$ earthquakes (FLETCHER and CHRISTENSEN, 1996). PAGE et al. (1995) proposed a block rotation model for east-central Alaska, which is based on the NW-SE striking lineations seen in the crustal seismicity and in air-magnetic surveys. The largest earthquake recorded in the Interior of Alaska over the past ten years was the 1995 M_w 6.2 Tatalina River earthquake, located approximately 50 km north of Fairbanks (Fig. 5a). To investigate spatial and temporal homogeneity of the earthquake reporting, we sample the seismicity using overlapping cylindrical volumes, each containing 1000 earthquakes. This large sample size is chosen to enhance robustness of the analysis and because we are mostly interested in larger scale changes in catalog homogeneity rather than a high-resolution study. As in the Guerrero case, we compute the standard deviate z as a function of time by comparing the overall seismicity rate with the rate, only that here we used a 2-year window. The maps of z-values are inspected for all times. The highest z-value

($z = 4$), corresponding to the most significant rate change, is found around 1992.5, and we show the corresponding spatial distribution of z-values in Figure 5b. The red colors north of the Denali fault indicate a rate decrease starting mid-1992. Blue and green colors indicate a slight increase in the number of events detected. The cumulative number of events as a function of time plot for the high z-value region (Fig. 6a) shows a stable rate of events for the period 1989–1992 (~ 300 events per year), then a drop of the rate by over 50% for the period 1992.5–1995.5. This rate decrease is followed by the October 1995 Tatalina River earthquake and its subsequent aftershock activity, which produces a rate increase despite the declustering of the catalog.

One may propose that this rate decrease is causally linked to the upcoming mainshock. However, if we compare the number of events reported in each magnitude bin (Fig. 6c) we find that the rate of magnitude 1.5 and above earthquakes is essentially unchanged, and only the number of $M < 1.5$ events decreased. The b-value remains unchanged (Fig. 6b), and the most logical explanation for the observed rate change around mid-1992 is a decrease in the detection ability of the network in the Interior of Alaska. A plot of the magnitude of completeness as a function of time supports this interpretation (Fig. 7): We find a step-like increase in the magnitude of completeness around 1992.5 from $Mc \sim 1.4$ to $Mc \sim 1.6$.

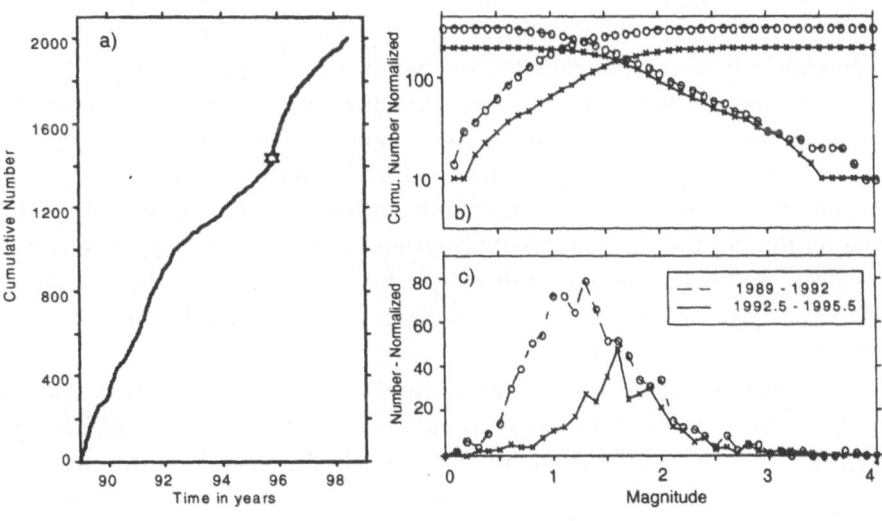

Figure 6

(a) Cumulative number of events as a function of time for the seismicity in the Interior of Alaska. The 1995 M 6.2 Tatalina river earthquake is marked by a star. Note the rate decrease that started in 1992.5. (b) Cumulative and (c) non-cumulative number of events as a function of magnitude. Two periods are compared in each frame; 1989–1992 (○) and 1992.5–1995.5 (×). The numbers have been normalized by the duration of each period.

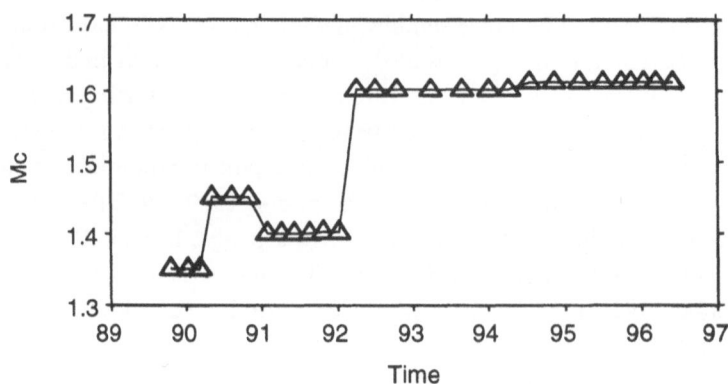

Figure 7
Magnitude of completeness as a function of time for the seismicity in the Interior of Alaska. The completeness is computed based on the frequency-magnitude distribution and using overlapping windows containing 500 earthquakes each.

Discussion

Seismicity rates vary with the applied stress and can serve as a remote stress sensor in the earth's crust. However, seismicity rates are also highly susceptible to changes in operating procedure of seismic networks. In this paper, we demonstrate how by using two simple techniques artificially introduced rate changes can easily be detected. The first technique is the systematic comparison of the cumulative and non-cumulative frequency-magnitude distribution for two periods (Figs. 3 and 6). Adopting a common sense approach, an investigator is often able to reconstruct and explain the history of seismicity rate changes based on these graphs. The second technique is the quantitative mapping of seismicity rates as a function of space and time. Most efforts of quantitative seismicity rate mapping have been focused on the identification of possible earthquake precursors (e.g., WIEMER and WYSS, 1994; WYSS et al., 1996). In this paper we demonstrate for the first time the usefulness of spatial information to unravel the reporting history of seismic network.

The change in seismicity rate observed around 1993.5 in the Guerrero case study (Figs. 2, 3 and 4) is likely to be a result of different operative practices which induced an ASR. It is difficult to think of a natural change in the earthquake production that would produce a drastic reporting change as seen in Figure 3. Such a change could, for example, consist of a decrease in medium size events ($2 < M < 4$) for unknown reasons, while simultaneously the number of larger events remains unchanged and an increase in the detection capability of the network increases the number of small events detected ($M < 2$). Although not impossible per se, it is highly unlikely that such a complex change would occur.

A much simpler explanation can be found when considering the operating procedures of the seismic network. As aforementioned, magnitudes have been determined from the duration of the signal from the beginning of cataloguing. Around 1990 the signals were sent through an AD filter before being recorded in memory. Records were automatically stored in blocks. The operator had to retrieve the blocks from memory and display them in order to measure the duration. A truncation of the signal would sometimes prevent measuring the complete coda. Different operators used different ways to deal with the problem. One operator employed the decay trend and estimated the end of the signal. Another operator estimated the length by comparing it with other records. The change in operators coincides with the time for which the seismicity variation is observed in the cumulative curves (Fig. 2). Therefore, we believe that there is a direct connection between the two events. Even though the different operative practices might have introduced a change in the number of reported events (if, for instance, the operator had decided to exclude those events which appeared truncated) and not affect the determination of magnitudes, the curves and histogram in Figure 3 demonstrate that this was not the case, since such a change would not effectively produce the shift observed between both curves as well as in the histograms. The difference between the curves and histograms can only be attained by a change in magnitude determination, that being a simple shift or a stretch. It is important to mention that some of the problems detected in this study might have been avoided if magnitudes had been determined based on amplitude and not on duration.

The change in rates is most pronounced for the central region of the catalog (Fig. 4g,h) and less obvious for outlying regions. One might assume that the center of a seismic network is less susceptible to ASR than outlying areas. However, we can understand this at first counterintuitive behavior by analyzing the magnitude of completeness as a function of space (Fig. 8). As one would expect, the magnitude of completeness is with $Mc \sim 2.0$ lowest in the central part of the network, and increases to $Mc > 2.5$ in outlying areas. The change in reporting procedures has the greatest impact for smaller events ($2 < M < 2.5$), because the change in magnitude reporting moves these events below the cut-off magnitude of $M = 2$. Consequently, the area with the lowest magnitude of completeness manifests the most significant artificial rate change.

We have shown that correcting a catalog after a suspicion of spurious changes may provide some indication of the nature of the change. For the cases analyzed, there are instances where apparent anomalies remain after the correction or are even enhanced by it. Other examples reveal that some anomalies are no longer significant or are not seen altogether after the correction. Thus, even if the correction is subjected to later refinements, preliminary results can be used as a tool to identify ASR.

The Alaska case study also clearly demonstrates the usefulness of spatial and temporal mapping of the seismicity rates as a tool to investigate the homogeneity of earthquake reporting. The overall seismicity rate for crustal activity in Alaska shows little change, because the rate decrease in the Interior is at least partially compensated by detection capability increases in South Central Alaska (blue areas in Fig. 5b). Consequently, the network operators were not aware that a change in the detection ability had occurred until our z-value mapping revealed this increase-decrease pattern. Subsequently it was discovered that in order to reduce the workload during the eruption of Mt. Spurr in the summer of 1992, the decision was reached (but unfortunately not documented) to change the triggering setup for the Interior subnet. One additional triggering station was subsequently required to declare an event, causing the number of small earthquakes detected to be drastically reduced (Fig. 6c). This change did not influence the detection ability south of the Denali fault, which belongs to a different triggering subnet, thus we find no rate reduction (Fig. 5b). The detection ability in the Interior was never restored and remains below the 1989–1992 level despite the introduction of a new processing system in early 1997 (the Iceworm system; LINDQUIST, 1998). We conclude that our analysis of the homogeneity of reporting was able to detect and quantify a major change in the network capabilities previously unknown, and we suggest that a routine monitoring of the seismicity rates, using the simple tools applied in this study, can help to minimize the unnecessary loss of data by offering an early detection capability to identify rate changes. In these as in other cases studied, experience indicates that for long-term experiments, best results are obtained when

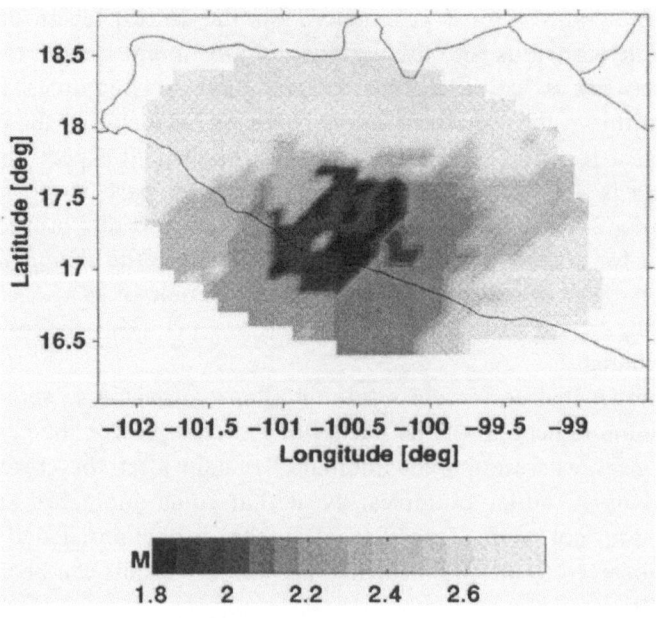

Figure 8
Map of the Guerrero region contouring the overall magnitude of completeness of the Guerrero network.

a conservative approach is employed, i.e., introducing as few changes in the operation of a network as possible.

Acknowledgements

The authors would like to thank Paul Reasenberg and an anonymous reviewer for their comments and suggestions which improved the presentation of this study. Gerardo Suárez' effort in keeping the Guerrero network alive for such a long time is gratefully acknowledged. Jaime Domínguez was very helpful in providing information related to operative practices of the Guerrero network. We acknowledge the Alaska Earthquake Information Center and the Institute of Geophysics-UNAM for providing seismicity catalogs used in this study. Support for this work has been provided by NSF Grant EAR-9505837, the Science and Technology Agency of Japan and ETH Zurich (SW) and by UNAM through its DGAPA-PAPIIT program (RZ). This paper is contribution number 1074 of the Institute of Geophysics ETHZ.

REFERENCES

BUFE, C. G., NISHENKO, S. P., and VARNES, D. J. (1994), *Seismicity Trends and Potential for Large Earthquakes in the Alaska-Aleutian Region*, Pure appl. geophys. *142*, 83–99.

DIETERICH, J. H. (1994), *A Constitutive Law for Rate of Earthquake Production and its Application to Earthquake Clustering*, J. Geophys. Res. *99*, 2601–2618.

DIETERICH, J. H., and OKUBO, P. G. (1996), *An Unusual Pattern of Seismic Quiescence at Kalapana, Hawaii*, Geophys. Res. Letts. *23*, 447–450.

FLETCHER, H. J., and CHRISTENSEN, D. H. (1996), *A Determination of Source Properties of Large Intraplate Earthquakes in Alaska*, Pure appl. geophys. *146*, 21–41.

FROHLICH, C., and DAVIS, S. D. (1993), *Teleseismic b-values; or, much ado about 1.0*, J. Geophys. Res. *98*, 631–644.

HABERMANN, R. E., and CREAMER, F. (1994), *Catalog Errors and the M 8 Earthquake Prediction Algorithm*, Bull. Seismol. Soc. Am. *84*, 1551–1559.

HABERMANN, R. E. (1982), *Consistency of Teleseismic Reporting since 1963*, Bull. Seismol. Soc. Am. *72*, 93–112.

HABERMANN, R. E. (1983), *Teleseismic Detection in the Aleutian Island Arc*, J. Geophys. Res. *88*, 5056–5064.

HABERMANN, R. E. (1987), *Man-made Changes of Seismicity Rates*, Bull. Seismol. Soc. Am. *77*, 141–159.

HABERMANN, R. E. (1991), *Seismicity Rate Variations and Systematic Changes in Magnitudes in Teleseismic Catalogs*, Tectonophys. *193*, 277–289.

LINDQUIST, K. (1998), *Seismic Array Processing and Computational Infrastructure for Improved Monitoring of Alaskan and Aleutian Seismicity and Volcanoes*, University of Alaska, Fairbanks.

KATO, N., OHTAKE, M., and HIRASAWA, T. (1997), *Possible Mechanism of Precursory Seismic Quiescence: Regional Stress Relaxation due to Preseismic Sliding*, Pure appl. geophys. *150*, 249–267.

KEILIS-BOROK, V. I., and KOSSOBOKOV, V. G. (1990), *Premonitory Activation of Earthquake Flow: Algorithm M 8*, Phys. Earth Plan. Int. *61*, 73–83.

KOSSOBOKOV, V. G., and CARLSON, J. M. (1995), *Active Zone Size versus Activity: A Study of Different Seismicity Patterns in the Context of the Prediction Algorithm M 8*, J. Geophys. Res. *100*, 6431–6441.

LAHR, J. C. (1989), *HYPOELLIPSE/version 2.00: A Computer Program for Determining Local Earthquakes Hypocentral Parameters, Magnitude, and First Motion Pattern*, U.S. Geol. Surv. Open-File Rep. 89–116.

PAGE, R. A., PLAFKER, G., and PULPAN, H. (1995), *Block Rotation in East-central Alaska: A Framework for Evaluation Earthquake Potential*, Geology *23*, 629–632.

PEREZ, O. J., and SCHOLZ, C. H. (1984), *Heterogeneity of the Instrumental Seismicity Catalog (1904–1980) for Strong Shallow Earthquakes*, Bull. Seismol. Soc. Am. *74*, 669–686.

REASENBERG, P. A. (1985), *Second-order Moment of Central California Seismicity*, J. Geophys. Res. *90*, 5479–5495.

REASENBERG, P. A., and MATTHEWS, M. V. (1988), *Precursory Seismic Quiescence: A Preliminary Assessment of the Hypothesis*, Pure appl. geophys. *126*, 373–406.

SUÁREZ, G., MONFRET, T., WITTLINGER, G., and DAVID, C. (1990), *Geometry of Subduction and Depth of the Seismogenic Zone in the Guerrero Gap, Mexico*, Nature *345*, 336–338.

VARNES, D. J. (1989), *Predicting Earthquakes by Analyzing Accelerating Precursory Seismic Activity*, Pure appl. geophys. *130*, 661–686.

WIEMER, S., and WYSS, M. (1994), *Seismic Quiescence before the Landers (M = 7.5) and Big Bear (M = 6.5) 1992 Earthquakes*, Bull. Seismol. Soc. Am. *84*, 900–916.

WYSS, M. (1991), *Reporting History of the Central Aleutians Seismograph Network and the Quiescence Preceding the 1986 Andreanof Island Earthquake*, Bull. Seismol. Soc. Am. *81*, 1231–1254.

WYSS, M. (1997), *Nomination of Precursory Seismic Quiescence as a Significant Precursor*, Pure appl. geophys. *149*, 79–114.

WYSS, M., SHIMAZAKI, K., and URABE, T. (1996), *Quantitative Mapping of a Precursory Quiescence to the Izu-Oshima 1990 (M = 6.5) Earthquake, Japan*, Geophys. J. Int. *127*, 735–743.

WYSS, M., and WIEMER, S. (1997), *Two Current Seismic Quiescences within 40 km of Tokyo*, Geophys. J. Int. *128*, 459–473.

ZÚÑIGA, F. R. (1989), *A Study of the Homogeneity of the NOAA Earthquake Data File in the Mid-America Region by the Magnitude Signature Technique*, Geofísica Internacional *28*, 103–119.

ZÚÑIGA, F. R., and WYSS, M. (1995), *Inadvertent Changes in Magnitude Reported in Earthquake Catalogs: Their Evaluation Through b-value Estimates*, Bull. Seismol. Soc. Am. *85*, 1858–1866.

(Received September 1, 1998, revised January 16, 1999, accepted January 20, 1999)

 To access this journal online:
http://www.birkhauser.ch

GeoSciences with Birkhäuser

http://www.birkhauser.ch

1999. 168 pages. Hardcover
sFr. 128.– / DM 148.– / öS 1081.–
ISBN 3-7643-6041-0

(Prices are subject to change. 4/99)

ISNM 127 • International Series of Numerical Mathematics

Gerke, H.H., ZALF, Müncheberg, Germany / **Hornung, U.**, (†) / **Kelanemer, Y.**, Montreal, Canada / **Slodicka, M.**, Comenius Univ., Bratislava, Slovakia / **Schumacher, S.**, Bayer. Landesbank, München, Germany

Optimal Control of Soil Venting: Mathematical Modeling and Applications

The task to develop a practicable simulation tool for optimal control of soil vapor extraction was attacked by an interdisciplinary group consisting of mathematicians, soil physicists, chemists and in cooperation with a remediation company. The purpose of this book is to describe the latest and most appropriate mathematical and numerical methods for optimizing soil venting and to demonstrate the application beginning with a conceptual mathematical model, the derivation and estimation of model parameters, generation of parameter distributions, sensitivity analysis, model calibration and numerical analysis of two spillage test cases, and stochastic optimization. The monograph considers mathematical, numerical, technical aspects as well as the practical significance. This book will be of interest to applied mathematicians, geophysicists, geoecologist, soil physicists, and environmental engineers.

For orders originating from all over the world except USA and Canada:
Birkhäuser Verlag AG
P.O Box 133
CH-4010 Basel/Switzerland
Fax: +41/61/205 07 92
e-mail: orders@birkhauser.ch

For orders originating in the USA and Canada:
Birkhäuser
333 Meadowland Parkway
USA-Secaurus, NJ 07094-2491
Fax: +1 201 348 4033
e-mail: orders@birkhauser.com

Birkhäuser